BOLYAI SOCIETY
MATHEMATICAL STUDIES

22

BOLYAI SOCIETY MATHEMATICAL STUDIES

1. **Combinatorics, Paul Erdős is Eighty, Vol. 1**
D. Miklós, V.T. Sós, T. Szőnyi (Eds.)
2. **Combinatorics, Paul Erdős is Eighty, Vol. 2**
D. Miklós, V.T. Sós, T. Szőnyi (Eds.)
3. **Extremal Problems for Finite Sets**
P. Frankl, Z. Füredi, G. O. H. Katona, D. Miklós (Eds.)
4. **Topology with Applications**
A. Császár (Ed.)
5. **Approximation Theory and Function Series**
P. Vértesi, L. Leindler, Sz. Révész, J. Szabados, V. Totik (Eds.)
6. **Intuitive Geometry**
I. Bárány, K. Böröczky (Eds.)
7. **Graph Theory and Combinatorial Biology**
L. Lovász, A. Gyárfás, G. Katona, A. Recski (Eds.)
8. **Low Dimensional Topology**
K. Böröczky, Jr., W. Neumann, A. Stipsicz (Eds.)
9. **Random Walks**
P. Révész, B. Tóth (Eds.)
10. **Contemporary Combinatorics**
B. Bollobás (Ed.)
11. **Paul Erdős and His Mathematics I+II**
G. Halász, L. Lovász, M. Simonovits, V. T. Sós (Eds.)
12. **Higher Dimensional Varieties and Rational Points**
K. Böröczky, Jr., J. Kollár, T. Szamuely (Eds.)
13. **Surgery on Contact 3-Manifolds and Stein Surfaces**
B. Ozbagci, A. I. Stipsicz
14. **A Panorama of Hungarian Mathematics in the Twentieth Century, Vol. 1**
J. Horváth (Ed.)
15. **More Sets, Graphs and Numbers**
E. Győri, G. O. H. Katona, L. Lovász (Eds.)
16. **Entropy, Search, Complexity**
I. Csiszár, G. O. H. Katona, G. Tardos (Eds.)
17. **Horizons of Combinatorics**
E. Győri, G. O. H. Katona, L. Lovász (Eds.)
18. **Handbook of Large-Scale Random Networks**
B. Bollobás, R. Kozma, D. Miklós (Eds.)
19. **Building Bridges**
M. Grötschel, G. O. H. Katona (Eds.)
20. **Fete of Combinatorics and Computer Science**
G.O.H. Katona, A. Schrijver, T. Szőnyi (Eds.)
21. **An Irregular Mind, Szemerédi is 70**
I. Bárány, J. Solymosi (Eds.)

Hajnal Andréka
Miklós Ferenczi
István Németi
(Eds.)

Cylindric-like Algebras and Algebraic Logic

Springer

JÁNOS BOLYAI MATHEMATICAL SOCIETY

Hajnal Andréka
Alfréd Rényi Institute of Mathematics
Hungarian Academy of Sciences
Reáltanoda u. 13-15
Budapest 1053
Hungary
e-mail: andreka.hajnal@mta.renyi.hu

Miklós Ferenczi
Department of Algebra
Budapest University of Technology and Economics
Egry J. u.1
Budapest 1111
Hungary
e-mail: ferenczi@math.bme.hu

István Németi
Alfréd Rényi Institute of Mathematics
Hungarian Academy of Sciences
Reáltanoda u. 13-15
Budapest 1053
Hungary
e-mail: nemeti.istvan@mta.renyi.hu

Mathematics Subject Classification (2010):
03G15, 06E25, 03B10, 03B45, 03B48, 03B25, 03C05, 03G27, 03C40, 03H05, 03B15

Library of Congress Control Number: 2012952598

ISSN 1217-4696
ISBN 978-3-642-35024-5 Springer Berlin Heidelberg New York
ISBN 978-963-9453-15-9 János Bolyai Mathematical Society, Budapest

Springer is a part of Springer Science+Business Media
springer.com

Cover photo of Alfred Tarski made by Steven Givant

Cover design: WMXDesign GmbH, Heidelberg

Printed on acid-free paper 44/3142/db – 5 4 3 2 1 0

Contents

III. Notions of Logic Related to Cylindric-Like Algebras

IV. Applications of Cylindric-Like Algebras

V. Other Algebraic Versions of Logic

VI. Connections with Abstract Algebraic Logic and Universal Logic

Dedicated to the memory of

Leon Henkin,

friend and co-author

Introduction

The theory of cylindric algebras (CAs) and Tarskian algebraic logic have been immensely successful in, e.g., applications to such a diversity of fields as logic, mathematics, computer science, artificial intelligence, linguistics and even relativity theory. In some of the applications one has to transcend (a little) the original definition of CAs but *not its spirit.* For example, the emphasis shifted to relativized CAs and to adding extra operations such as, e.g., substitutions (s_τs). To keep the original Tarskian spirit of CAs, in this volume we consider only quasi-polyadic operations (e.g., substitutions where τ is finite) as opposed to Halmos' polyadic ones (e.g., substitutions with all infinite τs). These and similar generalizations are the reason for "cylindric-like" algebras in the title in place of simply CAs.

In the literature of cylindric-like algebras, the following three books are regarded as the main source of the theory: [Hen-Mon-Tar,85], Part I, [Hen-Mon-Tar-And-Nem,81] and [Hen-Mon-Tar,85], Part II. Sometimes, they are referred to as [CA1], [CA1.5], and [CA2], respectively. We will use this notation ([CA1] etc.) in the present introduction. We will refer to the three of them together as [CA]. Since those volumes were completed before 1985 (almost 30 years ago) and since the theory grew and diversified much in the meantime, a general consensus formed that it would be timely to edit a new book [CA3] summarizing new trends, new directions of development/applications in a coherent volume, using the terminology, notation and overall structure of the original [CA]. For completeness, we note that after [CA2] important volumes did appear on the topic, some of them being [And-Mon-Nem,91b], [Ber-Mad-Pig,90], [Cra,06], [Csir-Gab-Rij,95], [Gab-Kur-Wol-Zak,03], [Hir-Hod,02a], [Mad,06], [Mar-Pol-Mas,96], [Mar-Ven,97], [Tar-Giv,87], but neither of them had the just described ambition.

The present volume [CA3] intends to be a guide to the field of cylindric-like algebras and algebraic logic as it developed after the completion of [CA2]. It is also the purpose of the present volume to highlight new horizons, new research directions found after [CA2], promising new possibilities for research and application of cylindric-like algebras. Let us take an example. The application of Boolean algebras (BAs) to logic was spectacularly successful. At the time [CA1] and [CA1.5] were written, this success story of BAs could not be repeated for CA-theory, e.g., because of the nonexistence of a Stone-type representation theorem for CAs. Actually, such a representation theorem was known to be impossible by the non-finitizability results of J. Donald Monk, Roger Maddux, James Johnson, Hajnal Andréka, Balázs Bíró and others. This mercilessly unfriendly picture started to soften when the Resek–Thompson result with Andréka's simple proof appeared in 1988 [And-Tho,88]. This meant that shifting emphasis to representing CAs by *relativized* cylindric set algebras in place of "square" cylindric set algebras yielded positive representability (hence axiomatizability) results. Indeed, this insight (together with [CA1.5] and [Nem,86]) led eventually to a flurry of applications of cylindric-like algebras to logic as summarized in Johan van Benthem's paper in this volume [Ben,thisVol]. (For completeness we note that a distant hope for such a positive turn was found already in [CA2] in the form of Remarks 3.2.88, p. 101.) Representation theory of cylindric-like algebras is nicely developing in the direction of using more and more natural variants of relativised set algebras, see, e.g., by Maddux, Andréka and Ferenczi: [Mad,89b], [And,01] and [Fer,12a]; developments in this field are surveyed in Miklós Ferenczi's paper in this volume [Fer,thisVol,a].

We dedicate this volume to the memory of Leon Henkin, our friend, collaborator, co-author and co-teacher, who passed away in 2006. Leon had a noble unselfish passion (and insistence) on looking for positive results (as a reaction to discovering negative ones). His positive-oriented influence can be felt throughout logic, not just in algebraic logic. Henkin-type semantics and his completeness theorem for higher-order logics is an example of this approach. Diane Resek was a student of Henkin, and their work on relativized cylindric algebras and merry-go-round equations led to the just discussed Resek–Thompson–Andréka turn. A further example of Henkin's striving for positive results (in the face of negative ones) is his method of "twisting elements" which he designed for non-standard representation of non-representable CAs ([CA2], 3.2.71, p. 89). Indeed, Henkin conjectured that by his methods of relativization and twisting, all CAs might be representable (in terms of square set-CAs). This conjecture was confirmed by András Simon in 1999; it is summarized in his paper for this volume [Sim,thisVol]. In the spirit of Leon's work, we would like to pass on his dedication for advancing science by striving for positive contributions, to our readers, our students, and to future generations.

The papers in this volume are survey papers expanded by proofs or by ideas of proofs. All of them were invited by the editors. The subjects were carefully selected in order to provide a balanced representation of the development of the field after completion of [CA2], and we are pleased with the result. There are subjects, though, we would have liked to devote separate papers to, but which were omitted from the list of papers for one reason or another. Below we list some of these subjects with some pointers to the literature:

– relation algebras (RAs) (two monographs appeared recently about these [Hir-Hod,02a], [Mad,06] and [Giv,12] by Steven R. Givant is to appear; further, paying proper homage to RAs would have overthrown the balance of [CA3]),

– translation of logic to algebra and back, the process of algebraization of logics and the "bridge theorems" (development after [CA2] in this important subject is discussed in detail in [And-Nem-Sai,01], [Nem,87a], [Nem,90] and also, in less detail, in the PhD dissertations of Eva Hoogland, Szabolcs Mikulás and others in Amsterdam from around 2000; György Serény's paper [Ser,thisVol] discusses the model theoretic part or aspects of this "translation of logic to algebra and back" subject under the name "cylindric algebraic model theory", and this subject is also touched upon in this volume in [Sai,thisVol], [Say,thisVol,a]),

– the finitization problem and its positive solutions (by Ildikó Sain and Viktor Gyuris [Sai-Gyu,97], Vera Stebletsova and Yde Venema [Ste,00], [Ste-Ven,98], and via non-well-founded set theory by Ági Kurucz and István Németi [Kur-Nem,00]; this finitization problem/issue is discussed (to some extent) in the updated version of [Nem,91]),

– the related weakly higher-order algebraic logic approach (CA $\uparrow$) (by Németi, Simon, and Gábor Sági [Nem-Sim,09], [Sag,00]),

– Galois theory of CAs by, e.g., Stephen Comer [Com,84],

– connection between CAs and RAs (by Maddux, Simon, Robin Hirsch, Ian Hodkinson and Németi [Nem-Sim,97], [Hir-Hod,02a]).

In this volume, some familiarity with the central notions of [CA] is assumed. [CA3] uses the notation and terminology of [CA]. Further, [CA3] follows the logical structure (chapter titles) of [CA] with some minor changes. Relative to [CA], there are two new kinds of chapters: Chapter 4 (Applications of cylindric-like algebras) and Chapter 6 (Connections with abstract algebraic logic and universal logic). Both topics, Chapters 4 and 6, are regarded as "hot topics" recently, e.g., there is an article on the latter in the Stanford Encyclopedia of Philosophy. The bibliography of [CA3] intends to cover most publications, in the subject, which appeared after [CA2]. An

Internet site devoted to open problems in cylindric-like algebras as well as following the status of problems published in, e.g., [CA] is under preparation.

We want to express our grateful thanks to J. Donald Monk and Leon Henkin whose consistent help and support throughout the years beginning with the conception of [CA1.5] made this volume possible. We want to express our thanks to all the authors of this volume for accepting our invitation and contributing excellent papers. Finally, warm thanks go to Ildikó Miklós who typed the volume, trying to coordinate the typographies of the papers originating from 19 authors.

Budapest, July 2012. The Editors

I. Algebraic Notions Applied to Cylindric-Like Algebras

BOLYAI SOCIETY
MATHEMATICAL STUDIES, 22

**Cylindric-like Algebras
and Algebraic Logic**
pp. 15–35.

Reducing First-order Logic to Df_3, Free Algebras

HAJNAL ANDRÉKA and ISTVÁN NÉMETI

Alfred Tarski in 1953 formalized set theory in the equational theory of relation algebras [Tar,53a, Tar,53b]. Why did he do so? Because the equational theory of relation algebras (RA) corresponds to a logic without individual variables, in other words, to a propositional logic. This is why the title of the book [Tar-Giv,87] is "Formalizing set theory without variables". Tarski got the surprising result that a propositional logic can be strong enough to "express all of mathematics", to be the arena for mathematics. The classical view before this result was that propositional logics in general were weak in expressive power, decidable, uninteresting in a sense. By using the fact that set theory can be built up in it, Tarski proved that the equational theory of RA is undecidable. This was the first propositional logic shown to be undecidable.

From the above it is clear that replacing RA in Tarski's result with a "weaker" class of algebras is an improvement of the result and it is worth doing. For more on this see the open problem formulated in Tarski–Givant ([Tar-Giv,87, p. 89, line 2 bottom up – p. 90, line 4 and footnote 17 on p. 90]).

A result of J. D. Monk says that for every finite n there is a 3-variable first-order logic (FOL) formula which is valid but which can be proved (in FOL) with more than n variables only (cf. [Hen-Mon-Tar,85, 3.2.85]). Intuitively this means that during any proof of this formula there are steps when we have to use n independent data (stored in the n variables as in n machine registers). For example, the associativity of relation composition of binary relations can be expressed with 3 variables but 4 variables are needed for any of its proofs.

Tarski's main idea in [Tar-Giv,87] is to use pairing functions to form ordered pairs, and so to store two pieces of data in one register. He used this technique to translate usual infinite-variable first-order logic into the three-variable fragment of it. From then on, he used the fact that any three-variable-formula about binary relations can be expressed by an RA-equation, [Hen-Mon-Tar,85, 5.3.12]. He used two registers for storing the data belonging to a binary relation and he had one more register available for making computations belonging to a proof.

The finite-variable fragment hierarchy of FOL corresponds to the appropriate hierarchy of cylindric algebras (CA_n's). The n-variable fragment $\mathcal{L}_n$ of FOL consists of all FOL-formulas which use only the first n variables. By Monk's result, $\mathcal{L}_n$ is essentially incomplete for all $n \geq 3$, it cannot have a finite Hilbert-style complete and strongly sound inference system. We get a finite Hilbert-style inference system $\vdash_{\overline{n}}$ for $\mathcal{L}_n$ by restricting a usual complete one for infinite-variable FOL to the first n variables (see [Hen-Mon-Tar,85, Sec. 4.3]). This inference system $\vdash_{\overline{n}}$ belonging to $\mathcal{L}_n$ is a translation of an equational axiom system for CA_n, it is strongly sound but not complete: $\vdash_{\overline{n}}$ is much weaker than validity $\models_n$ (which is the restriction of $\models$ to the formulas in $\mathcal{L}_n$).

Relation algebras are halfway between CA_3 and CA_4, the classes of 3-dimensional and 4-dimensional cylindric algebras, respectively. We sometimes jokingly say that RA is $\mathsf{CA}_{3.5}$. Why is RA stronger than CA_3? Because, the so-called relation-algebra-type reduct of a CA_3 is not necessarily an RA, e.g., associativity of relation composition can fail in the reduct. See [Hen-Mon-Tar,85, Sec. 5.3], and for more in this line see [Nem-Sim,97]. Why is CA_4 stronger than RA? Because not every RA can be obtained, up to isomorphism, as the relation-algebra-type reduct of a CA_4, and consequently not every 4-variable sentence can be expressed as an RA-term. However, the same equations are true in RA and in the class of all relation-algebra-type reducts of CA_4's (Maddux's result, see [Hen-Mon-Tar,85, Sec. 5.3]). Thus Tarski formulated set theory, roughly, in CA_4, i.e., in $\mathcal{L}_4$ with $\vdash_{\overline{4}}$, or in $\mathcal{L}_3$ with validity $\models_3$.

Németi ([Nem,85b], [Nem,86a]) improved this result by formalizing set theory in CA_3, i.e., in $\mathcal{L}_3$ with $\vdash_{\overline{3}}$ in place of validity $\models_3$. The main idea for this improvement was using the pairing functions to store all data always, during every step of a proof, in one register only, so as to get two registers to work with in the proofs. In this approach one represents binary relations as unary ones (of ordered pairs).

First-order logic has equality as a built-in relation. One of the uses of equality in FOL is that it can be used to express (simulate) substitutions of variables, thus to "transfer" content of one variable to the other. The reduct SCA_3 of CA_3 "forgets" equality d_{ij} but retains substitution in the form of the term-definable operations s^i_j. The logic belonging to SCA_3 is weaker than 3-variable fragment of FOL. Zalán Gyenis [Gye,11] improved parts of Németi's result by extending them from CA_3 to SCA_3.

We get a much weaker logic by forgetting substitutions, too, this is the logic corresponding to Df_3 in which FOL and set theory were formalized in [And-Nem,11a].

Three-dimensional diagonal-free cylindric algebras, Df_3's, are Boolean algebras with 3 commuting complemented closure operators, see [Hen-Mon-Tar,85, 1.1.2] or [Kur,thisVol]. The logic $\mathcal{L}df_3$ corresponding to Df_3 has several intuitive forms, one is 3-variable equality- and substitution-free fragment of first-order logic with a rather weak proof system $\vdash^{df}$, another form of this same logic is modal logic $[\mathbf{S5}, \mathbf{S5}, \mathbf{S5}]$, see [Kur,thisVol] and [Gab-Kur-Wol-Zak,03]. Not only set theory but the whole of FOL is recaptured in $\mathcal{L}df_3$. This is a novelty w.r.t. previous results in this line. All the formalizability theorems mentioned above follow from this last result.

In Section 1 we define our weak "target logic" $\mathcal{L}df_3$ and we state the existence of a structural translation mapping of FOL with countably many relation symbols of arbitrary ranks, $\mathcal{L}_\omega$, into $\mathcal{L}df_3$ with a single ternary relation symbol, see Theorem 1.1.6. If equality is available in our target logic, then we can do with one binary relation symbol, we do not need a ternary one, see Theorem 1.1.7. For theories in which a conjugated pair of quasi-pairing functions can be defined, such as most set theories, we can define a similar translation function which preserves meaning of formulas a bit more closely, see Theorem 1.1.6(ii), Theorem 1.1.7(ii). Theorem 1.1.6(ii) is a very strong version of Tarski's main result in [Tar-Giv,87, Theorem (xxxiv), p. 122], which states roughly the same for the logic corresponding to RA in place of Df_3. After Theorem 1.1.7 we discuss the conditions in both Theorem 1.1.6 and Theorem 1.1.7, and we obtain that almost all of them are needed and that they cannot be substantially weakened.

In Sections 2 and 3 we concentrate on the applications of the theorems stated in Section 1. In Section 2 we show that our translation functions are useful in proving properties for n-variable logics as well as for other "weak" logics. In particular, we prove a partial completeness theorem for the n-variable fragment of FOL ($n \geq 3$) and we prove that Gödel's incompleteness property holds for it.

In Section 3 we review some results and problems on free cylindric-like algebras from the literature since 1985. As an application of the theorems in

Section 1, we show that the free cylindric algebras are not atomic (solution for [Hen-Mon-Tar,85, Problem 4.14]) and that these free algebras are not "wide", i.e., the $k+1$-generated free cylindric algebra cannot be embedded into the k-generated one, but these free algebras have many k-element irredundant non-free generator sets (solution for [Hen-Mon-Tar,85, Problem 2.7]).

1. Interpreting FOL in its Small Fragments

Instead of Df_3 and CA_3, we will work with fragments of FOL which are equivalent to them because this will be convenient when stating our theorems. We treat FOL as [Hen-Mon-Tar,85] does, i.e., with equality and with no operation symbols. We deviate from [Hen-Mon-Tar,85] in that our connectives are $\vee$, $\neg$, $\exists v_i$, $v_i = v_j$, $i, j \in \omega$ and we treat the rest as derived ones, by defining $\varphi \wedge \psi \stackrel{d}{=} \neg(\neg\varphi \vee \neg\psi)$, $\bot \stackrel{d}{=} (v_0 = v_0 \wedge \neg v_0 = v_0)$, $\top \stackrel{d}{=} \neg\bot$. We will use the derived connectives $\forall$, $\rightarrow$, $\leftrightarrow$, too, as abbreviations: $\forall v\varphi \stackrel{d}{=} \neg\exists v\neg\varphi$, $\varphi \rightarrow \psi \stackrel{d}{=} \neg\varphi \vee \psi$, $\varphi \leftrightarrow \psi \stackrel{d}{=} (\varphi \rightarrow \psi) \wedge (\psi \rightarrow \varphi)$. We will use x, y, z to denote the first three variables v_0, v_1, v_2. Sometimes we will write, e.g., $\exists xy$ or $\forall xyz$ in place of $\exists x\exists y$ or $\forall x\forall y\forall z$, respectively.

We begin with defining the fragment $\mathcal{L}df_3(P, 3)$ of FOL. It contains three variables and one ternary relation symbol P. It is a fragment of FOL in which we omit the equality, quantifiers $\exists v$ for v distinct from x, y, z, and atomic formulas $P(u, v, w)$ for $uvw \neq xyz$; and we omit all relation symbols distinct from P.

Definition 1.1.1 (3-variable restricted FOL without equality $\mathcal{L}df_3(P, 3)$).

(i) The *language* of our system contains one atomic formula, namely $P(x, y, z)$. (E.g., the formula $P(y, x, z)$ is not available in this language, this feature is what the adjective *"restricted"* refers to.) The logical connectives are $\vee$, $\neg$, $\exists x$, $\exists y$, $\exists z$. Thus, the set Fdf_3 of formulas of $\mathcal{L}df_3(P, 3)$ is the smallest set F containing $P(x, y, z)$ and such that $\varphi \vee \psi$, $\neg\varphi$, $\exists x\varphi$, $\exists y\varphi$, $\exists z\varphi \in F$ whenever $\varphi, \psi \in F$.

(ii) The *proof system* $\vdash^{df}$ which we will use is a Hilbert-style one with the following logical axiom schemes and rules.

The logical axiom schemes are the following. Let $\varphi, \psi \in Fdf_3$ and $v, w \in \{x, y, z\}$.

((1)) φ, if φ is a propositional tautology.

((2)) $\forall v(\varphi \to \psi) \to (\exists v\varphi \to \exists v\psi)$.

((3)) $\varphi \to \exists v\varphi$.

((4)) $\exists v\exists v\varphi \to \exists v\varphi$.

((5)) $\exists v(\varphi \lor \psi) \leftrightarrow (\exists v\varphi \lor \exists v\psi)$.

((6)) $\exists v\neg\exists v\varphi \to \neg\exists v\varphi$.

((7)) $\exists v\exists w\varphi \to \exists w\exists v\varphi$.

The inference rules are Modus Ponens ((MP), or detachment), and Generalization ((G)).

(iii) We define $\mathcal{L}df_3(P, 3)$ as the logic with formulas Fdf_3 and with proof system $\models^{df}$.

(iv) We define $\mathcal{L}df_n(\mathcal{R}, \rho)$ where n is an ordinal, $\mathcal{R}$ is a sequence of relation symbols and ρ is the sequence of their ranks (i.e., numbers of arguments), all $\leq n$, analogously to $\mathcal{L}df_3(P, 3)$. When we do not indicate $\mathcal{R}$, ρ in $\mathcal{L}df_n(\mathcal{R}, \rho)$, we mean to have infinitely many n-place relation symbols. ■

The fragment $\mathcal{L}ca_3$ is similar to the above fragment $\mathcal{L}df_3$, except that we do not omit equality from the language, hence we will have $u = v$ as formulas for $u, v \in \{x, y, z\}$, and we will have two more axiom schemes concerning equality in the proof system. Since we have equality, our "smallest interesting" language will be when we have one binary relation symbol E.

Definition 1.1.2 (3-variable restricted FOL with equality $\mathcal{L}ca_3(E, 2)$).

(i) The *language* of our system contains one atomic formula, namely $E(x, y)$. The logical connectives are $\lor$, $\neg$, $\exists x$, $\exists y$, $\exists z$ together with $u = v$ for $u, v \in \{x, y, z\}$ as zero-place connectives. Thus, $x = x$, $x = y$, etc are formulas of $\mathcal{L}ca_3$. We denote the set of formulas (of $\mathcal{L}ca_3$) by Fca_3.

(ii) The *proof system* $\models^{ca}$ which we will use is a Hilbert-style one with the logical axiom schemes and rules of $\mathcal{L}df_3$ (understood as schemes for $\mathcal{L}ca_3$) extended with the following two axiom schemes:

Let $\varphi \in Fca_3$ and $u, v, w \in \{x, y, z\}$.

((8)) $(u = v \to v = u) \land (u = v \land v = w \to u = w) \land \exists v \; u = v$.

((9)) $u = v \land \exists v(u = v \land \varphi) \to \varphi$, where u, v are distinct.

(iii) We define $\mathcal{L}ca_3(E,2)$ as the logic with formulas Fca_3 and with proof system $\vdash^{ca}$.

(iv) We define $\mathcal{L}ca_n(\mathcal{R},\rho)$ where n is an ordinal, $\mathcal{R}$ is a sequence of relation symbols and ρ is the sequence of their ranks, all $\leq n$, analogously to $\mathcal{L}ca_3(E,2)$. When we do not indicate $\mathcal{R}$, ρ in $\mathcal{L}ca_n(\mathcal{R},\rho)$, we mean to have infinitely many n-place relation symbols. ■

Remark 1.1.3 (On the fragment $\mathcal{L}df_3$ of FOL).

(i) The proof system $\vdash^{df}$ is a direct translation of the equational axiom system of Df_3. Axiom ((2)) is needed for ensuring that the equivalence relation defined on the formula algebra by $\varphi \equiv \psi \Leftrightarrow \vdash^{df} \varphi \leftrightarrow \psi$ be a congruence with respect to (w.r.t.) the operation $\exists v$. It is congruence w.r.t. the Boolean connectives $\vee$, $\neg$ by axiom ((1)). Axiom ((1)) expresses that the formula algebra factorized with $\equiv$ is a Boolean algebra, axiom ((5)) expresses that the quantifiers $\exists v$ are operators on this Boolean algebra (i.e., they distribute over $\vee$), axioms ((3)), ((4)) express that these quantifiers are closure operations, axiom ((6)) expresses that they are complemented closure operators (i.e., the negation of a closed element is closed again). Together with ((5)) they imply that the closed elements form a Boolean subalgebra, and hence the quantifiers are normal operators (i.e., the Boolean zero is a closed element). Finally, axiom ((7)) expresses that the quantifiers commute with each other. We note that ((1)) is not an axiom scheme in the sense of [And-Nem-Sai,01] since it is not a formula scheme, but it can be replaced with three formula schemes, see [Hen-Mon-Tar,85, Problem 1.1] (solved in [McC,97]).

(ii) The logic $\mathcal{L}df_3$ corresponds to Df_3 in the sense of [Hen-Mon-Tar,85, Sec. 4.3], as follows. What is said in (i) above immediately implies that the proof-theoretic (Lindenbaum–Tarski) formula algebra of $\mathcal{L}df_3$ (which is just the natural formula-algebra factorized by the equivalence relation $\equiv$ defined in (i) above) is the infinitely generated Df_3-free algebra, and that of $\mathcal{L}df_3(P,3)$ is the one-generated Df_3-free algebra. Moreover, valid formulas of $\mathcal{L}df_3$ correspond to equations valid in Df_3, namely we claim that $\mathcal{L}df_3 \vdash^{df} \varphi \Leftrightarrow \mathsf{Df}_3 \models \tau\mu(\varphi) = 1$ for all $\varphi \in Fdf_3$, where $\tau\mu(\varphi)$ is as defined in [Hen-Mon-Tar,85, 4.3.55].

(iii) The logic $\mathcal{L}df_3$ inherits a natural semantics from first-order logic (namely Mod, the class of models of FOL, and $\models_3$, the validity relation restricted to 3-variable formulas). The proof system $\vdash^{df}$ is strongly sound with respect to this semantics, but it is not complete, for more on this see Section 2. We note that just as Df_3 corresponds to the logic $\mathcal{L}df_3 = \langle Fdf_3, \models^{df} \rangle$, the class of algebras corresponding to $\langle Fdf_3, \models_3 \rangle$ is the class RDf_3 of representable diagonal-free cylindric algebras ([Hen-Mon-Tar,85, 5.1.33(v)]). For more on connections between logics and classes of algebras, besides [Hen-Mon-Tar,85, Sec. 4.3], see [And-Nem-Sai,01], or [Sai,thisVol].

The expressive power of $\mathcal{L}df_3$ is seemingly very small. It's not only that "we cannot count" due to lack of the equality, we cannot transfer any information from one variable to the other by the use of the equality, so all such transfer must go through an atomic formula. Hence if we have only binary relation symbols, in the restricted language there is just no way of meaningfully using the third variable z, and we basically have two-variable logic which is decidable. However, Theorem 1.1.6 below says that if we have at least one ternary relation symbol and we are willing to express formulas in a more complicated way (than the most natural one would be), then we can express any sentence that we can in FOL.

(iv) In the present paper we will use $\mathcal{L}df_3$ as introduced above because it will be convenient to consider it a fragment of FOL. However, $\mathcal{L}df_3$ has several different but equivalent forms, each of which has advantages and disadvantages. Some of the different forms are reviewed in [And-Nem,11a, Sec. 2]. We mention two of the equivalent forms. One is modal logic $[\mathbf{S5}, \mathbf{S5}, \mathbf{S5}]$, this is equivalent to $\mathcal{L}df_3$ (while the modal logic $\mathbf{S5} \times \mathbf{S5} \times \mathbf{S5}$ introduced in [Kur,thisVol] is equivalent with $\langle Fdf_3, \models_3 \rangle$), see [Gab-Kur-Wol-Zak,03, p. 379]. The other equivalent form is just equational logic with the defining equations of Df_3 as extra axioms. ■

Remark 1.1.4 (On the fragment $\mathcal{L}ca_3$ of FOL).

(i) The logic $\mathcal{L}ca_3$ corresponds to CA_3 just the way $\mathcal{L}df_3$ corresponds to Df_3. The proof system $\models^{ca}$ is a direct translation of the equational axiom system of CA_3. ((8)) expresses that $=$ is an equivalence relation and ((9)) expresses that formulas do not distinguish equivalent (equal) elements. Take the Hilbert-style proof system with axiom schemes ((1))–((9)) and rules as (MP) and (G). Add the axioms

((0)) $R(v_{i1}, \ldots, v_{in}) \leftrightarrow \exists v_j R(v_{i1}, \ldots, v_{in})$ for R an n-place relation symbol and $j \notin \{i1, \ldots, in\}$.

Then the so obtained proof system is complete for FOL (with usual semantics Mod, $\models$). Hence, $\mathcal{L}df_3$ and $\mathcal{L}ca_3$ are "proof-theoretic" fragments of FOL when taking this complete proof system for FOL.

(ii) The expressive power of $\mathcal{L}ca_3$ is much greater than that of $\mathcal{L}df_3$, due to the presence of equality. E.g., one can express that a binary relation is actually a function, one can express composition of binary relations, one can express (simulate) substitution of variables. However, the proof system $\models^{ca}$ is still very weak, e.g., one can express but cannot prove the following: the composition of two functions is a function again, composition of binary relations is associative, converse of the converse of a binary relation is the original one, interchanging the variables x, y in two different ways by using z as "auxiliary register" results in an equivalent formula (this is the famous merry-go-round equation [Hen-Mon-Tar,85, 3.2.88], see also [Fer,thisVol,a], [Sim,thisVol]). More precisely, one cannot prove these statements if one expresses them the most natural ways. Our theorems below say that if we express the same statements in more involved ways, they become $\models^{ca}$-provable. ■

Let $\mathcal{L}_\omega$ denote usual FOL with countably many variables and with countably many relation symbols for each rank, i.e., we have countably many n-place relation symbols for all positive n. Let L_ω denote the set of formulas of $\mathcal{L}_\omega$. Thus $\mathcal{L}_\omega = \langle L_\omega, \vdash\rangle$ where $\vdash$ is either the proof system outlined in Remark 1.1.4(i) above, or just the usual semantic consequence relation $\models$. We assume that E is a binary and P is a ternary relation symbol in $\mathcal{L}_\omega$. Then $L_\omega(E, 2)$ denotes the set of formulas in $\mathcal{L}_\omega$ in which only E occurs from the relation symbols. Zermelo–Fraenkel set theory written up in $L_\omega(E, 2)$ is denoted by ZF. A formula of $\mathcal{L}_\omega$ is called a *sentence* if it does not contain free variables.

Definition 1.1.5 (Structural translations). Let $\mathcal{L} = \langle F, \vdash\rangle$ be a logic (in the sense of Remark 1.1.8(i) below). Assume that $\vee, \neg$ are connectives in $\mathcal{L}$, and let $\rightarrow$ denote the corresponding derived connective in $\mathcal{L}$, too. Let $f : L_\omega \rightarrow F$ be an arbitrary function. We say that f is *structural* iff the following (i)–(ii) hold for all sentences $\varphi, \psi \in L_\omega$.

(i) $\vdash f(\varphi \vee \psi) \leftrightarrow \left[f(\varphi) \vee f(\psi)\right]$,

(ii) $\vdash f(\varphi \rightarrow \psi) \rightarrow \left[f(\varphi) \rightarrow f(\psi)\right]$. ■

The following is proved in [And-Nem,11a] together with [And-Nem,11b], by defining concrete translations tr. For the role of $\neg\,\mathsf{tr}(\bot)$ in (ii) below see Remark 1.1.8(iii), (iv).

Theorem 1.1.6 (Formalizability of FOL in $\mathcal{L}df_3$).

(i) *There is a structural computable translation function* $\mathsf{tr} : \mathcal{L}_\omega \to \mathcal{L}df_3(P,3)$ *such that* tr *has a decidable range and the following* (a), (b) *are true for all sets of sentences* $Th \cup \{\varphi\}$ *in* $\mathcal{L}_\omega$:

(a) $Th \models \varphi$ *iff* $\mathsf{tr}(Th) \models^{df} \mathsf{tr}(\varphi)$.

(b) $Th \models \varphi$ *iff* $\mathsf{tr}(Th) \models \mathsf{tr}(\varphi)$.

(ii) *There is a structural computable translation function* $\mathsf{tr} : \mathcal{L}_\omega(E,2) \to \mathcal{L}df_3(P,3)$ *such that* tr *has a decidable range and the following* (c), (d) *are true, where* Δ *denotes the set of the following two formulas:*

$$E(x,y) \leftrightarrow \forall z P(x,y,z) \wedge \exists xyz\big[P(x,y,z) \wedge \neg\forall z P(x,y,z)\big],$$

$$x = y = z \leftrightarrow P(x,y,z) \wedge \neg E(x,y).$$

(c) *Statements* (a) *and* (b) *in* (i) *above hold for all sets of sentences* $Th \cup \{\varphi\}$ *in* $\mathcal{L}_\omega(E,2)$ *such that* $Th \cup \Delta \models \neg\mathsf{tr}(\bot)$. *Further,* $ZF \cup \Delta \models \neg\mathsf{tr}(\bot)$.

(d) $\Delta \cup \{\neg\mathsf{tr}(\bot)\} \models \varphi \leftrightarrow \mathsf{tr}(\varphi)$ *for all sentences* $\varphi \in L_\omega(E,2)$. ■

The following is proved in [Nem,85b], [Nem,86a] (taken together with [And-Nem,11b]), by constructing concrete tr's. It says that we can replace the ternary relation symbol P with a binary one in Theorem 1.1.6 if we have equality.

Theorem 1.1.7 (Formalizability of FOL in $\mathcal{L}ca_3$).

(i) *There is a computable, structural translation function* $\mathsf{tr} : \mathcal{L}_\omega \to \mathcal{L}ca_3(E,2)$ *such that* tr *has a decidable range and the following* (a), (b) *are true for all sets of sentences* $Th \cup \{\varphi\}$ *in* $\mathcal{L}_\omega$:

(a) $Th \models \varphi$ *iff* $\mathsf{tr}(Th) \models^{ca} \mathsf{tr}(\varphi)$.

(b) $Th \models \varphi$ *iff* $\mathsf{tr}(Th) \models \mathsf{tr}(\varphi)$.

(ii) *There is a computable, structural translation function* $\mathsf{tr} : \mathcal{L}_\omega(E,2) \to \mathcal{L}ca_3(E,2)$ *such that* tr *has a decidable range and the following* (c), (d) *are true:*

(c) *Statements* (a) *and* (b) *in* (i) *above hold for all sets of sentences* $Th \cup \{\varphi\}$ *in* $\mathcal{L}_\omega(E,2)$ *such that* $Th \models \neg\mathsf{tr}(\bot)$. *Further,* $ZF \models \neg\mathsf{tr}(\bot)$.

(d) $\neg\mathsf{tr}(\bot) \models \varphi \leftrightarrow \mathsf{tr}(\varphi)$, *for all sentences* $\varphi \in L_\omega(E,2)$. ■

On the proof of Theorem 1.1.7. The most difficult part of this theorem is proving $\models \varphi \Rightarrow \models^{ca} \mathsf{tr}(\varphi)$, for all $\varphi \in L_\omega$, therefore we will outline the ideas for proving this part. So, we want to prove a kind of completeness theorem for $\models^{ca}$.

Formulas $\varphi(x,y)$ with two free variables x, y represent binary relations and then the natural way of expressing relation composition of binary relations is the following:

$$(\varphi \circ \psi)(x,y) \stackrel{d}{=} \exists z\big(\varphi(x,z) \wedge \psi(z,y)\big), \quad \text{where}$$

$$\varphi(x,z) \stackrel{d}{=} \exists y\big(y = z \wedge \varphi(x,y)\big) \quad \text{and} \quad \psi(z,y) \stackrel{d}{=} \exists x\big(x = z \wedge \psi(x,y)\big).$$

Now, assume that we have two unary *partial* functions, p, q which form pairing functions, i.e. for which the following formula π holds:

$$\pi \stackrel{d}{=} \forall xy \exists z(\mathsf{p}(z) = x \wedge \mathsf{q}(z) = y).$$

For supporting intuition, let us write $z_0 = x$ and $z_1 = y$ in place of $\mathsf{p}(z) = x$ and $\mathsf{q}(z) = y$, and let $\langle x, y\rangle$ denote an arbitrary z for which $z_0 = x$ and $z_1 = y$. Now, we can "code" binary relations as unary ones, i.e., if $\varphi(x)$ is a formula with one free variable x, then we can think of it as representing the binary relation $\{\langle x_0, x_1\rangle : \varphi(x)\}$. With this in mind then a natural way of representing relation composition is the following

$$(\varphi \odot \psi)(x) \stackrel{d}{=} \exists y(\varphi(y_0) \wedge \psi(y_1) \wedge x_0 = y_{00} \wedge y_{01} = y_{10} \wedge y_{11} = x_1).$$

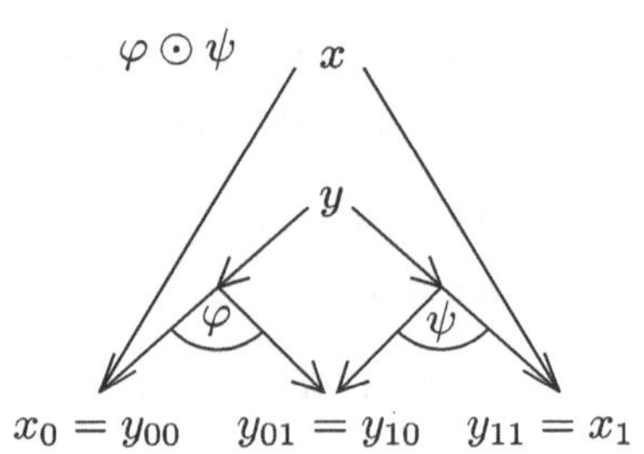

Fig. 1. Illustration of $\varphi \odot \psi$

As we said in Remark 1.1.4(ii), associativity of $\circ$ cannot be proved by $\models^{ca}$, i.e., there are formulas φ, ψ, η in Fca_3 such that

$$\not\models^{ca} \big((\varphi \circ \psi) \circ \eta\big)(x,y) \leftrightarrow \big(\varphi \circ (\psi \circ \eta)\big)(x,y).$$

However, associativity of relation composition expressed in the unary form can be proved, by assuming a formula $\mathsf{Ax} \in Fca_3$ which is semantically equivalent with π but proof-theoretically stronger:

$$\mathsf{Ax} \models^{ca} \big((\varphi \odot \psi) \odot \eta\big)(x) \leftrightarrow \big(\varphi \odot (\psi \odot \eta)\big)(x)$$

for all formulas φ, ψ, η with one free variable x such that $\models^{ca} \varphi(x) \to \mathsf{pair}(x)$, etc, where $\mathsf{pair}(x) \overset{d}{=} \exists y x_0 = y \wedge \exists y x_1 = y$. (I.e., $\mathsf{pair}(x)$ holds for x if both p and q are defined on it.) We note that π is not strong enough for proving associativity of $\odot$, and even Ax is not strong enough for proving associativity of $\circ$, see [Nem,86a, 15T(ii), (iv)]. We mentioned already that $\models^{ca}$ cannot prove that composition of functions is a function again. Roughly, Ax is π together with stating that composition of at most three "copies" of p, q is a function again (i.e., $\mathsf{p} \circ \mathsf{p} \circ \mathsf{q}$, $\mathsf{p} \circ \mathsf{q}$ etc are all functions). Similarly to the above, we can express converse of binary relations and the identity relation (coded for their unary form) and prove for these by $\models^{ca}$ all the relation algebraic equations, from Ax of course. Thus we defined relation-algebra-type operations on the set of formulas of form $\varphi(x) \wedge \mathsf{pair}(x)$, and we can prove from Ax that these operations form an RA. If p, q can be expressed as above, then we have a so-called quasi-projective RA, a QRA, which are representable by [Tar-Giv,87, 8.4(iii)], and we know that representation theorems help us to get provability from validity (i.e., the hard direction of completeness theorems). It remains to get suitable pairing formulas p, q (see (1) below) and to translate all FOL-formulas, in a meaning-preserving way, to the above QRA-fragment of $\mathcal{L}ca_3$ (see (2) below).

(1) We can get p, q by "brute force": we add a new binary relation symbol E to our language, intuitively we will think of it as the element-of relation $\in$. Then we express ordered pairs the way usually done in set theory (i.e., $\langle x, y\rangle \overset{d}{=} \{\{x\}, \{x, y\}\}$), and realize that we can write up the two projection functions belonging to these using only three variables. By using these projection functions we can convert every FOL-formula to one in Fca_3 so that we preserve validity (we can use the pairing-technique to code all the relations into one binary one, and then we can code up all the variables into the first three ones). This part is not so difficult because we may think "semantically".

(2) It remains now to translate all 3-variable formulas $\varphi \in Fca_3$ into the QRA-fragment of $\mathcal{L}ca_3$ we obtained above. The paper Simon [Sim,07] comes to our aid. In [Sim,07], to every QRA a CA_3-type subreduct is defined which is representable, i.e., which is in RCA_3. Let $\mathfrak{C}$ be this subreduct of our above-defined QRA, then the universe of $\mathfrak{C}$ is a subset of Fca_3 and the operations

of $\mathfrak{C}$ are defined in terms of formulas of Fca_3, too. Let $f : Fca_3 \to \mathfrak{C}$ be a homomorphism (where the CA_3-type operations of Fca_3 are the natural ones), and then we define $\mathsf{tr} : Fca_3 \to Fca_3$ by $\mathsf{tr}(\varphi) \stackrel{d}{=} \mathsf{Ax} \to f(\varphi)$. Now, one can check that $\models \varphi \Rightarrow \models^{ca} \mathsf{tr}(\varphi)$. For the details of the proof outlined above and for the proof of Theorem 1.1.6 we refer the reader to [And-Nem,11a, And-Nem,11b, Nem,85b, Nem,86a]. ∎

Remark 1.1.8 (Discussion of the conditions in Theorems 1.1.6, 1.1.7).

(i) In Abstract Algebraic Logic, AAL, and/or in Universal Logic the key concept is a logical system (logic in short) $\langle F, \vdash \rangle$ where F is a set (thought of as the set of formulas) and $\vdash \subseteq \mathsf{Sb}(F) \times F$ (thought of as a consequence relation), where $\mathsf{Sb}(F)$ denotes the powerset of F. If $f : F \to F'$ is a function between two logics $\mathcal{L} = \langle F, \vdash \rangle$ and $\mathcal{L}' = \langle F', \vdash' \rangle$, then f is called a *translation* iff it preserves $\vdash$, i.e., iff $Th \vdash \varphi \Rightarrow f(Th) \vdash' f(\varphi)$ holds for all $Th \cup \{\varphi\} \subseteq F$, and f is called a *conservative translation* if $\Rightarrow$ can be replaced by $\Leftrightarrow$ in the above. Jeřábek [Jer,11] proved that FOL can be conservatively translated even to classical propositional calculus (CPC); and moreover, every countable logic can be conservatively translated to CPC. In this sense, the existence of conservative translations does not mean much in itself. However, if we require the translation to be computable in addition, then undecidability is preserved along the translation, and so FOL can be translated to undecidable logics only (i.e., where $\{\varphi \in F : \emptyset \vdash \varphi\}$ is undecidable), and so it cannot be translated to CPC. For this reason, the conditions that we have at least one at least ternary relation symbol, we have at least 3 quantifiers (closure operators), and that they commute are all necessary conditions for our target logic in Theorem 1.1.6 since without these conditions we get decidable logics. (We have seen that $\mathcal{L}df_3(P, 2)$ is basically 2-variable logic which is decidable [Hen-Mon-Tar,85, 4.2.7], and the logic we get from $\mathcal{L}df_n$ by omitting the axiom scheme ((7)) requiring that the quantifiers commute is proved to be decidable in [Nem,86a, Chap. III], [Nem,95, Theorem 1.1]). If we require more properties for the translation function to hold, then more properties are preserved along them. E.g., structural computable conservative translations preserve Gödel's incompleteness property from one logic to the other, see Theorem 1.2.4.

(ii) The achievement (of Theorems 1.1.6, 1.1.7) that the range of the translation is decidable can be omitted, since if we have a translation function then by using the trick in [Cra-Vau,58] we can modify this function so that its range becomes decidable and keep all the other good properties, at least in our case when our logics are extensions of CPC.

(iii) There are sentences in $L_\omega(E,2)$ which are not equivalent semantically to any formula in $L_3(E,2)$, hence there is no function $f : L_\omega \to L_3(E,2)$ for which $\models \varphi \leftrightarrow f(\varphi)$ would hold for all sentences $\varphi \in L_\omega$. For this reason, $\neg\,\mathsf{tr}(\bot)$ cannot be omitted in Theorems 1.1.6(ii)(d), 1.1.7(ii)(d). For example, such a 4-variable sentence is exhibited in [Nem,86a, p. 39]. We note that Δ in Theorem 1.1.6(ii) is an explicit "definition" of E and $=$ from P.

(iv) Our translation functions are not Boolean homomorphisms in general, e.g., the translations tr we define in the proofs of Theorems 1.1.6, 1.1.7 do not preserve negation in the way they preserve disjunction. Consequently, $\neg\,\mathsf{tr}(\bot)$ is not the same as $\mathsf{tr}(\top)$, and more importantly, $\neg\,\mathsf{tr}(\bot)$ is not a valid formula. From the fact that tr is structural, it can be proved that $\vdash \mathsf{tr}(\varphi) \leftrightarrow \left[\neg\,\mathsf{tr}(\bot) \to \mathsf{tr}(\varphi)\right]$. Hence, $\neg\,\mathsf{tr}(\bot)$ seems to be the weakest assumption under which one can expect semantical equivalence of φ with $\mathsf{tr}(\varphi)$. Intuitively, $\neg\,\mathsf{tr}(\bot)$ is the "background knowledge" we assume for the translation function tr to preserve meaning. This is the role of $\neg\,\mathsf{tr}(\bot)$ in Theorems 1.1.6(ii), 1.1.7(ii).

(v) A logic $\langle F,\vdash\rangle$ is defined to be a *propositional logic* (or sentential logic) in AAL if F is built up from some set, called propositional variables, by using connectives and $\vdash$ is *substitutional*, i.e., $\vdash$ is preserved by substitution of arbitrary formulas for propositional variables. In this sense, $\mathcal{L}df_3$ and $\mathcal{L}ca_3$ are propositional logics, but $\mathcal{L}_\omega$ is not, see e.g., [And-Nem-Sai,01] or [Sai,thisVol].

(vi) Any logic $\mathcal{L} = \langle F,\vdash\rangle$ which is between $\mathcal{L}df_3(P,3)$ and $\mathcal{L}_\omega$ can be taken in Theorem 1.1.6 in place of $\mathcal{L}df_3(P,3)$. (We say that a logic $\mathcal{L} = \langle F,\vdash\rangle$ is contained in another one, $\mathcal{L}' = \langle F',\vdash'\rangle$, if $F \subseteq F'$ and $\vdash \subseteq \vdash'$.) This is easy to check. ∎

Problem 1.1.9 (Interpreting FOL in weaker fragments). Can the requirement of the closure operators being complemented be omitted in our theorems? I.e., is there a computable (structural) conservative translation from $\mathcal{L}_\omega$ to the equational theory EqBf_3 of Bf_3 where Bf_3 denotes the class of all Boolean algebras with three commuting (not necessarily complemented) closure operators? Is EqBf_3 undecidable?

2. Applications of the Interpretation

There are many applications, of different flavors, of the interpretability theorems of which Theorem 1.1.6 is presently the strongest one. In this and the next sections we state some of these applications. We will concentrate on consequences for cylindric algebras, CA_n, RCA_n and their logical counterparts, but analogous results hold for all their variants, e.g., for diagonal-free cylindric algebras Df_n, RDf_n, substitution-cylindrification algebras SCA_n, RSCA_n, polyadic equality algebras PEA_n, RPEA_n, polyadic algebras PA_n, RPA_n, for relation algebras SA, RA, RRA and their logical counterparts. For the definition of these classes of algebras see, e.g., [Hen-Mon-Tar,85], [Mad-Say,thisVol], [Mad,82].

The first applications we talk about here concern completeness theorems. Tarski used his translation in [Tar-Giv,87] to transfer the completeness theorem for $\mathcal{L}_\omega$ into a kind of completeness theorem for his target logic, which in algebraic form is stated as a representation theorem, namely that every quasi-projective relation algebra is representable. (Later Maddux [Mad,78] gave a purely algebraic proof for this.) It is shown in [Nem,85b, 3.7–3.10], [Nem,86a, 17T(viii)] that RA cannot be replaced with CA_3 in this consequence, namely, quasi-projective CA_3's are rather far from being representable (and the same is true for the class SA of semi-associative relation algebras, in place of CA_3). So, in this respect, Tarski's result cannot be improved.

Yet, we can use our translations in Theorems 1.1.6, 1.1.7 to prove completeness results for our target logics, but in a different way. We begin with recalling some definitions from [And-Nem-Sai,01, D. 33, D. 48]. A proof system is called *Hilbert-style* if it is given by finitely many axiom schemes and rules where the rules are of form $\varphi_1, \dots, \varphi_k \vdash \varphi_0$ for some formula schemes $\varphi_0, \dots, \varphi_k$. A proof system $\vdash$ is called *sound* w.r.t. the semantics $\models$ iff $\vdash \varphi$ implies $\models \varphi$, *strongly sound* if $Th \vdash \varphi$ implies $Th \models \varphi$, *complete*, *strongly complete* when "implies" is replaced with "implied by" in the above, for all sets $Th \cup \{\varphi\}$ of formulas. Finally, we define $\mathcal{L}_n = \langle L_n, \vdash_n \rangle$, the usual n-variable fragment of $\mathcal{L}_\omega$, as restricting $\mathcal{L}_\omega$ to those formulas of L_ω which contain only the first n variables. (E.g., $R(y, x, x, z) \in L_3$ when R is a 4-place relation symbol in $\mathcal{L}_\omega$.) More precisely, $\vdash_n$ is the provability relation we get from the axiom schemes ((0))–((9)) understood as schemes for L_n and rules (MP), (G), cf. Remark 1.1.4(i). Throughout this section, we assume that E is a binary relation symbol in $\mathcal{L}_\omega$ and $n \geq 3$ is finite. Hence, in Theorem 1.1.7 we can replace $\mathcal{L}ca_3(E,2)$ with $\mathcal{L}_n$ (see Remark 1.1.8(vi)).

We know that $\mathcal{L}_n$ is inherently incomplete, i.e., there is no complete and strongly sound Hilbert-style proof system for the "standard" validity $\models$ restricted to L_n. In the literature, there are approaches aimed at getting around this inherent incompleteness of $\mathcal{L}_n$. One goes by replacing "standard" models and validity with "non-standard" models and validity which one can obtain from CA_n. This approach originates with Leon Henkin. The other approach is keeping the standard semantics and using new complete inference systems which are sound but not strongly sound. Such inference systems are introduced, e.g., in [Ven,91], [Ven,thisVol] and in [Sim,91]. Problem 7.2 in [Sai,thisVol], as well as [Hen-Mon-Tar,85, Problem 4.16], and [And-Mon-Nem,91b, Problem 1(a) (p. 730), Problems 49, 50 (p. 740)] are strongly related to this direction.

Let $\vdash_{\overline{nt}}$ denote the proof system we obtain from $\vdash_{\overline{n}}$ by adding the rule which infers φ from $\mathsf{tr}(\varphi)$ when φ is a sentence in $\mathcal{L}_n$, where tr is the translation in Theorem 1.1.7(i). This last rule is sound but not strongly sound, i.e., $\vdash_{\overline{nt}} \varphi$ implies $\models \varphi$, but it is not true that $Th \vdash_{\overline{nt}} \varphi$ implies $Th \models \varphi$ (namely, $\mathsf{tr}(\varphi) \vdash_{\overline{nt}} \varphi$ for all φ, but $\mathsf{tr}(\varphi) \models \varphi$ is not true for all φ).

Our first theorem in this section is an immediate corollary of Theorem 1.1.7(i). It says that the "standard" Hilbert-style proof system $\vdash_{\overline{n}}$ is strongly complete and strongly sound within a large enough subset of L_n; and the "non-standard" proof system $\vdash_{\overline{nt}}$ is complete and sound for the whole of $\mathcal{L}_n$.

In more detail, the first part of Corollary 1.2.1 below says that we can select a subset G of formulas, call it the set of "formulas of good shape", such that the natural Hilbert-style proof system $\vdash_{\overline{n}}$ is strongly complete within this subset; moreover we can decide whether a formula is in good shape, and every formula φ can be algorithmically converted to one in a good-shape such that meaning is preserved in the sense described in Theorem 1.1.7(i).

Corollary 1.2.1. *Let* G *denote the range of* tr *in Theorem* 1.1.7(i) *and let* $n > 2$ *be finite. Then* (i)–(iii) *below hold:*

(i) $\vdash_{\overline{n}}$ *is strongly complete within* $G \subseteq L_n$, *i.e., for all* $Th \cup \{\varphi\} \subseteq G$ *we have that* $Th \models \varphi \Leftrightarrow Th \vdash_{\overline{n}} \varphi$.

(ii) G *is large enough in the sense that* $\mathsf{tr}(\varphi) \in G$ *for all* $\varphi \in L_n$ *and* $\models \varphi \Leftrightarrow \models \mathsf{tr}(\varphi)$.

(iii) $\models_{\overline{nt}}$ *is complete and sound in the whole of* L_n, *i.e., for all formulas* $\varphi \in L_n$ *we have* $\models \varphi \Leftrightarrow \models_{\overline{nt}} \varphi$. ■

The next corollary concerns connections between RCA_n and its finitary approximation, CA_n. Let $\otimes$ denote the complement of the symmetric difference, i.e., $x \otimes y \stackrel{d}{=} (x \cdot y) + (-x \cdot -y)$. We note that $\otimes$ is the algebraic counterpart of $\leftrightarrow$.

Corollary 1.2.2. *There is a computable function* f *mapping* CA_n*-terms to* CA_n*-terms such that for all* CA_n*-terms* τ, σ *we have*

(i) $\mathsf{RCA}_n \models \tau = \sigma$ *iff* $\mathsf{CA}_n \models f(\tau \otimes \sigma) = 1$, *or, in an equivalent form*

(ii) $\mathsf{RCA}_n \models \tau = 1$ *iff* $\mathsf{CA}_n \models f(\tau) = 1$. ■

The above corollary of Theorem 1.1.7 justifies, in a way, the introduction of CA_n. Namely, CA_n was devised in order to "control", have a firm grasp on equations true in RCA_n. Nonfinite axiomatizability of RCA_n implies that this firm grasp cannot be attained in the form of $\mathsf{Eq}\,\mathsf{CA}_n = \mathsf{Eq}\,\mathsf{RCA}_n$ where EqK denotes the equational theory of the class K of algebras. By contrast, the above theorem says that a firm grasp can be obtained by using the computable function f; the axioms of CA_n together with the definition of f provide a finitary tool that captures (reconstructs completely) $\mathsf{Eq}\,\mathsf{RCA}_n$.

A corollary of Theorem 1.1.6 says that the computational complexity of FOL is the same as that of the equational theory of Df_3. We recall from [Coo,03], informally, that the Turing-degree of $S \subseteq \omega$ is less than or equal to that of $Z \subseteq \omega$, in symbols $S \leq_T Z$, if by using a decision procedure for Z we can decide S. The Turing-degrees of S and Z are the same, in symbols $S \equiv_T Z$, if $S \leq_T Z$ and $Z \leq_T S$. The same notion can be applied to the equational theories of various classes of algebras, and to various FOL-theories. Let $Th(\emptyset)$ denote the set of valid formulas of $\mathcal{L}_\omega$.

The following corollary says that if we have a decision procedure for any one of $Th(\emptyset)$, EqK with K one of Df_n, RDf_n, CA_n, RCA_n, SCA_n, RSCA_n, …, RA, RRA, $3 \leq n < \omega$ then we can decidable any other of the same list. In short, the Turing-degrees of all these classes are the same. This corollary follows from Theorems 1.1.6, 1.1.7.

Corollary 1.2.3. *Let* $3 \leq n < \omega$ *and let* K *be any one of* Df_n, RDf_n, CA_n, RCA_n, SCA_n, RSCA_n, PA_n, RPA_n, PEA_n, RPEA_n, SA, RA, RRA. *Then* (i) *and* (ii) *below hold.*

(i) $\mathsf{Eq}\,\mathsf{Df}_3 \equiv_T \mathsf{Eq}\mathsf{K}$.

(ii) $\mathsf{Eq}\,\mathsf{Df}_3 \equiv_T Th(\emptyset)$. ■

The above applications are all relevant to Problem 4.1 of [Hen-Mon-Tar,85]. Indeed, from Corollary 1.2.2 we can get a decidable equational base for $\mathsf{Eq}\,\mathsf{RCA}_n$ similar to that in [Hen-Mon-Tar,85, 4.1.9], and $\models_{nt}$ gives a kind of solution for [Hen-Mon-Tar,85, Problem 4.1] similar to [Hen-Mon-Tar,85, 4.1.20].

Now we turn to other kinds of applications. Tarski introduced and used translation functions from a logic $\mathcal{L}$ into a logic $\mathcal{L}'$ in order to transfer some properties of $\mathcal{L}$ to $\mathcal{L}'$. For example, if the translation function is computable, then undecidability of the valid formulas of $\mathcal{L}$ implies the same for $\mathcal{L}'$. This is how Tarski proved that $\mathsf{Eq}\,\mathsf{RA}$ was undecidable. The same way, Theorem 1.1.6 immediately implies that the sets of validities of $\mathcal{L}df_n$, $\mathcal{L}ca_n$ as well as the equational theories of Df_n, CA_n for $n \geq 3$ are undecidable. These have been known and have been proved by using Tarski's translation of set theory into RA for $n \geq 4$, and for $n = 3$ it is a result of Maddux [Mad,80], proved by an algebraic method.

The next theorem says that structural computable translations are capable of transferring Gödel's incompleteness property. For $\mathcal{L}ca_3$ this is proved in [Nem,85b, Theorem 1.6], and for $\mathcal{L}df_3$ it is proved in [And-Nem,11a, Theorem 2.3].

Theorem 1.2.4 (Gödel-style incompleteness theorem for $\mathcal{L}df_3$). *There is a formula $\varphi \in Fdf_3$ such that no consistent decidable extension T of φ is complete, and moreover, no decidable extension of φ separates the $\models^{df}$-consequences of φ from the φ-refutable sentences (where ψ is φ-refutable iff $\varphi \models^{df} \neg\psi$). The same is true for $\mathcal{L}ca_3$ and $\mathcal{L}_n$ in place of $\mathcal{L}df_3$.*

The proof of Theorem 1.2.4 goes by showing that the translation of an inseparable formula which is consistent with $\neg\,\mathsf{tr}(\bot)$ by a structural computable translation function tr is inseparable again.

In algebraic logic, the algebraic property corresponding to the logical property of Gödel's incompleteness is atomicity of free algebras (see [Hen-Mon-Tar,85, 4.3.32] and [Nem,85b, Proposition 1.8]). Indeed, Theorem 1.2.4 above implies non-atomicity of free cylindric algebras, this way providing a solution for [Hen-Mon-Tar,85, Problem 4.14]. We devote the next section entirely to free cylindric algebras, because of their importance.

3. Structure of Free Cylindric Algebras

In general, the free algebras of a variety are important because they show, in a sense, the structure of the different "concepts" (represented by terms) of the variety. In algebraic logic, the free algebras of a variety corresponding to a logic $\mathcal{L}$ are even more important, because they correspond to the so-called Lindenbaum–Tarski algebras of $\mathcal{L}$. This implies that the structures of free cylindric algebras are quite rich, since these reflect the whole of FOL, in a sense. Thus proving properties about free cylindric algebras is not easy in general. Often, one proves properties of free algebras by applying logical results to algebras, and it is a task then to find purely algebraic proofs, too. Chapter 2.5 of [Hen-Mon-Tar,85] is devoted to free cylindric algebras. Most of what we say here about free cylindric algebras generalizes to its variants such as $\mathsf{Df}_n, \mathsf{RDf}_n, \mathsf{PA}_n, \ldots$ by using Theorem 1.1.6 in place of Theorem 1.1.7 and keeping Remark 1.1.8(vi) in mind.

Atoms in the Lindenbaum–Tarski algebras of sentences correspond to finitely axiomatizable complete theories, while atoms in the Lindenbaum–Tarski algebras of formulas with n free variables of FOL are related to the Omitting Type theorems and prime models, see [Cha-Kei,90, Sec. 2.3].

$\mathfrak{Fr}_k\, \mathsf{K}$ denotes the k-generated K-free algebra, see [Hen-Mon-Tar,85, 0.4.19]. We usually assume $k \neq 0$, just for simplicity. $\mathfrak{Fr}_k\, \mathsf{CA}_n$ is atomless if k is infinite (by Pigozzi, [Hen-Mon-Tar,85, 2.5.13]). Assume k is finite, nonzero. If $n < 2$ then $\mathfrak{Fr}_k\, \mathsf{CA}_n$ is finite ([Hen-Mon-Tar,85, 2.5.3(i)]), hence atomic. $\mathfrak{Fr}_k\, \mathsf{CA}_2$ is infinite but still atomic (by Henkin, [Hen-Mon-Tar,85, 2.5.3(ii), 2.5.7(ii)]). If $2 \leq n < \omega$ then $\mathfrak{Fr}_k\, \mathsf{CA}_n$ has infinitely many atoms (by Tarski, [Hen-Mon-Tar,85, 2.5.9]), and it was asked in [Hen-Mon-Tar,85] as Problem 4.14 whether it is atomic or not. The following solution is proved in [Nem,85b], [Nem,86a].

Theorem 1.3.1. *Let $0 < k < \omega$ and $n \geq 3$. Then $\mathfrak{Fr}_k\, \mathsf{CA}_n$ is not atomic.* ■

A proof of the above theorem is based on Theorem 1.2.4. This is a metalogical proof, "transferring" Gödel's incompleteness theorem for FOL to three-variable logic. [Hen-Mon-Tar,85, Problem 4.14] also raised the problem of finding purely algebraic proofs for these properties of free algebras. Németi ([Nem,84]) contains direct, purely algebraic proofs showing that $\mathfrak{Fr}_k\, \mathsf{CA}_n$ is not atomic, for $n \geq 4$. However, those proofs do not work for $n = 3$ (counterexamples show that the crucial lemmas fail for $n = 3$), and they are longer than the present metalogical proof.

So, in particular, it remains open to find a direct, algebraic proof for non-atomicity of $\mathfrak{Fr}_k \mathsf{Df}_3$, $k > 0$.

In the proof of Theorem 1.3.1 we had to show that there is an element in the free algebra below which there is no atom. Problem 2.5 in [Hen-Mon-Tar,85], still open, asks if the sum of all atoms in $\mathfrak{Fr}_k \mathsf{CA}_n$ exists for finite k and $3 \leq n < \omega$. This problem is equivalent to asking if there is a biggest element in the free algebra below which there is no atom.

Remark 1.3.2 (On zero-dimensional atoms). $\mathfrak{Fr}_k \mathsf{CA}_n$ has exactly 2^k zero-dimensional atoms (by Pigozzi, [Hen-Mon-Tar,85, 2.5.11]). It was conjectured that these are all the atoms if $n \geq \omega$ (see [Hen-Mon-Tar,85, 2.5.12, Problem 2.6]). We note that there may be many more atoms in $\mathfrak{Zd}\,\mathfrak{Fr}_k \mathsf{CA}_n$, the zero-dimensional part of $\mathfrak{Fr}_k \mathsf{CA}_n$, than the zero-dimensional atoms of $\mathfrak{Fr}_k \mathsf{CA}_n$. I.e., the atoms of $\mathfrak{Zd}\,\mathfrak{Fr}_k \mathsf{CA}_n$ usually are not atoms in $\mathfrak{Fr}_k \mathsf{CA}_n$.

The metalogical proof of Theorem 1.3.1 automatically proves that $\mathfrak{Zd}\,\mathfrak{Fr}_k \mathsf{CA}_n$ is not atomic either, if $2 < n < \omega$. In [Mad-Nem,01], the locally finite part of $\mathfrak{Fr}_k \mathsf{CA}_\alpha$ for infinite α is characterized, and this solves [Hen-Mon-Tar,85, Problem 2.10]. This implies that $\mathfrak{Zd}\,\mathfrak{Fr}_k \mathsf{CA}_\alpha$ is atomic if $\alpha \geq \omega > n$.

On the other hand, the algebraic proofs in [Nem,84] show that there is an atom of $\mathfrak{Zd}\,\mathfrak{Fr}_k \mathsf{CA}_n$ (for $0 < k < \omega$ and $4 \leq n < \omega$) below which there is no atom of $\mathfrak{Fr}_k \mathsf{CA}_n$. We do not know whether this holds for $n = 3$ or not. As for the conjecture in [Hen-Mon-Tar,85] about the nonzero-dimensional atoms in the case $\alpha \geq \omega$, in [Nem,84] we prove that it is true for the free representable CA_α ($\alpha \geq \omega$), and we have some partial results that might point into the opposite direction for the free CA_α. Namely, in [Nem,84] we show that there is a nonzero element in $\mathfrak{Fr}_k \mathsf{CA}_\alpha$ which is below d_{ij} for all $i, j \in \alpha$, $i, j \notin 2$. This cannot happen in the representable case. ■

The proof that $\mathfrak{Fr}_k \mathsf{CA}_2$ for finite k is atomic relies on the fact that CA_2 is a discriminator variety and the equational theory of CA_2 is the same as the equational theory of finite CA_2's. See [Hen-Mon-Tar,85, 2.5.7] and [And-Jon-Nem,91, Theorem 4.1]. Let n be finite. Then Crs_n, the class of cylindric-relativized set algebras of dimension n [Hen-Mon-Tar,85, 3.1.1(iv)], satisfies the second condition, i.e., it is generated by its finite members as a variety (see [And-Hod-Nem,99]) but it is not a discriminator variety. The same holds for the variety NCA_n of non-commutative cylindric algebras (see [Nem,86a, 5T, p. 112]) and for the varieties WA and NA of weakly associative and non-associative relation algebras (by [And-Hod-Nem,99], [Nem,87b]));

for the definitions of NA, WA, SA see [Mad,82] or, e.g., [And-Jon-Nem,91]. It is proved in [Nem,85b] that neither one of $\mathfrak{Fr}_k$ RA and $\mathfrak{Fr}_k$ SA is atomic.

Problem 1.3.3. Let k, n be finite. Are the k-generated free Crs_n's and NCA_n's atomic? Is $\mathfrak{Fr}_k$ WA atomic? Is $\mathfrak{Fr}_k$ NA atomic? ■

Our next subject is generating and free subsets of free algebras. $\mathfrak{Fr}_k \mathsf{CA}_n$ cannot be generated by fewer than k elements by [Hen-Mon-Tar,85, 2.5.20] and all free generator sets have cardinality k. By [Hen-Mon-Tar,85, 2.5.23], every k-element generator set of $\mathfrak{Fr}_k \mathsf{CA}_n$ is a free generator set, if $n \leq 2$ and k is finite. Problem 2.7 of [Hen-Mon-Tar,85] asks if this continues to hold for $3 \leq n$ and finite k. The following theorem, proved as [Nem,86a, Theorem 19, p. 100], gives a negative answer. Its proof essentially uses the translation mapping in the present Theorem 1.1.7.

Theorem 1.3.4. *There is a k-element irredundant non-free generator set in $\mathfrak{Fr}_k \mathsf{CA}_n$, for every $0 < k$ and $3 \leq n$.* ■

The proof of the above theorem goes by finding such generator sets in $\mathfrak{Fr}_k \mathsf{RCA}_\omega$, which allows us to think in a model theoretical way, and then using the translation function (in a non-trivial way) of Theorem 1.1.7 to translate the idea from RCA_ω to CA_n.

Andréka–Jónsson–Németi ([And-Jon-Nem,91, Theorem 9.1]) generalizes the existence of non-free generator sets in Theorem 1.3.4 from CA_n to many subvarieties of SA. We note that Jónsson–Tarski ([Jon-Tar,61]) proves that if a variety is generated by finite algebras then any k-element generator set of the k-generated free algebra generates it freely. Thus CA_n in Theorem 1.3.4 cannot be replaced with Crs_n or NCA_n or WA or NA.

Are there big, not necessarily generating but free subsets in $\mathfrak{Fr}_k \mathsf{CA}_n$? A way of formalizing this question is whether $\mathfrak{Fr}_{k+1} \mathsf{CA}_n$ can be embedded in $\mathfrak{Fr}_k \mathsf{CA}_n$. This question is investigated thoroughly in [And-Jon-Nem,91] and the following negative answer is proved as part of [And-Jon-Nem,91, Theorem 10.3].

Theorem 1.3.5. *$\mathfrak{Fr}_{k+1} \mathsf{CA}_n$ is not embeddable into $\mathfrak{Fr}_k \mathsf{CA}_n$, for $0 < k < \omega$.* ■

Many properties of free cylindric algebras are proved in [And-Jon-Nem,91]. E.g., [And-Jon-Nem,91, Theorem 10.1] gives a complete structural description of the free CA_1's, i.e., of the free monadic algebras.

The number of elements of this free algebra is given in [Hen-Mon-Tar,85, 2.5.62]. The cardinality of finitely generated free monadic Tarski algebras (which is a reduct of CA_1) is given in [Mon-Aba-Sav-Sew,97].

We had many illuminating discussions in the subject of this paper, during writing it, with Barry Cooper, Steven Givant, Ági Kurucz, András Simon and with the members of the Logic and Philosophy of Mathematics student and faculty seminar of the Eötvös University Budapest. This research was supported by the Hungarian Scientific Research Fund grant No. T81188.

BOLYAI SOCIETY
MATHEMATICAL STUDIES, 22

**Cylindric-like Algebras
and Algebraic Logic**
pp. 37–59.

Varieties of Two-Dimensional Cylindric Algebras

NICK BEZHANISHVILI*

In this chapter we survey recent developments in the theory of two-dimensional cylindric algebras. In particular, we will deal with varieties of two-dimensional diagonal-free cylindric algebras and with varieties of two-dimensional cylindric algebras with the diagonal. It is well known that two-dimensional diagonal-free cylindric algebras correspond to the two variable equality-free fragment of classical first-order logic FOL, whereas two-dimensional cylindric algebras with the diagonal correspond to the two variable fragment of FOL with equality. It is also well known that one-dimensional cylindric algebras, also called Halmos monadic algebras, provide algebraic completeness for the one variable fragment of FOL. For a systematic discussion on the connection between various fragments of FOL and classes of (cylindric) algebras that correspond to these fragments we refer to [And-Nem-Sai,01].

The variety Df_1 of one-dimensional cylindric algebras has a lot of 'nice' properties: Df_1 is finitely axiomatizable, it is generated by its finite algebras, and has a decidable equational theory. Moreover, the lattice of subvarieties of Df_1 is rather simple: it is an $(\omega+1)$-chain. Every proper subvariety of Df_1 is finitely generated, finitely axiomatizable and has a decidable equational theory (see [Hal,62], [Hen-Mon-Tar,85] and [Scr,51]). In contrast to this, the three variable fragment of FOL corresponding to three-dimensional cylindric algebras is much more complicated and no longer has 'nice' properties. It has been shown by Maddux [Mad,80] that the equational theory of three-dimensional cylindric algebras is undecidable. Moreover, every subvariety in between the variety of all representable three-dimensional cylindric algebras

*The author is grateful to Ian Hodkinson for his comments on the earlier version of this chapter. Special thanks go to the referee for valuable suggestions and pointers to the literature that substantially improved the presentation of this chapter.

and the variety of all three-dimensional cylindric algebras is undecidable. Kurucz [Kur,02b] strengthened this by showing that none of these varieties are generated by their finite algebras. It follows from Monk [Mon,69] and Johnson [Joh,69] that varieties of all representable three-dimensional cylindric algebras with and without diagonals are not finitely axiomatizable.

Our aim is to show that the two-dimensional case is not as complicated as the three-dimensional one, but is not as simple as the one-dimensional case. It has been known for a long time that the variety Df_2 of two-dimensional diagonal-free cylindric algebras is generated by its finite algebras and has a decidable equational theory. Moreover, every Df_2-algebra is representable. The variety CA_2 of two-dimensional cylindric algebras with the diagonal is also generated by its finite algebras and has a decidable equational theory. However, not every CA_2-algebra is representable. Representable CA_2-algebras form a subvariety of CA_2 denoted by RCA_2. Unlike the three-dimensional case, RCA_2 can be axiomatized by adding only one axiom to the axiomatization of CA_2. In this chapter we will mainly concentrate on the lattices of subvarieties of Df_2, CA_2 and RCA_2, respectively. As we will see below, the lattice of subvarieties of Df_2, although more complicated than the one-dimensional case, is still countable. Moreover, every subvariety of Df_2 is finitely axiomatizable and has a decidable equational theory. The lattices of subvarieties of CA_2 and RCA_2, respectively, have a more complex structure. In particular, it is known that unlike Df_2, CA_2 and RCA_2 have continuum many subvarieties. As a result, there is a continuum of non-finitely axiomatizable and undecidable subvarieties of CA_2 and RCA_2. We will show that every proper subvariety of Df_2 is locally finite and every subvariety of Df_2 is generated by its finite algebras. On the other hand, we will prove that there are continuum many non-locally finite subvarieties of CA_2 and RCA_2. It is still an open problem whether every subvariety of CA_2 and RCA_2 is generated by its finite algebras.

Our main technique in studying varieties of two-dimensional cylindric algebras will be the duality between two-dimensional cylindric algebras and Stone spaces enriched with two commuting equivalence relations. The topology-free analogue of this duality is thoroughly discussed in [Hen-Mon-Tar,85, Section 2.7]. Stone-like topological dualities are extensively used to investigate modal logics (see, e.g, [Bla-Rij-Ven,01, Cha-Zak,97]). In fact, many of the results discussed in this chapter are proved using techniques developed in modal logic. These techniques apply to Df_2 and CA_2-algebras since they are algebraic models of cylindric modal logic (see [Ven,thisVol]).

The chapter is organized as follows in Section 2: we recall the definition of Df_2 and CA_2-algebras and their topological representation. We also recall the functor from CA_2 to Df_2 that forgets the diagonal and discuss some of its basic properties. In Section 3 we recall representable cylindric algebras and a topological characterization of representable CA_2-algebras. We also construct rather simple non-representable CA_2-algebras. In Section 4 we discuss the cardinality of the lattices of subvarieties of CA_2 and RCA_2. Section 5 reviews a criterion of local finiteness for subvarieties of Df_2 and Section 6 discusses a classification of subvarieties of Df_2. In Section 7 we review local finiteness of subvarieties of CA_2 and RCA_2. Section 8 is devoted to finitely generated varieties of Df_2 and CA_2-algebras, and finally, we close the chapter by discussing some open problems.

1. Cylindric Algebras and Cylindric Spaces

In this section we recall the basic duality for two-dimensional cylindric algebras that will be used throughout this chapter.

Definition 2.1.1. A triple $\mathcal{B} = \langle B, \mathsf{c}_1, \mathsf{c}_2 \rangle$ is said to be a *two-dimensional diagonal-free cylindric algebra,* or a Df_2*-algebra* for short, if B is a Boolean algebra and $\mathsf{c}_i : B \to B$, $i = 1, 2$, are unary operations satisfying the following axioms for all $a, b \in B$:

(C_1) $\mathsf{c}_i 0 = 0$,

(C_2) $a \leq \mathsf{c}_i a$,

(C_3) $\mathsf{c}_i(\mathsf{c}_i a \cdot b) = \mathsf{c}_i a \cdot \mathsf{c}_i b$,

(C_4) $\mathsf{c}_1 \mathsf{c}_2 a = \mathsf{c}_2 \mathsf{c}_1 a$.

Let Df_2 denote the variety of all two-dimensional diagonal-free cylindric algebras [Hen-Mon-Tar,85, Definition 1.1.2].

Definition 2.1.2. A quadruple $\mathfrak{B} = \langle B, \mathsf{c}_1, \mathsf{c}_2, \mathsf{d} \rangle$ is said to be a *two-dimensional cylindric algebra,* or a CA_2*-algebra* for short, if $\langle B, \mathsf{c}_1, \mathsf{c}_2 \rangle$ is a Df_2-algebra and $\mathsf{d} \in B$ is a constant satisfying the following axioms for all $a \in B$ and $i = 1, 2$.

(C_5) $\mathsf{c}_i(\mathsf{d}) = 1$,

(C_6) $\mathsf{c}_i(\mathsf{d} \cdot a) \leq -\mathsf{c}_i(\mathsf{d} \cdot -a)$.

Let CA_2 denote the variety of all two-dimensional cylindric algebras [Hen-Mon-Tar,85, Definition 1.1.1].

Cylindric algebras were first introduced in [Chi-Tar,48] without explicitely mentioning the term 'cylindric algebra'. The full definition later appeared in [Jon-Tar,51], [Tar-Tho,52] and [Tar,52].

We recall that a Df_1-algebra or a monadic algebra [Hal,62] is a pair $\langle B, \mathsf{c}\rangle$ such that B is a Boolean algebra and c is a unary operator on B satisfying conditions (C_1)–(C_3) of Definition 2.1.1; see, e.g., [Hal,62, p. 40]. The unary operator c is called a *monadic operator,* and Df_1-algebras are widely known as *monadic algebras.* A systematic investigation of monadic algebras has been carried out by Halmos [Hal,62], Bass [Bas,58], Monk [Mon,70a], and Kagan and Quackenbush [Kag-Qua,76].

Now we turn to a topological representation of two-dimensional cylindric algebras. This duality (see [Bez,02, Section 2.2] and [Bez,04b, Section 2]) is based on a standard Jónsson–Tarski duality for modal algebras [Jon-Tar,51] (see also [Bla-Rij-Ven,01, Section 5.5] and [Cha-Zak,97, Section 8.2]). A topological duality for Df_1-algebras was developed by Halmos [Hal,62]. We recall that a *Stone space* is a compact, Hausdorff space with a basis of *clopen* (simultaneously closed and open) sets. For a Stone space X, we denote by $\mathsf{CP}(X)$ the set of all clopen subsets of X. For an arbitrary binary relation R on X, $x \in X$ and $A \subseteq X$, we let $R(x) = \{y \in X : xRy\}$, $R^{-1}(x) = \{y \in X : yRx\}$, $R(A) = \bigcup_{x\in A} R(x)$ and $R^{-1}(A) = \bigcup_{x\in A} R^{-1}(x)$. A relation R on a Stone space X is said to be *point-closed* if for each $x \in X$ the set $R(x)$ is closed, and R is called a *clopen* relation, if $A \in \mathsf{CP}(X)$ implies $R^{-1}(A) \in \mathsf{CP}(X)$. Note that if R is an equivalence relation, then $R(x) = R^{-1}(x)$ and $R(A) = R^{-1}(A)$. In such a case we call $R(A)$ the *R-saturation* of A.

A Df_2-*space* is a triple $\mathcal{X} = \langle X, E_1, E_2\rangle$, where X is a Stone space, and E_1 and E_2 are point-closed and clopen equivalence relations on X such that

$$(\forall\, x, y, z \in X)\big((xE_1y \wedge yE_2z) \to (\exists\, u \in X)(xE_2u \wedge uE_1z)\big).$$

Given two Df_2-spaces $\mathcal{X} = \langle X, E_1, E_2\rangle$ and $\mathcal{X}' = \langle X', E_1', E_2'\rangle$, a map $f : X \to X'$ is said to be a Df_2-*morphism* if f is continuous, and for each $x \in X$ and $i = 1, 2$ we have $fE_i(x) = E_i'f(x)$. We denote by **DS** the category of Df_2-spaces and Df_2-morphisms. Then Df_2 is dual (dually equivalent) to **DS**. In particular, every Df_2-algebra $\langle B, \mathsf{c}_1, \mathsf{c}_2 \rangle$ can be represented as $\big\langle \mathsf{CP}(X), E_1, E_2 \big\rangle$, for the corresponding Df_2-space $\langle X, E_1, E_2 \rangle$. We recall that $\langle X, E_1, E_2 \rangle$ is constructed as follows: X is the set of all ultrafilters of B, $\varphi(a) = \{x \in X : a \in x\}$, $\big\{\varphi(a) : a \in B\big\}$ is a basis for the topology on X, and xE_iy if ($\mathsf{c}_i a \in x \Leftrightarrow \mathsf{c}_i a \in y$), for each $a \in B$ and $i = 1, 2$. We call $\langle X, E_1, E_2 \rangle$ the *dual* of $\langle B, \mathsf{c}_1, \mathsf{c}_2 \rangle$. As an easy corollary of this duality, we obtain that the category FinDf_2 of finite Df_2-algebras is dual to the category $\mathsf{Fin}\mathbf{DS}$ of finite Df_2-spaces with the discrete topology. Hence, every finite Df_2-algebra is represented as the algebra $\big\langle \mathcal{P}(X), E_1, E_2 \big\rangle$ for the corresponding finite Df_2-space $\langle X, E_1, E_2 \rangle$, where $\mathcal{P}(X)$ denotes the powerset of X (see [Hen-Mon-Tar,85, Theorem 2.7.34]). For $i = 1, 2$ we call the E_i-equivalence classes of X the *E_i-clusters.*

The dual spaces of CA_2-algebras are obtained as easy extensions of Df_2-spaces. A quadruple $\langle X, E_1, E_2, D \rangle$ is said to be a *CA_2-space* if $\langle X, E_1, E_2 \rangle$ is a Df_2-space and D is a clopen subset of X such that for each $i = 1, 2$, each E_i-cluster of X contains a unique point from D. This implies that in each CA_2-space $\langle X, E_1, E_2, D \rangle$, unlike Df_2-spaces, the cardinality of the set of all E_1-clusters of X is always equal to the cardinality of the set of all E_2-clusters of X. Given two CA_2-spaces $\langle X, E_1, E_2, D \rangle$ and $\langle X', E_1', E_2', D' \rangle$, a map $f : X \to X'$ is said to be a *CA_2-morphism* if f is a Df_2-morphism and in addition $f^{-1}(D') = D$. We denote the category of CA_2-spaces and CA_2-morphisms by **CS**. Then CA_2 is dual to **CS**. In particular, every CA_2-algebra $\mathfrak{B} = \langle B, \mathsf{c}_1, \mathsf{c}_2, \mathsf{d} \rangle$ can be represented as $\big\langle \mathsf{CP}(X), E_1, E_2, D \big\rangle$ for the corresponding CA_2-space $\langle X, E_1, E_2, D \rangle$. The construction of $\langle X, E_1, E_2, D \rangle$ is the same as in the Df_2-case with the addition that we let $D = \varphi(\mathsf{d}) = \{x \in X : \mathsf{d} \in x\}$. We call $\langle X, E_1, E_2, D \rangle$ the *dual* of $\langle B, \mathsf{c}_1, \mathsf{c}_2, \mathsf{d} \rangle$. As an easy corollary of this duality we obtain that the category FinCA_2 of finite CA_2-algebras is dual to the category $\mathsf{Fin}\mathbf{CS}$ of finite CA_2-spaces with the discrete topology. In particular, every finite CA_2-algebra is represented as the algebra $\big\langle \mathcal{P}(X), E_1, E_2, D \big\rangle$ for the corresponding finite CA_2-space $\langle X, E_1, E_2, D \rangle$ (see [Hen-Mon-Tar,85, Theorem 2.7.34]).

Remark 2.1.3. We point out a close connection between Df_2 and CA_2-spaces and cylindric atom structures defined in [Hen-Mon-Tar,85]. Recall from [Hen-Mon-Tar,85, Definition 2.7.32] that if $\mathfrak{B} = \langle B, \mathsf{c}_1, \mathsf{c}_2, \mathsf{d} \rangle$ is a CA_2-algebra (the case of Df_2-algebras is similar) such that B is an atomic Boolean algebra, then the cylindric atom structure of $\mathfrak{B}$ is defined as the quadruple $\mathfrak{At}\,\mathfrak{B} = \langle At\,B, E_1, E_2, D \rangle$, where $At\,B$ is the set of all atoms of B; E_i is defined by setting: xE_iy if $\mathsf{c}_i x = \mathsf{c}_i y$, for $x, y \in At\,B$, $i = 1, 2$; and $D = \{x \in At\,B : x \leq \mathsf{d}\}$.

Now suppose $\mathfrak{B} = \langle B, \mathsf{c}_1, \mathsf{c}_2, \mathsf{d}\rangle$ is an arbitrary CA_2-algebra. Let $\mathfrak{B}^\sigma = \langle B^\sigma, \mathsf{c}_1^\sigma, \mathsf{c}_2^\sigma, \mathsf{d}^\sigma\rangle$ be the canonical extension of $\mathfrak{B}$, and let $i : B \to B^\sigma$ be the canonical embedding [Hen-Mon-Tar,85, Definition 2.7.4]. Then it is well known that $\mathfrak{B}^\sigma$ is complete and atomic. Let $\mathfrak{At}\,\mathfrak{B}^\sigma$ be the cylindric atom structure of $\mathfrak{B}^\sigma$. For $a \in B$ let $O_a = \{x \in At\, B^\sigma : x \leq i(a)\}$. We make $\mathfrak{At}\,\mathfrak{B}^\sigma$ into a topological space by letting $\{O_a : a \in B\}$ to be a basis for the topology τ. Then it can be shown that $\langle \mathfrak{At}\,\mathfrak{B}^\sigma, \tau\rangle$ is a CA_2-space, and that $\langle \mathfrak{At}\,\mathfrak{B}^\sigma, \tau\rangle$ is isomorphic to the dual of $\mathfrak{B}$. We note that this connection applies not only to cylindric algebras, but, in general, to any Boolean algebra with operators; see, e.g, [Ven,07, Section 5].

Having this duality at hand, we can obtain dual descriptions of important algebraic concepts of Df_2 and CA_2-algebras. A subset U of a Df_2-space (resp. of a CA_2-space) is said to be *saturated* if $E_1(U) = E_2(U) = U$. For each Df_2-algebra $\mathcal{B}$ (resp. CA_2-algebra $\mathfrak{B}$), the lattice of all congruences of $\mathcal{B}$ (resp. $\mathfrak{B}$) is dually isomorphic to the lattice of all closed saturated subsets of its dual space. A Df_2-space $\langle X, E_1, E_2\rangle$ (resp. a CA_2-space $\langle X, E_1, E_2, D\rangle$) is called *rooted* if

$$(\forall x, y \in X)(\exists\, z \in X)(xE_1z \wedge zE_2y).$$

(By the commutativity of E_1 and E_2, if such a z exists, then there exists $u \in X$ such that xE_2u and uE_1y). Rooted spaces correspond to subdirectly irreducible and simple cylindric algebras. In particular, a Df_2-algebra (resp. a CA_2-algebra) is subdirectly irreducible iff it is simple iff its dual space is rooted [Hen-Mon-Tar,85, Theorems 2.4.43 and 2.4.14]. As a result of the above, we obtain that both Df_2 and CA_2 are semi-simple and congruence distributive varieties with the congruence extension property. As shown in [Nem,91], these results also follow from the fact that Df_2 and CA_2 are discriminator varieties. The aforementioned properties are also discussed in [Ven,07, Section 4] in a wider context of Boolean algebras with operators.

Now we are ready to recall the definition of the reduct functor $\mathfrak{Df} : \mathsf{CA}_2 \to \mathsf{Df}_2$ [Hen-Mon-Tar,85, Definition 1.1.2] (see also [Bez,04b, Section 2.2]). For each CA_2-algebra $\langle B, \mathsf{c}_1, \mathsf{c}_2, \mathsf{d}\rangle$ we set

$$\mathfrak{Df}\langle B, \mathsf{c}_1, \mathsf{c}_2, \mathsf{d}\rangle = \langle B, \mathsf{c}_1, \mathsf{c}_2\rangle.$$

Thus, $\mathfrak{Df}$ forgets the diagonal element d from the signature of CA_2-algebras.

Next we will give a simple argument showing that $\mathfrak{Df}$ is not onto (see [Hen-Mon-Tar,85, Corollary 5.1.4(ii)]). In fact, the set of isomorphism types of $\mathsf{Df}_2 - \mathfrak{Df}(\mathsf{CA}_2)$ is infinite. For this, we define the reduct functor $\mathfrak{Rd}$: $\mathbf{CS} \to \mathbf{DS}$. For each CA_2-space $\langle X, E_1, E_2, D\rangle$ we set

$$\mathfrak{Rd}\langle X, E_1, E_2, D\rangle = \langle X, E_1, E_2\rangle.$$

Suppose a Df_2-space $\langle Y, E_1, E_2\rangle$ is rooted. We call $\langle Y, E_1, E_2\rangle$ a *quasi-square* if the cardinalities of the sets of all E_1 and E_2-clusters of Y coincide. It is not hard to see that a rooted $\langle Y, E_1, E_2\rangle$ is a reduct of some CA_2-space iff it is a quasi-square. Note that not every rooted space from $\mathbf{DS}$ is a quasi-square. The simplest examples of rooted Df_2-spaces that are not quasi-squares are rectangle Df_2-spaces (for the definition of a rectangle consult the next section). Since there are infinitely many rectangle Df_2-spaces (as we will see below), the set $\mathbf{DS} \setminus \mathfrak{Rd}(\mathbf{CS})$ is infinite.

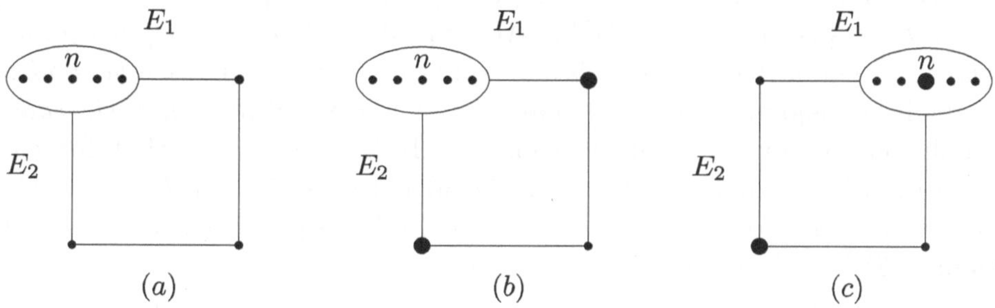

Fig. 1. Some CA_2-spaces and their reducts

We call a Df_2-algebra $\mathcal{B}$ a *quasi-square algebra* if its dual space is a quasi-square. As follows from the above, for each simple CA_2-algebra $\mathfrak{B}$, its Df_2-reduct is a quasi-square algebra. Therefore, the set $\mathsf{Df}_2 \setminus \mathfrak{Df}(\mathsf{CA}_2)$ is infinite. Moreover, one Df_2-algebra can be the reduct of many non-isomorphic CA_2-algebras. For instance, a Df_2-algebra whose dual space is shown in Fig. 1(a) is the reduct of the CA_2-algebras whose dual CA_2-spaces are shown in Figs. 1(b) and 1(c), where dots represent points of the spaces, while big dots represent the points belonging to the (diagonal) set D.

More algebraic properties of Df_2 are discussed in [Hen-Mon-Tar,85, Section 5.1] and [Bez,02]. In particular, a characterization of finitely approximable Df_2-algebras, projective and injective Df_2-algebras, and absolute retracts in Df_2 is given in [Bez,02, Sections 3.1 and 3.2].

2. Representable Cylindric Algebras

In this section we recall some basic facts about representable two-dimensional cylindric algebras.

Let W and W' be sets. We define on the Cartesian product $W \times W'$ two equivalence relations E_1 and E_2 by setting

$$(w, w')E_1(v, v') \text{ if } w' = v',$$

$$(w, w')E_2(v, v') \text{ if } w = v,$$

for $w, v \in W$ and $w', v' \in W'$. We call $\langle W \times W', E_1, E_2\rangle$ a *rectangle.* If $W = W'$ then we call $\langle W \times W, E_1, E_2\rangle$ a *square.* For a square $\langle W \times W, E_1, E_2\rangle$ we set $D = \{(w, w) : w \in W\}$. We call $\langle W \times W, E_1, E_2, D\rangle$ a *cylindric square.* A *generalized rectangle* is a disjoint union of rectangles, a *generalized square* is a disjoint union of squares, and a *generalized cylindric square* is a disjoint union of cylindric squares. It is easy to see that for each (generalized) rectangle $\langle U, E_1, E_2\rangle$, the algebra $\langle \mathcal{P}(U), E_1, E_2\rangle$ is a Df_2-algebra and that for each (generalized) cylindric square $\langle U, E_1, E_2, D\rangle$ the algebra $\langle \mathcal{P}(U), E_1, E_2, D\rangle$ is a CA_2-algebra.

We call a Df_2-algebra $\langle \mathcal{P}(U), E_1, E_2\rangle$ a *(generalized) rectangular* Df_2-*algebra,*[1] if $\langle U, E_1, E_2\rangle$ is a (generalized) rectangle and we call $\langle \mathcal{P}(U), E_1, E_2\rangle$ a (generalized) *square* Df_2-*algebra* if $\langle U, E_1, E_2\rangle$ is a (generalized) square. We call the CA_2-algebra $\langle \mathcal{P}(U), E_1, E_2, D\rangle$ a *(generalized) square* CA_2-*algebra,* if $\langle U, E_1, E_2, D\rangle$ is a (generalized) square. Let ($\mathbb{GRECT}$) $\mathbb{RECT}$ denote the class of all (generalized) rectangular Df_2-algebras, let ($\mathbb{GSQ}$) $\mathbb{SQ}$ denote the class of all (generalized) square algebras, and let ($\mathbb{GCSQ}$) $\mathbb{CSQ}$ denote the class of all (generalized) square CA_2-algebras. Finally, we let $\mathsf{Fin}\mathbb{RECT}$, $\mathsf{Fin}\mathbb{SQ}$ and $\mathsf{Fin}\mathbb{CSQ}$ denote the classes of all finite rectangular, finite square and finite cylindric square algebras, respectively.

For a class K of algebras, we denote by $\mathbf{H}(\mathsf{K})$, $\mathbf{S}(\mathsf{K})$ and $\mathbf{P}(\mathsf{K})$ the closure of K under homomorphic images, subalgebras and products, respectively. We say that a variety V is generated by a class K of algebras if $\mathsf{V} = \mathbf{HSP}(\mathsf{K})$. The classes $\mathbf{S}(\mathbb{RECT})$, $\mathbf{S}(\mathbb{SQ})$, $\mathbf{S}(\mathbb{CSQ})$, $\mathbf{S}(\mathbb{GRECT})$, and $\mathbf{S}(\mathbb{GCSQ})$ in [Hen-Mon-Tar,85, Definitions 3.1.1 and 5.1.33] are denoted by Csdf_2, Csudf_2, Cs_2, Gsdf_2, and Gs_2, respectively. The algebras

[1] Note that the concept of a 'rectangular algebra' is different from the one of a 'rectangular element' defined in [Hen-Mon-Tar,85, Definition 1.10.6].

in these classes are called *two-dimensional: diagonal-free cylindric set algebras, diagonal-free uniform cylindric set algebras, cylindric set algebras, diagonal-free generalized cylindric set algebras,* and *generalized cylindric set algebras,* respectively. Since we only work with two-dimensional cylindric algebras, we find our terminology and notations more suggestive.

Definition 2.2.1.

(i) A Df_2-algebra $\mathcal{B}$ is said to be *representable* if $\mathcal{B} \in \mathbf{S}(\mathbb{GRECT})$.

(ii) A CA_2-algebra $\mathfrak{B}$ is said to be *representable* if $\mathfrak{B} \in \mathbf{S}(\mathbb{GCSQ})$.

([Hen-Mon-Tar,85, Definitions 5.1.33(v), 3.1.1(vii) and Remark 1.1.13]).

The classes of representable Df_2 and CA_2-algebras are usually denoted by RDf_2 and RCA_2, respectively. For the proof of the next theorem we refer to [Hen-Mon-Tar,85, Corollary 5.1.35, Theorems 5.1.43 and 5.1.47] for (i) (see also [Gab-Kur-Wol-Zak,03, Corollary 5.10]), to [Hen-Mon-Tar,85, Corollary 3.1.108] for (ii), and to [Hen-Mon-Tar,85, Lemmas 2.6.41 and 2.6.42] for (iii).

Theorem 2.2.2.

(i) $\mathsf{RDf}_2 = \mathsf{Df}_2 = \mathbf{HSP}(\mathbb{RECT}) = \mathbf{HSP}(\mathbb{SQ}) = \mathbf{SP}(\mathbb{RECT}) = \mathbf{SP}(\mathbb{SQ}) = \mathbf{S}(\mathbb{GSQ})$.

(ii) $\mathsf{RCA}_2 = \mathbf{HSP}(\mathbb{CSQ}) = \mathbf{SP}(\mathbb{CSQ})$.

(iii) $\mathsf{RCA}_2 \subsetneq \mathsf{CA}_2$.

We call (H) and (V) below the *Henkin* and *Venema axioms,* respectively.

$$\text{(H)} \qquad \mathsf{c}_i\big(a \cdot -b \cdot \mathsf{c}_j(a \cdot b)\big) \leq \mathsf{c}_j(-\mathsf{d} \cdot \mathsf{c}_i a), \quad i \neq j, \quad i,j = 1,2,$$

and

$$\text{(V)} \qquad \mathsf{d} \cdot \mathsf{c}_i(-a \cdot \mathsf{c}_j a) \leq \mathsf{c}_j(-\mathsf{d} \cdot \mathsf{c}_i a), \quad i \neq j, \quad i,j = 1,2.$$

Then RCA_2 is axiomatized by adding either of these axioms to the axiomatization of CA_2; see, e.g., [Hen-Mon-Tar,85, Theorem 3.2.65(ii)] or [Ven,91, Proposition 3.5.8]). We denote by $\mathsf{V} + (\text{Ax})$ the addition of the axiom (Ax) to the axiomatization of a variety V. Then

Theorem 2.2.3. $\mathsf{RCA}_2 = \mathsf{CA}_2 + (\text{H}) = \mathsf{CA}_2 + (\text{V})$.

Next we recall from [Ven,91] and [Bez,04b] a dual characterization of representable CA_2-algebras, and construct rather simple finite non-representable CA_2-algebras. For this purpose we recall that on each Df_2-space $\langle X, E_1, E_2\rangle$ and on each CA_2-space $\langle X, E_1, E_2, D\rangle$ we can define yet another equivalence relation that naturally arises from E_1 and E_2. We define E_0 by: xE_0y if xE_1y and xE_2y, for each $x, y \in X$. In other words, $E_0 = E_1 \cap E_2$. We call E_0-equivalence classes *E_0-clusters.* Suppose $\langle X, E_1, E_2, D\rangle$ is a CA_2-space. We call $x \in D$ a *diagonal point,* and we call $x \in X \setminus D$ a *non-diagonal point.* We also call an E_0-cluster C a *diagonal E_0-cluster* if it contains a diagonal point. Otherwise we call C a *non-diagonal E_0-cluster.*

Definition 2.2.4. A CA_2-space $\langle X, E_1, E_2, D\rangle$ is said to satisfy $(*)$ if there exists a diagonal point $x_0 \in D$ such that $E_0(x_0) = \{x_0\}$ and there exists a non-singleton E_0-cluster C whose elements are either E_1 or E_2-related to x_0 ([Bez,04b, Definition 3.3]).

The $(*)$ condition is equivalent to Venema's condition NH7 of [Ven,91, Definition 3.2.5] (see also [Ven,thisVol, Definition 1.3]). In the terminology of [Hen-Mon-Tar,85] a CA_2-space satisfies the condition $(*)$ of Definition 2.2.4 iff the corresponding CA_2-algebra has at least one so-called defective atom (for details see [Hen-Mon-Tar,85, Lemma 3.2.59]). Next we recall a dual characterization of representable CA_2-algebras. There are at least three different proofs of Theorem 2.2.5. The one given in [Hen-Mon-Tar,85, Lemma 3.2.59] uses Henkin's axioms, the proof of [Ven,91, Theorem 3.2.6] (see also [Ven,thisVol, Proposition 1.5]) is based on a powerful technique of modal logic called Sahlqvist correspondence and [Bez,04b, Theorem 3.4] applies Venema's axioms and order-topological methods.

Theorem 2.2.5. *A CA_2-algebra $\mathfrak{B}$ is representable iff its dual CA_2-space $\mathcal{X}$ does not satisfy $(*)$.*

Using this criterion it is easy to see that the CA_2-algebras corresponding to the CA_2-spaces shown in Fig. 1(c) (big spots denote the diagonal points) are representable, while the CA_2-algebras corresponding to the CA_2-spaces shown in Fig. 1(b) are not. Moreover, the smallest non-representable CA_2-algebra is the algebra corresponding to the CA_2-space shown in Fig. 1(b), where the non-singleton E_0-cluster contains only two points.

3. The Finite Model Property and Cardinality of Lattices of Varieties

There is a wide variety of proofs available for the decidability of classical first-order logic with two variables. Equivalent results were stated and proved using quite different methods in first-order, modal and algebraic logic. We present a short historic overview.

Decidability of the validity of equality-free first-order sentences in two variables was proved by Scott [Sco,62]. The proof uses a reduction to the set of prenex formulas of the form $\exists^2\forall^n\varphi$, whose validity is decidable by Gödel [God,33]. The result was stated with equality in the language, because at that time it was still believed that the validity problem for $\exists^2\forall^n$ formulas containing equality is decidable. This belief was however refuted in Goldfarb [Gol,84]. Scott's result was extended by Mortimer [Mor,75], who included equality in the language and showed that such sentences cannot enforce infinite models, obtaining decidability as a corollary. A simpler proof was provided in Grädel et al. [Gra-Kol-Var,97]. They showed that any satisfiable formula can actually be satisfied in a model whose size is single exponential in the length of the formula. Adding two unary function symbols to the language with only one variable leads to undecidability, as shown in Gurevich [Gur,76]. Segerberg [Seg,73] proved the finite model property and decidability for so-called 'two-dimensional modal logic', which is essentially cylindric modal logic (see [Ven,thisVol]) enriched with the operation of involution. For an algebraic proof we refer to [Hen-Mon-Tar,85, Lemma 5.1.24 and Theorem 5.1.64]. A mosaic type proof can be found in Marx and Mikulás [Mar-Mik,99]. A proof using quasimodels is provided in [Gab-Kur-Wol-Zak,03, Theorem 5.22], and [Mar-Ven,07, Proposition 7.4.3] and [Bez,06, Theorem 6.1.1] give simple proofs via the filtration method.

The fact that CA_2 is generated by its finite algebras and has a decidable equational theory was first proved by Henkin [Hen-Mon-Tar,85, Lemma 2.5.4 and Theorem 4.2.7] (see also [Bez,06, Theorem 7.1.1] for a proof using the filtration method). That RCA_2 is generated by its finite algebras and has a decidable equational theory follows from Mortimer [Mor,75] (see also [Hen-Mon-Tar,85, Theorem 4.2.9], [Mar-Ven,97, Theorem 2.3.5] and [Mar,97]). Summing all this up we arrive at the following result.

Theorem 2.3.1. Df_2, CA_2 *and* RCA_2 *are generated by their finite algebras and have decidable equational theories.*

However, Df_2 is not only generated by its finite algebras, but it is also generated by its finite rectangular and finite square algebras. This result follows from Segerberg [Seg,73]. A short algebraic proof can be found in Andréka and Németi [And-Nem,94]. For a frame-theoretic proof using quasimodels see [Gab-Kur-Wol-Zak,03, Theorem 5.25]. Another frame-theoretic proof is given in [Bez,06, Theorem 6.1.11]. All these proofs can be adjusted to show that RCA_2 is generated by finite cylindric square algebras. This result also follows from Mortimer [Mor,75]. Thus, we arrive at the following theorem.

Theorem 2.3.2.

(i) $\mathsf{Df}_2 = \mathbf{HSP}(\mathsf{Fin}\mathbb{RECT}) = \mathbf{HSP}(\mathsf{Fin}\mathbb{SQ})$.

(ii) $\mathsf{RCA}_2 = \mathbf{HSP}(\mathsf{Fin}\mathbb{CSQ})$.

Now we turn to lattices of subvarieties of two-dimensional cylindric algebras. Let $\Lambda(\mathsf{Df}_2)$ denote the lattice of subvarieties of Df_2, $\Lambda(\mathsf{CA}_2)$ denote the lattice of subvarieties of CA_2 and $\Lambda(\mathsf{RCA}_2)$ denote the lattice of subvarieties of RCA_2. We also let $\Lambda(\mathsf{Df}_1)$ denote the lattice of subvarieties of Df_1-algebras. This lattice is easy to describe. The lattice of all subvarieties of Df_1 is an $(\omega+1)$-chain that converges to Df_1 (see [Scr,51, Theorem 4] and [Hen-Mon-Tar,85, Theorem 4.1.22]). As we will see below, the lattice $\Lambda(\mathsf{Df}_2)$ is also countable, although much more complex than $\Lambda(\mathsf{Df}_1)$. The lattices $\Lambda(\mathsf{CA}_2)$ and $\Lambda(\mathsf{RCA}_2)$, however, are not countable [Hen-Mon-Tar,85, Theorem 4.1.27 and Remark 4.1.28]. We will sketch the proof of this fact by using the technique of Jankov–Fine formulas, which is a standard tool in modal logic. For an overview on Jankov–Fine formulas we refer to [Cha-Zak,97, Section 9.4], [Bla-Rij-Ven,01, Section 3.4] or [Bez,06, Section 3.4]. This proof will also underline the difference between the finite square Df_2 and CA_2-algebras.

Let $\mathfrak{B}$ and $\mathfrak{B}'$ be simple CA_2-algebras. We write

$$\mathfrak{B} \leq \mathfrak{B}' \text{ iff } \mathfrak{B} \text{ is a subalgebra of } \mathfrak{B}'.$$

Next we construct simple $\leq$-antichains of finite simple RCA_2-algebras. It follows from [Bez-Hod,04] that there do not exist any infinite $\leq$-antichains of finite simple Df_2-algebras. For the proof of the next lemma we refer to [Hen-Mon-Tar,85, Theorem 4.1.27] (see also [Bez,04b, Lemma 4.1] or [Bez,06, Lemma 7.1.13]).

Lemma 2.3.3. *Every two non-isomorphic finite cylindric square algebras are $\leq$-incomparable.*

Applying the standard technique of Jankov–Fine formulas to two-dimensional cylindric algebras [Bez,06, Section 7.1.2], as an immediate consequence of Lemma 2.3.3, we obtain the following result ([Hen-Mon-Tar,85, Theorem 4.1.27], (see also [Bez,04b, Theorem 4.2] or [Bez,06, Theorem 7.1.14]).

Theorem 2.3.4. *The cardinality of* $\Lambda(\mathsf{RCA}_2)$ *is that of the continuum.*

Moreover, by replacing a singleton non-diagonal E_0-cluster with a two-element E_0-cluster in each finite cylindric square, we obtain an infinite $\leq$-antichain of finite rooted CA_2-spaces satisfying $(*)$. Therefore, the corresponding algebras do not belong to RCA_2. Applying Jankov–Fine formulas again we obtain the following result [Hen-Mon-Tar,85, Remark 4.1.28], (see also [Bez,04b, Theorem 4.4, Corollary 4.5] or [Bez,06, Theorem 7.16, Corollary 7.1.17]).

Theorem 2.3.5.

(i) *The cardinality of* $\Lambda(\mathsf{CA}_2) \setminus \Lambda(\mathsf{RCA}_2)$ *is that of the continuum.*

(ii) *There exists a continuum of varieties in between* RCA_2 *and* CA_2.

Note that there are only countably many finitely axiomatizable varieties and there are only countably many varieties with a decidable equational theory. Therefore, Theorem 2.3.4 also implies that there exists a continuum of non-finitely axiomatizable subvarieties of CA_2 (resp. of RCA_2), and there exists a continuum of subvarieties of CA_2 (resp. of RCA_2) with an undecidable equational theory. We also remark that for an uncountable α, CA_α has 2^α many subvarieties [Nem,87a, p. 246]. It is still an open problem whether the same result holds for RCA_α [Hen-Mon-Tar,85, Problem 4.2] and [Mon,91b]. We will see in the next section that $\Lambda(\mathsf{Df}_2)$ is countable.

4. Locally Finite Subvarieties of Df_2

In this section we investigate locally finite subvarieties of Df_2. We recall that a variety V is called *locally finite* if every finitely generated V-algebra is finite. It is well known (see e.g., Halmos [Hal,62]) that Df_1 is locally finite.

It was Tarski who first noticed that Df_2 is not locally finite. Detailed proofs of this fact can be found in [Hen-Mon-Tar,85, Theorem 2.1.11], [Hal,62, p. 92], [Erd-Fab-Lar,81], [Bez,06, Example 6.2.1].

Let $\mathcal{B}$ be a simple Df_2-algebra and $\mathcal{X}$ its dual rooted Df_2-space. Let also $i = 1, 2$ and $n > 0$. We say that $\mathcal{X}$ is of *E_i-depth n* if the number of E_i-clusters of $\mathcal{X}$ is exactly n. The E_i-depth of $\mathcal{X}$ is said to be *infinite* if $\mathcal{X}$ has infinitely many E_i-clusters. $\mathcal{B}$ is said to be of *E_i-depth $n < \omega$* if the E_i-depth of $\mathcal{X}$ is n. The E_i-depth of $\mathcal{B}$ is said to be *infinite* if $\mathcal{X}$ is of infinite E_i-depth. $\mathsf{V} \subseteq \mathsf{Df}_2$ is said to be of *E_i-depth $n < \omega$* if n is the maximal E_i-depth of the simple members of V, and V is of *E_i-depth ω* if there is no bound on the E_i-depth of simple members of V. For a simple Df_2-algebra $\mathcal{B}$ and its dual $\mathcal{X}$, let $d_i(\mathcal{B})$ and $d_i(\mathcal{X})$ denote the E_i-depth of $\mathcal{B}$ and $\mathcal{X}$, respectively. Similarly, let $d_i(\mathsf{V})$ denote the E_i-depth of a variety $\mathsf{V} \subseteq \mathsf{Df}_2$. We note that there exists a formula measuring the depth of a subvariety of Df_2 (see [Bez,02, Theorem 4.2] or [Bez,06, Theorem 6.2.4]).

Definition 2.4.1. For a variety V, let $\mathsf{SI}(\mathsf{V})$ and $\mathsf{S}(\mathsf{V})$ denote the classes of all subdirectly irreducible and simple V-algebras, respectively. Let also $\mathsf{FinSI}(\mathsf{V})$ and $\mathsf{FinS}(\mathsf{V})$ denote the classes of all finite subdirectly irreducible and simple V-algebras, respectively.

We recall from [Bez,01] a criterion of local finiteness.

Theorem 2.4.2. *A variety V of a finite signature is locally finite iff the class $\mathsf{SI}(\mathsf{V})$ is* uniformly locally finite; *that is, for each natural number n there is a natural number $M(n)$ such that for each n-generated $\mathcal{A} \in \mathsf{SI}(\mathsf{V})$ we have $|\mathcal{A}| \leq M(n)$.*

The next theorem is an important tool in characterizing locally finite subvarieties of Df_2. Its proof, which can be found in [Bez,02, Lemma 4.4] or [Bez,06, Lemma 6.2.7], relies on the fact that the variety of Df_1-algebras is locally finite.

Theorem 2.4.3. *Every subvariety $\mathsf{V} \subseteq \mathsf{Df}_2$ such that $d_1(\mathsf{V}) < \omega$ or $d_2(\mathsf{V}) < \omega$ is locally finite.*

Now we are in a position to prove that every proper subvariety of Df_2 is locally finite.

Theorem 2.4.4. *If a variety $\mathsf{V} \subseteq \mathsf{Df}_2$ is not locally finite, then $\mathsf{V} = \mathsf{Df}_2$.*

Proof. We sketch the main idea of the proof. Suppose V is not locally finite. Then there exists a finitely generated infinite V-algebra $\mathcal{B}$. Let $\mathcal{X}$ be the dual of $\mathcal{B}$. Then either there exists an infinite rooted saturated subset of $\mathcal{X}$, or $\mathcal{X}$ consists of infinitely many finite rooted saturated subsets.

First suppose that $\mathcal{X}$ contains an infinite rooted saturated subset $\mathcal{X}_0$. If either the E_1 or E_2-depth of $\mathcal{X}_0$ is finite, then the Df_2-algebra, call it $\mathcal{B}_0$, corresponding to $\mathcal{X}_0$ belongs to some variety $\mathsf{V}' \subseteq \mathsf{Df}_2$ of a finite E_1 or E_2-depth. Then $\mathcal{B}_0$ is a homomorphic image of $\mathcal{B}$ and is finitely generated. Moreover, by our assumption, $\mathcal{X}_0$ and hence $\mathcal{B}_0$ is infinite. This is a contradiction, since by Theorem 2.4.3, V' is locally finite. Thus, both the E_1 and E_2-depths of $\mathcal{X}_0$ are infinite. Next we can show (see [Bez,02, Claim 4.7] or [Bez,06, Claim 6.2.10]) that for each $n \in \omega$, the square $\langle W \times W, E_1, E_2 \rangle$ with $|W| = n$ is a Df_2-morphic image of $\mathcal{X}_0$. By duality, this means that the algebra $\big\langle \mathcal{P}(W \times W), E_1, E_2 \big\rangle$ is a subalgebra of $\big\langle \mathsf{CP}\,(\mathcal{X}_0), E_1, E_2 \big\rangle$ for each $n < \omega$. Since Df_2 is generated by finite square algebras (see Theorem 2.2.2) this implies that $\mathsf{V} = \mathsf{Df}_2$.

Now suppose that $\mathcal{X}$ consists of infinitely many finite rooted spaces which we denote by $\{\mathcal{X}_j\}_{j \in J}$. If either the E_1 or E_2-depth of the members of $\{\mathcal{X}_j\}_{j \in J}$ is bounded by some integer n, then their corresponding algebras belong to some variety $\mathsf{V}' \subseteq \mathsf{Df}_2$ with $d_1(\mathsf{V}') < n$ or $d_2(\mathsf{V}') < n$. This means that there is an infinite finitely generated algebra in V'. By Theorem 2.4.3, this is a contradiction. Therefore, we can assume that neither the E_1 nor E_2-depth of $\{\mathcal{X}_j\}_{j \in J}$ is bounded by any integer. Then we can again show that every finite square algebra is a subalgebra of $\big\langle \mathsf{CP}\,(\mathcal{X}_0), E_1, E_2 \big\rangle$. This, by Theorem 2.2.2, means that $\mathsf{V} = \mathsf{Df}_2$.

Thus, if V is not locally finite, then $\mathsf{V} = \mathsf{Df}_2$, which completes the proof of the theorem. ■

Recall that a variety V is called *pre-locally finite* if V is not locally finite but every proper subvariety of V is locally finite. We also recall that every locally finite variety is generated by its finite algebras. Therefore, we arrive at the following theorem.

Corollary 2.4.5.

(i) $\mathsf{V} \in \Lambda(\mathsf{Df}_2)$ *is locally finite iff* V *is a proper subvariety of* Df_2.

(ii) Df_2 *is the only pre-locally finite subvariety of* Df_2.

(iii) *Every variety* $\mathsf{V} \subseteq \mathsf{Df}_2$ *is generated by its finite algebras.*

In fact, Corollary 2.4.5(iii) can be significantly strengthened. It is proved in [Bez-Mar,03] (see also [Bez,06, Section 8.2]) that the (bi-)modal logic corresponding to every proper subvariety of Df_2 has the poly-size model property. Moreover, using combinatorial set theory, namely, the theory of better-quasi-orderings, [Bez-Hod,04] (see also [Bez,06, Section 8.1]) proves that every subvariety of Df_2 is finitely axiomatizable. Combining this with Corollary 2.4.5(iii) gives us that the equational theory of every subvariety of Df_2 is decidable. However, even more is true. It is proved in [Bez-Hod,04] (see also [Bez,06, Section 8.4]) that the equational theory of every subvariety of Df_2 is NP-complete.

We finish this section by mentioning the analogy of these results with those obtained in Monk [Mon,70a] for two-dimensional polyadic algebras. Two-dimensional polyadic algebras are obtained by adding four extra unary operations to the signature of Df_2-algebras (see [Hen-Mon-Tar,85, Definition 5.4.1] or [Mon,70a]). Using the methods very similar to ours Monk [Mon,70a] proves that the variety PA_2 of two-dimensional polyadic algebras has only countably many subvarieties, each subvariety is finitely axiomatizable, is determined by its finite members and has a decidable equational theory. Moreover, similarly to Df_2, each proper subvariety of PA_2 is locally finite. As was noted in [Hen-Mon-Tar,85, Theorem 5.4.5 and Remark 5.4.6], the results on subvarieties of PA_2 do not immediately transfer to subvarieties of Df_2. For example, the subvarieties of Df_2 axiomatized by the equations $(\mathsf{c}_1 x = x)$ and $(\mathsf{c}_2 x = x)$, respectively, are distinct, while the subvarieties of PA_2 axiomatized by these equations coincide. This also means that PA_2 is not a conservative extension of Df_2.

5. Classification of Subvarieties of Df_2

In this section we will see that Corollary 2.4.5 enables us to give a classification of subvarieties of Df_2 in terms of E_1 and E_2-depths (see [Bez,02, Section 4] or [Bez,06, Section 6.3]). It follows from Corollary 2.4.5(iii) that every subvariety V of Df_2 is generated by $\mathsf{FinS}\,(\mathsf{V})$.

Theorem 2.5.1. *For every proper subvariety* V *of* Df_2 *there exists a natural number* n *such that* $\mathsf{FinS}\,(\mathsf{V})$ *can be divided into three disjoint sets* $\mathsf{FinS}\,(\mathsf{V}) = \mathsf{F}_1 \uplus \mathsf{F}_2 \uplus \mathsf{F}_3$, *where* $d_2(\mathsf{F}_1), d_1(\mathsf{F}_2) \leq n$ *and* $d_1(\mathsf{F}_3), d_2(\mathsf{F}_3) \leq n$. *(Note that any two of the sets* $\mathsf{F}_1, \mathsf{F}_2$ *and* F_3 *may be empty.)*

Proof. We sketch the proof. Suppose V is a proper subvariety of Df_2. By Theorem 2.2.2, Df_2 is generated by finite square algebras. Therefore, there exists $n \in \omega$ and a square $\langle W \times W, E_1, E_2 \rangle$ such that $|W| = n$ and $\langle \mathcal{P}(W \times W), E_1, E_2 \rangle \notin \mathsf{FinS}(\mathsf{V})$. Let n be the minimal such number. We consider three subclasses of $\mathsf{FinS}(\mathsf{V})$: $\mathsf{F}_1 = \{\mathcal{B} \in \mathsf{FinS}(\mathsf{V}) : d_1(\mathcal{B}) > n\}$, $\mathsf{F}_2 = \{\mathcal{B} \in \mathsf{FinS}(\mathsf{V}) : d_2(\mathcal{B}) > n\}$ and $\mathsf{F}_3 = \{\mathcal{B} \in \mathsf{FinS}(\mathsf{V}) : d_1(\mathcal{B}), d_2(\mathcal{B}) \leq n\}$. It is obvious that $\mathsf{FinS}(\mathsf{V}) = \mathsf{F}_1 \cup \mathsf{F}_2 \cup \mathsf{F}_3$. We prove that F_1, F_2 and F_3 are disjoint.

Let us show that if $\mathcal{B} \in \mathsf{F}_1$, then $d_2(\mathcal{B}) \leq n$ and if $\mathcal{B} \in \mathsf{F}_2$, then $d_1(\mathcal{B}) \leq n$. Suppose $\mathcal{B} \in \mathsf{F}_1 \cup \mathsf{F}_2$, $d_1(\mathcal{B}) = k$, $d_2(\mathcal{B}) = m$ and both $k, m > n$. Let $\mathcal{X}$ be the dual of $\mathcal{B}$. Then we can show that a finite square $\langle W \times W, E_1, E_2 \rangle$ such that $|W| = n$ is a Df_2-morphic image of $\mathcal{X}$. By duality, this means that the square algebra $\langle \mathcal{P}(W \times W), E_1, E_2 \rangle$ is a subalgebra of $\mathcal{B}$ and therefore belongs to $\mathsf{FinS}(\mathsf{V})$, which is a contradiction. Thus, $\mathcal{B} \in \mathsf{F}_1$ implies $d_1(\mathcal{B}) > n$ and $d_2(\mathcal{B}) \leq n$, and $\mathcal{B} \in \mathsf{F}_2$ implies $d_1(\mathcal{B}) \leq n$ and $d_2(\mathcal{B}) > n$. Also, if $\mathcal{B} \in \mathsf{F}_3$, then $d_1(\mathcal{B}), d_2(\mathcal{B}) \leq n$. This shows that all the three sets are disjoint. ■

From this theorem we obtain the following classification of subvarieties of Df_2 (see [Bez,02, Theorem 4.10] or [Bez,06, Theorem 6.3.4]).

Theorem 2.5.2. *For each* $\mathsf{V} \in \Lambda(\mathsf{Df}_2)$, *either* $\mathsf{V} = \mathsf{Df}_2$, *or* $\mathsf{V} = \bigvee_{i \in S} \mathsf{V}_i$ *for some* $S \subseteq \{1, 2, 3\}$, *where* $d_1(\mathsf{V}_1), d_2(\mathsf{V}_2), d_1(\mathsf{V}_3), d_2(\mathsf{V}_3) < \omega$.

Proof. The proof follows from Theorem 2.5.1, by setting $\mathsf{V}_i = \mathbf{HSP}(\mathsf{F}_i)$ for $i = 1, 2, 3$. ■

We close this section by recalling from [Bez,02, Section 6] a characterization of rectangularly and square representable subvarieties of Df_2. First we give a general definition of representability for varieties of Df_2-algebras.

Definition 2.5.3. A variety $\mathsf{V} \subseteq \mathsf{Df}_2$ is called *representable by (algebras from class)* $\mathsf{K} \subseteq \mathsf{Df}_2$ if $\mathsf{V} = \mathbf{SP}(\mathsf{K} \cap \mathsf{V})$.

For a variety $\mathsf{V} \subseteq \mathsf{Df}_2$, we denote by $(\mathbb{GRECT}_\mathsf{V})$ $\mathbb{RECT}_\mathsf{V}$ and $(\mathbb{GSQ}_\mathsf{V})$ $\mathbb{SQ}_\mathsf{V}$ the classes of (generalized) rectangular and (generalized) square V-algebras, respectively. By Definition 2.5.3, $\mathsf{V} \subseteq \mathsf{Df}_2$ is *rectangularly representable* if $\mathsf{V} = \mathbf{SP}(\mathbb{RECT}_\mathsf{V})$ and V is *square representable* if $\mathsf{V} = \mathbf{SP}(\mathbb{SQ}_\mathsf{V})$. Moreover, we have that $\mathsf{V} \subseteq \mathsf{Df}_2$ is *rectangularly representable* iff $\mathsf{V} = \mathbf{S}(\mathbb{GRECT}_\mathsf{V})$ and V is *square representable* iff $\mathsf{V} = \mathbf{S}(\mathbb{GSQ}_\mathsf{V})$. (This is a consequence of a general result of modal logic concerning a duality between products of complete and atomic modal algebras and disjoint unions of corresponding frames; see e.g., [Ven,07, Section 5.6].). Thus, square and rectangular representability in the subvarieties of Df_2 are restricted versions of the general definition of representability (Definition 2.2.1).

For positive integers m and n let $\mathcal{P}(m \times n)$ denote the rectangular algebra $\langle \mathcal{P}(W \times W'), E_1, E_2 \rangle$ such that $|W| = m$ and $|W'| = n$. Let n_1 and n_2 be positive integers. We let $\mathsf{V}_{(\omega,n_1)} = \mathbf{HSP}\big(\{\mathcal{P}(m \times n_1)\}_{m\in\omega}\big)$ and $\mathsf{V}_{(n_2,\omega)} = \mathbf{HSP}\big(\{\mathcal{P}(n_2 \times m)\}_{m\in\omega}\big)$. The next theorem provides a full characterization of rectangular and square representable subvarieties of Df_2.

Theorem 2.5.4. *Let* $\mathsf{V} \in \Lambda(\mathsf{Df}_2)$.

(i) V *is square representable iff* $\mathsf{V} = \mathsf{Df}_2$ *or* $\mathsf{V} = \mathbf{HSP}\big(\mathcal{P}(n \times n)\big)$ *for some* $n \in \omega$.

(ii) V *is rectangularly representable iff* $\mathsf{V} = \mathsf{Df}_2$ *or* $\mathsf{V} = \mathsf{V}_{(\omega,n_1)} \vee \mathsf{V}_{(n_2,\omega)} \vee \mathsf{V}'$, *where* $\mathsf{V}' = \bigvee_{i=1}^{r} \mathbf{HSP}\big(\mathcal{P}(m_i \times k_i)\big)$ *for some* $m_i, k_i, r \in \omega$.

6. Locally Finite Subvarieties of CA_2

In the previous section we proved that Df_2 is pre-locally finite. It is known (see, e.g., [Hen-Mon-Tar,85, Theorem 2.1.11]) that RCA_2, and hence every variety in the interval $[\mathsf{RCA}_2, \mathsf{CA}_2]$, is not locally finite. In this section, we present a criterion of local finiteness for varieties of CA_2-algebras (see [Bez,04b, Section 5] or [Bez,06, Section 7.2]). We also show that there exists exactly one pre-locally finite subvariety of CA_2. The E_1 and E_2-depths of simple CA_2-algebras are defined as in the Df_2-case. Since the number of the E_1 and E_2-clusters in every CA_2-space is the same, for each simple CA_2-algebra $\mathfrak{B}$, we have $d_1(\mathfrak{B}) = d_2(\mathfrak{B})$. We denote it by $d(\mathfrak{B})$ and call it the *depth* of $\mathfrak{B}$. The depth of a variety of CA_2-algebras we denote by $d(\mathsf{V})$. Our goal is to show that a variety V of CA_2-algebras is locally finite iff $d(\mathsf{V}) < \omega$. For this we need the following definition.

Definition 2.6.1.

(i) Call a rooted CA_2-space $\mathcal{X}$ *uniform* if every non-diagonal E_0-cluster of $\mathcal{X}$ is a singleton set, and every diagonal E_0-cluster of $\mathcal{X}$ contains only two points.

(ii) Call a simple CA_2-algebra $\mathfrak{B}$ *uniform* if its dual rooted CA_2-space $\mathcal{X}$ is uniform.

Finite uniform rooted spaces are shown in Fig. 2, where big dots denote the diagonal points. Let $\mathcal{X}_n$ denote the uniform rooted space of depth n. Also let $\mathfrak{B}_n$ denote the uniform CA_2-algebra of depth n. It is obvious that $\mathcal{X}_n$ is (isomorphic to) the dual CA_2-space of $\mathfrak{B}_n$. Let U denote the variety generated by all finite uniform CA_2-algebras; that is $\mathsf{U} = \mathbf{HSP}(\{\mathfrak{B}_n\}_{n\in\omega})$. Applying the criterion of Theorem 2.2.5, it is easy to check that $\mathsf{U} \subseteq \mathsf{RCA}_2$. For the proof of the next lemma we refer to [Bez,04b, Lemma 5.2] or [Bez,06, Lemma 7.2.4].

Lemma 2.6.2.

(i) *If $\mathfrak{B}$ is a simple cylindric algebra of infinite depth, then each $\mathfrak{B}_n$ is a subalgebra of $\mathfrak{B}$.*

(ii) *If $\mathfrak{B}$ is a simple cylindric algebra of depth $2n$, then $\mathfrak{B}_n$ is a subalgebra of $\mathfrak{B}$.*

Now we characterize varieties of CA_2-algebras of infinite depth in terms of U.

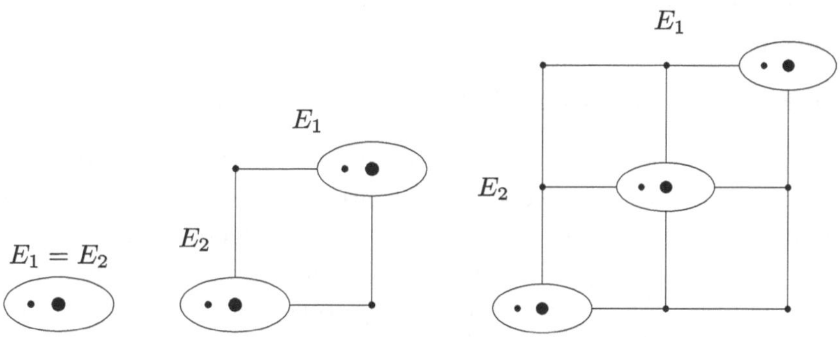

Fig. 2. Uniform rooted CA_2-spaces

Theorem 2.6.3. *For a variety V of CA_2-algebras, $d(\mathsf{V}) = \omega$ iff $\mathsf{U} \subseteq \mathsf{V}$.*

Proof. It is obvious that $d(\mathsf{U}) = \omega$. So, if $\mathsf{U} \subseteq \mathsf{V}$, then obviously $d(\mathsf{V}) = \omega$. Conversely, suppose $d(\mathsf{V}) = \omega$. We want to show that every finite uniform CA_2-algebra belongs to V. Since $d(\mathsf{V}) = \omega$, the depth of the simple members of V is not bounded by any integer. So, either there exists a family of simple V-algebras of increasing finite depth, or there exists a simple V-algebra of infinite depth. In either case, it follows from Lemma 2.6.2 that $\{\mathfrak{B}_n\}_{n\in\omega} \subseteq \mathsf{V}$. Therefore, $\mathsf{U} \subseteq \mathsf{V}$, since $\{\mathfrak{B}_n\}_{n\in\omega}$ generates U. ■

Our next task is to show that U is not locally finite. For this we first need to observe that every finite uniform algebra is 1-generated (see [Bez,04b, Lemma 5.6] or [Bez,06, Lemma 7.2.6]). Now Theorem 2.4.2 immediately implies that U is not locally finite. Next, using the fact that the variety of Boolean algebras is locally finite, we show that varieties of CA_2-algebras of finite depth are locally finite. For the proof of this result we refer to [Bez,04b, Theorem 5.10] or [Bez,06, Theorem 7.2.9].

Theorem 2.6.4. *If* $d(\mathsf{V}) < \omega$, *then* V *is locally finite.*

We note that Theorem 2.6.4 is a CA_2-analogue of Theorem 2.4.3. Its proof, however, relies on local finiteness of Boolean algebras, whereas the proof of Theorem 2.4.3 uses the fact that Df_1 is locally finite. Finally, combining Theorems 2.6.3 and 2.6.4, we obtain the following characterization of locally finite varieties of CA_2-algebras.

Theorem 2.6.5.

(i) *For* $\mathsf{V} \in \Lambda(\mathsf{CA}_2)$ *the following conditions are equivalent:*

  (a) V *is locally finite,*

  (b) $d(\mathsf{V}) < \omega$,

  (c) $\mathsf{U} \not\subseteq \mathsf{V}$.

(ii) U *is the only pre-locally finite subvariety of* CA_2.

Therefore, in contrast to the diagonal-free case, there exist uncountably many subvarieties of CA_2 (RCA_2) which are not locally finite. Since every locally finite variety is generated by its finite algebras we obtain from Theorem 2.6.5 that every subvariety of CA_2 of finite depth is generated by its finite algebras. We leave it as an open problem whether each subvariety of CA_2 is generated by its finite algebras.

7. Finitely Generated Varieties of Cylindric Algebras

Recall that a variety is called *finitely generated* if it is generated by a single finite algebra, and that a variety is called *pre-finitely generated* if it is not finitely generated, but all its proper subvarieties are finitely generated. In studying a lattice of subvarieties of a given variety, finitely generated and pre-finitely generated varieties play an important role. Finitely generated varieties constitute the 'lower' part of this lattice, whereas pre-finitely generated ones are borderlines between the finitely generated and non-finitely generated ones. Pre-finitely generated varieties are minimal among non-finitely generated ones. Moreover, an explicit description of pre-finitely generated varieties provides a criterion for characterizing finitely generated varieties. For varieties of modal and Heyting algebras there are few well-known characterizations of pre-finitely generated varieties. Maksimova [Mak,72] showed that there are exactly three pre-finitely generated varieties of Heyting algebras. Maksimova [Mak,75] and Esakia and Meskhi [Esa-Mes,77] proved that there are exactly five pre-finitely generated varieties of **S4**-algebras. Blok [Blo,80] showed that there is a continuum of pre-finitely generated varieties of **K4**-algebras. On the other hand, Df_1 is the only pre-finitely generated variety in the lattice of subvarieties of Df_1. For Df_2 and CA_2 the picture is more complex than for Df_1. As follows from [Bez,02, Theorem 5.4] and [Bez,06, Corollary 6.4.7] there are exactly six pre-finitely generated varieties in $\Lambda(\mathsf{Df}_2)$, there are exactly fifteen pre-finitely generated varieties in $\Lambda(\mathsf{CA}_2)$, and six of them belong to $\Lambda(\mathsf{RCA}_2)$ (see [Bez,04b, Corollary 6.6] and [Bez,06, Corollary 7.3.7]). These results yield a characterization of finitely generated varieties of Df_2 and CA_2-algebras. A variety V is finitely generated iff none of the pre-finitely generated varieties is a subvariety of V. Another characterization of finitely generated varieties of Df_2 and CA_2-algebras can be found in [Bez,02, Section 5] and [Bez,04b, Section 7].

In Section 5 we gave a classification of subvarieties of Df_2. We close this section with a very rough description of $\Lambda(\mathsf{CA}_2)$. We need the following notation: Let FG denote the class of all finitely generated subvarieties of CA_2. Also let D_F denote the class of varieties of CA_2-algebras of finite depth which are not finitely generated varieties and let D_ω denote the class of varieties of CA_2-algebras of infinite depth.

It follows from the results discussed in Sections 7 and 8 that the variety V_{tr} generated by the trivial CA_2-algebra is the least element of FG, that FG does not have maximal elements, that D_F has precisely fifteen minimal elements, that D_F does not have maximal elements, and that U and CA_2 are the least and greatest elements of D_ω, respectively.

The detailed investigation of the 'lower' part of $\Lambda(\mathsf{CA}_2)$ can be found in [Bez,04b, Section 7]. In particular, a complete characterization of the lattice structure of the extensions of CA_2 of depth one is given in [Bez,04b, Section 7.1]. Using the reduct functor $\mathfrak{Df} : \mathsf{CA}_2 \to \mathsf{Df}_2$, we can define a reduct functor from the lattice $\Lambda(\mathsf{CA}_2)$ into the lattice $\Lambda(\mathsf{Df}_2)$. This reduct functor and the properties that are preserved and reflected by it are investigated in [Bez,04b, Section 7].

8. Problems

We close this chapter by listing some open problems.

(i) As we saw in Section 4, every subvariety of Df_2 is generated by its finite members. Moreover, corresponding logical systems have the poly-size model property and NP-complete satisfiability problem. The same question for CA_2-algebras remains open. Every subvariety of CA_2 of finite depth is locally finite and therefore is generated by its finite algebras. However, it is still an open (and rather complicated) question whether every subvariety of CA_2 and RCA_2 of infinite depth is generated by its finite algebras.

(ii) Subvarieties of CA_α (for both infinite and finite α) are investigated in [Nem,87a, Section 1.1]. In particular, a characterization of subvarieties of CA_ω with decidable equational theories is given in [Nem,87a]. An existence of such a characterization for finite α (especially in the case $\alpha = 2$) remains an open problem.

(iii) That Df_2 (resp. CA_2 and RCA_2) does not have the amalgamation property was first noticed by Comer [Com,69] (see also Sain [Sai,90] and Marx [Mar,00]). In particular, it follows from the proof of this result that every subvariety V of Df_2 (resp. of CA_2 and RCA_2) such that $d_1(\mathsf{V}) > 2$ or $d_2(\mathsf{V}) > 2$ lacks the amalgamation property. We leave it as an open problem to give a full characterization of subvarieties of Df_2 (resp. CA_2 and RCA_2) with the amalgamation property.

(iv) In this chapter we considered two types of two-dimensional algebras: diagonal-free cylindric algebras and cylindric algebras with the diagonal. However, in order to get the full two-variable fragment of FOL (with substitution), we could have added to our signature four extra unary operations (analogous to substitutions of first-order variables) used in polyadic algebras. Then in addition to Df_2-algebras and CA_2-algebras we would have two more similarity types of two-dimensional cylindric-like algebras: *two-dimensional polyadic algebras*

(PA_2-algebras) and *two-dimensional polyadic equality algebras* (PEA_2-algebras), see [Hen-Mon-Tar,85, Section 5.4]. A PA_2-like similarity type was also considered in [Seg,73].

As was pointed out earlier, it was shown by Monk [Mon,70a] that subvarieties of the variety of PA_2-algebras have very similar (good) properties as subvarieties of Df_2-algebras. We leave it as an open problem to investigate the lattice of varieties of PEA_2-algebras and compare it with the lattices of varieties of PA_2-algebras, Df_2-algebras and CA_2-algebras. We also suggest studying the obvious reduct functors arising between these lattices. We conjecture that in the same way most of the properties of (varieties of) Df_2-algebras hold for (varieties of) PA_2-algebras, most of the properties of (varieties of) CA_2-algebras would hold for (varieties of) PEA_2-algebras.

BOLYAI SOCIETY
MATHEMATICAL STUDIES, 22

**Cylindric-like Algebras
and Algebraic Logic**
pp. 61–89.

Completions and Complete Representations

ROBIN HIRSCH and IAN HODKINSON

The title of this chapter indicates a rather technical topic, but it can also be thought of as a foundational issue in logic. The question to be considered is this: to what extent can we use an abstract mathematical language to express and reason about relations? Going back at least as far as Augustus de Morgan [Mor,60], a relation can be *defined* explicitly, as a set of tuples of some fixed length. This allows us to focus on the mathematical aspects of relations and ignore other more problematic features that might arise from other approaches, such as a linguistic analysis of the use of relations in natural language. In order to treat relations algebraically, we consider them abstractly, identify certain relational operations (e.g., the operation of taking the *converse* of a binary relation) and write down some equational axioms which are sound for the chosen kind of relations (e.g., a binary relation is equal to the converse of its converse). Ideally, our set Γ of equations will be *equationally complete,* so that any equation valid over fields of relations of a certain rank equipped with the chosen set-theoretically definable operators will be entailed by Γ.

The finite set of equational axioms for Boolean algebra is very successful in this respect, for handling *unary relations.* The chosen operators are union and complementation together with constants 0 denoting the empty set and 1 for the unit of the Boolean algebra. Other operators, like intersection, can be defined within this signature. The finite set of axioms defining a Boolean algebra is complete and every Boolean algebra is isomorphic to a genuine field of sets.

Similarly, for binary relations, we treat the boolean operators plus some additional operators and try to write down some equational axioms that are equationally complete for binary relations. Different choices are possible for a set of operators for binary relations – for relation algebras we use the boolean operators together with the unary operator of taking the converse, the binary operator of composition and a constant for the identity. Axiomatising binary relations with the relation algebra operators turns out to be more difficult than was the case for unary relations, and we know that any complete set of axioms is necessarily infinite [Mon,64], but recursively enumerable, complete, equational axiomatisations are known [Lyn,56, Hir-Hod,02a]. For relations of higher finite rank, different choices of algebras can be considered – cylindric algebra, polyadic algebra, diagonal-free algebra – but for ranks at least three, the situation is largely similar to the relation algebra case. All of the operators of these algebras are additive in each argument, and normal, meaning that their value is 0 whenever any argument is 0. All algebras mentioned above are *Boolean algebras with operators* (BAOs).

Let $\mathcal{F}$ be one of the following: (i) the class of fields of sets equipped with the boolean operators, (ii) the class of fields of binary relations with the relation algebra operators, (iii) the class of fields of n-ary relations (for some n) with the cylindric algebra operators, (iv) the class of fields of n-ary relations with the polyadic operators, (v) the class of fields of n-ary relations with the diagonal free operators. Let Γ be a set of equations of the appropriate signature equationally complete over $\mathcal{F}$. Since the closure of $\mathcal{F}$ under isomorphism is known to be a variety, every model of Γ is isomorphic to a member of $\mathcal{F}$. A *representation* is an isomorphism from an algebra to a field of relations, and its *base* is the underlying set of objects that the relations relate.

But the correspondence between algebras and fields of relations may not be quite as close as we had hoped. By completeness, any equation valid in $\mathcal{F}$ is entailed by Γ, and since $\mathcal{F}$ is a variety, Γ entails all first-order sentences valid over $\mathcal{F}$. But there might be other true properties of $\mathcal{F}$, not expressible by equations or even first-order sentences, that do not follow from Γ. At least some second-order properties do follow from Γ in these cases. Since each k-ary operator f of each algebra in $\mathcal{F}$ is *conjugated,* it follows that f is *completely additive* in each argument [Jon-Tar,51], meaning that if an arbitrary non-empty set S of elements of some $\mathfrak{A} \in \mathcal{F}$ has a supremum $\sup(S)$, and if $i < k$ and $a_0, \ldots, a_{i-1}, a_{i+1}, \ldots a_{k-1} \in \mathfrak{A}$, then $b = \sup\{f(a_0, \ldots, a_{i-1}, s, a_{i+1}, \ldots, a_{k-1}) : s \in S\}$ exists and

$$f(a_0, a_1, \ldots, a_{i-1}, \sup(S), a_{i+1}, \ldots, a_{k-1}) = b.$$

But there are other second-order properties of $\mathcal{F}$ that might not be properly captured in our algebraic framework. The first problem is that a model $\mathfrak{A}$ of Γ might be *incomplete* – there could be a set S of elements of $\mathfrak{A}$ that has no supremum in $\mathfrak{A}$. With a field of concrete relations, we can always extend the field to include a supremum of any set of relations, simply by taking the set-theoretic union of each set of relations, and generating a field of relations. A construction of Monk [Mon,70b] gives us, for any completely additive BAO $\mathfrak{A}$, a complete extension $\mathsf{Com}(\mathfrak{A})$ in which $\mathfrak{A}$ is dense, and which respects all existing suprema in $\mathfrak{A}$. Such an extension is unique up to isomorphism, and is called the *completion* of $\mathfrak{A}$. However, a potential problem is that $\mathfrak{A}$ could be a model of Γ, so $\mathfrak{A}$ is isomorphic to a field of relations, but $\mathsf{Com}(\mathfrak{A})$ could fail some of the axioms in Γ and have no representation. For binary and higher order relations, this problem is real, as we will see.

The second problem is that even if h is a representation of $\mathfrak{A}$, so that h is an isomorphism from the algebra to a field of relations, there are certain operators definable in second order logic that might not be preserved by h. We say that h is a *complete representation* of $\mathfrak{A}$ if

$$h\big(\sup(S)\big) = \bigcup_{s\in S} h(s)$$

for any subset S of $\mathfrak{A}$ where the supremum $\sup(S)$ exists in $\mathfrak{A}$. By the De Morgan Laws, a complete representation also preserves arbitrary infima wherever they are defined. Every representation of a finite algebra is of course complete. A saturation argument shows that all infinite algebras, even boolean ones, have incomplete representations. So the main question is when an algebra has some complete representation. Complete representability is connected to the omitting types theorem for the corresponding logic: see Chapter 3.4 for more on this. We will devise an infinite game to characterise when an algebra has a complete representation, and we will use this game to analyse the class of completely representable algebras.

1. Boolean Algebra

We start with the easiest case: algebras of unary relations. We can define an ordering in a Boolean algebra by $x \leq y \iff x + y = y$. An *atom* of a Boolean algebra is a $\leq$-minimal non-zero element and the algebra is *atomic* if every non-zero element of the algebra is above some atom. For a Boolean algebra $\mathfrak{B}$ write $At\,\mathfrak{B}$ for the set of all atoms of $\mathfrak{B}$. All non-trivial finite Boolean algebras are atomic but there are Boolean algebras with no atoms at all. (For example, let X be an infinite set and define the equivalence relation over the subsets of X by $S \sim T$ iff the symmetric difference of S and T is finite. The boolean operators have a well-defined action on the equivalence classes yielding a Boolean algebra with no atoms.)

A representation h of a Boolean algebra $\mathfrak{B}$ is called *atomic* if for all $x \in h(1)$ there is an atom $b \in At\,\mathfrak{B}$ with $x \in h(b)$.

Theorem 3.1.1. *Let h be a representation of the Boolean algebra $\mathfrak{B}$. The following are equivalent.*

(i) *h is a complete representation.*

(ii) *h is an atomic representation* ([Hir-Hod,97c]).

Proof. If h is an atomic representation then for all $b \in \mathfrak{B}$, $h(b) = \bigcup \{ h(a) : a \in At\,\mathfrak{B},\ a \leq b \}$. Let S be a set of elements of $\mathfrak{B}$ with a supremum $\sup(S) \in \mathfrak{B}$. An atom a is below $\sup(S)$ iff there is $s \in S$ with $a \leq s$. So

$$\begin{aligned} x \in h\big(\sup(S)\big) &\iff x \in \bigcup \{ h(a) : a \in At\,\mathfrak{B},\ \exists s \in S,\ a \leq s \} \\ &\iff \exists s \in S\ \exists a \in At\,\mathfrak{B},\ \big(a \leq s,\ x \in h(a)\big) \\ &\iff \exists s \in S\ \big(x \in h(s)\big) \\ &\iff x \in \bigcup \{ h(s) : s \in S \} \end{aligned}$$

so h is a complete representation.

Conversely, suppose h is complete. Let $x \in h(1)$. The set $\gamma = \{ b \in \mathfrak{B} : x \in h(b) \}$ is an ultrafilter of $\mathfrak{B}$. 0 is a lower bound of γ. 0 cannot be the greatest lower bound of γ, else

$$x \in \bigcap \{ h(b) : b \in \gamma \} \setminus h\big(\inf(\gamma)\big)$$

contradicting the assumed completeness of h. So there must be a non-zero lower bound of γ, say a. Since $a \not\leq -a$, $-a \notin \gamma$, so $a \in \gamma$. Since if $b + c \in \gamma$ then $b \in \gamma$ or $c \in \gamma$, it follows that a is an atom. Hence $x \in h(a)$ for some atom a, and h is an atomic representation. ∎

Corollary 3.1.2. *The class of completely representable Boolean algebras is the same as the class of atomic Boolean algebras* ([Hir-Hod,97c]).

Other potential problems that we mentioned earlier do not arise for Boolean algebras. Since every Boolean algebra is representable it follows trivially that the completion of a Boolean algebra is always representable.

2. Completely Representable Relation Algebras

The main focus of this article is about completions and complete representations of n-dimensional cylindric algebras for finite $n \geq 3$. Historically the main results were all established first for relation algebra, and we outline these results here, without including any proofs. In the following sections we will go through the corresponding material for cylindric algebras in more detail.

In 1950 Roger Lyndon published a set LC of axioms (now called the *Lyndon conditions*) and proved that a finite relation algebra satisfies the conditions iff it is representable [Lyn,50]. He also claimed to prove that his conditions were valid over complete, representable atomic relation algebras, but in fact his proof only works for finite relation algebras. His main result was to construct a finite relation algebra which failed some of his conditions and was therefore not representable. This showed that Tarski's set of equations for relation algebra [Chi-Tar,51] was not complete. He then defined two infinite atomic relation algebras $\mathfrak{M}$, $\mathfrak{M}'$, and showed that (i) $\mathfrak{M}$ was representable, (ii) $\mathfrak{M}'$ failed one of the first Lyndon conditions, and (iii) every finitely generated subalgebra of either relation algebra was isomorphic to a finitely generated subalgebra of the other. He concluded from (ii) that $\mathfrak{M}'$ was not representable and from (iii) that there could be no equational axiomatization of the class of representable relation algebras (RRA). But in 1955 Tarski proved that RRA was closed under homomorphic image, subalgebra and direct product and was therefore an equational variety [Tar,55]. The situation appeared contradictory.

In fact, by Tarski's result, both algebras were representable, but the fact that $\mathfrak{M}'$ failed a Lyndon condition did not prove it to be unrepresentable, but only that it had no complete representation. The mistake in Lyndon's paper turned out to be a very fruitful one, mainly because it led him to publish a second paper with the first correct axiomatization of RRA [Lyn,56], but also because it led to a thorough investigation of the relationship between representability, the Lyndon conditions, complete representability etc. It was shown in [Hir,95] that the class of completely representable relation algebras is non-elementary. In [Hir-Hod,97c] this was extended to RCA_n for all $n \geq 3$. [Hod,97c] showed that RRA (and RCA_n for finite $n \geq 3$) is not closed under completions.

3. Complete Representations of Cylindric Algebras and Games

We now consider complete representations of n-dimensional cylindric algebras, for $3 \leq n < \omega$. It is clear that determining whether a cylindric algebra has a complete representation or not can be tricky. (Indeed we will see that the class of completely representable cylindric algebras of dimension n is not even elementary.) We saw in Theorem 3.1.1 that a representation of a Boolean algebra is complete if and only if it is atomic. This theorem generalises to algebras of higher order relations, since their representations are, *inter alia,* boolean representations. It follows that only atomic algebras can have complete representations, although Lyndon's relation algebra $\mathfrak{M}'$ shows that not every representable atomic relation algebra need have a complete representation, and similarly (as it turns out), not every atomic representable cylindric algebra need have a complete representation.

We will introduce a two-player game that tests complete representability of an atomic cylindric algebra, but we have some preliminaries concerning networks first.

The dimension n (where $3 \leq n < \omega$) remains fixed until section 8. RCA_n denotes the class of representable n-dimensional cylindric algebras and CCA_n denotes the class of completely representable n-dimensional cylindric algebras. In the following, we often suppress references to n, so it is implicit that all cylindric algebras are n-dimensional. To avoid unnecessary checking, it will often be convenient to consider a slightly wider class of algebras: by a *cylindric-type algebra* we will mean a completely additive BAO of the signature of n-dimensional cylindric algebras. Note that every n-dimensional cylindric algebra is such an algebra (because it is conjugated: see the introductory part), and every representable cylindric-type algebra is a cylindric algebra. A cylindric-type algebra $\mathfrak{A}$ is said to be *atomic* if its boolean reduct is atomic, and in that case we let $At\,\mathfrak{A}$ denote the set of atoms of its boolean reduct.

We consider functions from n to A, where A is a set. (The set of all functions from a set X to a set Y is as usual denoted by ${}^XY$.) We identify the function $x \in {}^nA$ with the sequence $\big(x(0), x(1), \ldots, x(n-1)\big)$, and we sometimes write x as $\bar{x} = (x_0, \ldots, x_{n-1})$. Given $x, y \in {}^nA$ and $i < n$, we write $x \equiv_i y$ if for all $j < n$, if $j \neq i$ then $x(j) = y(j)$. For $i < n$ and $a \in A$, we write $x[i/a]$ for the function that is identical to x except $x[i/a]$ maps i to a.

Definitions 3.3.1 and 3.3.2 below appeared first as [Hir-Hod,97c, Definition 27].

Definition 3.3.1 (Network). Let $\mathfrak{A}$ be an atomic cylindric-type algebra. An *$\mathfrak{A}$-pre-network* $N = (N_1, N_2)$ consists of a set of nodes N_1 and a 'labelling' function $N_2 : {}^nN_1 \to At\,\mathfrak{A}$. N is said to be a *network* if it satisfies, for all $x, y \in {}^nN_1$ and $i, j < n$,

- $N_2(x) \leq \mathsf{d}_{ij} \iff x(i) = x(j)$,
- if $x \equiv_i y$ then $N_2(x) \leq \mathsf{c}_i N_2(y)$.

Write $(M_1, M_2) \subseteq (N_1, N_2)$ if (M_1, M_2), (N_1, N_2) are networks, $M_1 \subseteq N_1$ and $M_2 = N_2 \restriction_{M_1}$. For a limit ordinal λ and a sequence of networks $(N_1^0, N_2^0) \subseteq (N_1^1, N_2^1) \subseteq \cdots \subseteq (N_1^\mu, N_2^\mu) \subseteq \cdots$ ($\mu < \lambda$), define the *limit* of the sequence to be the network $(N_1, N_2) = \bigcup_{\mu<\lambda} (N_1^\mu, N_2^\mu)$ with nodes $N_1 = \bigcup_{\mu<\lambda} N_1^\mu$ and labelling $N_2 = \bigcup_{\mu<\lambda} N_2^\mu$: that is, $N_2(m, n) = N_2^\mu(m, n)$ for any $\mu < \lambda$ such that $m, n \in N_1^\mu$.

The elements of ${}^nN_1$ are called n-dimensional hyperedges (or simply hyperedges) of the network. We will frequently drop the suffices and let N denote the network (N_1, N_2), the set of nodes N_1 and the labelling function N_2, distinguishing cases by context.

A complete representation of an atomic cylindric-type algebra $\mathfrak{A}$ can be identified with a set $\{N_a : a \in At\,\mathfrak{A}\}$ of $\mathfrak{A}$-networks such that

(3.3.1) for each $a \in At\,\mathfrak{A}$ there is $x \in {}^nN_a$ with $N_a(x) = a$, and whenever $x \in {}^nN_a$, $b \in At\,\mathfrak{A}$, $i < n$, and $N_a(x) \leq \mathsf{c}_i b$, there is $y \in {}^nN_a$ with $x \equiv_i y$ and $N_a(y) = b$.

By dint of Theorem 3.1.1, such a set of networks can easily be constructed from a complete representation. Conversely, by renaming the nodes of the networks, we can arrange that the nodes of N_a and N_b are disjoint, when a and b are distinct atoms. An atomic (hence complete) representation h of $\mathfrak{A}$ whose base is the union of the sets of nodes of the N_a, for $a \in At\,\mathfrak{A}$, is defined by

$$h(b) = \{x : \exists a \in At\,\mathfrak{A},\ x \in {}^nN_a,\ N_a(x) \leq b\},$$

for each element b of $\mathfrak{A}$.

Definition 3.3.2 (Atomic Game). Let $\mathfrak{A}$ be an atomic cylindric-type algebra and let $\kappa > 0$ be a cardinal. The two player game $G^\kappa(\mathfrak{A})$ is defined as follows. A *play* of the game is a sequence $N_0 \subseteq N_1 \subseteq \cdots \subseteq N_t \subseteq \cdots$ of $\mathfrak{A}$-networks ($t < \kappa$). In round 0, $\forall$ picks an atom $a \in \mathfrak{A}$ and $\exists$ plays a network N_0. If there is no $x \in {}^n(N_0)$ such that $N_0(x) = a$ then $\forall$ wins the play.

For a limit ordinal $\lambda < \kappa$ let $N_\lambda = \bigcup_{t<\lambda} N_t$. $\forall$ does not win in the round of a limit ordinal.

For successor ordinals, suppose the play has proceeded $N_0 \subseteq \cdots \subseteq N_t$ for some t with $t + 1 < \kappa$. In the $(t+1)$th round, $\forall$ picks $i < n$, $x \in {}^nN_t$, and an atom $a \in At\,\mathfrak{A}$ such that $N_t(x) \leq \mathsf{c}_i a$. Such a move by $\forall$ is denoted (i, x, a). $\exists$ responds with a network $N_{t+1} \supseteq N_t$. If there is no node $l \in N_{t+1}$ such that $N_{t+1}\big(x[i/l]\big) = a$ then $\forall$ wins.

The *limit* of the play is defined to be $\bigcup_{t<\kappa} N_t$. If $\forall$ does not win in any round then $\exists$ wins the play.

The next theorem generalises [Hir-Hod,97c, Theorem 28] to the uncountable case.

Theorem 3.3.3. *Let $\mathfrak{A}$ be an atomic cylindric-type algebra with κ atoms. The following are equivalent.*

(i) *$\mathfrak{A}$ is completely representable.*

(ii) *$\exists$ has a winning strategy in $G^{\kappa+\omega}(\mathfrak{A})$.*

Proof. If $\mathfrak{A}$ has a complete representation then by Theorem 3.1.1 it has an atomic representation and $\exists$'s winning strategy is to maintain an embedding of the current network in a play of the game into the base of the atomic representation. Conversely, if $\exists$ has a winning strategy in $G^{\kappa+\omega}(\mathfrak{A})$, then for each $a \in At\,\mathfrak{A}$ consider a play of the game in which $\exists$ plays networks with fewer than $\kappa + \omega$ nodes, and $\forall$ picks the atom a initially and picks all possible $i < n$, all hyperedges and all legitimate atoms eventually. Let the limit of the play be N_a. Then $\big\{ N_a : a \in At\,\mathfrak{A}\big\}$ satisfies (3.3.1). ∎

For finite m, we can define a first-order sentence ρ_m such that for any atomic cylindric-type algebra $\mathfrak{A}$, $\exists$ has a winning strategy in $G^m(\mathfrak{A})$ iff $\mathfrak{A} \models \rho_m$. These formulas ρ_m correspond, roughly, to the Lyndon conditions that we mentioned in the section on Relation Algebra. By König's lemma, a *finite* n-dimensional cylindric algebra $\mathfrak{A}$ satisfies $\{\rho_m : m < \omega\}$ iff $\exists$

has a winning strategy in $G^\omega(\mathfrak{A})$, iff $\mathfrak{A}$ is representable. This can fail for infinite algebras, but still there is a generalisation to arbitrary algebras (Corollary 3.3.5 below).

Theorem 3.3.4. *If $\mathfrak{A}$ is an atomic cylindric-type algebra and $\exists$ has a winning strategy in $G^m(\mathfrak{A})$ (all $m < \omega$), then $\exists$ has a winning strategy in the game $G^\omega\big(\prod_U \mathfrak{A}\big)$ on the ultrapower $\prod_U \mathfrak{A}$, for any non-principal ultrafilter U over ω.*

Proof. (See [Hir-Hod,97c, Theorem 28(2)] for details.) Let X be a finite set and suppose N^i is an $\mathfrak{A}$-pre-network with nodes X, for all $i < \omega$. The *ultraproduct* of $\langle N^i : i < \omega\rangle$ is defined to be the $\prod_U \mathfrak{A}$-pre-network N with nodes X and labelling defined by $N(\bar{x}) = \big[\big\langle N^i(\bar{x}) : i < \omega\big\rangle\big] \in \prod_U \mathfrak{A}$. Łoš's theorem can be used to prove that this is a network iff $\{i < \omega : N^i \text{ is a network}\} \in U$.

Consider a play $N_0 \subseteq N_1 \subseteq \ldots \subseteq N_m \subseteq \ldots$ of $G^\omega\big(\prod_U \mathfrak{A}\big)$. For each $m < \omega$, $\exists$ maintains a sequence of $\mathfrak{A}$-pre-networks $\big\langle N^j_m : j < \omega\big\rangle$, each with the same nodes as N_m, such that N_m is the ultraproduct of the N^j_m's. Inductively, she also arranges that there is a set $X_m \in U$ such that for all $j \in X_m$, $j \geq m$ and the sequence $N^j_0 \subseteq N^j_1 \subseteq \ldots \subseteq N^j_m$ is the initial segment of a play of $G^j(\mathfrak{A})$ in which $\exists$ uses her winning strategy. Let $X_0 = \omega$. In round m, suppose $\forall$ plays $i < n$, $\bar{x}$ and an atom $\big[\langle a_j : j < \omega\rangle\big]$ of $\prod_U \mathfrak{A}$. By Łoš's theorem, $L = \big\{j < \omega : j > m, (i, \bar{x}, a_j) \text{ is a legal } \forall\text{-move}\big\} \in U$, so $X_m \cap L \in U$. Fix a new node x_m. For each $j \in X_m \cap L$, $\exists$ uses her winning strategy to determine a network $M^j_{m+1} \supseteq N^j_m$. We can assume that M^j_{m+1} extends N^j_m by at most the single node x_m. If $\big\{j \in X_m \cap L : M^j_{m+1} \text{ has the same nodes as } N^j_m\big\} \in U$ then $\exists$ plays $N_{m+1} = N_m$ in the main game. Otherwise, $Y = \big\{j \in X_m \cap L : M^j_{m+1} \text{ extends } N^j_m \text{ by a single node } x_m\big\} \in U$, and she lets N_{m+1} extend N_m by the single new node x_m. For all $j \in Y$ she lets $N^j_{m+1} = M^j_{m+1}$ and for $j \notin Y$ she lets N^j_{m+1} be an arbitrary pre-network with the same nodes as N_{m+1}. By Łoš's theorem again, this maintains the induction hypothesis and defines a valid move for $\exists$ in round m. Since she can do this in all rounds she will win the play. ■

Corollary 3.3.5. *Let $\mathfrak{A}$ be an atomic cylindric-type algebra. Then $\exists$ has a winning strategy in $G^m(\mathfrak{A})$ for all finite m, iff $\mathfrak{A}$ is elementarily equivalent to a completely representable cylindric algebra.*

Proof. If $\mathfrak{B}$ is a completely representable cylindric algebra, then by Theorem 3.3.3, $\exists$ has a winning strategy in $G^m(\mathfrak{B})$ for all finite m, and hence $\mathfrak{B} \models \{\rho_m : m < \omega\}$. If $\mathfrak{A}$ is elementarily equivalent to $\mathfrak{B}$ then $\mathfrak{A} \models \{\rho_m : m < \omega\}$ as well, so $\exists$ has a winning strategy in $G^m(\mathfrak{A})$ for all finite m.

Conversely, if $\exists$ has a winning strategy in $G^m(\mathfrak{A})$ for all finite m, then by Theorem 3.3.4, $\exists$ has a winning strategy in $G^\omega(\prod_U \mathfrak{A})$ where $\prod_U \mathfrak{A}$ is a non-principal ultrapower of $\mathfrak{A}$. It can be checked using elementary chains that $\prod_U \mathfrak{A}$ has a countable elementary subalgebra $\mathfrak{B}$ where $\exists$ still has a winning strategy in $G^\omega(\mathfrak{B})$. By Theorem 3.3.3, $\mathfrak{B}$ is completely representable, and plainly, $\mathfrak{A}$ and $\mathfrak{B}$ are elementarily equivalent. ■

4. Atom Structures

Duality has been important in the theory of BAOs since [Jon-Tar,51], and much earlier for boolean and other algebras. In the rest of the chapter we will consider representations from the dual perspective of atom structures.

The action of the non-boolean operators in a completely additive atomic BAO is determined by their behaviour over the atoms, and this in turn is encoded by the *atom structure* of the algebra.

Definition 3.4.1 (Atom Structure). Let $\mathfrak{A} = \langle A, 0, 1, +, -, \Omega_i : i \in I\rangle$ be an atomic Boolean algebra with operators $\Omega_i : i \in I$. Let the rank of Ω_i be $\rho(i)$. The atom structure $\mathfrak{At}\,\mathfrak{A}$ of $\mathfrak{A}$ is a relational structure

$$\langle At\,\mathfrak{A}, R_{\Omega_i} : i \in I\rangle$$

where $At\,\mathfrak{A}$ is the set of atoms of $\mathfrak{A}$ as before, and R_{Ω_i} is a $\big(\rho(i)+1\big)$-ary relation over $At\,\mathfrak{A}$ defined by

$$R_{\Omega_i}\big(a_0, \ldots, a_{\rho(i)}\big) \iff \Omega_i\big(a_1, \ldots, a_{\rho(i)}\big) \geq a_0.$$

Similar 'dual' structures arise in other ways, too. For any not necessarily atomic BAO $\mathfrak{A}$ as above, its *ultrafilter frame* is the structure

$$\mathfrak{A}_+ = \big\langle \mathsf{Uf}\,(\mathfrak{A}), R_{\Omega_i} : i \in I\big\rangle,$$

where $\mathsf{Uf}\,(\mathfrak{A})$ is the set of all ultrafilters of (the boolean reduct of) $\mathfrak{A}$, and for $\mu_0, \ldots, \mu_{\rho(i)} \in \mathsf{Uf}\,(\mathfrak{A})$, we put $R_{\Omega_i}\big(\mu_0, \ldots, \mu_{\rho(i)}\big)$ iff $\big\{\Omega_i\big(a_1, \ldots, a_{\rho(i)}\big) : a_j \in \mu_j \text{ for } 0 < j \leq \rho(i)\big\} \subseteq \mu_0$.

Definition 3.4.2 (Complex algebra). Conversely, if we are given an arbitrary structure $\mathcal{S} = (S, r_i : i \in I)$ where r_i is a $(\rho(i)+1)$-ary relation over S, we can define its *complex algebra*

$$\mathfrak{Cm}\,(\mathcal{S}) = \big\langle \wp(S), \emptyset, S, \cup, \setminus, \Omega_i : i \in I \big\rangle,$$

where $\wp(S)$ is the power set of S, and Ω_i is the $\rho(i)$-ary operator defined by

$$\Omega_i\big(X_1, \ldots, X_{\rho(i)}\big) \\ = \big\{ s \in S : \exists s_1 \in X_1 \ldots \exists s_{\rho(i)} \in X_{\rho(i)},\ r_i\big(s, s_1, \ldots, s_{\rho(i)}\big) \big\},$$

for each $X_1, \ldots, X_{\rho(i)} \in \wp(S)$.

It is easy to check that, up to isomorphism, $\mathfrak{At}\,\mathfrak{Cm}\,(\mathcal{S}) \cong \mathcal{S}$ always (we identify the two), and $\mathfrak{A} \subseteq \mathfrak{Cm}(\mathfrak{At}\,\mathfrak{A}) \cong \mathsf{Com}\,(\mathfrak{A})$ for any completely additive atomic BAO $\mathfrak{A}$. If $\mathfrak{A}$ is finite then of course $\mathfrak{A} \cong \mathfrak{Cm}(\mathfrak{At}\,\mathfrak{A})$.

Atom structures of cylindric-type algebras have the form $\langle S, R_{\mathsf{c}_i}, R_{\mathsf{d}_{ij}} : i, j < n \rangle$, where the R_{c_i} and $R_{\mathsf{d}_{ij}}$ are binary and unary relations on S, respectively. We call such objects *cylindric-type atom structures.* One can construct from the standard axiomatization of cylindric algebras [Hen-Mon-Tar,85, Definition 1.1.1] a *Sahlqvist correspondent:* a first-order sentence true in all atom structures of atomic cylindric algebras, and such that the complex algebra of any atom structure in which it is true is a cylindric algebra. We call any model of this sentence a *cylindric algebra atom structure* cf. [Hen-Mon-Tar,85, Theorem 2.7.40].

It turns out that if $\mathfrak{A}$ is any cylindric algebra, $\mathfrak{A}_+$ is a cylindric algebra atom structure. Its complex algebra $\mathfrak{Cm}\,(\mathfrak{A}_+)$ is often written $\mathfrak{A}^\sigma$, and is called the *canonical extension* of $\mathfrak{A}$ [Jon-Tar,51]. $\mathfrak{A}$ is isomorphic to a subalgebra of $\mathfrak{A}^\sigma$ and the isomorphism is $a \mapsto \{f \in \mathfrak{A}_+ : a \in f\}$. This $\mathfrak{A}^\sigma$ is a different kind of complete extension of $\mathfrak{A}$ to the Monk completion $\mathsf{Com}\,(\mathfrak{A})$ that we mentioned in the introduction. Whereas suprema and infima are preserved from $\mathfrak{A}$ to $\mathsf{Com}\,(\mathfrak{A})$, this is not the case for $\mathfrak{A}^\sigma$ if $\mathfrak{A}$ is infinite. On the other hand, $\mathfrak{A}^\sigma$ is always complete and atomic, while $\mathsf{Com}\,(\mathfrak{A})$ will be atomic iff $\mathfrak{A}$ is. Monk proved that RCA_n is *canonical* (closed under taking canonical extensions): see [Hen-Mon-Tar,85, Theorem 2.7.23]. In fact,

Theorem 3.4.3. *An n-dimensional cylindric algebra $\mathfrak{A}$ is representable iff $\mathfrak{A}^\sigma$ has a* complete *representation.*

See [Hir-Hod,02a, Theorem 3.36] for the analogous result for relation algebras.

5. Representability and Atom Structures

Given an atomic cylindric-type algebra $\mathfrak{A}$, the games $G^{\kappa}(\mathfrak{A})$ are effectively played on the atom structure $\mathfrak{At}\,\mathfrak{A}$, so by Theorem 3.3.3, *whether* $\mathfrak{A}$ *has a complete representation or not depends only on its atom structure.* It follows that if a cylindric algebra $\mathfrak{A}$ has a complete representation then any cylindric algebra with the same atom structure as $\mathfrak{A}$ is completely representable, and in particular the completion $\mathsf{Com}\,(\mathfrak{A})$ of $\mathfrak{A}$ is also completely representable. At any rate, the atom structures of completely representable cylindric algebras form an important class, which we would like to characterise, perhaps by first-order sentences.

But whether the plain representability of $\mathfrak{A}$ is determined by $\mathfrak{At}\,\mathfrak{A}$ is not so clear. On the one hand, $\mathfrak{A}$ is determined by its boolean structure and by $\mathfrak{At}\,\mathfrak{A}$, and since Boolean algebras are easy to represent, one might surmise that impediments to representing $\mathfrak{A}$ reside in its atom structure. On the other hand, the boolean and atom structure of $\mathfrak{A}$ may interact, perhaps allowing two atomic cylindric algebras with the same atom structure, one being representable, the other not. This happens iff there is a representable atomic cylindric algebra whose completion is not representable. It would lead to two different kinds of 'representability' for a cylindric algebra atom structure, depending on whether *some* or *all* atomic cylindric algebras with that atom structure are representable. This turns out to be the case: it is possible to construct a weakly but not strongly representable cylindric algebra atom structure [Hod,97c], as we will see below.

In this section, we examine these issues (see [Hir-Hod,02b] for the corresponding definitions and results for relation algebra).

Definition 3.5.1. Let $\mathcal{S}$ be an n-dimensional cylindric algebra atom structure.

(i) $\mathcal{S}$ is *completely representable* if some (equivalently, every) atomic n-dimensional cylindric algebra $\mathfrak{A}$ with $\mathfrak{At}\,\mathfrak{A} = \mathcal{S}$ has a complete representation. CRAS_n denotes the class of completely representable (n-dimensional) cylindric algebra atom structures.

(ii) $\mathcal{S} \in \mathsf{LCAS}_n$ if $\exists$ has a winning strategy in $G^m(\mathfrak{Cm}\,\mathcal{S})$ for all $m < \omega$ – i.e., $\mathfrak{A} \models \{\rho_n : n < \omega\}$, for some (equivalently, all) $\mathfrak{A}$ where $\mathfrak{At}\,\mathfrak{A} = \mathcal{S}$.

(iii) $\mathcal{S}$ is *strongly representable* if every atomic cylindric algebra $\mathfrak{A}$ with $\mathfrak{At}\,\mathfrak{A} = \mathcal{S}$ is representable. We write SRAS_n for the class of strongly representable (n-dimensional) cylindric algebra atom structures.

(iv) $\mathcal{S}$ is *weakly representable* if there is a representable, atomic cylindric algebra $\mathfrak{A}$ with $\mathfrak{At}\,\mathfrak{A} = \mathcal{S}$. We let WRAS_n denote the class of weakly representable (n-dimensional) cylindric algebra atom structures.

Note that for any n-dimensional cylindric algebra $\mathfrak{A}$ and atom structure $\mathcal{S}$, if $\mathfrak{At}\,\mathfrak{A} = \mathcal{S}$ then $\mathfrak{A}$ embeds into $\mathfrak{Cm}\,\mathcal{S}$, and hence $\mathcal{S}$ is strongly representable iff $\mathfrak{Cm}\,\mathcal{S}$ is representable.

We want to investigate these classes, and the relationships between them. It is easily seen that

$$\text{(3.5.1)} \qquad \mathsf{CRAS}_n \subseteq \mathsf{LCAS}_n \subseteq \mathsf{SRAS}_n \subseteq \mathsf{WRAS}_n.$$

The last inclusion is trivial, and the first is immediate from the proof of Theorem 3.3.3: $\exists$ may use a complete representation of a cylindric algebra to guide her to victory in any atomic game played on the algebra. For the middle inclusion, let $\mathcal{S}$ be an atom structure in LCAS_n. To show $\mathcal{S} \in \mathsf{SRAS}_n$ we must show that an arbitrary atomic cylindric algebra $\mathfrak{A}$ with $\mathfrak{At}\,\mathfrak{A} = \mathcal{S}$ is representable. By Corollary 3.3.5, $\mathfrak{A}$ is elementarily equivalent to some (completely) representable algebra, and since RCA_n is an elementary class, $\mathfrak{A}$ is representable too. This shows that $\mathsf{LCAS}_n \subseteq \mathsf{SRAS}_n$. (Also, by Corollary 3.3.5, LCAS_n is the elementary closure of CRAS_n.)

We now ask *which of the inclusions in (3.5.1) are strict,* and *which of the classes are elementary.* LCAS_n is elementary and it is defined by $\{\rho_m : m < \omega\}$. The fact that WRAS_n is elementary is a special case of a more general result: given any variety V of completely additive BAOs, Venema showed in [Ven,97b] that the class $\mathsf{At}\,\mathsf{V}$ of atom structures of atomic algebras in V is elementary. The idea of the proof is as follows. Any atom structure $\mathcal{S}$ of a completely additive atomic BAO is also the atom structure of the subalgebra, say $\mathfrak{A}$, generated by the atoms. Then $\mathcal{S} \in \mathsf{At}\,\mathsf{V}$ iff $\mathfrak{A} \in \mathsf{V}$, and this holds iff each equation ε defining V is valid in $\mathfrak{A}$. But each element of $\mathfrak{A}$ is the value (in $\mathfrak{A}$) of some term $t(\bar{x})$ of the signature of $\mathfrak{A}$, whose variables $\bar{x}$ are instantiated by atoms. So the statement that ε is valid in $\mathfrak{A}$ is equivalent to the truth in $\mathcal{S}$ of an infinite set T_ε of first-order sentences in the signature of α, obtained by replacing the $\bar{x}$ by arbitrary terms, rewriting all function symbols into first-order formulas over $\mathcal{S}$ (using complete additivity), and then taking the universal ($\forall$) closure. The union of the T_ε, taken over all equations ε defining V, is then a set of first-order axioms defining $\mathsf{At}\,\mathsf{V}$.

This leaves the classes CRAS_n and SRAS_n. It turns out that they are not elementary [Hir-Hod,97c, Hir-Hod,09]. (Hence, all inclusions in (3.5.1) are strict.)

6. Monk and Rainbow Algebras

How are these non-elementary results proved? The games introduced earlier are potentially powerful tools for problems like this, since they can be used to determine when an atom structure lies in one of the classes. But to take advantage of them, we need a source of examples of atom structures whose game-theoretic properties we can control.

We will give two types of example, both obtained from graphs. Aspects of our constructions can be traced back to [Mon,69, Hir,95, Hod,97c, Hir-Hod,02b, Hir-Hod,09]. A *graph* is a structure $\Gamma = (V, E)$ where V is a non-empty set of 'nodes' or 'vertices', and $E \subseteq V \times V$ is a symmetric binary 'edge' relation on V. Note that our graphs can have 'loops': a *reflexive node* is a node $x \in V$ such that $(x, x) \in E$. A set $X \subseteq V$ is a *clique* if $(x, y) \in E$ for all distinct $x, y \in X$, and *independent* if $(X \times X) \cap E = \emptyset$. The *chromatic number* $\chi(\Gamma)$ of Γ is the least natural number k such that V is the union of k independent sets, and ∞ if no such k exists. For economy's sake, we often identify (notationally) Γ with V. In the same way, we identify (notationally) a model-theoretic structure M with its domain, the cardinality of which we write as $|M|$. We will write $M \subseteq N$ to mean that M is a substructure of N.

Proofs in this section are only sketched, owing to lack of space. More details can be found in the references. Recall that n is fixed ($3 \leq n < \omega$).

6.1. Strong homomorphisms

Before we proceed, a little more duality will be helpful.

Definition 3.6.1. Let $\mathcal{S} = \langle S, R_{\mathsf{c}_i}, R_{\mathsf{d}_{ij}} : i, j < n\rangle$ and $\mathcal{S}' = \langle S', R'_{\mathsf{c}_i}, R'_{\mathsf{d}_{ij}} : i, j < n\rangle$ be cylindric-type atom structures. A map $\theta : S \to S'$ is said to be a *strong homomorphism from $\mathcal{S}$ to $\mathcal{S}'$* if for each $x, y \in S$ and $i, j < n$ we have

(i) $(x, y) \in R_{\mathsf{c}_i} \iff \big(\theta(x), \theta(y)\big) \in R'_{\mathsf{c}_i}$,

(ii) $x \in R_{\mathsf{d}_{ij}} \iff \theta(x) \in R'_{\mathsf{d}_{ij}}$.

Lemma 3.6.2. *Let $\mathcal{S}$, $\mathcal{S}'$ be cylindric-type atom structures and let $\theta : \mathcal{S} \to \mathcal{S}'$ be a strong homomorphism.*

(i) *Let N_1 be a set and let $N_2 : {}^nN_1 \to \mathcal{S}$. Then $N = (N_1, N_2)$ is a $\mathfrak{Cm}\,\mathcal{S}$-network iff $\theta(N) = (N_1, \theta \circ N_2)$ is a $\mathfrak{Cm}\,\mathcal{S}'$-network.*

(ii) *If θ is surjective and the cylindric-type algebra $\mathfrak{Cm}\,\mathcal{S}$ is a completely representable cylindric algebra then $\mathfrak{Cm}\,\mathcal{S}'$ is a completely representable cylindric algebra.*[1]

Proof. The first part is a consequence of definitions 3.3.1 and 3.6.1. For the second part, θ induces a map $\theta^{-1} : \mathfrak{Cm}\,\mathcal{S}' \to \mathfrak{Cm}\,\mathcal{S}$ by $\theta^{-1}(X) = \{s \in S : \theta(s) \in X\}$, for $X \subseteq S'$. This can be checked to be an algebra embedding that preserves all meets and joins. Hence, $\mathfrak{Cm}\,\mathcal{S}'$ is a cylindric algebra, and if h is a complete representation of $\mathfrak{Cm}\,\mathcal{S}$ then $h \circ \theta^{-1}$ is a complete representation of $\mathfrak{Cm}\,\mathcal{S}'$. ■

6.2. Algebras from classes of structures

Both of our examples will be based on an underlying class of structures.

Definition 3.6.3. Let L be a first-order signature consisting of relation symbols of arity $< n$, and let K be a non-empty class of L-structures with the property that an L-structure M is in K iff every substructure of M with at most n elements is in K.

(i) Let X be a set, $M, N \in \mathsf{K}$, $f \in {}^XM$, and $g \in {}^XN$. We write $f \sim g$ if $\{(f(x), g(x)) : x \in X\}$ is a well defined partial isomorphism from M to N.

(ii) Let $\mathcal{F} = \mathcal{F}(\mathsf{K}) = \bigcup\{{}^nM : M \in \mathsf{K}\}$. For each $f \in \mathcal{F}$, we fix some $M_f \in \mathsf{K}$ with $f \in {}^n(M_f)$. The class relation $\sim$ induces an equivalence relation on $\mathcal{F}$, and we write the equivalence class of $f \in \mathcal{F}$ as $[f]$. For $f \in \mathcal{F}$, if we write $\ker f = \{(x, y) \in n \times n : f(x) = f(y)\}$,

[1] A slightly weaker property suffices for this than θ being a strong homomorphism, namely, being a 'bounded morphism', but we will need the stronger version in Lemma 3.6.4.

we may identify $[f]$ with the L-structure induced on the set $n/\ker f$ of $(\ker f)$-equivalence classes by pulling back from M_f in the obvious way. Therefore, we may treat each equivalence class $[f]$, and $\mathcal{F}$ itself, as a set. Then, $\mathcal{F}/{\sim}$ denotes the set of $\sim$-equivalence classes.

(iii) We now define a structure $\rho(\mathsf{K}) = \langle \mathcal{F}/{\sim}, R_{\mathsf{c}_i}, R_{\mathsf{d}_{ij}} \ : \ i,j < n\rangle$, which will be the atom structure of the cylindric-type algebra $\mathfrak{Cm}\,\rho(\mathsf{K})$, as follows:

- $R_{\mathsf{c}_i} = \big\{ \big([f],[g]\big) \ : \ f,g \in \mathcal{F},\ f \restriction \big(n \setminus \{i\}\big) \sim g \restriction \big(n \setminus \{i\}\big) \big\}$,
- $R_{\mathsf{d}_{ij}} = \big\{ [f] \ : \ f \in \mathcal{F},\ f(i) = f(j) \big\}$,

where $i, j < n$.

(iv) As usual, we will identify any $[f] \in \mathcal{F}/{\sim}$ with the singleton $\big\{ [f] \big\} \in \mathfrak{Cm}\,\rho(\mathsf{K})$.

Fix L, K as in Definition 3.6.3, and write $\mathfrak{A}$ for $\mathfrak{Cm}\,\rho(\mathsf{K})$. Notions to do with $\mathfrak{A}$-networks and complete representations of $\mathfrak{A}$ have analogues in terms of structures in K. This can be seen as follows. We leave the reader to check the (quite standard) details.

There is a one-one correspondence between $\mathfrak{A}$-networks and structures in K. In one direction, we may view any $M \in \mathsf{K}$ as an $\mathfrak{A}$-network $\operatorname{Net} M$ via $\operatorname{Net} M(\bar{a}) = [\bar{a}]$, for each $\bar{a} \in {}^nM$. Conversely, let N be an $\mathfrak{A}$-network. We define an L-structure $\operatorname{Str} N$ on the same domain as N. For each k-ary $R \in L$ and $a_0, \ldots, a_{k-1} \in N$, we define $\operatorname{Str} N \models R(a_0, \ldots, a_{k-1})$ iff

(3.6.1)
$$N(a_0, \ldots, a_{k-1}, \underbrace{a_0, \ldots, a_0}_{n-k \text{ times}}) = [f] \quad \text{and} \quad M_f \models R\big(f(0), \ldots, f(k-1)\big).$$

This is independent of the choice of f in (3.6.1). Using the networkhood of N, it can be checked that for every $\bar{a} = (a_0, \ldots, a_{n-1}) \in {}^nN$, if $N(\bar{a}) = [f]$ then the partial map $\big\{ \big(a_i, f(i)\big) \ : \ i < n \big\} \ : \ \operatorname{Str} N \to M_f$ is a partial isomorphism. This is a useful property to bear in mind. Among other things, it implies that (i) every substructure of $\operatorname{Str} N$ with at most n elements is in K, and hence $\operatorname{Str} N \in \mathsf{K}$ too, and (ii) $\bar{a} \ : \ n \to \operatorname{Str} N$ and $\bar{a} \sim f$, so $\operatorname{Net}(\operatorname{Str} N)(\bar{a}) = [\bar{a}] = [f] = N(\bar{a})$, and hence $\operatorname{Net}(\operatorname{Str} N) = N$. Similarly, for $M \in \mathsf{K}$ we have $\operatorname{Str}(\operatorname{Net} M) = M$.

Taking account of (3.3.1), this correspondence allows us to view a complete representation of $\mathfrak{A}$ as a set[2] $\{N_{[f]} : [f] \in \rho(\mathsf{K})\} \subseteq \mathsf{K}$ of structures such that $f : n \to N_{[f]}$ for each $[f]$, and:

(3.6.2) whenever $F \subseteq N_{[f]}$, $F \subseteq A \in \mathsf{K}$, and $|A| \leq n$, the inclusion map $\iota : F \to N_{[f]}$ extends to an embedding $\iota' : A \to N_{[f]}$.

The correspondence also allows us to construe the game $G^\kappa(\mathfrak{A})$ of Definition 3.3.2 as a game played to build a chain of structures $M_t \in \mathsf{K}$ ($t < \kappa$) as follows. In the initial round, $\forall$ picks a structure $M_0 \in \mathsf{K}$ with $|M_0| \leq n$. In successor rounds $t + 1 < \kappa$, supposing that $M_t \in \mathsf{K}$ is the structure at the start of the round, $\forall$ picks a substructure $F \subseteq M_t$ and $A \in \mathsf{K}$ with $|A| \leq n$ and $F \subseteq A$. $\exists$ must respond by finding $M_{t+1} \in \mathsf{K}$ with $M_t \subseteq M_{t+1}$ and $\forall$ wins unless the identity map on F extends to an embedding of A into M_{t+1}. At limit rounds $\lambda < \kappa$ we take unions and define $M_\lambda = \bigcup_{t<\lambda} M_t$.

6.3. Algebras over graphs

We now present our first specific example of this construction. The algebras we construct are related to ones in [Hir-Hod,09] and have some affinity to algebras devised by Monk [Mon,69]. We will use them to study SRAS_n. Let Γ be a graph. We write $\Gamma \times n$ for the graph consisting of n pairwise disjoint copies of Γ, and with an edge added between every two nodes lying in different copies. Formally, if $\Gamma = (V, E)$,

$$\Gamma \times n = \Big(V \times n,\ \big\{\big((x, i), (y, j)\big) : i, j < n,\ (x, y) \in E \vee i \neq j\big\}\Big).$$

We regard $\Gamma \times n$ as a signature by regarding each node of it as an $(n-1)$-ary relation symbol. Let $\mathcal{I}(\Gamma)$ be the class of $(\Gamma \times n)$-structures M satisfying:

M1. all relations in M are irreflexive and symmetric: whenever $p \in \Gamma \times n$, $a_0, \ldots, a_{n-2} \in M$, and $M \models p(a_0, \ldots, a_{n-2})$, then $a_0, \ldots, a_{n-2}$ are pairwise distinct and $M \models p\big(a_{\pi(0)}, \ldots, a_{\pi(n-2)}\big)$ for each permutation π of $(n-1)$,

M2. whenever $a_0, \ldots, a_{n-2} \in M$ are pairwise distinct, $M \models p(a_0, \ldots, a_{n-2})$ for some unique $p \in \Gamma \times n$,

[2] In the case where $\mathfrak{A}$ is simple, this set may be taken to be a singleton. Cf. [Hen-Mon-Tar,85, Corollary 3.1.81].

M3. whenever $a_0, \ldots, a_{n-1} \in M$, $p_0, \ldots, p_{n-1} \in \Gamma \times n$, and $M \models \bigwedge_{i<n} p_i(a_0, \ldots, a_{i-1}, a_{i+1}, \ldots, a_{n-1})$, there are $i < j < n$ such that (p_i, p_j) is an edge of $\Gamma \times n$.[3]

Note that $\mathcal{I}(\Gamma)$ satisfies the conditions in Definition 3.6.3. We will write $\mathfrak{M}(\Gamma)$ for the algebra $\mathfrak{Cm}\, \rho(\mathcal{I}(\Gamma))$. Our aim is to prove (Proposition 3.6.8) that if Γ is infinite then $\mathfrak{M}(\Gamma) \in \mathsf{RCA}_n$ iff $\chi(\Gamma) = \infty$.

Lemma 3.6.4. *Let Γ be a graph that contains a reflexive node. Let $\mathcal{S} = (S, R_{\mathsf{c}_i}, R_{\mathsf{d}_{ij}} : i, j < n)$ be a cylindric-type atom structure, and suppose that $\theta : \mathcal{S} \to \rho(\mathcal{I}(\Gamma))$ is a surjective strong homomorphism. Then $\mathfrak{Cm}\, \mathcal{S}$ is a completely representable cylindric algebra.*

Proof. By Theorem 3.3.3, it is enough to show that $\exists$ has a winning strategy in the game $G^{|S|+\omega}(\mathfrak{Cm}\, \mathcal{S})$. Let N be a $\mathfrak{Cm}\, \mathcal{S}$-network. By Lemma 3.6.2(i), $\theta(N)$ is an $\mathfrak{M}(\Gamma)$-network. By the equivalence between networks and structures in $\mathcal{I}(\Gamma)$, $\theta(N)$ can be identified in a well defined way with a structure $N{\downarrow} \in \mathcal{I}(\Gamma)$ with the same domain as N and with the following property: for each $x_0, \ldots, x_{n-1} \in N$ with $\theta(N(x_0, \ldots, x_{n-1})) = [f]$, say, each $i < n$, and each $p \in \Gamma \times n$, we have $N{\downarrow} \models p(x_0, \ldots, x_{i-1}, x_{i+1}, \ldots, x_{n-1})$ iff $M_f \models p(f(0), \ldots, f(i-1), f(i+1), \ldots, f(n-1))$. This identification will provide $\exists$ with a winning strategy in the game.

The initial round and rounds indexed by limit ordinals pose no problems for her. In some successor round, suppose that the current $\mathfrak{Cm}\, \mathcal{S}$-network is N, say. Let $\forall$ choose $x \in {}^nN$, $i < n$, and $a \in \mathcal{S}$ with $a \leq \mathsf{c}_i N(x)$. If $N(x[i/x_j]) = a$ for some $j < n$ then $\exists$ simply responds to $\forall$'s move with the current network N.

Assume from now on that no such j exists. It follows that $a \leq -\mathsf{d}_{ij}$ for each $j \in n \setminus \{i\}$. Let $\theta(a) = [f]$, say. So $f(i) \neq f(j)$ for all $j \neq i$. Let $z \notin N$ be a new node, and write

$$y = x[i/z] \in {}^n(N \cup \{z\}), \tag{3.6.3}$$

$$Y = \{y_0, \ldots, y_{n-1}\}.$$

Then $f(j) = f(k)$ iff $y_j = y_k$ for each $j, k < n$.

We now extend $N{\downarrow}$ to a structure $M \in \mathcal{I}(\Gamma)$ defined as follows. Its domain is the domain of $N{\downarrow}$ together with z. We specify that

[3]The conclusion is a stronger condition than '$\{p_0, \ldots, p_{n-1}\}$ is not independent' in the case where Γ has reflexive nodes.

S1. $N\!\downarrow \subseteq M$, and $f \sim y$.

To complete the specification, we first define elements $q_j \in \Gamma \times n$ ($j \in n \setminus \{i\}$) as follows. If $y_0, \dots, y_{j-1}, y_{j+1}, \dots, y_{n-1}$ (as in (3.6.3)) are pairwise distinct, we let $q_j \in \Gamma \times n$ be the unique element satisfying $M \models q_j(y_0, \dots, y_{j-1}, y_{j+1}, \dots, y_{n-1})$ according to S1. If they are not all distinct, we choose $q_j \in \Gamma \times n$ arbitrarily. Next, recalling that $\Gamma \times n$ consists of n pairwise disjoint copies of Γ, choose one of these copies that does not contain any of $q_0, \dots, q_{i-1}, q_{i+1}, \dots, q_{n-1}$. Let d be a reflexive node in this copy. We now specify that

S2. $M \models d(t_0, \dots, t_{n-2})$ whenever $t_0, \dots, t_{n-2} \in M$ are distinct and $z \in \{t_0, \dots, t_{n-2}\} \nsubseteq Y$,

where Y is as in (3.6.3). It can be checked that this specifies a well defined $(\Gamma \times n)$-structure M. We check that $M \in \mathcal{I}(\Gamma)$. Properties M1 and M2 are easy to verify. We pass to M3. Let $t_0, \dots, t_{n-1} \in M$ be distinct, let $p_0, \dots, p_{n-1} \in \Gamma \times n$, and suppose that $M \models p_j(t_0, \dots, t_{j-1}, t_{j+1}, \dots, t_{n-1})$ for each $j < n$. We need to show that (p_j, p_k) is an edge of $\Gamma \times n$, for some $j < k < n$. Using S1, it can be seen that this holds if $t_0, \dots, t_{n-1} \in N\!\downarrow$, since $N\!\downarrow \in \mathcal{I}(\Gamma)$, and it holds if $\{t_0, \dots, t_{n-1}\} = Y$, because $M_f \in \mathcal{I}(\Gamma)$. So assume that $z \in \{t_0, \dots, t_{n-1}\} \nsubseteq Y$. Clearly, there are at least $n-2$ indices $j < n$ with

$$z \in \{t_0, \dots, t_{j-1}, t_{j+1}, \dots, t_{n-1}\} \nsubseteq Y.$$

By S2, there are at least $n-2$ indices $j < n$ with $p_j = d$. Since $n \geq 3$, there is at least one such j. There are now two cases.

1. If there are $j < k < n$ with $p_j = p_k = d$, then as d is a reflexive node, (p_j, p_k) is an edge of $\Gamma \times n$ as required.

2. If there is a unique $j < n$ with $p_j = d$, we plainly must have $n = 3$. Let $k, l \in 3 \setminus \{j\}$ satisfy $z = t_k$ and $t_l \notin Y$. Then $M \models p_l(t_j, t_k)$ by the above, and $t_j, t_k \in Y$. So $p_l \in \{q_j : j \in n \setminus \{i\}\}$. But d lies in a copy of Γ that does not contain p_l. As there are edges of $\Gamma \times n$ connecting all nodes in distinct copies of Γ, (p_j, p_l) and (p_l, p_j) are edges of $\Gamma \times n$ as required.

We now extend N to a network $N' \supseteq N$ whose set of nodes is the domain of M, with $N'(y) = a$, and with $N'\!\downarrow = M$, in any way at all; Lemma 3.6.2(i) guarantees that the result will be a $\mathfrak{Cm}\,\mathcal{S}$-network. $\exists$ responds to $\forall$ with this network N', and thus has the capability to win the game. ■

Definition 3.6.5. Let Γ be a graph. We write $\mathfrak{Ue}\,\Gamma$, the *ultrafilter extension* of Γ, for the graph whose nodes are the ultrafilters on Γ (i.e., ultrafilters of the Boolean algebra of subsets of Γ), and such that (μ, ν) is an edge of $\mathfrak{Ue}\,\Gamma$ iff for every $X \in \mu$, $Y \in \nu$, there are $p \in X$, $q \in Y$ such that (p, q) is an edge of Γ.

Lemma 3.6.6. *Let Γ be any graph.*

(i) $\mathfrak{Ue}\,(\Gamma \times n) \cong (\mathfrak{Ue}\,\Gamma) \times n$ *(we will identify the two).*

(ii) $\chi(\mathfrak{Ue}\,\Gamma) = \chi(\Gamma)$.

(iii) $\chi(\Gamma) = \infty$ *iff* $\mathfrak{Ue}\,\Gamma$ *has a reflexive node.*

Proof. Write Δ for $\mathfrak{Ue}\,\Gamma$ in the proof. The first part is easy and we leave it to the reader. Let $k < \omega$. If $\Gamma = \bigcup_{i<k} I_i$ for independent $I_i \subseteq \Gamma$, then for each i, $I_i^\Delta = \{\nu \in \Delta : I_i \in \nu\}$ is an independent subset of Δ, and $\bigcup_{i<k} I_i^\Delta = \Delta$. Similarly, suppose $\Delta = \bigcup_{j<k} J_j$ for independent $J_j \subseteq \Delta$. For $p \in \Gamma$ let $\langle p \rangle \in \Delta$ be the principal ultrafilter generated by $\{p\}$. Then for each j, the set $J_j^\Gamma = \big\{p \in \Gamma : \langle p \rangle \in J_j\big\}$ is an independent subset of Γ, and $\Gamma = \bigcup_{j<k} J_j^\Gamma$.

For the last part, if there is finite k with $\Gamma = \bigcup_{i<k} I_i$ for independent $I_i \subseteq \Gamma$, then each ultrafilter on Γ contains some I_i and so cannot be reflexive. Conversely, if $\chi(\Gamma) = \infty$ then the set of independent subsets of Γ generates a proper ideal of subsets of Γ. Any ultrafilter on Γ disjoint from this ideal contains no independent sets and is therefore a reflexive node of Δ. ∎

Recall that $\mathfrak{A}_+$ (for a $\mathsf{BAO}\,\mathfrak{A}$) was defined in §4.

Lemma 3.6.7. *For any graph Γ, there is a surjective strong homomorphism $\theta : \mathfrak{M}(\Gamma)_+ \to \rho\big(\mathcal{I}(\mathfrak{Ue}\,\Gamma)\big)$.*

Proof. First, for any $x_0, \ldots, x_{n-2} < n$ and $X \subseteq \Gamma \times n$, define the following element of $\mathfrak{M}(\Gamma)$:

$$X^{(x_0,\ldots,x_{n-2})} = \big\{[f] \in \rho\big(\mathcal{I}(\Gamma)\big) \ : \exists p \in X\big[M_f \models p\big(f(x_0), \ldots, f(x_{n-2})\big)\big]\big\}.$$

Now let μ be an ultrafilter of $\mathfrak{M}(\Gamma)$. Define an equivalence relation $\sim$ on n by $i \sim j \iff \mathsf{d}_{ij} \in \mu$. Let $g : n \to n/{\sim}$ be given by $g(i) = i/{\sim}$. Define a $(\mathfrak{Ue}\,\Gamma \times n)$-structure M_μ with domain $n/{\sim}$ as follows. For each $\nu \in \mathfrak{Ue}\,\Gamma \times n$ and $x_0, \ldots, x_{n-2} < n$, we let

(3.6.4)
$$M_\mu \models \nu\big(g(x_0), \ldots, g(x_{n-2})\big) \iff X^{(x_0,\ldots,x_{n-2})} \in \mu \text{ for each } X \in \nu.$$

It is straightforward (though lengthy) to check that this is well defined and that $M_\mu \in \mathcal{I}(\mathfrak{Ue}\,\Gamma)$. So $g : n \to M_\mu$ and hence $g \in \mathcal{F}(\mathcal{I}(\mathfrak{Ue}\,\Gamma))$ (see Definition 3.6.3(ii)). Then we define $\theta(\mu) = [g] \in \rho(\mathcal{I}(\mathfrak{Ue}\,\Gamma))$.

Using M1 and M2, it can now be checked that θ is a strong homomorphism. We show it is surjective. Let $[g] \in \rho(\mathcal{I}(\mathfrak{Ue}\,\Gamma))$ be given, so $g : n \to M_g \in \mathcal{I}(\mathfrak{Ue}\,\Gamma)$. Let $D_g = \{\mathsf{d}_{ij} : i,j < n,\ g(i) = g(j)\} \cup \{-\mathsf{d}_{ij} : i,j < n,\ g(i) \neq g(j)\} \subseteq \mathfrak{M}(\Gamma)$. There are three cases.

Case 1: g is one-one. By M2, for each $i < n$ there is a unique $\nu_i \in \mathfrak{Ue}\,\Gamma$ with $M_g \models \nu_i(g(0), \dots, g(i-1), g(i+1), \dots, g(n-1))$, and by M3, there are $i < j < n$ such that (ν_i, ν_j) is an edge of $\mathfrak{Ue}\,\Gamma$. We show that

$$(3.6.5) \qquad \mu_0 = D_g \cup \{X^{(0,\dots,l-1,l+1,\dots,n-1)} : l < n, X \in \nu_l\} \subseteq \mathfrak{M}(\Gamma)$$

has the finite intersection property. As the ν_l are ultrafilters, it suffices to check that whenever $X_l \in \nu_l$ $(l < n)$, we have

$$G = D_g \cap \bigcap_{l<n} X_l^{(0,\dots,l-1,l+1,\dots,n-1)} \neq \emptyset.$$

We may choose $p_l \in X_l$ (each l) such that (p_i, p_j) is an edge of $\Gamma \times n$. Then we can define a $(\Gamma \times n)$-structure $M \in \mathcal{I}(\Gamma)$ with domain n by specifying that $M \models p_l(0, \dots, l-1, l+1, \dots, n-1)$ for each l. Because (p_i, p_j) is an edge, M3 is satisfied. If $f : n \to M$ is the identity map on n, then $[f] \in G$, which is therefore non-empty as required. So μ_0 extends to an ultrafilter μ of $\mathfrak{M}(\Gamma)$. By (3.6.4) and (3.6.5) we have $\theta(\mu) = [g]$.

Case 2: there are unique $i < j < n$ with $g(i) = g(j)$. Using M2, let $\nu \in \mathfrak{Ue}\,\Gamma$ be such that $M_g \models \nu(g(0), \dots, g(i-1), g(i+1), \dots, g(n-1))$. Using the irreflexivity condition in M1, it can be verified that

$$D_g \cup \{X^{(0,\dots,i-1,i+1,\dots,n-1)} : X \in \nu\}$$

has the finite intersection property and so extends to a (unique) ultrafilter μ of $\mathfrak{M}(\Gamma)$, and $\theta(\mu) = [g]$.

Case 3: otherwise. By irreflexivity M1, all structures in $\mathcal{I}(\Gamma)$ with fewer than $n-1$ elements are isomorphic, so $\bigwedge D_g$ is an atom of $\mathfrak{M}(\Gamma)$. Let μ be the unique ultrafilter of $\mathfrak{M}(\Gamma)$ containing D_g. Then $\theta(\mu) = [g]$.

■

Combining these lemmas and with a little more work, we reach our goal:

Proposition 3.6.8. *For any infinite graph* Γ, *we have* $\mathfrak{M}(\Gamma) \in \mathsf{RCA}_n$ *iff* $\chi(\Gamma) = \infty$.

Proof. Suppose that $\mathfrak{M}(\Gamma)$ is representable. By Theorem 3.4.3, $\mathfrak{M}(\Gamma)^\sigma$ is completely representable. By Lemmas 3.6.7 and 3.6.2(ii), so is $\mathfrak{M}(\mathfrak{Ue}\,\Gamma)$, and hence, choosing any $[f] \in \rho\big(\mathcal{I}(\mathfrak{Ue}\,\Gamma)\big)$, there is $M = M_{[f]} \in \mathcal{I}(\mathfrak{Ue}\,\Gamma)$ satisfying (3.6.2). Since Γ is infinite, it can be checked that M is also infinite.

Suppose for contradiction that $\chi(\Gamma) < \infty$. By lemma 3.6.6, $\chi(\mathfrak{Ue}\,\Gamma) < \infty$, and it is clear that $\chi(\mathfrak{Ue}\,\Gamma \times n) = \chi(\mathfrak{Ue}\,\Gamma) \cdot n < \infty$ as well. So there are $k < \omega$ and independent sets $I_0, \ldots, I_{k-1} \subseteq \mathfrak{Ue}\,\Gamma \times n$ with $\mathfrak{Ue}\,\Gamma \times n = \bigcup_{i<k} I_i$. Choose pairwise distinct $x_0, x_1, \ldots \in M$. By M2, for each $i_0 < \cdots < i_{n-2} < \omega$ there is $c(i_0, \ldots, i_{n-2}) < k$ such that $M \models p(x_{i_0}, \ldots, x_{i_{n-2}})$ for some $p \in I_{c(i_0,\ldots,i_{n-2})}$. By Ramsey's theorem, we may assume that c has constant value c_0, say. Then for each $i < n$ there is some $p_i \in I_{c_0}$ with $M \models p_i(x_0, \ldots, x_{i-1}, x_{i+1}, \ldots, x_{n-1})$. Since $\{p_i : i < n\} \subseteq I_{c_0}$, an independent set, this contradicts M3.

Conversely, suppose that $\chi(\Gamma) = \infty$. By Lemma 3.6.6, $\mathfrak{Ue}\,\Gamma$ contains a reflexive node. By Lemmas 3.6.7 and 3.6.4, $\mathfrak{M}(\Gamma)^\sigma$ is completely representable. By Theorem 3.4.3, $\mathfrak{M}(\Gamma)$ is representable. ■

6.4. 'Rainbow algebras' over graphs

Our second specific instance of the construction of section 6.2 is so-called 'rainbow algebras'. They are similar to algebras constructed in [Hir-Hod,97c] and will be used to study CRAS_n.

Definition 3.6.9. Let Γ be a graph.

(i) Let $L = L(\Gamma)$ be the signature

$$\Gamma \cup \{\mathsf{g}_0^j : j < \omega\} \cup \{\mathsf{g}_i : 1 \le i \le n-2\} \cup \{\mathsf{w}_i : i \le n-2\}$$
$$\cup \big\{\mathsf{y}_S : S \subseteq \omega,\ |S| < \omega\big\}.$$

Each y_S is an $(n-1)$-ary relation symbol, regarded as yellow. All the others are binary relation symbols. We regard the g_0^j and g_i as green and the w_i as white. We define the following formulas:

- $G(x,y) = \bigvee_{j<\omega} \mathsf{g}_0^j(x,y) \vee \bigvee_{1\le i\le n-2} \mathsf{g}_i(x,y)$ (an $L_{\omega_1\omega}$-formula),
- $\chi^j(x_0,\ldots,x_{n-2},y) = \mathsf{g}_0^j(x_0,y) \wedge \bigwedge_{1\le i\le n-2} \mathsf{g}_i(x_i,y)$, for $j<\omega$.

(ii) We let $\mathsf{K} = \mathsf{K}(\Gamma)$ be the class of L-structures M such that:

R1. all relations in M are irreflexive: if $R \in L$ is k-ary, $a_0,\ldots,a_{k-1} \in M$, and $M \models R(a_0,\ldots,a_{k-1})$, then $a_0,\ldots,a_{k-1}$ are pairwise distinct,

R2. all non-yellow binary relations are symmetric,

R3. exactly one non-yellow binary relation holds on each pair of distinct elements of M,

R4. M has no green triangles: $M \models \neg\exists xyz\big(G(x,y) \wedge G(y,z) \wedge G(x,z)\big)$,

R5. M has no green-green-white triangles with equal lower indices:

$$M \models \neg\exists xyz\big(\mathsf{g}_0^j(x,y) \wedge \mathsf{g}_0^k(y,z) \wedge \mathsf{w}_0(x,z)\big)$$

for each $j,k<\omega$, and

$$M \models \neg\exists xyz\big(\mathsf{g}_i(x,y) \wedge \mathsf{g}_i(y,z) \wedge \mathsf{w}_i(x,z)\big)$$

for $1 \le i \le n-2$,

R6. $M \models \neg\exists xyz\big(p(x,y) \wedge q(y,z) \wedge r(x,z)\big)$ whenever $p,q,r \in \Gamma$ and $\{(p,q),(q,r),(p,r)\} \not\subseteq E$,

R7. $M \models \neg\exists x_0\ldots x_{n-2}\big(\mathsf{y}(x_0,\ldots,x_{n-2}) \wedge \bigvee_{i<j\le n-2} G(x_i,x_j)\big)$ for each yellow $\mathsf{y} \in L$,

R8. if $S \subseteq \omega$ is finite and $j \in \omega \setminus S$ then

$$M \models \neg\exists x_0\ldots x_{n-2}y\big(\mathsf{y}_S(x_0,\ldots,x_{n-2}) \wedge \chi^j(x_0,\ldots,x_{n-2},y)\big).$$

Note that K satisfies the conditions in Definition 3.6.3.

(iii) We write $\mathcal{R}(\Gamma)$ for the 'rainbow' algebra $\mathfrak{Cm}\,\rho\big(\mathsf{K}(\Gamma)\big)$.

6.5. Complete representability of the algebras $\mathcal{R}(\Gamma)$

Proposition 3.6.10. *Let Γ be any graph. If $\mathcal{R}(\Gamma)$ is completely representable then Γ has a reflexive node or an infinite clique.*

Proof. Assume that $\mathcal{R}(\Gamma)$ has a complete representation, viewed as a structure $M \in \mathsf{K} = \mathsf{K}(\Gamma)$ satisfying (3.6.2) above. By (3.6.2), we can find elements $a_0, \ldots, a_{n-2} \in M$ such that $M \models \neg G(a_i, a_j)$ for each $i, j < n-1$, and $M \models \neg \mathsf{y}(a_0, \ldots, a_{n-2})$ for each yellow $\mathsf{y} \in L$. By (3.6.2) again, for each $j < \omega$ there is $b_j \in M$ such that $M \models \chi^j(a_0, \ldots, a_{n-2}, b_j)$. Since $M \in \mathsf{K}$, M satisfies conditions R3–R5 of Definition 3.6.9. It follows that for each $j < k < \omega$ there is $p_{jk} \in \Gamma$ with $M \models p_{jk}(b_i, b_j)$. Considering triangles (b_0, b_j, b_k) and using the definition of $\mathsf{K}(\Gamma)$, we see that (p_{0j}, p_{0k}) is an edge of Γ for all $0 < j < k < \omega$. So $\{p_{0j} : 1 \le j < \omega\}$ is either an infinite clique in Γ or contains a reflexive node. ■

Proposition 3.6.11. *If Γ is a countable graph containing a reflexive node or an infinite clique, then $\mathcal{R}(\Gamma)$ is completely representable.*

Proof. Assume that $C \subseteq \Gamma$ is an infinite clique or a singleton consisting of a reflexive node of Γ. By Theorem 3.3.3, it suffices to show how $\exists$ can win $G^\omega(\mathcal{R}(\Gamma))$, construed as above as a game on structures in $\mathsf{K} = \mathsf{K}(\Gamma)$. Let $M \in \mathsf{K}$ be the structure at the start of some round t ($1 \le t < \omega$). Suppose inductively that M is finite. In round t, suppose that $\forall$ chooses $F \subseteq M$ with $|F| < n$, and an extension $A \in \mathsf{K}$ of F with $|A| \le n$. We can assume without loss of generality that $|A \setminus F| = 1$ and that $A \setminus F = \{d\}$, say, where $d \notin M$. $\exists$ must extend M to some $M^\sharp \in \mathsf{K}$ in such a way that the inclusion map $\iota : F \to M$ extends to an embedding $\iota^\sharp : A \to M^\sharp$.

If there is already such an $\iota^\sharp : A \to M$, then $\exists$ lets $M^\sharp = M$. So assume not. $\exists$ defines an extension $M^\sharp$ of M with domain $M \cup \{d\}$ as follows. Let $M^\flat$ be the union of M and A over F.[4]

- For each $a_0, \ldots, a_{n-2} \in M^\flat$ such that $d \in \{a_0, \ldots, a_{n-2}\} \not\subseteq A$ and $M^\flat \models \neg G(a_i, a_j)$ for each $i < j \le n-2$, $\exists$ defines

$$M^\sharp \models \mathsf{y}_S(a_0, \ldots, a_{n-2}),$$

$$\text{where} \quad S = \big\{ j < \omega : M^\flat \models \exists x \chi^j(a_0, \ldots, a_{n-2}, x) \big\}.$$

Then, for each $b \in M \setminus F$, $\exists$ chooses a binary relation symbol x_b and lets $M^\sharp \models x_b(b, d) \wedge x_b(d, b)$. In each case she chooses $x_b \in \{\mathsf{w}_i : i \le n-2\} \cup C$. She chooses these elements in turn, as follows.

[4] That is, we assume that $M \cap A = F$; for each k-ary $R \in L$ and k-tuple $\bar{a}$ of elements of $M \cup A$, we define $M^\flat \models R(\bar{a})$ iff the elements of $\bar{a}$ lie in M and $M \models R(\bar{a})$ or the elements of $\bar{a}$ lie in A and $A \models R(\bar{a})$.

- If there are no $a_0 \in F$ and $j, j' < \omega$ such that $M^\flat \models \mathsf{g}_0^j(a_0, d) \wedge \mathsf{g}_0^{j'}(a_0, b)$, then $\exists$ defines $M^\sharp \models \mathsf{w}_0(b, d) \wedge \mathsf{w}_0(d, b)$.
- Otherwise, if there is $1 \leq i \leq n-2$ such that for no $a_i \in F$ do we have $M^\flat \models \mathsf{g}_i(d, a_i) \wedge \mathsf{g}_i(a_i, b)$, then $\exists$ chooses the least such i and defines $M^\sharp \models \mathsf{w}_i(b, d) \wedge \mathsf{w}_i(d, b)$.
- Otherwise, there must be $\bar{a} \in {}^{n-1}F$ and $j, j' < \omega$ with $M^\flat \models \chi^j(\bar{a}, b) \wedge \chi^{j'}(\bar{a}, d)$. If C consists of a single reflexive node ρ, say, she lets $M^\sharp \models \rho(b, d) \wedge \rho(d, b)$. If C is an infinite clique then she picks some $x \in C$ that has not been used as a label so far (either in a previous round or for some other 'b' in the current round) and lets $M^\sharp \models x(b, d) \wedge x(d, b)$.

Note that $\exists$ never defines any green relations, so

$$M^\sharp \models \neg G(d, b) \quad \text{for every} \quad b \in M \setminus F. \tag{3.6.6}$$

This strategy can be checked to be winning for $\exists$. We have no space for a full proof, but the chief point to check is that $M^\sharp \in \mathsf{K}$, and in particular that $M^\sharp$ satisfies Condition R6 of Definition 3.6.9. This boils down to checking that whenever $b, c \in M \setminus F$, $\exists$ defines $M^\sharp \models p(b, d) \wedge q(c, d)$ for $p, q \in C$ as per her strategy, and also $M \models r(b, c)$ for some $r \in \Gamma$, then (p, q), (p, r), (q, r) are edges of Γ. Certainly (p, q) is an edge, since $p, q \in C$ are chosen successively by $\exists$ as already outlined. So it is sufficient to show that $r \in C$.

By $\exists$'s strategy, this will certainly be the case if $\exists$ defined $M \models r(b, c)$ herself in an earlier round of the game. We will show that she did. $\exists$ is currently defining $M^\sharp \models p(b, d) \wedge q(c, d)$, so according to her strategy there must be $\bar{a}, \bar{a}' \in {}^{n-1}F$ and $j, k, l, l' < \omega$ with $M^\flat \models \chi^j(\bar{a}, b) \wedge \chi^l(\bar{a}, d) \wedge \chi^k(\bar{a}', c) \wedge \chi^{l'}(\bar{a}', d)$. As $|F| \leq n-1$, by R3 we have $\bar{a} = \bar{a}'$ and $l = l'$. So

$$M^\flat \models \chi^j(\bar{a}, b) \wedge \chi^k(\bar{a}, c) \wedge \chi^l(\bar{a}, d). \tag{3.6.7}$$

Now M has been built by the game: its elements were added one at a time in earlier rounds. Let $\bar{a} = (a_0, \ldots, a_{n-2})$. Clearly, $a_0, \ldots, a_{n-2}, b, c$ are pairwise distinct. Consider the round in which the final element among them, say d', was added. In his move in that round, suppose that $\forall$ chose $F' \subseteq M$ with $|F'| < n$.

Suppose for contradiction that $d' = a_i$ for some $i \leq n-2$. By (3.6.7), $M \models G(a_i, b) \wedge G(a_i, c)$, and by (3.6.6) applied to the earlier round we must have $b, c \in F'$. As $M \in \mathsf{K}$, by (3.6.7) and R4 we have $M^\flat \models \neg G(a_i, a_j)$ for

$i < j \leq n-2$. As $|F'| < n$, there is $i' \leq n-2$ with $i' \neq i$ and $a_{i'} \notin F'$. Referring to the strategy shows that $\exists$ defined $M \models \mathsf{y}_S(\bar{a})$, where S was the set of all $m < \omega$ such that $\exists x \chi^m(a_0, \ldots, a_{n-2}, x)$ was true in the structure existing at the start of that round. This structure is a substructure of M, so $S \subseteq \{m < \omega : M \models \exists x \chi^m(a_0, \ldots, a_{n-2}, x)\}$. Now we return our attention to the current round. Since $A \in \mathsf{K}$ and $A \models \mathsf{y}_S(\bar{a}) \wedge \chi^l(\bar{a}, d)$ (see (3.6.7)), by R8 we must have $l \in S$, so there must be some $d' \in M$ with $M \models \chi^l(\bar{a}, d')$. It follows by Condition R7 of the definition of K that $\iota^\sharp = \iota \cup \{(d, d')\}$ embeds A into M, contradicting our assumption that there is no such embedding.

So $d' \in \{b, c\}$. Suppose that $d' = b$ (the case where $d' = c$ is similar). For each $i \leq n-2$, $M \models G(b, a_i)$, and by (3.6.6) applied to the earlier round, $a_i \in F'$. Since $|F'| < n$, we have $c \notin F'$, and by (3.6.7), $\exists$'s strategy would have defined $M \models r(b, c)$ for $r \in C$, as required. ■

7. Consequences

It is now easy to derive several corollaries. We will use a few common graph constructions. The *disjoint union* of graphs $\Gamma_i = (V_i, E_i)$ $(i \in I)$ is the graph

(3.7.1)

$$\bigoplus_{i \in I} \Gamma_i = \Big(\bigcup \{ V_i \times \{i\} : i \in I \}, \{ ((x, i), (y, i)) : i \in I,\ (x, y) \in E_i \} \Big).$$

For a cardinal $\kappa > 0$, we write K_κ for the complete irreflexive graph $(\kappa, \{ (i, j) : i, j < \kappa,\ i \neq j \})$. For finite $k > 0$, we have $\chi(K_k) = k$. Also, $\chi(\bigoplus_{i \in I} \Gamma_i) = \max \{ \chi(\Gamma_i) : i \in I \}$ if this exists, and ∞ otherwise.

Corollary 3.7.1. CRAS_n *is not an elementary class* ([Hir-Hod,97c]).

Proof. Write $\Gamma = \bigoplus_{1 \leq k < \omega} K_k$. We know from Proposition 3.6.10 that $\mathcal{R}(\Gamma)$ is not completely representable. Therefore, its atom structure $\rho(\mathsf{K}(\Gamma))$ is not in CRAS_n.

However, since Γ has arbitrarily large finite cliques, there is a countable graph Δ that is elementarily equivalent to Γ and has an infinite clique. By Proposition 3.6.11, $\mathcal{R}(\Delta)$ is completely representable, so $\rho(\mathsf{K}(\Delta)) \in \mathsf{CRAS}_n$.

It can be checked that $\rho\big(\mathsf{K}(\Gamma)\big)$ is elementarily equivalent to $\rho\big(\mathsf{K}(\Delta)\big)$. This shows that CRAS_n is not closed under elementary equivalence and so cannot be elementary. ■

In fact, CRAS_n is pseudo-elementary, and so closed under ultraproducts [Cha-Kei,90, Exercise 4.1.17, corollary 6.1.16]. Hence [Cha-Kei,90, Theorems 4.1.12 and 6.1.15], it is not closed under ultraroots.

In contrast, SRAS_n is closed under ultraroots [Gol,89, 3.8.1(1)], but not ultraproducts, and hence is not elementary:

Corollary 3.7.2. SRAS_n *is not an an elementary class* ([Hir-Hod,09]).

Proof. We use a celebrated theorem of Erdős [Erd,59] stating that for each finite k, there exists a finite graph Γ_k with chromatic number at least k and with no cycles of length at most k. (For our purposes, a *cycle of length* k in a graph is a sequence $v_1, \ldots, v_k$ of distinct nodes such that $(v_1, v_2), \ldots, (v_{k-1}, v_k)$, and (v_k, v_1) are edges.) Let $\Delta_k = \bigoplus_{k<m<\omega} \Gamma_m$. Then $\chi(\Delta_k) = \infty$, and Δ_k is countably infinite and has no cycles of length at most k. Therefore, by Proposition 3.6.8, $\mathfrak{M}(\Delta_k)$ is representable, and so $\rho\big(\mathcal{I}(\Delta_k)\big) \in \mathsf{SRAS}_n$.

Now let Δ be a non-principal ultraproduct of the Δ_k. It follows from Łoš's theorem that Δ has no cycles of any finite length. So by a well known result from graph theory (cf. [Die,97, Proposition 1.6.1]), $\chi(\Delta) \leq 2$. By Proposition 3.6.8 again, $\mathfrak{M}(\Delta)$ is not representable, so $\rho\big(\mathcal{I}(\Delta)\big) \notin \mathsf{SRAS}_n$.

But it is easily seen that $\rho\big(\mathcal{I}(\Delta)\big)$ is isomorphic to an ultraproduct of the $\rho\big(\mathcal{I}(\Delta_k)\big)$. As elementary classes are closed under ultraproducts, it follows that SRAS_n is non-elementary. ■

Corollary 3.7.3. RCA_n *is not closed under completions* ([Hod,97c]).

Proof. In the notation of the preceding proof, let $\mathfrak{A}$ be a non-principal ultraproduct of the $\mathfrak{M}(\Delta_k)$. For each k we know $\mathfrak{M}(\Delta_k) \in \mathsf{RCA}_n$, so as this class is elementary, by Łoš's theorem we have $\mathfrak{A} \in \mathsf{RCA}_n$ as well. But $\mathfrak{A}$ is atomic with atom structure $\rho\big(\mathcal{I}(\Delta)\big)$, so its completion is $\mathfrak{M}(\Delta)$, which is not representable. ■

It follows that RCA_n is not Sahlqvist-axiomatizable [Ven,97a]. As $\mathfrak{At}\mathfrak{A} \in \mathsf{WRAS}_n \setminus \mathsf{SRAS}_n$, or as only one of them is elementary, we see that these classes are indeed distinct.

We conclude that:

Theorem 3.7.4. *For finite* $n \geq 3$, *we have* $\mathsf{CRAS}_n \subset \underline{\mathsf{LCAS}_n} \subset \mathsf{SRAS}_n \subset \underline{\mathsf{WRAS}_n}$, *the elementary classes being underlined.*

Related results for relation algebras are proved in [Hir,95, Hir-Hod,02b, Hod-Ven,05].

8. Cylindric Algebras of Low or High Dimension

We end by considering RCA_n for $n \leq 2$ and the infinite dimensional case. For $n \leq 2$ there are analogues of Corollary 3.1.2 for these classes:

Proposition 3.8.1. *For* $n \leq 2$, *an* n*-dimensional cylindric algebra is completely representable iff it is representable and atomic.*

Proof. '⇒' is immediate from Corollary 3.1.2, since a cylindric representation is *inter alia* a boolean representation. We sketch the proof of '⇐'. The case $n = 0$ follows from Corollary 3.1.2, as 0-dimensional cylindric algebras are just Boolean algebras. Let $\mathfrak{A} \in \mathsf{RCA}_1$ be atomic. Consider the equivalence relation on $At\,\mathfrak{A}$ defined by $x \sim y \iff \mathsf{c}_0 x = \mathsf{c}_0 y$. Let E be the set of $\sim$-equivalence classes, and for $e \in E$ write $\mathcal{C}(e)$ for the full 1-dimensional cylindric set algebra with base e. Then $f : \mathfrak{A} \to \prod_{e \in E} \mathcal{C}(e)$ given by $f(a) = \langle a \cap e : e \in E \rangle$ is an embedding preserving all meets and joins that exist in $\mathfrak{A}$.

Let $\mathfrak{A} \in \mathsf{RCA}_2$ be atomic. As RCA_2 is conjugated and defined by (algebraic versions of) Sahlqvist equations given in [Ven,95b, Definition 2.2], it is closed under completions [Giv-Ven,99]. So the equations are valid over the frame (atom structure) $\mathfrak{At}\,\mathfrak{A}$. By [Ven,95b, Theorem 2.4], $At\,\mathfrak{A}$ is a bounded morphic image of a disjoint union of square frames $\mathcal{F}_i$ $(i \in I)$. Each $\mathfrak{Cm}\,\mathcal{F}_i$ is a full 2-dimensional cylindric set algebra. By duality, the inverse of the bounded morphism is an embedding from $\mathfrak{A}$ into $\prod_{i \in I} \mathfrak{Cm}\,\mathcal{F}_i$ that can be checked to preserve all meets and joins existing in $\mathfrak{A}$. ■

By [Giv-Ven,99], for $n \leq 2$, since RCA_n is a conjugated variety defined by Sahlqvist equations, it is closed under completions.

For $n \geq \omega$, a simple cardinality argument will show that the class of completely representable n-dimensional cylindric algebras is not elementary [Hir-Hod,97c, Corollary 26]. Other results established in this chapter for the finite dimensional case have not yet been considered for $n \geq \omega$.

Problem 3.8.2. Which parts of Theorem 3.7.4 remain true when n is infinite?

Problem 3.8.3. Does the canonical extension of a representable ω-dimensional cylindric algebra necessarily have a complete, ω-dimensional representation? ([Hir-Hod,02a], Problem 5.18).

Problem 3.8.4. Are the classes CRAS_n, SRAS_n definable by sets of sentences in $L_{\infty\omega}$, or in first-order logic plus a least fixed point operator?

BOLYAI SOCIETY
MATHEMATICAL STUDIES, 22

Cylindric-like Algebras and Algebraic Logic
pp. 91–103.

Amalgamation, Interpolation and Epimorphisms in Algebraic Logic

JUDIT X. MADARÁSZ* and TAREK SAYED AHMED

The amalgamation property (for classes of models), since its discovery, has played a dominant role in algebra and model theory. Algebraic logic is the natural interface between universal algebra and logic (in our present context a variant of first order logic). Indeed, in algebraic logic amalgamation properties in classes of algebras are proved to be equivalent to interpolation results in the corresponding logic. In algebra, the properties of epimorphisms (in the categorial sense) being surjective is well studied. In algebraic logic, this property is strongly related to the famous Beth definability property [Hen-Mon-Tar,85] 5.6.10. Pigozzi [Pig,71] is a milestone for working out such equivalences for cylindric algebras, see also [Mad-Say,07], [And-Nem-Sai,01]. In first order logic, Craig interpolation theorem is equivalent to Beth definability theorem; however this equivalence no longer holds for certain modifications of first order logic, be it reducts or expansions, like L_n (first order logic restricted to finitely many n variables) and the extensions of first order logic studied in [Hen-Mon-Tar,85] Sec. 4.3, which are essentially finitary but have an infinitary flavour.

Roughly speaking, algebraic logic has three important branches: (1) Algebraization of logical calculi (deductive systems) which is known as abstract algebraic logic and is initiated by Blok and Pigozzi [Blo-Pig,89], and Andréka and Németi [And-Kur-Nem-Sai,95]. While the first approach is only syntactical, the second approach is more general since it deals with semantics in terms of "meaning" functions [And-Nem-Sai,01]. (2) algebraic approach to first order logic largely due to Tarski, but it's history can be traced back to Pierce and Schröder. (3) algebraic approach to non-classical propositional logics. The work of Maksimova [Mak,97] on amalgamation and interpolation belongs to (3). One could say that we use algebraic logic

*The work of Madarász was supported by a Bolyai grant and by OTKA grant T81188.

in the present exposition to mean exclusively the second branch. The reason is – our algebras are algebraic counterparts of variants, fragments and modifications of first order logic. However, we deal with concepts that were first (historically) formulated for non-classical logics, namely the super amalgamation property. Furthermore, results relating amalgamation and interpolation, though historically originate in connection with and cylindric algebras, now belong to abstract algebraic logic.

Diagneault [Dai,64] was the first to explicitly make the connection between the Craig interpolation theorem and amalgamation in the context of polyadic algebras. In [Pig,71] Pigozzi studied various amalgamation properties for classes of cylindric algebras. Some questions were left open in [Pig,71]. [Mad-Say,07] answers all these questions together with other related ones. However, in [Mad-Say,07] new questions arose. In this paper we collect and answer all open questions in [Pig,71] and [Mad-Say,07].

We will study various forms of the amalgamation property and related concepts (like surjectiveness of epimorphisms) for several algebraizations of first order logic. Our investigation of course includes cylindric algebras, the prime inspiration for introducing such algebras. We start by recalling the concrete versions of our algebras. The algebras dealt with throughout consist of sets of sequences, and the operations are set-theoretic operations on such sets. Let α be an ordinal. Let U be a set. Let $\Gamma \subseteq \alpha$, $E \subseteq \alpha \times \alpha$, $\tau : \alpha \to \alpha$ and $X \subseteq {}^{\alpha}U$:

$$\mathsf{c}_{(\Gamma)}X = \left\{ s \in {}^{\alpha}U : \exists t \in X,\ t(i) = s(i) \text{ for all } i \notin \Gamma \right\},$$

$$\mathsf{s}_{\tau}X = \{ s \in {}^{\alpha}U : s \circ \tau \in X \},$$

$$\mathsf{d}_E = \left\{ s \in {}^{\alpha}U : s_i = s_j \text{ for } (i,j) \in E \right\}.$$

The operations are called *cylindrifications, substitutions,* and *diagonal elements,* respectively. We use simplified notation for some of these operations $\mathsf{c}_i = \mathsf{c}_{(\{i\})}$ and $\mathsf{d}_{ij} = \mathsf{d}_{\{(i,j)\}}$. A full set algebra is an algebra $\mathfrak{A}$ with universe the powerset of a Cartesian space and with the operations the Boolean ones and cylindrifications c_i $i \in \alpha$ together with possibly some others from the above operations. $\mathfrak{Rd}_{df}\,\mathfrak{A}$ is the reduct of $\mathfrak{A}$ which contains only the Boolean operations and cylindfrications c_i. Variations on the above operations cover a wide range of concrete set algebras studied in algebraic logic starting from diagonal free cylindric set algebras to polyadic equality set algebras [Hen-Mon-Tar,85] Sec. 5. We give some concrete examples. $[i|j]$ is the replacement on α that takes i to j and leaves every other thing fixed, while $[i, j]$ is the transposition interchanging i and j. For set X, let $\mathfrak{B}(X) = \langle \wp(X), \cup, \cap, \sim, \emptyset, X \rangle$ be the full Boolean set algebra with universe $\wp(X)$. Let $\mathbf{S}$ be the operation of forming subalgebras, $\mathbf{P}$ be that of forming products, and $\mathbf{H}$ be the operation of forming homomorphic images. Then

(i) $\mathrm{RSC}_\alpha = \mathbf{SP}\{(\mathfrak{B}({}^\alpha U), \mathsf{c}_i, \mathsf{s}_{[i|j]})_{i,j<\alpha} : U \text{ is a set}\}$.

(ii) $\mathrm{RQA}_\alpha = \mathbf{SP}\{(\mathfrak{B}({}^\alpha U), \mathsf{c}_i, \mathsf{s}_{[i|j]}, \mathsf{s}_{[i,j]})_{i,j<\alpha} : U \text{ is a set}\}$.

(iii) $\mathsf{RCA}_\alpha = \mathbf{SP}\{(\mathfrak{B}({}^\alpha U), \mathsf{c}_i, \mathsf{d}_{ij})_{i,j<\alpha} : U \text{ is a set}\}$.

(iv) $\mathrm{RQEA}_\alpha = \mathbf{SP}\{(\mathfrak{B}({}^\alpha U), \mathsf{c}_i, \mathsf{d}_{ij}, \mathsf{s}_{[i,j]})_{i,j<\alpha} : U \text{ is a set}\}$.

(v) $\mathsf{RPA}_\alpha = \mathbf{SP}\{(\mathfrak{B}({}^\alpha U), \mathsf{c}_{(\Gamma)}, \mathsf{s}_\tau)_{\Gamma\subseteq\alpha,\ \tau\in{}^\alpha\alpha} : U \text{ is a set}\}$.

(vi) $\mathrm{RPEA}_\alpha = \mathbf{HSP}\{(\mathfrak{B}({}^\alpha U), \mathsf{c}_{(\Gamma)}, \mathsf{s}_\tau, \mathsf{d}_{ij})_{\Gamma\subseteq\alpha,\ i,j<\alpha,\ \tau\in{}^\alpha\alpha} : U \text{ is a set}\}$.

SC stands for the class of Pinter's substitution algebras, QA (QEA) for quasi-polyadic (equality) algebras and CA stands for the class of cylindric algebras. PA and PEA stand for polyadic and polyadic equality algebras, respectively. These algebras are defined by finite schema of equations in [And-Giv-Mik-Nem-Sim,98]. For $\mathsf{K} \in \{\mathrm{SC}, \mathsf{QA}, \mathsf{QEA}, \mathsf{CA}, \mathsf{PA}, \mathsf{PEA}\}$, RK_α as defined above stands for the class of representable algebras in K_α. It is not hard to show that RK_α as defined above is a variety. However, it is not finite schema axiomatizable, except when $\alpha \leq 2$ and $\mathsf{K} = \mathsf{PA}_\alpha$ with α infinite, cf. [Hen-Mon-Tar,85], [Sai-Tho,91].

We recall that a class K of models (in our case, algebras) has the amalgamation property, if for all $\mathfrak{A}$, $\mathfrak{B}$ and $\mathfrak{C} \in \mathsf{K}$, and all monomorphisms f and g of $\mathfrak{A}$ into $\mathfrak{B}$, $\mathfrak{C}$, respectively, there exist $\mathfrak{D} \in \mathsf{K}$, a monomorphism m from $\mathfrak{B}$ into $\mathfrak{D}$ and a monomorphism n from $\mathfrak{C}$ into $\mathfrak{D}$ such that $m \circ f = n \circ g$. $\mathfrak{D}$ is referred to an *amalgam* of $\mathfrak{B}$ and $\mathfrak{C}$ over the base algebra $\mathfrak{A}$. Since $\mathfrak{A}$ embeds in both $\mathfrak{B}$ and $\mathfrak{C}$ then it is sometimes referred to as the common subalgebra. Discovered by Roland Fraïssé, the idea of amalgamation proved very powerful. Subtly different kinds of amalgamation produce preservation theorems (like the Łoś–Tarski Theorem on the syntactic description of formulas preserved in substructures) and definability theorems (like the classical Beth-Definability Theorem.) In Model Theory we see Fraïssé's method of amalgamation to construct universal models, cf. [Hod,93a] Chapter 6. One is to build up a structure by taking smaller parts, extending them, and then amalgamating the extensions. Fraïssé's method of constructing universal (and homogeneous) models, is a typical instance of amalgamating small parts of a model to get the desired one. Another way of using amalgamation is to *classify* not to construct. In favourable cases this leads to a structural classification of all models of a given (first order) theory. Stability theory follows this path, cf. [Hod,93a].

In algebraic logic, the notion of amalgamation occurs in essentially two forms. One of them is to construct representations in a step-by-step fashion in the spirit of Fraïssé. See for example the mosaic method of Németi [Nem,96a], [Mar-Pol-Mas,96] and the work of Maddux, cf. [Mad,83], [Mad,89a], on relation algebras with n-dimensional basis and n-dimensional cylindric basis. Roughly, a basis consists of (finite) approximations to a representation which Maddux calls matrices. These matrices can be glued together or amalgamated in a step-by-step manner to give relativized representations, see e.g. [Mad,83]. This approach was further pursued by Hirsch and Hodkinson generalizing Maddux's basis to hyperbasis [Hir-Hod,02a], which can be amalgamated to give rise to cylindric algebras.

The other form of amalgamation occurring in algebraic logic is typically of the following form: Which classes frequently studied in algebraic logic – like for example the class of representable cylindric algebras or the class of polyadic algebras – have the amalgamation property [Kis-Mar-Pro-Tho,83]?

It is this form of amalgamation that we will be addressing in what follows.

This form of amalgamation is important (from the logical point of view) for it is in the context of algebraic logic that the equivalence of the interpolation property on the logic side and the amalgamation property (for the corresponding class of algebras) becomes explicit [And-Nem-Sai,01]. The connection is strongest in equational logic because here the algebras are built directly from the theories, and algebraic logic can be viewed as the process of formulating richer logical systems, or simply logics, as systems of equational logic, cf. [Blo-Pig,89]. The first systematic use of the link to obtain results about interpolation properties from results of amalgamation, or vice versa, can be found in Pigozzi's landmark paper [Pig,71] that appeared in Algebra Universalis in 1971. The principal context of [Pig,71] is the class of infinite dimensional cylindric algebras, an equational formalism of first order logic. In this paper Pigozzi deals basically with the following question: Which subclasses of infinite dimensional cylindric algebras, other than the class of locally finite dimensional ones, still have the (strong) amalgamation property. The fact that the class of locally finite dimensional cylindric algebras has the strong amalgamation property, proved earlier by

Diagneualt [Dai,64], is equivalent to the fact that first order logic has the Craig interpolation property. The classes that Pigozzi deals with consist solely of algebras that are infinite dimensional and we assume, to simplify notation, that such classes of algebras are ω-dimensional, where ω is the least infinite ordinal. These classes include the class of ω-dimensional locally finite algebras (Lf_ω), the class of dimension complemented algebras (Dc_ω), the class of ω-dimensional diagonal algebras (Di_ω) (exact definitions will be recalled below), and the class of ω dimensional semisimple algebras (Ss_ω .) Here a semisimple algebra is a subdirect product of simple algebras. All of these classes consist exclusively of algebras that are representable, but unlike RCA_ω none of these classes is first order axiomatizable, least a variety. While the amalgamation property speaks about amalgamating algebras in such a way that the amalgam only agrees on the common subalgebra, the amalgamation is said to be *strong* if the common subalgebra is the *only* overlap between the two algebras in the amalgam. The positive results of Section 2.2 in combination with the negative ones of Section 2.3 of [Pig,71] answer most of the natural questions one could ask about amalgamation for cylindric algebras. In particular, Pigozzi proves that in the (strictly) increasing sequence

$$\mathsf{Lf}_\omega \subset \mathsf{Dc}_\omega \subset \mathsf{Di}_\omega \subset \mathsf{RCA}_\omega$$

the first and third classes have the amalgamation property while the second and fourth fail to have it. However, most questions concerning the strong amalgamation property for several classes of cylindric algebras were posed as open questions, and other closely related ones appeared after Pigozzi's paper was published.

Pigozzi summarized his results and questions on amalgamation of several subclasses of (representable) cylindric algebras in tabular form consisting of two tables. The first table (Table 2.4.1) on p. 346 of [Pig,71] deals with the global property of amalgamation in a given class K, while the second table (Table 2.4.2) on the opposite page, deals with the corresponding local interpolation property on the K-independent free algebras (an instance of free algebras subject to certain defining relations). When K is a variety, the K independent algebras are just the free algebras. When it comes to cylindric algebras, the K-independent algebras Pigozzi deals with are what Henkin, Monk and Tarski call free algebras with dimension restricting functions, or simply dimension-restricted free algebras (cf. [Hen-Mon-Tar,85] Definition 2.5.31). When Pigozzi's paper was published, the two tables looked like Figs. 1 and 2. We start with Table 2.4.1 in [Pig,71] p. 346, which is represented in Fig. 1.

	strong AP	strong AP w.r.t RCA_ω	AP	AP w.r.t RCA_ω	AP for simple algebras	strong EP	EP	EP for simple algebras
Lf_ω	yes	yes	yes	yes	yes	yes	yes	yes
Dc_ω	no	yes	no	yes	no	no	no	no
Semisimple CA_ω's	?	?	yes	yes	yes	?	yes	yes
Diagonal CA_ω's	?	?	yes	yes	yes	?	yes	yes
RCA_ω	no	no	no	no	yes	yes	yes	yes
$\mathsf{RCA}_\omega \subset \mathsf{K} \subseteq \mathsf{CA}_\omega$	no	no	no	no	yes	?	?	yes
CA_ω's of positive characteristic	?	?	yes	yes	yes	yes	yes	yes
$\mathsf{RCA}_n \subseteq \mathsf{K} \subseteq \mathsf{CA}_n$	no	no	no	no	no	no	no	no
CA_n of positive characteristic	?	?	yes	yes	yes	yes	yes	yes
CA_1	?	?	yes	yes	yes	yes	yes	yes
CA_0	yes	yes	yes	yes	yes	yes	yes	yes

Fig. 1.

And Table 2.4.2 on the opposite page looked like Fig. 2.

In the top row of Fig. 1, we find a list of eight different amalgamation and embedding properties (AP and EP for short) and at the leftmost column we find a comprehensive list of classes of algebras. All these notions are defined in [Mad-Say,07], and of course in Pigozzi's original paper [Pig,71]. The first seven rows address infinite dimensional algebras and the last four address finite dimensional algebras. For the infinite dimensional case, we consider, without any loss of generality, only ω-dimensional algebras. This simplifies notation. Fig. 2 contains a summary of the results involving interpolation of the dimension-restricted free algebras on ω generators. Again this is a natural restriction, since in the (special) case of varieties, failure of amalgamation is met by failure of the corresponding interpolation property in the free algebra on ω generators. The rows addressing semisimple, diagonal and representable cylindric algebras in Fig. 1 collapse to just one row in Fig. 2, since the free algebra on ω generators coincide for all these classes[1]. Therefore we find only five rows addressing infinite dimensional algebras in Fig. 2. In more general contexts than first order logic, the Craig interpolation property ramifies into several different interpolation proper-

[1]More precisely, we have $\mathfrak{Fr}_\omega \mathsf{Sc}_\omega = \mathfrak{Fr}_\omega \mathsf{Di}_\omega = \mathfrak{Fr}_\omega \mathsf{RCA}_\omega$ since $\mathrm{Eq}\,\mathsf{Sc}_\omega = \mathrm{Eq}\,\mathsf{Di}_\omega = \mathsf{RCA}_\omega$, where $\mathrm{Eq}\,\mathsf{K}$ is the smallest variety containing K and $\mathfrak{Fr}_\omega\,\mathsf{K}$ is the K-free algebra on ω generators.

	strong IP	IP	weak IP	strong restricted IP	restricted IP	weak restricted IP
$\mathfrak{Fr}^{\rho}_{\omega}\,\mathsf{CA}_{\omega}$ which is in Lf_{ω}	yes	yes	yes	yes	yes	yes
$\mathfrak{Fr}^{\rho}_{\omega}\,\mathsf{CA}_{\omega}$ which is in Dc_{ω}	yes	yes	yes	yes	yes	yes
$\mathfrak{Fr}_{\omega}\,\mathsf{RCA}_{\omega}$	no	no	yes	?	?	yes
$\mathfrak{Fr}_{\omega}\,\mathsf{K}$, where	no	no	yes	?	?	yes
$\mathsf{RCA}_{\omega} \subset \mathsf{K} \subseteq \mathsf{CA}_{\omega}$	yes	yes	yes	yes	yes	yes
$\mathfrak{Fr}_{\omega}\,\mathsf{K}$, where $\mathsf{K} \subseteq \mathsf{CA}_{\omega}$ is the class of algebras of positive characteristic	?	yes	yes	?	yes	yes
$\mathfrak{Fr}_{\omega}\,\mathsf{K}$, where $\mathsf{RCA}_{n} \subset \mathsf{K} \subseteq \mathsf{CA}_{n}$	no	no	no	no	no	no
$\mathfrak{Fr}_{\omega}\,\mathsf{K}$, where $\mathsf{K} \subseteq \mathsf{CA}_{n}$ is the class of algebras of positive characteristic	?	yes	yes	?	yes	yes
$\mathfrak{Fr}_{\omega}\,\mathsf{CA}_{1}$	?	yes	yes	?	yes	yes
$\mathfrak{Fr}_{\omega}\,\mathsf{CA}_{0}$	yes	yes	yes	yes	yes	yes

Fig. 2.

ties (IP for short). These properties are summarized in the six columns of the uppermost row of Fig. 2. Most of the results listed in Fig. 2 are direct consequences of the results of Fig. 1 together with Theorems 1.2.8, 1.2.10, 2.1.11, 2.1.13 and 2.1.20 in [Pig,71] where the general relationships between the various amalgamation and interpolation properties are stated (cf. also 2.1.22 in [Pig,71]); the remaining results are, for their part, automatic consequences of results obtained this way. We will look at the two tables *without* the leftmost column and the uppermost row as matrices, and refer to an open question by its co-ordinates. For example, the question of whether or not the class of semisimple algebras in CA_{ω} has the strong amalgamation property, is given co-ordinates $(3, 1)$ because it is located in the third row and first column. At each matrix point of Fig. 1, the answer to whether the corresponding class has the amalgamation property in question is indicated by a yes or a no, while a question mark (?) indicates that the problem was open at the time. The same applies to Fig. 2. We add two columns to Fig. 1 corresponding to other amalgamation properties which, though discovered after Pigozzi's paper was published, are closely related to the subject matter of the present paper. These properties are the super amalgamation property, or SUPAP for short, introduced by Maksimova [Mak,91], and the property that epimorphisms in the categorial sense[2] are surjective, which we abbreviate by ES.

The link between ES and the Beth-definability property is discovered by Németi and the part of Németi's result relevant to this paper is fully proved in [Hen-Mon-Tar,85] Thm 5.6.10, see also [Hoo,00], [Hoo-Mar-Ott,99]. The name Beth definability Theorem is a generic title for assertions of the form "a Logic has the Beth definability property". What Beth himself proved is that first order logic has the Beth Definability property. The Beth Definability Theorem is one of the cornerstones of first order logic. Indeed, the Beth Definability Theorem together with the so called Downward Löwenheim–Skolem Theorem characterize first order logic. This, in turn, is known as Lindström Theorem. The Beth Definability Theorem for first order logic relates two notions of definability, implicit definability and explicit definability. A set Σ of formulas implicitly defines a relation symbol P if for any structure of the symbols in $\Sigma \sim P$ this structure has at most one expansion that is a model of Σ. On the other hand, Σ defines P explicitly if there is a formula built up of symbols distinct from P that turns out to be equivalent to P in any model of Σ. It is straightforward to check that explicit definability implies implicit definability. When the converse holds for a logic, then this logic is said to have the Beth definability property.

However, unlike Pigozzi, Németi did not settle the the corresponding algebraic question i.e whether epimorphisms are surjective in various classes of cylindric algebras. This appears as Open problem 10 of [Hen-Mon-Tar-And-Nem,81] p. 311. The finite dimensional case was settled in [And-Com-Mad-Nem-Say,09] and the infinite dimensional case was settled by Judit Madarász [Mad,12]. In fact, Madarász proved that ES fails for various subclasses of cylindric algebras, the smallest of which is the class of semisimple algebras. Sayed Ahmed used the result of Madarász, in conjuction with the sheaf-theoretic duality of cylindric algebras developed by Comer [Com,84], [Com,72] to show that ES fails even in the class of simple algebras. These result imply that the infinitary logics studied in [Hen-Mon-Tar,85] Sec. 4.3 fail to have the Beth definability property in a very strong sense.

[2] I.e. maps that are right cancellative.

Beth definability property BDP can be traced back to Padoa's method for showing definability of primitive notions in a language. Classical propositional logic, intuitionistic propositional logic, the minimal logic K and all normal extensions of the logic K4 have BDP. Failure of BDP for L_n (first order logic restricted to finitely many variables) was first announced in [And-Com-Nem,83] and proved algebraically in [And-Com-Mad-Nem-Say,09]. Sain [Sai,97], [Sai,90] proves that BDP fails for L_n without equality. Gurevich [Gur,84] shows that first order logic with only finite models fails BDP. This result was refined in [And-Com-Mad-Nem-Say,09] by showing that BDP fails when we consider finite models having $n+2$ elements (or more). If we allow only finite models having $< n+2$ elements, then BDP holds [And-Com-Mad-Nem-Say,09]. Finite model theory is intimately connected to finite variable logics, and there is a particularly strong connection between definability properties and complexity issues in computer science [Daw-Hel-Kol,95]. This is the main objective in descriptive complexity theory. BDP also fails for FOL with finitely many variables and finite models, even if enriched by infinite conjunctions [Hod,93b]. We should mention that for the highly relevant guarded fragments of first order logic the Beth definability property holds even for finite variable fragments [Hoo-Mar-Ott,99], [Hoo-Mar,02b].

Super amalgamation, on the other hand, is the algebraic equivalent of the property that the free algebras have the *strong interpolation property* in the sense of Pigozzi [Pig,71, Definition 2.1.16], which, in turn, is the algebraic equivalent of a strong form of interpolation in the corresponding logic. In Remark 2.1.21 in [Pig,71] p. 33, Pigozzi proposes the question to find the algebraic counterpart of the strong interpolation property. Madarász [Mad,98] found the answer; this property turns out to be SUPAP. This answer is sharp in view of [Sag-She,06] which gives an example of a variety of representable CA's that has the strong amalgamation property SAP but not SUPAP. While SUPAP is added as column 9 and ES as column 10 to Fig. 1, no new columns are added to Fig. 2. This is because the first added column addressing SUPAP in Fig. 1 corresponds to a column already existing in Fig. 2, namely that of strong interpolation. On the other hand, ES for a class K has no obvious counterpart on the K-free algebras. Note that CA_0, addressed in the last row in both tables, is just the class of Boolean algebras. Today Fig. 1 (with the additional two columns and with replacing the question marks by a yes or a no) looks like Fig. 3.

And Fig. 2 (without adding any new columns) today looks like Fig. 4

	strong AP	strong AP w.r.t RCA_ω	AP	AP w.r.t RCA_ω	AP for simple CA_ω	SEP	EP	EP for simple CA_ω	SUP AP	ES
Lf_ω	yes	yes	yes	yes	yes	yes	yes	yes	yes	yes
Dc_ω	no	yes	no	yes	no	no	no	no	no	yes
Semisimple CA's	no	no	yes	yes	yes	yes	yes	yes	no	no
Diagonal CA's	no	no	yes	yes	yes	yes	yes	yes	no	no
RCA_ω	no	no	no	no	yes	yes	yes	yes	no	no
$\mathsf{RCA}_\omega \subset \mathsf{K} \subseteq \mathsf{CA}_\omega$	no	no	no	no	yes	no for CA_ω	no for CA_ω	yes	no	no
CA_ω's of positive characteristic	yes	yes	yes	yes	yes	yes	yes	yes	yes	yes
RCA_n	no	no	no	no	no	no	no	no	no	no
CA_n of positive characteristic	yes	yes	yes	yes	yes	yes	yes	yes	yes	yes
CA_1	yes	yes	yes	yes	yes	yes	yes	yes	yes	yes
CA_0	yes	yes	yes	yes	yes	yes	yes	yes	yes	yes

Fig. 3.

Proofs for various amalgamation properties for finite dimensional CA's can be found in [Mar,00]. Now let $\mathsf{K} \in \{\mathsf{SC}, \mathsf{CA}, \mathsf{QA}, \mathsf{QEA}\}$. We define several distinguished classes of algebras.

$$\mathsf{Lf}\,\mathsf{K}_\omega = \{\mathfrak{A} \in \mathsf{K}_\omega \,:\, \Delta x \text{ is finite for all } x \in A\}$$

$$\mathsf{Dc}\,\mathsf{K}_\omega = \{\mathfrak{A} \in \mathsf{K}_\omega \,:\, \omega \sim \Delta x \text{ is infinite for all } x \in A\}$$

$$\mathsf{Ss}\,\mathsf{K}_\omega = \mathbf{SP}\{\mathfrak{A} \,:\, \mathfrak{A} \text{ is simple }\}$$

$$\mathsf{Re}\,\mathsf{K}_\omega = \big\{\mathfrak{A} \in \mathsf{K}_\omega \,:\, (\forall \Gamma \subseteq_\omega \omega)(\forall x \neq 0)(\exists i, j \in \omega \sim \Gamma)\big(i \neq j \wedge \mathsf{s}_i^j x \neq 0\big)\big\}.$$

$\mathsf{Lf}\,\mathsf{K}_\omega$ is the class of locally finite algebras, $\mathsf{Dc}\,\mathsf{K}_\omega$ is the class of dimension complemented algebras, $\mathsf{Ss}\,\mathsf{K}_\omega$ is the class of semisimple algebras and $\mathsf{Re}\,\mathsf{K}_\omega$ is the class of replacement algebras. It is known [Hen-Mon-Tar,85] that $\mathsf{Dc}\,\mathsf{K}_\omega \cup \mathsf{Ss}\,\mathsf{K}_\omega \subseteq \mathsf{Re}\,\mathsf{K}_\omega \subseteq \mathsf{RK}_\omega$.

	strong IP	IP	weak IP	strong restricted IP	restricted IP	weak restricted IP
$\mathfrak{Fr}^{\rho}_{\omega}\,\mathsf{CA}_{\omega}$ which is in Lf_{ω}	yes	yes	yes	yes	yes	yes
$\mathfrak{Fr}^{\rho}_{\omega}\,\mathsf{CA}_{\omega}$ which is in Dc_{ω}	yes	yes	yes	yes	yes	yes
$\mathfrak{Fr}_{\omega}\,\mathsf{RCA}_{\omega}$	no	no	yes	yes	yes	yes
$\mathfrak{Fr}_{\omega}\,\mathsf{K}$, where $\mathsf{RCA}_{\omega} \subset \mathsf{K} \subseteq \mathsf{CA}_{\omega}$	no	no	yes	no for CA_{ω}	no for CA_{ω}	yes
$\mathfrak{Fr}_{\omega}\,\mathsf{K}$, where $\mathsf{K} \subseteq \mathsf{CA}_{\omega}$ is the class of algebras of positive characteristic	yes	yes	yes	yes	yes	yes
$\mathfrak{Fr}_{\omega}\,\mathsf{K}$, where $\mathsf{RCA}_{n} \subset \mathsf{K} \subseteq \mathsf{CA}_{n}$	no	no	no	no	no	no
$\mathfrak{Fr}_{\omega}\,\mathsf{K}$, where $\mathsf{K} \subseteq \mathsf{CA}_{n}$ is the class of algebras of positive characteristic	yes	yes	yes	yes	yes	yes
$\mathfrak{Fr}_{\omega}\,\mathsf{CA}_{1}$	yes	yes	yes	yes	yes	yes
$\mathfrak{Fr}_{\omega}\,\mathsf{CA}_{0}$	yes	yes	yes	yes	yes	yes

Fig. 4.

The following table was published in [Mad-Say,07] with the question marks indicating open questions.

	strong AP	strong AP w.r.t K_{ω}	AP	AP w.r.t RK_{ω}	AP for simple algebras	strong EP	EP	EP for simple algebras	SUP AP	ES
$\mathsf{Lf}\,\mathsf{K}_{\omega}$	yes	yes	yes	yes	yes	yes	yes	yes	yes	yes
$\mathsf{Dc}\,\mathsf{K}_{\omega}$	no	yes	no	yes	no	no	no	no	no	yes
$\mathsf{Sc}\,\mathsf{K}_{\omega}$	no	no	?	?	?	yes	yes	yes	no	no
$\mathsf{Re}\,\mathsf{K}_{\omega}$	no	no	?	?	?	yes	yes	yes	no	no
RK_{ω}	no	no	?	?	?	yes	yes	yes	no	no
SA_{ω}	yes	yes	yes	yes	yes	yes	yes	yes	yes	yes
PA_{ω}	yes	yes	yes	yes	yes	yes	yes	yes	yes	yes

Fig. 5.

Today Fig. 5 appearing in [Mad-Say,07] without question marks with last row added for cylindric and quasipolyadic equality algebras of positive characterstic $\kappa > 0$ (by noting that SUPAP implies everything else) looks like Fig. 6.

	strong AP	strong AP w.r.t RK_ω	AP	AP w.r.t RK_ω	AP for simple algebras	strong EP	EP	EP for simple algebras	SUP AP	ES
$\mathsf{Lf\,K}_\omega$	yes	yes	yes	yes	yes	yes	yes	yes	yes	yes
$\mathsf{Dc\,K}_\omega$	no	yes	no	yes	no	no	no	no	no	yes
$\mathsf{Sc\,K}_\omega$	no	no	yes	yes	yes	yes	yes	yes	no	no
$\mathsf{Re\,K}_\omega$	no	no	yes	yes	yes	yes	yes	yes	no	no
RK_ω	no	no	no	no	yes	yes	yes	yes	no	no
SA_ω	yes	yes	yes	yes	yes	yes	yes	yes	yes	yes
PA_ω	yes	yes	yes	yes	yes	yes	yes	yes	yes	yes
$\mathsf{CA}'s$ QEA's of +ve chara.	yes	yes	yes	yes	yes	yes	yes	yes	yes	yes

Fig. 6.

SA_ω's are reducts of polyadic algebras introduced by Sain [Sai,00], [Sai-Gyu,97] to finitize first order logic. In [Say,04a] it is proved that such algebras have SUPAP. Generally, the amalgamation property for various reducts of polyadic algebras, possibly with equality, is investigated in [Say,09b] and [Say,08b]. In [Say,10a] it is proved that PA_α has SUPAP for infinite α.

We highlight the main new results in the above tables. In what follows, $\mathsf{K} \in \{\mathsf{SC}, \mathsf{QA}, \mathsf{CA}, \mathsf{QEA}\}$.

(1) Semisimple and replacement algebras do not have the strong amalgamation property, even if the amalgam is sought in the bigger class (of representable algebras in) K_ω [Mad-Say,09b]. For CA's this answers $(3, 1)$, $(3, 2)$, $(4, 1)$, and $(4, 2)$ of Fig. 1 negatively. In fact, Madarász [Mad,12], [Mad-Say,09b] proves something stronger, from which we can infer that ES fails for such classes of algebras even if the right-hand most algebra is in K_ω. For CA_ω, this appears as $(3, 10)$ and $(4, 10)$ in Fig. 3.

(2) In contrast, semisimple and diagonal K_ω's have the strong embedding property. For cylindric algebras this answers $(3,6)$ and $(4,6)$ of Fig. 1 positively. This is proved in [Say,06b].

(3) The class of cylindric algebras of positive characteristic and the class of the so-called monadic-generated cylindric algebras of any dimension, have the super amalgamation property. This is proved in [Say,06b], see also, [Mad-Say,09a] and [Mad-Say,09b]. In particular, CA_1 has SUPAP and consequently, we infer that $\mathfrak{Fr}_\omega\,\mathsf{CA}_1$ has the strong interpolation property. This answers $(7,1)$, $(7,2)$, $(9,1)$, $(9,2)$, $(10,1)$ and $(10,2)$ of Fig. 1 and $(5,1)$, $(5,4)$, $(7,1)$, $(7,4)$, $(8,1)$ and $(8,4)$ of Fig. 2 positively.

(4) CA_ω does not have the embedding property. This answers partially $(6,6)$ and $(6,7)$ of Fig. 1 negatively. Consequently, $\mathfrak{Fr}_\omega\,\mathsf{CA}_\omega$ does not have the restricted interpolation property, nor, *a fortiori*, the strong restricted interpolation property. This answers $(4,4)$ and $(4,5)$ of Fig. 2 negatively, but again only when $\mathsf{K} = \mathsf{CA}_\omega$. This is proved in [Sim,99]. Questions concerning embedding properties for classes strictly in between RCA_ω and CA_ω (see the fifth row in the leftmost column of Fig. 2) still involve some open problems.

(5) The free representable CA_ω on ω generators, or $\mathfrak{Fr}_\omega\,\mathsf{RCA}_\omega$ for short, has the strong restricted interpolation property. This is proved in [Mad-Say,09b].

(6) However, RK_ω does not have AP [Say,10b].

(1) is due to Madarász, (2), (3) and (5) are due to Madarász and Sayed-Ahmed, (4) is due to Madarász and Simon and (6) is due to Sayed Ahmed. We note that in Pigozzi's opinion, (cf. Remark 2.4.3, p. 347) $(3,4)$ together with $(4,5)$ of Fig. 2, were the most interesting remaining open questions on amalgamation. A relevant deep result of Németi is that Crs_α, the class of *relativized* cylindric algebras of dimension α, has SAP. Marx strenghthened this to SUPAP, but this is implicit in Németi's proof [Nem,85a].

Finally, we should mention that Sayed Ahmed [Say,09g] used the fact that the amalgamation property fails for RCA_ω to provide a solution to an open problem in [Hen-Mon-Tar,85], namely problem 2.13.

Open problem. Find an infinite dimensional variety of CA's with SAP and not SUPAP.

BOLYAI SOCIETY
MATHEMATICAL STUDIES, 22

**Cylindric-like Algebras
and Algebraic Logic**
pp. 105–131.

Neat Reducts and Neat Embeddings in Cylindric Algebras

TAREK SAYED AHMED

An important central concept introduced in [Hen-Mon-Tar,85] is that of neat reducts, and the related one of neat embeddings. The notion of neat reducts is due to Leon Henkin, and one can find that the discussion of this notion is comprehensive and detailed in [Hen-Mon-Tar,85] (closer to the end of the book). This notion proved useful in at least two respects. Analyzing the number of variables appearing in proofs of first order formulas [Hir-Hod,02c], and characterizing the class of representable algebras; those algebras that are isomorphic to genuine algebras of relations. In fact, several open problems that appeared in [Hen-Mon-Tar,85] are on neat reducts, some of which appeared in part 1, and (not yet resolved) appeared again in part 2. This paper, among other things, surveys the status of these problems 40 years after they first appeared. Long proofs are omitted, except for one, which gives the gist of techniques used to solve such kind of problems.

All the open problems in [Hen-Mon-Tar,85] on neat reducts are now solved. The most recent one was solved by the present author. This is Problem 2.13 in [Hen-Mon-Tar,85]. A solution of this problem on neat embeddings is presented in [Say,09g]. But the present paper also poses new problems related to this key notion in the representation theory of cylindric algebras, that have emerged in recent years.

Our notation is in conformity with the three monographs on the subject [Hen-Mon-Tar,85], [Hen-Mon-Tar-And-Nem,81]. Cylindric set algebras are algebras whose elements are relations of a certain pre-assigned arity, endowed with set-theoretic operations that utilize the form of elements of the algebra as sets of sequences. Our $\mathfrak{B}(X)$ denotes the boolean set algebra $\big(\wp(X), \cup, \cap, \sim, \emptyset, X\big)$. Let U be a set and α an ordinal. α will be the dimension of the algebra. For $\mathsf{s}, t \in {}^{\alpha}U$ write $\mathsf{s} \equiv_i t$ if $\mathsf{s}(j) = t(j)$ for all $j \neq i$.

For $X \subseteq {}^{\alpha}U$ and $i, j < \alpha$, let

$$\mathsf{C}_i X = \left\{ \mathsf{s} \in {}^{\alpha}U \, : \, \exists t \in X (t \equiv_i \mathsf{s}) \right\}$$

and

$$\mathsf{D}_{ij} = \{ \mathsf{s} \in {}^{\alpha}U \, : \, \mathsf{s}_i = \mathsf{s}_j \}.$$

$\left(\mathfrak{B}\left({}^{\alpha}U \right), \mathsf{C}_i, \mathsf{D}_{ij} \right)_{i,j<\alpha}$ is called the full cylindric set algebra of dimension α with unit (or greatest element) ${}^{\alpha}U$. Examples of subalgebras of such set algebras arise naturally from models of first order theories. Indeed if M is a first order structure in a first order language L with α many variables, and one sets

$$\phi^{\mathsf{M}} = \left\{ \mathsf{s} \in {}^{\alpha}\mathsf{M} \, : \, \mathsf{M} \models \phi[\mathsf{s}] \right\},$$

(here $\mathsf{M} \models \phi[\mathsf{s}]$ means that s satisfies ϕ in M), then the set $\{\phi^{\mathsf{M}} \, : \, \phi \in Fm^L\}$ is a cylindric set algebra of dimension α. Indeed

$$\phi^{\mathsf{M}} \cap \psi^{\mathsf{M}} = (\phi \wedge \psi)^{\mathsf{M}},$$

and

$${}^{\alpha}\mathsf{M} \sim \phi^{\mathsf{M}} = (\neg\phi)^{\mathsf{M}},$$

$$\mathsf{C}_i\left(\phi^{\mathsf{M}}\right) = \exists v_i \phi^{\mathsf{M}},$$

and finally

$$\mathsf{D}_{ij} = (x_i = x_j)^{\mathsf{M}}.$$

Cs_{α} denotes the class of all subalgebras of full set algebras of dimension α. Let RCA_{α} denote the variety generated by the class Cs_{α}. Algebras in RCA_{α} are said to be representable. An old problem in algebraic logic is ([Hen-Mon-Tar,85], [Hir-Hod,97a]): Can we describe RCA_{α} by a simple schema of equations? In other words, is there a simple set of equations Σ such that $\mathfrak{A} \models \Sigma$ if and only if $\mathfrak{A}$ is representable? Let us call this problem the *representation problem.* Andréka, Németi and Sain refer to the related problem of modifying the class of representable cylindric algebras to get a new variety that is still adequate for the algebraisation of first order logic, but is finitely axiomatizable, as the *finitizability* problem in algebraic logic [Nem,91]. Both of these problems are discussed at length in [Say,05a]. The representation problem, and for that matter the finitizability problem, have provoked extensive research and are still, in some sense, open! An approximation CA_{α} was introduced by Tarski. CA_{α} is defined by an indeed simple finite set of equations Σ that aims at capturing the algebraic essence of existential quantifiers and equality. But early on in the investigations of CA's it turned out that there are cylindric algebras that are not representable. The choice of Σ was motivated by the fact that it works in some special cases that are significant (like for example locally finite cylindric algebras). The locally finite cylindric algebras correspond in an exact sense [Hen-Mon-Tar,85] 4.3.28 (ii) to Lindenbaum–Tarski algebras of formulas in (ordinary) first order logic.

If $\mathfrak{C} \in \mathsf{Cs}_\beta$ with base U, then for any $\alpha < \beta$, the elements of $\mathfrak{C}$ that are fixed by C_i, $i \geq \alpha$, can be thought of as representations of α ary relations on U. In fact if we keep only these elements and those operations whose indices are all in α, then the resulting algebra is obviously isomorphic to a Cs_α (and in fact to one with base U). This observation carries over to abstract CA_β's in general yielding the concept of neat reducts.

A reduct of an algebra $\mathfrak{A}$ is another algebra $\mathfrak{B}$ obtained from $\mathfrak{A}$ by dropping some of the operations. $\mathfrak{B}$ thus has the same universe of $\mathfrak{A}$ but the operations defined on these elements constitute only a part of the original operations. In cylindric algebras, reducts are important because certain reducts of cylindric algebras are cylindric algebras (of a different dimension, though).

Let $\mathfrak{A} = (A, +, \cdot, -, \mathsf{c}_i, \mathsf{d}_{ij}) \in \mathsf{CA}_\beta$[1] *and* $\rho : \alpha \to \beta$ *be one to one. Then* $\mathfrak{Rd}^\rho \mathfrak{A} = \left(A, +, \cdot, -, \mathsf{c}_{\rho(i)}, \mathsf{d}_{\rho(i),\rho(j)}\right)_{i,j<\alpha}$ *is a* CA_α [Hen-Mon-Tar,85] 2.6.1.

Here a reduct is defined by renaming the operations. However, when $\alpha \subseteq \beta$ and ρ is the inclusion map then $\mathfrak{Rd}_\alpha \mathfrak{A}$ is just the algebra obtained by discarding the operations indexed by ordinals in $\beta \sim \alpha$. For $x \in \mathfrak{A}$, let $\Delta x = \{i \in \beta : \mathsf{c}_i x \neq x\}$. Then for $i, j < \alpha$ we have $\Delta \mathsf{d}_{ij} \subseteq \alpha$ and if $\Delta x \subseteq \alpha$ and $i < \alpha$ then $\Delta \mathsf{c}_i x \subseteq \alpha$. Also $\Delta(x + y) \subseteq \Delta x \cup \Delta y$ and $\Delta(-x) = \Delta x$.

The set $\mathfrak{Nr}_\alpha \mathfrak{B} = \{x \in B : \Delta x \subseteq \alpha\}$ *is a subuniverse of* $\mathfrak{Rd}_\alpha \mathfrak{B}$. *The algebra* $\mathfrak{Nr}_\alpha \mathfrak{B} \in \mathsf{CA}_\alpha$ *with universe* $\mathfrak{Nr}_\alpha \mathfrak{B}$ *is called the neat* α *reduct of* $\mathfrak{B}$ [Hen-Mon-Tar,85] 2.6.28.

If there is an embedding $e : \mathfrak{C} \to \mathfrak{Nr}_\alpha \mathfrak{B}$ then we say that $\mathfrak{C}$ neatly embeds in $\mathfrak{B}$. For $\mathsf{K} \subseteq \mathsf{CA}_\beta$ and $\alpha < \beta$, $\mathsf{Nr}_\alpha \mathsf{K} = \{\mathfrak{Nr}_\alpha \mathfrak{B} : \mathfrak{B} \in \mathsf{K}\}$. Now if $\alpha < \beta$ and $\theta : \wp({}^\alpha U) \to \wp({}^\beta U)$ is defined by

$$X \mapsto \{\mathsf{s} \in {}^\beta U : (\mathsf{s} \restriction \alpha) \in X\},$$

[1]We follow the conventions of [Hen-Mon-Tar,85], so that operations of abstract algebras are denoted by $+, \cdot, -, \mathsf{c}_i, \mathsf{d}_{ij}$, standing for join, meet, complementation, etc, while in set algebras these operations are denoted by $\cup, \cap, \sim \mathsf{C}_i, \mathsf{D}_{ij}$.

then θ maps $\wp({}^\alpha U)$ into $\mathsf{Nr}_\alpha \wp({}^\beta U)$. Thus set algebras can be neatly embedded into algebras in arbitary extra dimensions. But the converse is strikingly not true. If $\mathfrak{A} \in \mathsf{CA}_\alpha$ and there exists an embedding $e : \mathfrak{A} \to \mathfrak{Nr}_\alpha \mathfrak{B}$, with $\mathfrak{B} \in \mathsf{CA}_{\alpha+\omega}$, then $\mathfrak{A}$ is representable. So we have the following (neat) Neat Embedding theorem, or NET for short, of Henkin: $\mathsf{RCA}_\alpha = \mathbf{S}\mathsf{Nr}_\alpha \mathsf{CA}_{\alpha+\omega}$ for any α. Here $\mathbf{S}$ stands for the operation of forming subalgebras. This is fully proved in [Hen-Mon-Tar,85], cf. Theorem 3.2.10. This theorem has several incarnations in the literature, see e.g. [Sam-Say,07a], [Sai,00], [Sim,07] and [Fer,00], some of which are quite sophisticated. Following the conventions of [Hen-Mon-Tar,85], algebras in the class $\mathbf{S}\mathsf{Nr}_\alpha \mathsf{CA}_{\alpha+\omega}$ are said to have the Neat Embedding property (*NEP*). Monk proved that $\mathsf{RCA}_\alpha \subset \mathbf{S}\mathsf{Nr}_\alpha \mathsf{CA}_{\alpha+n}$ for every $\alpha > 2$ and $n \in \omega$, so that all ω extra dimensions are needed to enforce representability. However, we can dispense with some of the CA axioms, when we get to ω-extra dimensions as illustrated by Ferenczi [Fer,00]. We will return to such issues in some depth at the end of the paper. The non-finite axiomatizability result for RCA_α when $\alpha > 2$ is finite, follows from Monk's result. Indeed, let $\mathfrak{A}_k \in \mathbf{S}\mathsf{Nr}_\alpha \mathsf{CA}_{\alpha+k} \sim \mathsf{RCA}_\alpha$, then any non trivial ultraproduct of the $\mathfrak{A}_k$'s will be representable.

But why is the notion of neat reducts so important in cylindric algebra theory and related structures; in a nut shell: due to its intimate connection to the notion of representability, via Henkin's Neat Embedding Theorem.

However, this is not the end of the story, in fact this is where the fun begins. A new unexpected viewpoint can yield dividends, and indeed the notion of neat reducts has been revived lately, to mention a few references: [Say,02b], [Say,02c], [Say,02d], [Say,03], [Fer,00], [Fer,07b], [Say,08d], [Say,07d], [Say,01], [Hir-Hod,02c], [Hir,07], [Sim,07], [Sam-Say,07a], [And-Com-Mad-Nem-Say,09], [And-Nem-Say,08], and [Say,05a]. Indeed there has been a rise of interest in the study of neat embeddings for cylindric algebras, and related structures with pleasing progress. In this paper we intend to survey (briefly) such results on neat reducts putting them in a wider perspective.

Our first family of results will concern the class of neat reducts proper, that is the class $\mathsf{Nr}_\alpha \mathsf{CA}_\beta$, e.g. is it closed under homomorphic images, products; is it a variety, if not, is it perhaps an elementary class? But why address such questions on neat reducts? There are (at least) three possible answers to this question. First there are aesthetic reasons. Motivated by intellectual curiosity, the investigation of such questions is likely to lead to nice mathematics. The second reason concerns definability or classification.

Now that we have the class of neat reducts in front of us, the most pressing need is to try to classify it. Classifying is a kind of defining. Most mathematical classification is by axioms (preferably first order) or, even better, equations (if the class in question is a variety.) Now we come to the third reason, for studying such questions on neat reducts. Here we do not address neat embeddings as an end itself but rather discuss such notion in connection to the so called amalgamation property and the notion of complete representations.

Accordingly, the rest of the paper is divided into four parts. In the first part (Section 1), we discuss results on the class of neat reducts proper. In the second part (Section 2) we discuss neat embeddings in connection to the amalgamation property. Two open questions in the problem session paper of [And-Mon-Nem,91b] are answered. In Section 3 we discuss neat embeddings in connection to complete representations. In Section 4, we go back to the classical NET of Henkin and review several variations on this deep theorem introduced by Ferenczi. In the final section we comment on related results concerning relation algebras. We note that many other classes of algebras studied in algebraic logic enjoy a NET, like Pinter's substitution algebras, Halmos quasipolyadic algebras and Halmos's polyadic algebras. To keep the paper as short as possible, we discuss those very briefly.

1. The Class of Neat Reducts is not Elementary

Problems 2.11 and 2.12 in the monograph [Hen-Mon-Tar,85] are on neat reducts. Problem 2.12 is solved by Hirsch, Hodkinson and Maddux [Hir-Hod,02c]. Hirsch, Hodkinson and Maddux show that the sequence $\langle \mathsf{SNr}_n \mathsf{CA}_{n+k} : k \in \omega \rangle$ is strictly decreasing for $n > 2$ with respect to inclusion. Problem 2.11 which is relevant to our later discussion asks: For which pairs of ordinals $\alpha < \beta$ is the class $\mathsf{Nr}_\alpha \mathsf{CA}_\beta$ closed under forming subalgebras? Németi [Nem,83] proves that for any $1 < \alpha < \beta$ the class $\mathsf{Nr}_\alpha \mathsf{CA}_\beta$ though closed under forming homomorphic images and products is not a variety, i.e. it is not closed under forming subalgebras. The next natural question is whether this class is elementary? Andréka and Németi prove that the class $\mathsf{Nr}_2 \mathsf{CA}_\beta$ for $\beta > 2$ is not elementary. Their remarkable proof appears in [Hen-Mon-Tar-And-Nem,81]. Not resolved for higher dimensions, this problem re-appears in [Hen-Mon-Tar,85] Problem 4.4. Since this class is closed under ultraproducts this is equivalent to asking whether it is closed under forming elementary subalgebras. In [Say,01] it is proved that for any $2 < \alpha < \beta$, the class $\mathsf{Nr}_\alpha \mathsf{CA}_\beta$ is not elementary. Here we give a model theoretic proof of this result that has appeared in [Say,02b].

Definition 5.1.1.

(i) Let L be a signature and $\mathfrak{D}$ an L structure. The *age* of $\mathfrak{D}$ is the class K of all finitely generated structures that can be embedded in $\mathfrak{D}$.

(ii) A class K is the *age* of $\mathfrak{D}$ if the structures in K are *up to isomorphism*, exactly the finitely generated substructures of $\mathfrak{D}$.

(iii) Let K be a class of structures.

K has the *Hereditary Property,* HP *for short,* if whenever $\mathfrak{A} \in \mathsf{K}$ and $\mathfrak{B}$ is a finitely generated substructure of $\mathfrak{A}$ then $\mathfrak{B}$ is isomorphic to some structure in K.

K has the *Joint Embedding Property,* JEP *for short,* if whenever $\mathfrak{A}, \mathfrak{B} \in \mathsf{K}$ then there is a $\mathfrak{C} \in \mathsf{K}$ such that both $\mathfrak{A}$ and $\mathfrak{B}$ are embeddable in $\mathfrak{C}$.

K has *Amalgamation Property, or* AP *for short,* if $\mathfrak{A}, \mathfrak{B}, \mathfrak{C} \in \mathsf{K}$ and $e : \mathfrak{A} \to \mathfrak{B}$, $f : \mathfrak{A} \to \mathfrak{C}$ are embeddings, then there are $\mathfrak{D}$ in K and embeddings $g : \mathfrak{B} \to \mathfrak{D}$ and $h : \mathfrak{C} \to \mathfrak{D}$ such that $g \circ e = h \circ f$.

(iv) A structure $\mathfrak{D}$ is weakly homogeneous if it has the the following property if $\mathfrak{A}$, $\mathfrak{B}$ are finitely generated substructures of $\mathfrak{D}$, $A \subseteq B$ and $f : \mathfrak{A} \to \mathfrak{D}$ is an embedding, then there is an embedding $g : \mathfrak{B} \to \mathfrak{D}$ which extends f.

(v) We call a structure $\mathfrak{D}$ homogeneous if every isomorphism between finitely generated substructures extends to an automorphism of $\mathfrak{D}$.

Note that if $\mathfrak{D}$ is homogeneous, then it is weakly homogeneous. We recall Theorem 7.1.2 from [Hod,93a], a theorem of Fraisse that puts the above pieces together.

Theorem 5.1.2. *Let L be a countable signature and let K be a non-empty finite or countable set of finitely generated L-structures which has* HP, JEP *and* AP. *Then there is an L structure $\mathfrak{D}$, unique up to isomorphism, such that*

(i) *$\mathfrak{D}$ has cardinality $\leq \omega$*

(ii) K *is the age of* D, *and*

(iii) $\mathfrak{D}$ *is homogeneous.*

Using Theorem 5.1.2, we shall construct an algebra $\mathfrak{A} \in \mathsf{Nr}_3\,\mathsf{CA}_\beta$ that has an elementary equivalent algebra $\mathfrak{B} \notin \mathsf{Nr}_3\,\mathsf{CA}_\beta$. The proof for the finite dimensional case is the same. For infinite dimensions we refer to [Say,01].

S_3 denotes the set of all permutations of 3. ${}^XY$ denotes the set of functions from X to Y. For $u, v \in {}^33$, $i < 3$ we write u_i for $u(i) < 3$, and we write $u \equiv_i v$ if u and v agree off i, i.e. if $u_j = v_j$ for all $j \in 3 \smallsetminus \{i\}$. For a symbol R of the signature of M we write R^{M} for the interpretation of R in M.

Our algebras will be based on the model proven to exist in the next lemma.

Lemma 5.1.3. *Let L be a signature consisting of the unary relation symbols P_0, P_1, P_2 and uncountably many 3-ary predicate symbols. For $u \in {}^33$, let χ_u be the formula $\bigwedge_{i<3} P_{u_i}(x_i)$. Then there exists an L-structure M with the following properties:*

(i) M *has quantifier elimination, i.e. every L-formula is equivalent in* M *to a boolean combination of atomic formulas.*

(ii) *The sets P_i^{M} for $i < 3$ partition* M,

(iii) $\mathsf{M} \models \forall x_0x_1x_2\big(R(x_0, x_1x_2) \to \bigvee_{u\in S_3} \chi_u\big)$, *for all $R \in L$,*

(iv) $\mathsf{M} \models \exists x_0x_1x_2\big(\chi_u \wedge R(x_0, x_1, x_2) \wedge \neg S(x_0, x_1, x_2)\big)$ *for all distinct ternary $R, S \in L$, and $u \in S_3$,*

(v) *For $u \in S_3$, $i < 3$,* $\mathsf{M} \models \forall x_0x_1x_2\big(\exists x_i\chi_u \leftrightarrow \bigvee_{v\in{}^33, v\equiv_i u} \chi_v\big)$,

(vi) *For $u \in S_3$ and any L-formula $\phi(x_0, x_1, x_2)$, if* $\mathsf{M} \models \exists x_0x_1x_2(\chi_u \wedge \phi)$ *then* $\mathsf{M} \models \forall x_0x_1x_2\big(\exists x_i\chi_u \leftrightarrow \exists x_i(\chi_u \wedge \phi)\big)$ *for all $i < 3$.*

Sketch of Proof. The proof of this lemma is model theoretic. Let $\mathcal{L}$ be the relational signature containing unary relation symbols $P_0, \ldots, P_3$ and a 4-ary relation symbol X. Let K be the class of all finite $\mathcal{L}$-structures $\mathfrak{D}$ satisfying

(5.1.1) The P_i's are disjoint : $\forall x \bigvee_{i<j<4} \Big(P_i(x) \wedge \bigwedge_{j\neq i} \neg P_j(x)\Big)$.

$$(5.1.2) \qquad \forall x_0 \cdots x_3 \Big(X(x_0, \cdots, x_3) \longrightarrow P_3(x_3) \wedge \bigvee_{u \in \mathsf{S}_3} \chi_u \Big).$$

Then K contains countably many isomorphism types, because for each $n \in \omega$, there are countably many isomorphism types of finite L structures (satisfying (5.1.1) and (5.1.2)) having cardinality $\leq n$. Also it is easy to check that K is closed under substructures and that K has the AP. From the latter it follows that it has the JEP, since K contains the one element structure that is embeddable in any structure in K.[2] By Theorem 5.1.2 there is a countably infinite homogeneous $\mathcal{L}$-structure $\mathfrak{N}$ with age K. $\mathfrak{N}$ has quantifier elimination, and obviously, so does any elementary extension of $\mathfrak{N}$. K contains structures with arbitrarily large P_3-part, so $P_3^{\mathfrak{N}}$ is infinite. Let $\mathfrak{N}^*$ be an elementary extension of $\mathfrak{N}$ such that $\left|P_3^{\mathfrak{N}^*}\right| = |L|$, and fix a bijection $*$ from the set of ternary relation symbols of L to $P_3^{\mathfrak{N}^*}$. Define an L-structure M with domain $P_0^{\mathfrak{N}^*} \cup P_1^{\mathfrak{N}^*} \cup P_2^{\mathfrak{N}^*}$, by: $P_i^{\mathsf{M}} = P_i^{\mathfrak{N}^*}$ for $i < 3$ and for ternary $R \in L$,

$$\mathsf{M} \models R(a_0, a_1, a_2) \quad \text{iff} \quad \mathfrak{N}^* \models X(a_0, a_1, a_2, R^*).$$

If $\phi(\bar{x})$ is any L-formula, let $\phi^*(\bar{x}, \bar{R})$ be the $\mathcal{L}$-formula with parameters $\bar{R}$ from $\mathfrak{N}^*$ obtained from ϕ by replacing each atomic subformula $R(x, y, z)$ by $X(x, y, z, R^*)$ and relativizing quantifiers to $\neg P_3$, that is replacing $(\exists x)\phi(x)$ and $(\forall x)\phi(x)$ by $(\exists x)\big(\neg P_3(x) \to \phi(x)\big)$ and $(\forall x)\big(\neg P_3(x) \to \phi(x)\big)$, respectively. A straightforward induction on complexity of formulas gives that for $\bar{a} \in \mathsf{M}$

$$\mathsf{M} \models \phi(\bar{a}) \quad \text{iff} \quad \mathfrak{N}^* \models \phi^*(\bar{a}, \bar{R}).$$

Then M is as required. ■

Now we are going to prove that the class $\mathsf{Nr}_\alpha \mathsf{CA}_\beta$ is not elementary for $3 \leq \alpha < \beta$. We prove the result for $\alpha = 3$. The proof for higher finite dimensions is the same. For the infinite dimensional case, we refer to [Say,01].

Theorem 5.1.4. *For $\beta > 3$, the class $\mathsf{Nr}_3 \mathsf{CA}_\beta$ is not elementary.*

Proof. Fix L and M as in Lemma 5.1.3. Let $A_\omega = \{\phi^{\mathsf{M}} : \phi \in L\}$ and $A = \{\phi^{\mathsf{M}} : \phi \in L_3\}$ with operations defined as for set algebras. Then $\mathfrak{A} \cong \mathfrak{Nr}_3 \mathfrak{A}_\omega$, the isomorphism is given by

$$\phi^{\mathsf{M}} \mapsto \phi^{\mathsf{M}}.$$

[2]It is not always true that AP implies JEP; think of fields.

Quantifier elimination in M guarantees that this map is onto. For $u \in {}^3 3$, let $\mathfrak{A}_u$ denote the relativization of $\mathfrak{A}$ to χ_u^{M} i.e.

$$\mathfrak{A}_u = \{x \in A : x \le \chi_u^{\mathsf{M}}\}.$$

$\mathfrak{A}_u$ is a Boolean algebra. Also $\mathfrak{A}_u$ is uncountable for every $u \in S_3$ because by property (iv) of Lemma 5.1.3 the sets $\left(\chi_u \wedge R(x_0, x_1, x_2)\right)^{\mathsf{M}}$, for $R \in L$ are distinct elements of A_u. Define a map $f : \mathfrak{A} \to \prod_{u \in {}^3 3}(\mathfrak{A}_u)$, by

$$f(a) = \langle a \cdot \chi_u \rangle_{u \in {}^3 3}.$$

We will expand the language of the Boolean algebra $\prod_{u \in {}^3 3} \mathfrak{A}_u$ in such a way that the cylindric algebra $\mathfrak{A}$ becomes interpretable in the expanded structure. For this we need the following definition:

Let $\mathfrak{P}$ denote the following structure for the signature of Boolean algebras expanded by constant symbols 1_u for $u \in {}^3 3$ and d_{ij} for $i, j \in 3$:

(1) The boolean part of $\mathfrak{P}$ is the Boolean algebra $\prod_{u \in {}^3 3} \mathfrak{A}_u$,

(2) $1_u^{\mathfrak{P}} = f(\chi_u^{\mathsf{M}}) = \langle 0, \cdots 0, 1, 0, \cdots \rangle$ (with the 1 in the uth place) for each $u \in {}^3 3$,

(3) $\mathsf{d}_{ij}^{\mathfrak{P}} = f(\mathsf{d}_{ij}^{\mathfrak{A}})$ for $i, j < 3$.

We now show that $\mathfrak{A}$ is interpretable in $\mathfrak{P}$ [Hod,93a]. For this it is enough to show that f is one to one and that Rn f and the f-images of the graphs of the cylindric algebra functions in $\mathfrak{A}$ are definable in $\mathfrak{P}$. Since the χ_u^{M} partition the unit of $\mathfrak{A}$, each $a \in A$ has a unique expression in the form $\sum_{u \in {}^3 3}(a \cdot \chi_u^{\mathsf{M}})$, and it follows that f is boolean isomorphism: bool $(\mathfrak{A}) \to \prod_{u \in {}^3 3} \mathfrak{A}_u$. So the f-images of the graphs of the boolean functions on $\mathfrak{A}$ are trivially definable. f is bijective so Rng (f) is definable, by $x = x$. For the diagonals, $f(\mathsf{d}_{ij}^{\mathfrak{A}})$ is definable by $x = \mathsf{d}_{ij}$. Finally we consider cylindrifications. For $S \subseteq {}^3 3$, $i < 3$, let t_S be the closed term

$$\sum \{1_v : v \in {}^3 3,\ v \equiv_i u \text{ for some } u \in S\}.$$

Let

$$\eta_i(x, y) = \bigwedge_{S \subseteq {}^3 3} \Big(\bigwedge_{u \in S} x.1_u \neq 0 \wedge \bigwedge_{u \in {}^3 3 \smallsetminus S} x.1_u = 0 \to y = t_S \Big).$$

We claim that for all $a \in A$, $b \in P$, we have

$$\mathfrak{P} \models \eta_i\big(f(a), b\big) \quad \text{iff} \quad b = f\big(\mathsf{c}_i^{\mathfrak{A}} a\big).$$

To see this, let $f(a) = \langle a_u \rangle_{u \in {}^33}$, say. So in $\mathfrak{A}$ we have $a = \sum_u a_u$. Let u be given; a_u has the form $(\chi_i \wedge \phi)^{\mathsf{M}}$ for some $\phi \in L^3$, so $\mathsf{c}_i^A(a_u) = \big(\exists x_i(\chi_u \wedge \phi)\big)^{\mathsf{M}}$. By property (vi) of Lemma 5.1.3, if $a_u \neq 0$, this is $(\exists x_i \chi_u)^{\mathsf{M}}$; by property 5, this is $\big(\bigvee_{v \in {}^33, v \equiv_i u} \chi_v\big)^{\mathsf{M}}$. Let $S = \big\{u \in {}^33 : a_u \neq 0\big\}$. By normality and additivity of cylindrifications we have,

$$\mathsf{c}_i^A(a) = \sum_{u \in {}^33} \mathsf{c}_i^A a_u = \sum_{u \in S} \mathsf{c}_i^A a_u = \sum_{u \in S} \Big(\sum_{v \in {}^33, v \equiv_i u} \chi_v^{\mathsf{M}} \Big)$$

$$= \sum \big\{\chi_v^{\mathsf{M}} : v \in {}^33, v \equiv_i u \text{ for some } u \in S\big\}.$$

So $\mathfrak{P} \models f(\mathsf{c}_i^{\mathfrak{A}} a) = t_S$. Hence $\mathfrak{P} \models \eta_i\big(f(a), f(\mathsf{c}_i^{\mathfrak{A}} a)\big)$. Conversely, if $\mathfrak{P} \models \eta_i\big(f(a), b\big)$, we require $b = f(\mathsf{c}_i a)$. Now S is the unique subset of ^{33} such that

$$\mathfrak{P} \models \bigwedge_{u \in S} f(a) \cdot 1_u \neq 0 \wedge \bigwedge_{u \in {}^33 \smallsetminus S} f(a) \cdot 1_u = 0.$$

So we obtain

$$b = t_S = f\big(\mathsf{c}_i^A a\big).$$

We have proved that $\mathfrak{A}$ is interpretable in $\mathfrak{P}$. Furthermore it is easy to see that the interpretation is one dimensional and quantifier free. Next we extract an algebra $\mathfrak{B}$ elementary equivalent to $\mathfrak{A}$ that is not a neat reduct i.e. not in $\mathsf{Nr}_3\,\mathsf{CA}_4$. Let $Id \in {}^33$ be the identity map on 3. Choose any countable boolean elementary subalgebra of $\mathfrak{A}_{Id}$, $\mathfrak{B}_{Id}$ say. Thus $\mathfrak{B}_{Id} \preceq \mathfrak{A}_{Id}$. Then

$$Q = \Big(\Big(B_{Id} \times \prod_{u \in {}^33 \smallsetminus Id} \mathfrak{A}_u\Big), 1_u, \mathsf{d}_{ij}\Big)_{u \in {}^33,\ i,j<3}$$

$$\equiv \Big(\Big(\prod_{u \in {}^33} \mathfrak{A}_u\Big), 1_u, \mathsf{d}_{ij}\Big)_{u \in {}^33,\ i,j<3} = P.$$

Let $\mathfrak{B}$ be the result of applying the interpretation given above to Q. Then $\mathfrak{B} \equiv \mathfrak{A}$ as cylindric algebras. Now we show that $\mathfrak{B}$ cannot be a neat reduct, in fact we show that $\mathfrak{B} \notin \mathsf{Nr}_3\,\mathsf{CA}_\beta$ for any $\beta > 3$. Assume for contradiction that $\mathfrak{B} = \mathfrak{Nr}_3\,\mathfrak{D}$ for some $\mathfrak{D} \in \mathsf{CA}_\beta$; with $\beta > 3$. Note that $\mathfrak{D}$ may not be representable. It is only here that we deal with possibly non-representable algebras. Now $\chi_u^{\mathsf{M}} \in B$ for each $u \in {}^33$. Identifying functions with sequences we let $v = \langle 1, 0, 2\rangle \in {}^33$. Let $t(x)$ be the CA_2 term $\mathsf{s}_1^0\mathsf{c}_1 x \cdot \mathsf{s}_0^1\mathsf{c}_0 x$, where $\mathsf{s}_i^j(x) = \mathsf{c}_i(\mathsf{d}_{ij}\cdot x)$, for $i \neq j$. Then we claim that $t^B(\chi_v^{\mathsf{M}}) = \chi_{Id}^{\mathsf{M}}$. For the sake of brevity, denote χ_v^{M} by 1_{10} and χ_{Id}^{M} by 1_{01}. Then, by definition, we have

$$t^B(1_{01}) = \mathsf{c}_0(\mathsf{d}_{01}\cdot \mathsf{c}_1 1_{10})\cdot \mathsf{c}_1(\mathsf{d}_{01}\cdot \mathsf{c}_0 1_{10}).$$

Computing we get

$$\begin{aligned}\mathsf{c}_0(\mathsf{d}_{01}\cdot \mathsf{c}_1 1_{10}) &= \mathsf{c}_0\left(\mathsf{d}_{01}\cdot\left(\sum\{1_u : u \equiv_1 1_{10}\}\right)\right)\\ &= \mathsf{c}_0(\mathsf{d}_{01}\cdot 1_{112}) = 1_{01} + 1_{112}.\end{aligned}$$

Here 1_{112} denotes $\chi_{\langle 1,1,2\rangle}$. Note that we are using that the evaluation of the term $\mathsf{c}_1 1_{10}$ in $\mathfrak{B}$ is equal to its value in $\mathfrak{A}$. This is so, because $\mathfrak{B}$ inherits the interpretation given to $\prod A_u$. A similar computation gives

$$\mathsf{c}_1(\mathsf{d}_{01}\cdot \mathsf{c}_0 1_{01}) = 1_{002} + 1_{01},$$

where 1_{002} denotes $\chi_{\langle 0,0,2\rangle}$. Therefore as claimed

$$t^B(1_{10}) = 1_{01}.$$

Now let ${}_3\mathsf{s}(0,1)$ be the unary substitution term as defined in [Hen-Mon-Tar,85] 1.5.12, that is

$${}_3\mathsf{s}(0,1)x = \mathsf{s}_0^3\mathsf{s}_1^0\mathsf{s}_3^1(x).$$

Then for any $\beta > 3$ we have

$$\mathsf{CA}_\beta \models {}_3\mathsf{s}(0,1)\mathsf{c}_3 x \leq t(\mathsf{c}_3 x).$$

Indeed by [Hen-Mon-Tar,85] 1.5.12, 1.5.8 and 1.5.10 (ii), we get

$$\begin{aligned}{}_3\mathsf{s}(0,1)\mathsf{c}_3 x \leq {}_3\mathsf{s}(0,1)\mathsf{c}_1\mathsf{c}_3 x &= \mathsf{s}_0^3\mathsf{s}_1^0\mathsf{s}_3^1\mathsf{c}_1\mathsf{c}_3 x = \mathsf{s}_0^3\mathsf{s}_1^0\mathsf{c}_1\mathsf{c}_3 x\\ &= \mathsf{s}_0^3\mathsf{s}_1^0\mathsf{c}_3\mathsf{c}_1 x = \mathsf{s}_0^3\mathsf{c}_3\mathsf{s}_1^0\mathsf{c}_1 x = \mathsf{c}_3\mathsf{s}_1^0\mathsf{c}_1 x = \mathsf{s}_0^1\mathsf{c}_1\mathsf{c}_3 x.\end{aligned}$$

Similarly

$${}_3\mathsf{s}(0,1)\mathsf{c}_3 x \leq \mathsf{s}_1^0\mathsf{c}_0\mathsf{c}_3 x.$$

Therefore

$$_3\mathsf{s}(0,1)\mathsf{c}_3x \leq t(\mathsf{c}_3x).$$

It thus follows that

$$\mathfrak{D} \models {_3\mathsf{s}(0,1)}\big(\chi_u^{\mathsf{M}}\big) \leq \mathsf{s}_1^0\mathsf{c}_1\big(\chi_u^{\mathsf{M}}\big) \cdot \mathsf{s}_0^1\mathsf{c}_0\big(\chi_u^{\mathsf{M}}\big) = \chi_{Id}^{\mathsf{M}}.$$

Now $_3\mathsf{s}(0,1)$ preserves $\leq$ and is one to one $\mathfrak{Nr}_3\,\mathfrak{D}$. By [Hen-Mon-Tar,85], 1.5.12 and 1.5.1, we have:

$$_3\mathsf{s}(0,1)\mathsf{c}_3x = \mathsf{s}_0^n\mathsf{s}_1^0\mathsf{s}_3^1\mathsf{c}_3x = \mathsf{c}_3(\mathsf{d}_{30} \cdot \mathsf{c}_0\big(\mathsf{d}_{01} \cdot \mathsf{c}_1\big(\mathsf{d}_{01} \cdot \mathsf{c}_1(\mathsf{d}_{13} \cdot \mathsf{c}_3x)\big)\big).$$

By [Hen-Mon-Tar,85], 1.3.8, $0 < x$, implies $0 < \mathsf{d}_{ij} \cdot \mathsf{c}_jx$, for all $i, j \in \beta$. We have shown that if $x > 0 \in \mathsf{Nr}_3\, D$, then $_3\mathsf{s}(0,1)x > 0$, i.e. that $_3\mathsf{s}(0,1)$, being a boolean endomorphism, is one to one. Since $B_v = A_v$ it follows (by condition (iv) in Lemma 5.1.3) that $B_v = \big\{b \in B : b \leq \chi_v^{\mathsf{M}}\big\}$ is uncountable. Since $_3\mathsf{s}(0,1)$ is one to one, it follows that $_3\mathsf{s}(0,1)B_u$ is also uncountable. But by the above we have

$$_3\mathsf{s}(0,1)B_u \subseteq B_{Id} = \big\{b \in B : b \leq \chi_{Id}^{B}\big\},$$

and so B_{Id} is also uncountable. But by construction, we have $B_{Id} = \big\{b \in B : b \leq \chi_{Id}^{\mathsf{M}}\big\}$ is countable. This contradiction shows that $\mathfrak{B} \notin \mathsf{Nr}_3\,\mathsf{CA}_\beta$ for any $\beta > 3$. ■

We formulate a (new) theorem that further indicates that the class of neat reducts is really hard to characterize. But first some set-theoretic preparations. Let M denote the universe of sets and let $\mathfrak{C} \in \mathsf{M}$ be a complete Boolean algebra. (Note that $\mathfrak{C}$ is a Boolean algebra "from the outside as well" but not necessarily complete.) Form the Boolean valued extension $\mathsf{M}^{\mathfrak{C}}$ of M and let $\|\phi\|$ be the boolean value of a sentence ϕ of set theory containing parameters from $\mathsf{M}^{\mathfrak{C}}$. ϕ is valid in M if $\|\phi\| = 1$ in symbols $\mathsf{M}^{\mathfrak{C}} \models \phi$. Write $\mathfrak{C} : \mathfrak{A} \cong \mathfrak{B}$ if $\mathsf{M}^{\mathfrak{C}} \models \bar{\mathfrak{A}} \cong \bar{\mathfrak{B}}$. Here $\check{\mathsf{s}}$ is the canonical name of s in M. We say that $\mathfrak{A}$ and $\mathfrak{B}$ are Boolean isomorphic if there is such $\mathfrak{C}$, and (in which case) we write $\mathfrak{A} \cong_b \mathfrak{B}$. It turns out that boolean isomorphism lies somewhere between $\equiv$ (elementary equivalence) and $\cong$ (isomorphism). Such an equivalence relation, as it turns out, is purely structural and can characterized by games. (The idea is to look at isomorphisms between a finite number of elements at a time. In Model Theory this is expressed by back- and-forth systems.) Of course if $\mathfrak{A} \cong \mathfrak{B}$ then trivially $\mathfrak{A} \cong_b \mathfrak{B}$. Call a class of algebras K *boolean closed* if whenever $\mathfrak{A} \in \mathsf{K}$ and $\mathfrak{B} \cong_b \mathfrak{A}$ in some Boolean valued extension of the universe of sets, then $\mathfrak{B} \in \mathsf{K}$. We now have the following theorem proved in [Say,10c].

Theorem 5.1.5. *Let $1 < \alpha < \beta$. Then the following hold:*

(i) *The class* $\mathsf{Nr}_\alpha \mathsf{CA}_\beta$ *is not boolean closed*

(ii) *The classes* $\mathsf{Nr}_\alpha \mathsf{CA}_\beta$ *regarded as concrete categories are not finitely complete (that is closed under finite limits).*

(iii) *Let* L *denote the first order language of* CA_α*. There is no sentence* $\sigma \in L_{\infty\omega}$ *that characterizes* $\mathsf{Nr}_\alpha \mathsf{CA}_\beta$.

Here $L_{\infty\omega}$ is the logic obtained from first order logic by allowing infinite conjunctions without any restrictions on cardinality. Examples of finite limits are products and equalizers. In [Say,10c] it is shown that there are $\mathfrak{A}, \mathfrak{B} \in \mathsf{Nr}_\alpha \mathsf{CA}_\beta$ and morphisms f, g from $\mathfrak{A}$ to $\mathfrak{B}$ such that $\{x \in A : f(x) = g(x)\}$ is not the universe of an algebra in $\mathsf{Nr}_\alpha \mathsf{CA}_\beta$.

The closure of the class of neat reducts for other algebras under forming (elementary) subalgebras is investigated in [Say,02d], [Say,06c], [Say,07a].

2. Amalgamation and Neat Reducts

Let K *be a class of algebras having a boolean reduct.* $\mathfrak{A}_0 \in \mathsf{K}$ *is in the amalgamation base of* K *if for all* $\mathfrak{A}_1, \mathfrak{A}_2 \in \mathsf{K}$ *and monomorphisms* $i_1 : \mathfrak{A}_0 \to \mathfrak{A}_1$, $i_2 : \mathfrak{A}_0 \to \mathfrak{A}_2$ *there exist* $\mathfrak{D} \in \mathsf{K}$ *and monomorphisms* $m_1 : \mathfrak{A}_1 \to \mathfrak{D}$ *and* $m_2 : \mathfrak{A}_2 \to \mathfrak{D}$ *such that* $m_1 \circ i_1 = m_2 \circ i_2$. *If in addition,* $(\forall x \in A_j)(\forall y \in A_k)\big(m_j(x) \leq m_k(y) \implies (\exists z \in A_0)(x \leq i_j(z) \wedge i_k(z) \leq y)\big)$ *where* $\{j, k\} = \{1, 2\}$, *then we say that* $\mathfrak{A}_0$ *lies in the super amalgamation base of* K. *Here* $\leq$ *is the boolean order.* K *has the (super) amalgamation property* $\big((\text{SUP})AP\big)$, *if the (super) amalgamation base of* K *coincides with* K.

The amalgamation property (for classes of models), since its discovery, has played a dominant role in algebra and model theory [Hod,93a]. Algebraic logic is the natural interface between universal algebra and logic (in our present context a variant of first order logic). Indeed, in algebraic logic amalgamation properties in classes of algebras are proved to be equivalent to interpolation results in the corresponding logic. Pigozzi [Pig,71], is a milestone for working out such equivalences for cylindric algebras, see also [Mad-Say,07], [Mad-Say,thisVol]. The super amalgamation property was introduced by Maksimova [Mak,91] (for expansions of Boolean algebras) and it is studied extensively by Madarász in e.g. [Mad,98] and more recently (for cylindric algebras) by Sági and Shelah [Sag-She,06] and by the present author [Say,10a]. The super amalgamation property for a class of algebras corresponds to a strong form of interpolation in the corresponding logic [Say,04a], [Mad-Say,07], [Mad,98].

It is usually not an easy matter to characterize the amalgamation base (or for that matter the super amalgamation base) of K when K does not have the AP(SUPAP). An example is the case when $\mathsf{K} = \mathsf{RCA}_\alpha$ with $\alpha > 1$ [Com,69], [Pig,71]. We set out to determine both the amalgamation base and super amalgamation base of RCA_α for any ordinal α answering a question in the problem session paper in [And-Mon-Nem,91b]. (This question, formulated as Problem 45 in [And-Mon-Nem,91b] addresses only the amalgamation base case).

We start by giving a natural sufficient condition for an algebra $\mathfrak{A}$ to belong to the (super) amalgamation base of RCA_α. The conditions are formulated in terms of *neat embeddings.* This is indeed expected since we have a NET. For a cylindric algebra $\mathfrak{A}$ and $X \subseteq A$, $\mathfrak{Sg}^{\mathfrak{A}} X$ denotes the subalgebra of $\mathfrak{A}$ generated by X. $\mathfrak{Ig}^{\mathfrak{A}} X$ is the ideal generated by X.

Definition 5.2.1. Let $\mathfrak{A} \in \mathsf{RCA}_\alpha$. Then $\mathfrak{A}$ has the *UNEP* (short for unique neat embedding property) if for all $\mathfrak{A}' \in \mathsf{CA}_\alpha$, $\mathfrak{B}, \mathfrak{B}' \in \mathsf{CA}_{\alpha+\omega}$, isomorphism $i : \mathfrak{A} \to \mathfrak{A}'$, embeddings $e_A : \mathfrak{A} \to \mathfrak{Nr}_\alpha \mathfrak{B}$ and $e_{A'} : \mathfrak{A}' \to \mathfrak{Nr}_\alpha \mathfrak{B}'$ such that $\mathfrak{Sg}^{\mathfrak{B}} e_A(A) = \mathfrak{B}$ and $\mathfrak{Sg}^{\mathfrak{B}'} e_{A'}(A)' = \mathfrak{B}'$, there exists an isomorphism $\bar{i} : \mathfrak{B} \to \mathfrak{B}'$ such that $\bar{i} \circ e_A = e_{A'} \circ i$.

Definition 5.2.2. Let $\mathfrak{A} \in \mathsf{RCA}_\alpha$. Then $\mathfrak{A}$ has the *NS* property (short for neat reducts commuting with forming subalgebras) if for all $\mathfrak{B} \in \mathsf{CA}_{\alpha+\omega}$ if $\mathfrak{A} \subseteq \mathfrak{Nr}_\alpha \mathfrak{B}$ then for all $X \subseteq A$, $\mathfrak{Sg}^{\mathfrak{A}} X = \mathfrak{Nr}_\alpha \mathfrak{Sg}^{\mathfrak{B}} X$.

Let us examine closely the above conditions. At first glance Definition 5.2.1 might seem complicated, but in fact it is a slight generalization of a very simple and indeed "natural" property. Let $\mathfrak{A} \in \mathsf{CA}_\alpha$. Let $\mathfrak{A} \subseteq \mathfrak{Nr}_\alpha \mathfrak{B}$ with $\mathfrak{B} \in \mathsf{CA}_{\alpha+\omega}$. Call $\mathfrak{B}$ an ω dilation of $\mathfrak{A}$. If further A generates $\mathfrak{B}$ (using the $\alpha + \omega$ operations of $\mathfrak{B}$) call $\mathfrak{B}$ a *minimal* ω dilation of $\mathfrak{A}$. In this case, one might expect that $\mathfrak{A}$ has some control of $\mathfrak{B}$. In fact, Definition 5.2.1 implies that any two minimal ω dilations of $\mathfrak{A}$ are in fact *isomorphic.* Furthermore this isomorphism can be chosen to fix A. This follows from the special case when $\mathfrak{A} = \mathfrak{A}'$, and i and $e_A = e_{A'}$ are the inclusion maps. So, roughly, Definition 5.2.1 says that $\mathfrak{A}$ determines essentially the structure of its minimal ω dilations.

Now for Definition 5.2.2. Again let $\mathfrak{B}$ be an ω dilation of $\mathfrak{A} \in \mathsf{CA}_\alpha$ so that $\mathfrak{A} \subseteq \mathfrak{Nr}_\alpha \mathfrak{B}$, with $\mathfrak{B} \in \mathsf{CA}_{\alpha+\omega}$. Let $X \subseteq A$. Form the subalgebra of $\mathfrak{A}$ generated by X and form the subalgebra of $\mathfrak{B}$ generated by X. Then, in principal, in the second process of generation, new α dimensional elements can be generated. Definition 5.2.2 excludes this possibility. It says that if we take the set of α dimensional elements of $\mathfrak{Sg}^{\mathfrak{B}} X$ (i.e. we form $\mathfrak{Nr}_\alpha \mathfrak{Sg}^{\mathfrak{B}} X$), then we come back exactly to where we started namely to $\mathfrak{Sg}^{\mathfrak{A}} X$ (and not to a bigger algebra). No new α dimensional elements are generated (even in the presence of ω extra dimensions). Note that in this case, we have

$$\mathfrak{Sg}^{\mathfrak{A}} X = \mathfrak{Sg}^{\mathfrak{Nr}_\alpha \mathfrak{B}} X = \mathfrak{Nr}_\alpha \mathfrak{Sg}^{\mathfrak{B}} X.$$

Here two operations commute. Forming the subalgebra of the neat reduct is the same as taking the neat reduct of the subalgebra.

Let APbase(K) be the class of algebras that lie in the amalgamation base of K and SUPAPbase(K) the class of algebras that lie in the super amalgam base of K. Then the following is known

$$\mathsf{RCA}_\alpha = \text{APbase}(\mathsf{RCA}_\alpha) = \text{SUPAPbase}(\mathsf{RCA}_\alpha)$$

if and only if $\alpha \leq 1$, cf. [Com,69], [Pig,71], [Mad-Say,07]. Now we have

Theorem 5.2.3. *Let α be an ordinal. Let $\mathfrak{A} \in \mathsf{RCA}_\alpha$.*

(i) *If $\mathfrak{A}$ has UNEP, then $\mathfrak{A} \in$ APbase(RCA_α).*

(ii) *If $\mathfrak{A}$ has UNEP and NS, then $\mathfrak{A} \in$ SUPAPbase(RCA_α).*

We omit the proof that can be found in [Say,10c]. Let $\mathsf{Dc}_\alpha = \{\mathfrak{A} \in \mathsf{CA}_\alpha : \Delta x \neq \alpha, \text{ for all } x \in \mathfrak{A}\}$. These algebras are referred to as dimension complemented cylindric algebras. It is not hard to show for $\alpha \geq \omega$, Dc_α has *NS* and *UNEP*. Hence $\mathsf{Dc}_\alpha \subseteq$ SUPAPbase(RCA_α). Mn_α denotes the class of minimal cylindric algebras of dimension α. Since $\mathsf{Mn}_\alpha \subseteq \mathsf{Dc}_\alpha$ the latter class is also contained in the SUPAPbase of RCA_α. However, for $1 < n < \omega$, $\mathsf{Mn}_n \nsubseteq$ APbase(RCA_n) a result of Comer [Com,69]. Expressed differently RCA_n does not have the embedding property. Another comprehensive class of algebras that is contained in SUPAPbase(CA_α) for $\alpha > 1$ is the class of cylindric algebras of positive characteristic [Say,d]. It seems likely that the class of algebras having the unique neat embedding property coincides with APbase(RCA_α), and that for infinite α, SUPAPbase$(\mathsf{RCA}_\alpha) = \mathsf{Dc}_\alpha$ but further research is needed. The following theorem will be used to confirm some conjectures of Tarski on neat reducts. First we need:

For a cardinal $\beta > 0$, $L \subseteq \mathsf{CA}_\alpha$ *and* $\rho : \beta \to \wp(\alpha)$, $\mathfrak{Fr}^\rho_\beta L$ *stands for the dimension restricted* L *free algebra on* β *generators* [Hen-Mon-Tar,85] 2.5.31. *The sequence* $\langle \eta/Cr^\rho_\beta L : \eta < \beta \rangle$ *L-freely generates* $\mathfrak{Fr}^\rho_\beta L$, *cf.* [Hen-Mon-Tar,85] *Theorem* 2.5.35.

Theorem 5.2.4. *If* $\alpha < \beta$ *are any ordinals and* $L \subseteq \mathsf{CA}_\beta$, *then, in the sequence of conditions* (i)–(v) *below,* (i)–(iv) *implies the immediately following one:*

(i) *For any* $\mathfrak{A} \in L$ *and* $\mathfrak{B} \in \mathsf{CA}_\beta$ *with* $\mathfrak{A} \subseteq \mathfrak{Nr}_\alpha \mathfrak{B}$, *for all* $X \subseteq A$ *we have* $\mathfrak{Sg}^{\mathfrak{A}} X = \mathfrak{Nr}_\alpha \mathfrak{Sg}^{\mathfrak{B}} X$.

(ii) *For any* $\mathfrak{A} \in L$ *and* $\mathfrak{B} \in \mathsf{CA}_\beta$ *with* $\mathfrak{A} \subseteq \mathfrak{Nr}_\alpha \mathfrak{B}$, *if* $\mathfrak{Sg}^{\mathfrak{B}} A = \mathfrak{B}$, *then* $\mathfrak{A} = \mathfrak{Nr}_\alpha \mathfrak{B}$.

(iii) *For any* $\mathfrak{A} \in L$ *and* $\mathfrak{B} \in \mathsf{CA}_\beta$ *with* $\mathfrak{A} \subseteq \mathfrak{Nr}_\alpha \mathfrak{B}$, *if* $\mathfrak{Sg}^{\mathfrak{B}} A = \mathfrak{B}$, *then for any ideal* I *of* $\mathfrak{B}$, $\mathfrak{Ig}^{\mathfrak{B}}(A \cap I) = I$.

(iv) *If whenever* $\mathfrak{A} \in L$, *there exists* $x \in {}^{|A|}A$ *such that if* $\rho = \langle \Delta x_i : i < |A| \rangle$, $\mathfrak{D} = \mathfrak{Fr}^\rho_{|A|} \mathsf{CA}_\beta$ *and* $g_\xi = \xi/Cr^\rho_{|A|} \mathsf{CA}_\beta$, *then* $\mathfrak{Sg}^{\mathfrak{Rd}_\alpha \mathfrak{D}} \{ g_\xi : \xi < |A| \} \in L$, *then the following UNEP hold: For* $\mathfrak{A}, \mathfrak{A}' \in L$, $\mathfrak{B}, \mathfrak{B}' \in \mathsf{CA}_\beta$ *with embeddings* $e_A : \mathfrak{A} \to \mathfrak{Nr}_\alpha \mathfrak{B}$ *and* $e_{A'} : \mathfrak{A}' \to \mathfrak{Nr}_\alpha \mathfrak{B}'$ *such that* $\mathfrak{Sg}^{\mathfrak{B}} e_A(A) = \mathfrak{B}$ *and* $\mathfrak{Sg}^{\mathfrak{B}'} e_{A'}(A) = \mathfrak{B}'$, *whenever* $i : \mathfrak{A} \to \mathfrak{A}'$ *is an isomorphism, then there exists an isomorphism* $\bar{i} : \mathfrak{B} \to \mathfrak{B}'$*such that* $\bar{i} \circ e_A = e_{A'} \circ i$.

(v) *Assume that* $\beta = \alpha + \omega$. *Then* $L \subseteq$ APbase(RCA_α).

Proof. (i) implies (ii) is trivial. Now we prove (ii) implies (iii). The proof is similar to [Hen-Mon-Tar,85] 2.6.71. From the premise that $\mathfrak{A}$ is a generating subreduct of $\mathfrak{B}$ we easily infer that $|\Delta x \setminus \alpha| < \omega$ for all $x \in B$. We now have $\mathfrak{A} = \mathfrak{Nr}_\alpha \mathfrak{B}$. Now clearly $\mathfrak{Ig}^{\mathfrak{B}}(I \cap A) \subseteq I$. Conversely let $x \in I$. Then $\mathsf{c}_{(\Delta x \setminus \alpha)} x$ is in $\mathfrak{Nr}_\alpha \mathfrak{B}$, hence in $\mathfrak{A}$. Therefore $\mathsf{c}_{(\Delta x \setminus \alpha)} x \in A \cap I$. But $x \leq \mathsf{c}_{(\Delta x \setminus \alpha)} x$, hence the required. We now prove (iii) implies (iv). The proof is a generalization of the proof of [Hen-Mon-Tar,85] 2.6.72. Let $\mathfrak{A}, \mathfrak{A}' \in L$, $\mathfrak{B}, \mathfrak{B}' \in \mathsf{CA}_\beta$ and assume that $e_A, e_{A'}$ are embeddings from $\mathfrak{A}, \mathfrak{A}'$ into $\mathfrak{Nr}_\alpha \mathfrak{B}$, $\mathfrak{Nr}_\alpha \mathfrak{B}'$, respectively, such that $\mathfrak{Sg}^{\mathfrak{B}} (e_A(A)) = \mathfrak{B}$ and $\mathfrak{Sg}^{\mathfrak{B}'} (e_{A'}(A')) = \mathfrak{B}'$, and let $i : \mathfrak{A} \to \mathfrak{A}'$ be an isomorphism. We need to "lift" i to β dimensions. Let $\mu = |A|$. Let x be a bijection from μ onto A that satisfies the premise of (4). Let y be a bijection from μ onto A', such that $i(x_j) = y_j$ for all $j < \mu$. Let $\rho = \langle \Delta^{(\mathfrak{A})} x_j : j < \mu \rangle$,

$\mathfrak{D} = \mathfrak{Fr}_{\mu}^{(\rho)} \mathsf{CA}_{\beta}$, $g_{\xi} = \xi/Cr_{\mu}^{(\rho)} \mathsf{CA}_{\beta}$ for all $\xi < \mu$ and $\mathfrak{C} = \mathfrak{Sg}^{\mathfrak{Rd}_{\alpha}\mathfrak{D}}\{g_{\xi} : \xi < \mu\}$. Then $\mathfrak{C} \subseteq \mathfrak{Nr}_{\alpha}\mathfrak{D}$, C generates $\mathfrak{D}$ and by hypothesis $\mathfrak{C} \in L$. There exist $f \in \mathrm{Hom}(\mathfrak{D}, \mathfrak{B})$ and $f' \in \mathrm{Hom}(\mathfrak{D}, \mathfrak{B}')$ such that $f(g_{\xi}) = e_A(x_{\xi})$ and $f'(g_{\xi}) = e_{A'}(y_{\xi})$ for all $\xi < \mu$. Note that f and f' are both onto. We now have $e_A \circ i^{-1} \circ e_{A'}^{-1} \circ (f' \upharpoonright \mathfrak{C}) = f \upharpoonright \mathfrak{C}$. Therefore $\mathrm{Ker}\, f' \cap \mathfrak{C} = \mathrm{Ker}\, f \cap \mathfrak{C}$. Hence $\mathfrak{Ig}(\mathrm{Ker}\, f' \cap \mathfrak{C}) = \mathfrak{Ig}(\mathrm{Ker}\, f \cap \mathfrak{C})$. So by (iii), $\mathrm{Ker}\, f' = \mathrm{Ker}\, f$. Let $y \in B$, then there exists $x \in D$ such that $y = f(x)$. Define $\hat{i}(y) = f'(x)$. The map is well defined and is as required. The proof of (iv) $\Longrightarrow$ (v) follows from Theorem 5.2.3. ■

Since for $\alpha \geq \omega$, RCA_{α} does not have AP, a classical result of Pigozzi, it follows that *we cannot replace* Dc_{α} *in* 2.6.67(ii), 2.6.71–72 *of* [Hen-Mon-Tar,85] *by* RCA_{α} *when* $\alpha \geq \omega$. This answers a question of Monk and Henkin mentioned in the introduction of [Hen-Mon-Tar,85]. That this replacement cannot be made was mentioned in [Hen-Mon-Tar,85] with the proof deferred to the second part, but in the second the proof never appeared. Actually the co-authors Henkin and Monk admit in [Hen-Mon-Tar,85] p. (iv) that they could not reconstruct Tarski's proof. So the above theorem confirms three conjectures of Tarski, the proof of which could not be recovered by his co-authors Henkin and Monk. The first of those conjectures is confirmed more directly in [Nem-Say,01].

Let APbase (K) denote the amalgamation base of K and SUPAPbase (K) denote the super amalgamation base of K.

We say that $\mathfrak{A}_0 \in \mathsf{K}$ *is in the strong amalgamation base of* K, *briefly* $\mathfrak{A}_0 \in$ SAPbase (K) *if for all* $\mathfrak{A}_1, \mathfrak{A}_2 \in \mathsf{K}$ *and monomorphisms* $i_1 : \mathfrak{A}_0 \to \mathfrak{A}_1$, $i_2 : \mathfrak{A}_0 \to \mathfrak{A}_2$ *there exist* $\mathfrak{D} \in \mathsf{K}$ *and monomorphisms* $m_1 : \mathfrak{A}_1 \to \mathfrak{D}$ *and* $m_2 : \mathfrak{A}_2 \to \mathfrak{D}$ *such that* $m_1 \circ i_1 = m_2 \circ i_2$ *and* $m_1(A_1) \cap m_2(A_2) = m_2 \circ i_2(A_0)$.

Then, it is easy to see that

$$\text{SUPAPbase}(\mathsf{K}) \subseteq \text{SAPbase}(\mathsf{K}) \subseteq \text{APbase}(\mathsf{K}).$$

In [Sim,99], using the remarkable technique of twisting, it is shown that $\mathsf{Mn}_{\omega} \nsubseteq$ APbase (CA_{ω}), see also [Sim,thisVol]. In particular, we have

$$\text{APbase}(\mathsf{RCA}_{\omega}) \nsubseteq \text{APbase}(\mathsf{CA}_{\omega}),$$

and

$$\text{SAPbase}(\mathsf{RCA}_{\omega}) \nsubseteq \text{SAPbase}(\mathsf{CA}_{\omega}.)$$

This solves Problem 45 in the problem session paper of [And-Mon-Nem,91b]. The problem of finding the AP base for classes of algebras that does not have AP originates with Bjarni Jónsson. (This is mentioned in the problem session paper of [And-Mon-Nem,91b]). In [Say,09g] a solution to Problem 2.13 (using neat embeddings and the amalgamation property) in [Hen-Mon-Tar,85] is presented. Further results connecting neat embeddings to various amalgamation properties can be found in [Say,b], [Mad-Say,09a], [Mad-Say,09b].

3. Complete Representations

The topic of this section is also studied in [Hir-Hod,thisVol] and [Say,thisVol,b]. Unless otherwise specified, we assume that $n \leq \omega$. If $\mathfrak{A} \in \mathsf{RCA}_n = \mathbf{SP}\,\mathsf{Cs}_n$, then for all non-zero $a \in A$ there exist $\mathfrak{C}_a \in \mathsf{Cs}_n$ with base M and a homomorphism $f : \mathfrak{A} \to \mathfrak{C}_a$ such that $f(a) \neq 0$. (If $\mathfrak{A} \in \mathsf{Cs}_n$ has greatest element ${}^nM$, then M is called its base). This can be easily proved to be equivalent to the fact that $\mathfrak{A}$ has a representation on some set (in the sense of the coming definition).

Definition 5.3.1. A *representation* of $\mathfrak{A} \in \mathsf{CA}_n$ on a set V of n ary sequences, is an injective boolean homomorphism $h : \mathfrak{A} \to \wp(V)$ (the power set of V) such that

(i) $h(1) = V = \bigcup_{i\in I} {}^nX_i$ where the X_i's are disjoint. Here 1 is the greatest element of the boolean reduct of $\mathfrak{A}$ and I is an arbitrary non-empty set.

(ii) For all $i, j < n$, $\bar{x} \in h(\mathsf{d}_{ij})$ iff $x_i = x_j$.

(iii) For all $i < n$, $a \in A$ and $\bar{x} \in V$ we have $\bar{x} \in h(\mathsf{c}_i a)$ iff $\bar{x}^i_y \in h(a)$ for some $y \in X$. Here $\bar{x}^i_y$ is the sequence that agrees with $\bar{x}$ except for i where its value is y.

In this case $\mathfrak{A} \cong h(\mathfrak{A})$, and $h(\mathfrak{A})$ is a Gs_n in the sense of [Hen-Mon-Tar,85] with greatest element $V = \bigcup_{i\in I} {}^nX_i$. Then

$$h(\mathfrak{A}) \subseteq \left(\wp(V), \cup, \cap, \sim V, \emptyset, \mathsf{C}_i, \mathsf{D}_{ij}\right)_{i,j<n}$$

with the C_i's and D_{ij}'s defined as in set algebras. In this case we say that (h, V) is a representation of $\mathfrak{A}$. Let $\mathfrak{A} \in \mathsf{RCA}_n$ and (h, V) a representation of $\mathfrak{A}$. If $s \in V$, we let

$$f^{-1}(\mathsf{s}) = \left\{a \in \mathfrak{A} : \mathsf{s} \in f(a)\right\}.$$

Clearly $f^{-1}(\mathsf{s})$ is a boolean ultrafilter in $\mathfrak{A}$.

Definition 5.3.2.

(i) An *atomic representation* $f : \mathfrak{A} \to \wp(V)$ is an (injective cylindric) representation such that for each $\mathsf{s} \in V$, the ultrafilter $f^{-1}(\mathsf{s})$ is principal. Equivalently s is in the image of some atom of $\mathfrak{A}$. (Recall that an atom is a minimal non-zero element.)

(ii) A *complete representation* of $\mathfrak{A}$ is an injective representation $f : \mathfrak{A} \to \wp(V)$ satisfying

$$f\left(\prod X\right) = \bigcap f[X]$$

whenever $X \subseteq \mathfrak{A}$ and $\prod X$ is defined. Equivalently, $f\left(\sum X\right) = \bigcup f[X]$ whenever $\sum X$ is defined.

Lemma 5.3.3. *Let $\mathfrak{A} \in \mathsf{RCA}_n$. A representation f of $\mathfrak{A}$ is atomic if and only if it is complete.*

Proof. This is proved for Boolean algebras in [Hir-Hod,97c]. The proof lifts to the cylindric case with no modifications. ■

Lemma 5.3.4. *Assume that $\mathfrak{A}$ has a complete representation. Then $\mathfrak{A}$ is atomic. That is every non-zero element contains an atom.*

Proof. Let f be a complete representation. Then it is atomic. Let a be a non-zero element of $\mathfrak{A}$. Let $\mathsf{s} \in f(a)$. (Here we are using that f is injective). So there is an atom $b \in A$ with $\mathsf{s} \in f(b)$. Therefore $b \wedge a \neq 0$. Thus $b \leq a$. Hence $\mathfrak{A}$ is atomic ([Hir-Hod,97c]). ■

For an algebra $\mathfrak{A}$ with a boolean reduct, we write $At\mathfrak{A}$ for the set of atoms of $\mathfrak{A}$. By Lemma 5.3.4 a necessary condition for existence of complete representations is atomicity. However, representable atomic cylindric algebras may not be completely representable. In fact, the class of completely representable algebras is not even elementary [Hir-Hod,97c], [Kha-Say,09b]. But again we can characterize the class of completely representable algebras using neat embeddings: We start with a definition.

Definition 5.3.5. For K a class with a boolean reduct we define

$$S_{\mathsf{c}}\mathsf{K} = \Big\{\mathfrak{A} : \exists \mathfrak{B} \in \mathsf{K} : A \subseteq B, \text{ and whenever } \sum X = 1 \text{ in } \mathfrak{A}$$

$$\text{then } \sum X = 1 \text{ in } \mathfrak{B} \text{ for all } X \subseteq A\Big\}.$$

We sometimes refer to algebras in the class $\mathbf{S_c}\,\mathsf{Nr}_n\,\mathsf{CA}_{n+\omega}$ as algebras having the *strong* neat embedding property. We now have

Theorem 5.3.6. *Let $n < \omega$. Let $\mathfrak{A} \in \mathsf{CA}_n$ be countable. Then $\mathfrak{A}$ is completely representable if and only if $\mathfrak{A}$ is atomic and $\mathfrak{A} \in \mathbf{S_c}\,\mathsf{Nr}_n\,\mathsf{CA}_\omega$* ([Say,02c]).

One implication follows from Lemma 5.3.4. We sketch a proof of the non-trivial implication, the if part. However this follows from the stronger:

($*$). *If $\mathfrak{A} \in \mathbf{S_c}\,\mathsf{Nr}_n\,\mathsf{CA}_{n+\omega}$ is countable, $n \leq \omega$ (note that here n is allowed to be infinite) and $\{X_i : i < \omega\}$ is a family of subsets of $\mathfrak{A}$ such that $\prod X_i = 0$ for all $i < \omega$, then for every non-zero $a \in A$ there exists $\mathfrak{C} \in \mathsf{Ws}_n$, with countable base, and $f : \mathfrak{A} \to \mathfrak{C}$ a homomorphism such that $f(a) \neq 0$ and for all $i \in \omega$ we have $\bigcap_{x \in X_i} f(x) = \emptyset$.*

Here Ws_n stands for the class of weak set algebras of dimension n. ($*$) is proved in [Say,02c]. The proof is a Baire category argument at heart hence the condition of countability cannot be omitted [Say,07d]. We recall that a weak set algebra has unit of the form ${}^nU^{(p)} = \big\{\mathsf{s} \in {}^nU : \big|\{i \in n : \mathsf{s}_i \neq p_i\}\big| < \omega\big\}$, for some $p \in {}^nU$. U is called its base. To emphasize the connection with the omitting types Theorem, we refer to the X_i's as *non-principal types* and to $\mathfrak{B}$ as a representation *omitting* these types. Note that for $n < \omega$ we have $\mathsf{Ws}_n = \mathsf{Cs}_n$, so that a unit of a Ws_n is simply of the form ${}^nU$.

Sketch of proof of the non-trivial implication of Theorem 5.3.6. We show how the only if part of Theorem 5.3.6 follows from ($*$). Assume that ($*$) is proved and let $n < \omega$. Let $\mathfrak{A} \in \mathbf{S_c}\,\mathsf{Nr}_n\,\mathsf{CA}_\omega$ be countable and atomic. We will assume, to simplify matters, that $\mathfrak{A}$ is simple. The general case is not much harder. Then taking $X_i = Y = \{-b : b \text{ is an atom of } \mathfrak{A}\}$, and applying ($*$) for any non-zero a in A, upon noting that $\prod Y = 0$ since $\mathfrak{A}$ is atomic, we get an atomic representation, hence a complete representation of $\mathfrak{A}$. Note that since $\mathfrak{A}$ is simple, the representation is necessarily injective.
■

We note that the class of completely representable cylindric algebras of dimension > 2 is not elementary. When we consider $< {}^{\omega}2$ many types then $(*)$ becomes an instance of Martin's axiom restricted to countable Boolean algebras. Indeed, the notion of complete representations have been connected to Martin's axiom giving independent statements in set theory [Say,05c]. Our last result in this section unifies results on neat reducts and complete representations. Using the so called Rainbow construction for cylindric algebras the following is proved in [Say,e]:

Theorem 5.3.7. *Let $n > 2$. Then any K such that $\mathsf{Nr}_n\,\mathsf{CA}_\omega \subseteq \mathsf{K} \subseteq \mathbf{S}_{\mathsf{c}}\,\mathsf{Nr}_n\,\mathsf{CA}_{n+2}$ is not elementary.*

From which we readily get

Corollary 5.3.8. *For $n > 2$ and $k \geq 2$, the class $\mathsf{Nr}_n\,\mathsf{CA}_{n+k}$ and the class of completely representable algebras of dimension n are not elementary.*

4. Back to the Classical Neat Embedding Theorem

Let us go back to Henkin's classical Neat Embedding Theorem. We set out from the known property of cylindric algebras that the neat embedding property implies representability, i.e.

$$\text{(5.4.1)} \qquad \mathfrak{A} \in \mathbf{S}\mathsf{Nr}_\alpha\,\mathsf{CA}_{\alpha+\varepsilon} \quad \text{implies} \quad \mathfrak{A} \in \mathbf{I}\,\mathsf{Gs}_\alpha$$

where α, ε are infinite ordinals. The converse of the proposition is trivial, so the following is true: $\mathfrak{A} \in \mathbf{S}\mathsf{Nr}_\alpha\,\mathsf{CA}_{\alpha+\varepsilon}$ if and only if $\mathfrak{A} \in \mathbf{I}\,\mathsf{Gs}_\alpha$. Ferenczi went deeper into the analysis of property in (5.4.1) for various classes of algebras.

He introduced some new classes of cylindric like algebras (classes $\mathsf{K}^\alpha_{\alpha+\varepsilon}$, $\mathsf{M}^\alpha_{\alpha+\varepsilon}$, $\mathsf{F}^\alpha_{\alpha+\varepsilon}$ e.g.) and formulated a number of theorems that can be viewed as strengthenings of property (5.4.1). In fact, in [Fer,00] the following problem is investigated: Is it possible to replace in (5.4.1) the class CA by a larger class so that the implicitation in (5.4.1) still holds. Surprisingly, the answer is affirmative. Let (C_1)–(C_7) denote the cylindric axioms as defined in [Hen-Mon-Tar,85] 1.1.1. In [Fer,00] the following class $\mathsf{K}^\alpha_{\alpha+\varepsilon}$ is introduced: Suppose that $\mathsf{K}^\alpha_{\alpha+\varepsilon}$ is the class for which $\mathsf{K}^\alpha_{\alpha+\varepsilon} \models \left\{(\mathrm{C}_1), (\mathrm{C}_2), (\mathrm{C}_3), (\mathrm{C}_5), (\mathrm{C}_7), (\mathrm{C}_4)^-, (\mathrm{C}_6)^-\right\}$ where β denotes $\alpha + \varepsilon$

and $(C_4)^-$ denotes the pair of the following weakenings a) and b) of axiom (C_4):

$(C_4)^-$ a) $c_m s^j_n x = s^j_n c_m x$;

$(C_4)^-$ b) $c_m s^j_m c_m x = c_j c_m x$.

Here $j \in \alpha$ and $m, n \in \beta$, $m \neq n$, and $(C_6)^-$ denotes the following weakenings a), b) and c) of axiom (C_6):

$(C_6)^-$ a) $d_{mn} = d_{nm}$;

$(C_6)^-$ b) $d_{mn} \cdot d_{nj} \leq d_{mj}$;

$(C_6)^-$ c) $c_m d_{jn} = d_{jn} \quad m \notin \{j, n\}$.

Notice that $(C_4)^-$ and $(C_6)^-$ are restrictions of the β-dimensional axioms (C_4) and (C_6). The following theorem is proved in [Fer,00].

Theorem 5.4.1. *Suppose that $\mathfrak{A} \in \mathsf{CA}_\alpha$. Then $\mathfrak{A} \in \mathbf{S}\mathsf{Nr}_\alpha \mathsf{K}^\alpha_{\alpha+\varepsilon}$ if and only if $\mathfrak{A} \in \mathbf{I}\,\mathsf{Gws}_\alpha$ where α and ε are infinite fixed ordinals.*

So, although we need all ω dimensions for representability by Monk's classical result, we can dispense with some of the CA axioms when we get there (i.e. to ω extra dimensions)! We believe that this is an intriguing phenomenon. It can be proved that in some precise sense the class $\mathsf{K}^\alpha_{\alpha+\varepsilon}$ is the optimal extension of the class CA_α such that (5.4.1) remains valid, that is if L is a class having the appropriate similarity type, and $\mathsf{K}_{\alpha+\varepsilon} \subset L$ then $\mathbf{S}\mathsf{Nr}_\alpha L$ contains non representable algebras. Reflecting on the classical logical aspects of Theorem 5.4.1 we can draw the conclusion that in Gödel's completeness proof we use only a fragment of the complete calculus, i.e. it is possible to restrict the equality axioms, and to weaken the commutativity of the quantifiers $\exists x_i$ so that Gödel's theorem remains valid, but of course there is minimum set of logical axioms needed to achieve completeness, and those are expressed algebraically in the axioms of $\mathsf{K}_{\alpha+\epsilon}$. We note that commutativity of the quantifiers $\exists x_i$ cannot be just omitted by [Nem,95] Theorem 3.2.

In [Fer,10b], the generalization of (5.4.1), from the class Gws_α to the class $\mathsf{Crs}_\alpha \cap \mathsf{CA}_\alpha$ (i.e., to the class addressed in the famous Resek–Thompson representation Theorem) is investigated. The problem again is: Is there a class of cylindric like algebras instead of CA_α such that (5.4.1) remains true when the class Gs_α is replaced by the class $\mathsf{Crs}_\alpha \cap \mathsf{CA}_\alpha$? And, yet again, the answer is affirmative. A new class $\mathsf{M}^\alpha_{\alpha+\varepsilon}$ of cylindric like algebras is introduced in [Fer,10b]. The character of this class is similar to that of $\mathsf{K}^\alpha_{\alpha+\varepsilon}$ i.e. this class satisfies all the cylindric axioms except for (C_4) and (C_6). Instead of these last two axioms it satisfies some concrete weakenings of (C_4) and (C_6) together with the restrictions of the so called merry-go-round identities. The following theorem holds for $\mathsf{M}^{\alpha}_{\alpha+\varepsilon}$:

Theorem 5.4.2. *Suppose that* $\mathfrak{A} \in \mathsf{CA}_\alpha$. *Then* $\mathfrak{A} \in \mathbf{I}\,\mathsf{Crs}_\alpha \cap \mathsf{CA}_\alpha$ *if and only if* $\mathfrak{A} \in \mathbf{S}\mathsf{Nr}_\alpha\mathsf{M}^{\alpha}_{\alpha+\varepsilon}$ *where* α *and* ε *are infinite fixed ordinals.*

A similar theorem is formulated in [Fer,thisVol,b], Section 4.

Resek and Thompson's famous theorem says: $\mathfrak{A} \in \mathsf{CA}^M_\alpha$ if and only if $\mathfrak{A} \in \mathbf{I}\,\mathsf{Crs}_\alpha \cap \mathsf{CA}_\alpha$ where CA^M_α denotes the class of cylindric algebras satisfying the merry-go-round equations. Using Theorem 5.4.2 Ferenczi, basically using a NET, gives a new proof for this classical theorem that casts its shadow over the entire field. A proof of this theorem, due to Andréka can be found in [And-Tho,88]. It is worthy of note that the first proof of the Resek Thompson result, as mentioned in [Hen-Mon-Tar,85], was more than 100 pages long.

Theorem 5.4.2 has also some remarkable consequences for logic. In classical first order logic if the language is extended by new individual variables then the deduction system obtained is a conservative extension of the old one. This fails to be true for logics with infinitary predicates. But, as a consequence of Theorem 5.4.2 it can be proved that restricting the commutativity of quantifiers and the equality axioms in the expanded language and assuming the merry-go-round properties in the original language, the foregoing extension is already a conservative one (see [Fer,09]).

We give a sketch for the proof of Theorem 5.4.1:

The "only if" part is trivial, because every algebra in Gws_α can be neatly embedded to an algebra in $\mathsf{Gws}_{\alpha+\omega}$ ([Hen-Mon-Tar,85] II. p. 63).

The "if" part: Let $\mathfrak{A} \in \mathbf{S}\mathsf{Nr}_\alpha\mathsf{K}^{\alpha}_{\alpha+\varepsilon}$. We shall show that $\mathfrak{A}$ is representable by relativized set algebra.

The essence of the proof is similar to Tarski's classical proof for the representation of locally finite cylindric algebras (see [Hen-Mon-Tar,85] 3.2.9). Given $\mathfrak{A}$ and non-zero $a \in A$, we shall construct a weak set algebra $\mathfrak{B}$ and a CA homomorphism $h : \mathfrak{A} \to \mathfrak{B}$ such that $h(a) \neq \emptyset$. A proof similar to Tarski's original proof sketched in [Hen-Mon-Tar,85] but never published, is given by Andréka and Németi [And-Nem,75].

However there are some significant differences between our proof and Tarski's proof mentioned above. One is the definition of the ultrafilter occurring in Tarski's proof. To prove the theorem without (C_4) and (C_6) a very special ultrafilter which we refer to as a *specified perfect ultrafilter* is needed. The definition of such ultrafilters as we shall shortly see, is somewhat more complicated than Tarski's ultrafilter. Another important difference is the *definition of the foregoing homomorphism* – in the lack of (C_4) and (C_6). Afterall we have to work without the axioms (C_4) and (C_6) throughout the proof i.e. we have to show that *it is sufficient to use only certain weakened versions* of these axioms.

First we present the definition of the concept of a specified perfect ultrafilter.

Let T denote the set of all finite transformations on β such ($\beta = \alpha + \varepsilon$) that $\tau \in T$ if and only if $\big|\{i \in \beta : \tau i \neq i\}\big| < \omega$. $\tau \in T$ is *admitted* if $\tau = \big[i(n-1)/m(n-1)\big] \dots [i0/m0]$, $n \in \omega$, $\{i0, \dots, i(n-1)\} \subseteq \alpha$ and $\big\{m0, \dots, m(n-1)\big\} \cap \alpha = \emptyset$, when $n > 0$, $\tau \in T$. Here $[i/j]$ denotes the replacement that takes i to j and identity on $\beta - \{i\}$.

We refer to $\big\{ i0, \dots i(n-1)\big\}$ as the domain of τ ($Do\ \tau$), and $\big\{ m0, \dots, m(n-1)\big\}$ as the range of τ ($Rg\ \tau$). Let $G \subseteq T$ stand for the set of admitted substitutions. For $\tau \in G$, τ defines a unary operation s_τ on $\mathfrak{B}$ as follows:

$$\mathsf{s}_\tau = \mathsf{s}^{i0}_{m0} \dots \mathsf{s}^{i(n-1)}_{m(n-1)}.$$

A Boolean ultrafilter F in $\mathfrak{B}$ is specified perfect *if for any element of the form $\mathsf{s}_\tau \mathsf{c}_j x$ included in F, there exists an $m \notin \alpha \cup Rg\ \tau$ such that $\mathsf{s}_\tau \mathsf{s}^j_m x \in F$, where $j \in \alpha$, $x \in A$ and s_τ is any admitted substitution operator.*

Next, let a be an arbitrary, but fixed non zero element of A.

We build in a step by step fashion, a perfect ultrafilter F of $\mathfrak{B}$ containing a, that will be used to construct the required representation. However we do not provide details for this part of the proof.

But we provide some of the details in the next part of the proof, the *definition of the homomorphism h*.

Let F be a specified perfect ultrafilter containing the element a. Let

(5.4.2) $$\Gamma = \{m : m < \beta \text{ and } \mathsf{c}_i \mathsf{d}_{im} \in F \text{ for some } i \in \alpha\}.$$

The axiom (C_4) $\mathsf{c}_i \mathsf{d}_{ij} = 1$ implies that $\alpha \subseteq \Gamma$. Let us consider the following equivalence relation $\equiv$:

$$m \equiv n \quad (m, n \in \Gamma) \quad \text{iff} \quad \mathsf{d}_{mn} \in F.$$

The properties $(C_6)^-$ a. and b. ensure that $\equiv$ is an equivalence relation. Let Π denote the set of the equivalence classes, let $M, N, L, \dots$ denote the classes themselves and let $m, n, l, \dots$ be fixed representives in these classes, respectively. Γ is a proper superset of α, indeed for every $i \in \alpha$ there exists an $m \notin \alpha$, $m \in \Gamma$ such that $i \equiv m$. Let the support of V be an α-sequence y, for which $y_i = I$ if $i \in \alpha$, thus y_i is the equivalence class containing the ordinal i. Let the base of V be Π. Let $\mathcal{C}$ denote the full set algebra with unit V.

Instead of the notation y^i_M we write $f^i_M y$ where f^i_M is a substitution operator in the weak set algebra in question. Thus the notation of the substitution $\left(\left(\left(y^i_M\right)^j_N\right)\dots\right)^k_L$ is $f^k_L \dots f^j_N f^i_M y$, or $f_{\tau'} y$ for short. With the substitution $f^k_L \dots f^j_N f^i_M$, or with $f_{\tau'}$ for short, we associate the substitution $\mathsf{s}^i_m \dots \mathsf{s}^k_l$, or briefly s_τ in $\mathfrak{B}$ so that $l, \dots n, m$ are the prefixed representives of the classes $L, \dots M, N$, respectively. From now, we assume that in the substitutions s_τ's, the ordinals in α occur *at least once*, among $k, \dots, j, i$.

Now, we are ready to present the definition of h_F (briefly h) into $\mathcal{C}$:

$$hz = \{f_{\tau'} y \;:\; \mathsf{s}_\tau z \in F\}$$

where s_τ is an admitted substitution operator in $\mathfrak{B}$ associated with the substitution $f_{\tau'}$ in V, as above.

We omit the last part of the proof: to check that the definition of h is well defined and h is a homomorphism. ■

5. Some Further Final Comments

Robin Hirsch proved the analogous result of Theorem 5.3.6 for relation algebras (RA) [Hir,07]. Let RRA denote the class of representable RA's. For $\mathfrak{C} \in \mathsf{CA}_n$, $n \geq 4$, $\mathsf{Ra}\,\mathfrak{C}$, the relation algebra reduct of $\mathfrak{C}$, is defined as in [Hen-Mon-Tar,85] 5.3.7. For RA's we do have a NET to the effect that $\mathsf{RRA} = \mathbf{S}\,\mathsf{Ra}\,\mathsf{CA}_\omega = \mathbf{S}\,\mathsf{Ra}\,\mathsf{RCA}_\omega$. If a representable relation algebra $\mathfrak{A}$ generates at most one RCA_ω then $\mathfrak{A} \in$ APbase (RRA). This is another way of saying that an RA has the *UNEP*. In particular, QRA $\subseteq$ APbase (RRA). QRA defined in e.g. [Tar-Giv,87] p. 242 is the class of relation algebras with quasi-projections. In fact, we have QRA $\subseteq$ SUPAPbase (RRA) [Say,a]. A recent reference dealing with representability of QRA's via a NET for CA's is [Sim,07]. So for RA's, QRA is a "natural" class such that each of its members has *NS* and *UNEP*. The CA analogue of this class is the class of directed cylindric algebras invented by Németi, and studied by András

Simon and Gábor Sági [Sag,99a], [Nem-Sim,09]. These algebras are strongly related to higher order logics and they provide a solution to the finitizability problem (FP) in non-well founded set theories. The representablity of such algebras, providing a solution to the finite dimensional version of the (FP), can be also proved using a NET. Positive solution exists in non-well founded set theories, because one can generate infinitely many extra dimensions, forcing a neat embedding theorem, by digging "downwards" with nothing to stop him! This view comes across very much in the case of Németi's directed cylindric algebras. Furthermore for such algebras neat reducts commute with forming subalgebras, hence this class has SUPAP [Say,d]. A solution to the infinite dimensional version of the FP is provided by Sain [Sai,00] (in usual set theory) using also a NET. These algebras are obtained by expanding the language of quasi-polyadic algebras by finitely many infinitary substitutions and adding *finitely* many new axioms in the bigger language that enforces a NET. For those algebras neat reducts also commute with forming subalgebras, and so they have SUPAP [Say,04a]. The real technical difficulty that comes up here is that when we expand our languages and add axioms to code extra dimensions somehow, in the hope of obtaining a NET, then usually we succeed in representing the already existing operations; the difficult problem is that the new operations turn out representable as well! Sain succeeded to overcome this difficulty for first order logic *without* equality. Adding equality proves problematic so far. A sophisticated categorial formulation of the FP, is to look at *inverses* of the *Neat reduct* functor going from one category to another in ω extra dimensions, and try to reflect those in an adjoint situation. A solution to the finitizability problem is thereby presented as an equivalence of two categories.

In [Ame-Say,06] the NET of Henkin is likened to his completeness proof (the extra dimensions play the role of added witnesses to existential formulas); therefore it is not a coincidence that interpolation results and omitting types for variants of first order logic can be proved algebraically by using appropriate variations on the NET [Say,07c], [Sam-Say,07b]. We find it timely to make the following observation. There are algebras for which the NET does not hold, that is, neat embeddability in algebras in ω extra dimensions does not enforce representability. Surprisingly this occurs at the "end points". The NET fails for the class of diagonal free cylindric algebras ($\mathfrak{Df}$) and polyadic equality algebras of infinite dimension (PEA). In between, there is a whole stratum of proper reducts of PEA's that are proper expansions of $\mathfrak{Df}$'s (like CA's, Sain's algebras introduced in [Sai,00] and PA's) for which the NET holds. Finally (*) above after Theorem 5.3.6, which is basically a variation on a NET, is related to many statements from lattice theory and topology in [Say,08e].

6. Problems

We end this paper with the following two questions:

(i) Let $n > 2$ and $k \geq 2$. Is the class $\mathsf{SNr}_n\,\mathsf{CA}_{n+k}$ closed under completions?

(ii) Does the class of completely representable polyadic algebras of infinite dimension coincide with the atomic algebras?

Problem (i), attributed to the present author, appears in [Hir-Hod,02a] Problem 12 p. 627. For a partial result to (i), the reader is referred to [Kha-Say,09a].

II. Representation Theory

BOLYAI SOCIETY
MATHEMATICAL STUDIES, 22

Cylindric-like Algebras
and Algebraic Logic
pp. 135–162.

A NEW REPRESENTATION THEORY: REPRESENTING CYLINDRIC-LIKE ALGEBRAS BY RELATIVIZED SET ALGEBRAS

MIKLÓS FERENCZI*

Representation theorems. As is known, cylindric algebras are not representable in the classical sense (as a subdirect product of cylindric set algebras), in general. But, the Resek–Thompson theorem states that if the system of cylindric axioms is extended by a new axiom, by the merry-go-round property (MGR, for short), then the cylindric-like algebra obtained is representable by a cylindric relativized set algebra. Furthermore, if, in additional, axiom (C_4) is weakened (only the commutativity of the single substitutions is assumed) then it is representable by an algebra in D_α, (see [And-Tho,88] and [Fer,07a]). This style of representation theorem is closely connected with the completeness theorems based on the Henkin style semantics in mathematical logic. By an r-representation of a cylindric-like algebra we mean a representation by a cylindric-like *relativized* set algebra.

Analyzing the merry-go-round property, it turns out that this notation is underlain by the elementary operation of *transposition* (see [Fer,11]). The operation transposition cannot be introduced in every cylindric algebra (see [Fer,07b], [Say,09a]). This fact led to research into the representability of transposition algebras (TA_α). Transposition algebras are cylindric algebras extended by abstract transposition operations (p_{ij}) and single substitutions (s^i_j), where $i, j < \alpha$. They are weakening of the (so-called) finitary polyadic equality algebras introduced in [Sai-Tho,91]. They are definitionally equivalent to the so-called non-commutative quasi-polyadic equality algebras. Transposition algebras are not necessarily representable in the classical sense. However, it is proven that transposition algebras are r-representable, they are representable by relativized set algebras in Gwp_α (see [Fer,12a], Theorem 1).

*I dedicate this paper to the memory of Leon Henkin.

The next question is that whether polyadic equality algebras are r- representable? As is known, the behaviour of polyadic algebras without equality (PA_α) and with equality (PEA_α) are also essentially different. For example, PA_α is representable in the classical sense while PEA_α is not (except for the locally finite case, for example, see Halmos's [Hal,57], [Dai-Mon,63] or [Sag,thisVol]). Recall that polyadic algebras are essentially different from quasi-polyadic algebra in that the substitution operators are defined for real infinite transformations. In [Fer,12b], the problem of r-representability of polyadic equality algebras is answered for a large class: for polyadic equality algebras having ordinary cilindrifications (single cylindrifications), called *cylindric polyadic equality algebras* (class CPE_α – these algebras, in the rigorous sense, can no longer be considered as cylindric-like algebras). This class is r-representable by algebras in $\mathsf{Gp}_\alpha^{\mathrm{reg}}$. Furthermore, Halmos's result on the representability of locally finite quasi-polyadic algebras ([Hal,56]) can be generalized for m-quasi, locally-m cylindric polyadic algebras (m is infinite) and r-representability (see Section 3).

The representant structures (representant set algebras) related to the above r-representations are very simple. For example, the representant of a *transposition algebra* (TA_α) is a Boolean set algebra with unit V such that V is an arbitrary union of weak Cartesian products, i.e., $V = \bigcup_{k \in K} {}^{\alpha}U_k^{(p_k)}$ (class Gwp_α). Approaching our topic from set theory or geometry, the r-representation theorem says that the above Boolean set algebras are first order finite schema axiomatizable and the axioms can be the TA_α axioms. As is known if the disjointness of the members ${}^{\alpha}U_k^{(p_k)}$ is assumed in the above decompositions of V's, then the classes of set algebras obtained are no longer first order finite schema axiomatizable. For example, some additional non first order conditions for the algebra are needed (local finiteness). These representation theorems can be considered as an immediate generalizations of the Stone representation theorem for Boolean algebras, and their importance may be compared with that theorem.

A remarkable consequence of the investigation of the concept of r-representability is that *certain modifications* of the classical structures of algebraic logic came into the focus of research. A common feature of the abstract algebras occurring in these theorems is that the commutativity of cylindrifications is not required. Beyond this, additional axioms are assumed (e.g., the MGR property for cylindric algebras) or certain axioms are weakened (e.g., in the case of polyadic-style equality algebras, the last non-equality axiom).

Neat embedding. Some words about the techniques used in the proofs of the r-representation theorems are now in order. Two methods are used, both of them are known in the literature, but, several new ideas are needed for the applications of these techniques. The first one is the step-by-step method (it is closely related to the technique "games"). This technique is applied to prove the main r-representation theorems for cylindric algebras and transposition algebras. The other technique is the neat embedding technique (in this volume, see [Say,thisVol,a]).

Some words about neat embeddability, from the viewpoint of r-representability. In the theory of cylindric algebras the neat embedding theorem is of central importance. This theorem says that (classical) representability is equivalent to (classical) neat embeddability (see [Hen-Mon-Tar,85], 3.2.10). The importance of the theorem is that the concept of neat embeddability is more suitable for the abstract algebraic treatment than representability. The following question arises: is it possible to characterize the concept of r-representability of a cylindric algebra by the concept of neat embeddability? The answer is affirmative (see Theorem 6.4.2, by Ferenczi and [Fer,10b]). Of course, the concept of neat embeddability, obtained in this way is *different from the standard one.* The neat embedding happens here into a *many sorted* structure. This structure does not have the same axiomatization as the original one (but, the new axiomatization is similar to the original one, of course, it is the weakening of that). Thus, as embedding class, a *larger class* is allowed than in the case of the classical neat embedding.

The next question is whether this new kind of neat embedding theorem concerning cylindric algebras can be transfered to other structures: to transposition algebras, quasi-polyadic equality algebras, cylindric polyadic equality algebras. The answer depends on which particular class we are considering. It is obviously affirmative for transposition algebras and for quasi-polyadic equality algebras. But in the case of cylindric polyadic equality algebras, in the presence of infinite substitutions, the situation is essentially different. Here, as a consequence of the Daigneault–Monk–Keisler theorem (polyadic equality algebras have the classical neat embeddability property), there is *no* classical neat embedding theorem (i.e., neatly embeddability does not imply classical representability). The question whether there exists some kind of neat embedding theorem for polyadic equality algebras is a long standing problem. A neat embedding theorem is stated for m-quasi locally-m cylindric polyadic equality algebras (see Section 4). This theorem says that for this class, classical neat embeddability is equivalent to a kind of r-representability.

History. The pioneer of the researches discussed here is Leon Henkin. He introduced the concept of cylindric relativized set algebra (Crs_α), developed the merry-go-round properties, he was Resek's the doctoral advisor (Resek formulated her representation theorem concerning cylindric relativized set algebras in her PhD thesis [Res,75]) and, Henkin developed the famous completeness theorem in mathematical logic based on Henkin style semantics. Henkin style semantics and cylindric relativized set algebras are, closely related. The detailed research of the class Crs_α was initiated by István Németi ([Hen-Mon-Tar-And-Nem,81], [Fer,86a]), as regards Crs in this Volume, see van Benthem's paper [Ben,thisVol].

Some words on the first representation theorem concerning cylindric relativized set algebras, i.e., concerning r-representation. First, Resek proved, although never published, such a representation theorem in her PhD thesis, for simple, complete, atomic cylindric algebras satisfying infinitely many merry-go-round equalities. This result was improved, in a sense, by Thompson who reduced the infinitely many merry-go-round equalities merely to two and replaced the cylindric axiom (C_4) by a weaker axiom. Though the theorem was announced in [Hen-Mon-Tar,85], in Remark 3.2.88, a proof came out only in 1986 (by Andréka, [And-Tho,88]). Andréka also improved the theorem, using the class D_α as representant class. The theorem improved and proved in this way is called Resek–Thompson–Andréka theorem. That proof is relatively short and it is based on the step-by-step method (see Theorem 6.1.9). Later, some variants of the theorem have also found their way into the literature. Maddux has proved a somewhat stronger version of the theorem (see [Mad,89b] and Section 1). Ferenczi has published a simplification of the theorem, replacing the axiom (C_4) by the commutativity of single substitutions ([Fer,07a]). Andréka, applying the step-by-step method, has given a short axiomatization for finite dimensional locally square cylindric set algebras (class G_α) in [And,01].

The results concerning r-representation of *polyadic-style algebras* are due to the present author. In [Fer,11] the connection of the merry-go-round properties and the operation transposition is investigated and the class of partial transposition algebras is introduced. In [Fer,12a], the r-representation theorem is proven for transposition algebras (also for quasi-polyadic equality algebras). In [Fer,12b], the r-representation theorem is proven for cylindric polyadic equality algebras, and, in addition, a neat embedding theorem is proven for this class.

On the Sections. Sections 1–3 deal with the classes cylindric algebras (CA_α), transposition algebras (TA_α) and cylindric polyadic equality algebras (CPE_α), respectively, from the view point of r-representability. In Section 4 the connection of neat embeddability and r-representability is analyzed.

On the prerequisites of this paper: we suppose that the reader is familiar with the concepts of cylindric algebra, quasi-polyadic and polyadic algebra (with and without equality), with the most important kinds of cylindric set algebras (e.g., Gs_α, Gws_α, etc.), and with some basic properties of these algebras.

1. Cylindric Algebras

First, we recall the concepts of cylindric *relativized set* algebras:

Definition 6.1.1 (Crs_α). An algebra $\mathfrak{A}$ is a *cylindric relativized set algebra* of dimension α with unit V if it is of the form:

$$\left\langle A, \cup, \cap, \sim_V, 0, V, \mathsf{C}_i^V, \mathsf{D}_{ij}^V \right\rangle_{i,j<\alpha}$$

where V is a set of α-termed sequences, such that $V \subseteq {}^\alpha U$ for some set U, A is a non-empty set of subsets of V, closed under the Boolean operations $\cup$, $\cap$, $\sim_V$ and under the cylindrifications

$$\mathsf{C}_i^V X = \left\{ y \in V \,:\, y_u^i \in X \text{ for some } u \right\}$$

where $i < \alpha$, $X \in A$, and A contains the sets $\emptyset$, V and contains the diagonals

$$\mathsf{D}_{ij}^V = \{ y \in V \,:\, y_i = y_j \}$$

(see [Hen-Mon-Tar,85] 3.1.1).

Here the definition of y_u^i is $\left(y_u^i\right)_j = y_j$ if $j \neq i$, and $\left(y_u^i\right)_j = u$ if $j = i$. Another notation for y_u^i is $y(i/u)$. The superscript V is often omitted from the notation C_i^V and D_{ij}^V. We note that an algebra in Crs_α satisfies all the cylindric axioms, with the possible exception of the axioms (C_4) and (C_6) (see [Hen-Mon-Tar,85] 3.1.19).

Let us denote $\mathsf{C}_i^V(\mathsf{D}_{ij}^V \cap X)$ $(i \neq j)$ by ${}^V\mathsf{S}_j^i X$. Notice that ${}^V\mathsf{S}_j^i X = \{y \in V : y \circ [i/j] \in X\}$, where $X \in A$. Here $y \circ [i/j] = y_{y_j}^i$, by definition. In this sense if $\{y\} \in A$, then the elementary substitution $y_{y_j}^i$ can be defined in Crs_α in terms of ${}^V\mathsf{S}_j^i$.

Definition 6.1.2 (D_α). D_α is the subclass of Crs_α such that ${}^V\mathsf{S}_j^i V = V$ for every $i, j \in \alpha$, where V is the unit of the algebra (see [And-Tho,88]).

It is easy to check, that in Crs_α the equality ${}^V\mathsf{S}_j^i V = V$ and (C_6) are equivalent, thus D_α satisfies all the CA_α axioms with the possible exception of (C_4).

Definition 6.1.3 (G_α). G_α is a subclass of D_α, called the class of "*locally square*" cylindric set algebras, such that the unit V is of the form $\bigcup_{k\in K} {}^\alpha U_k$ for some sets U_k, $k \in K$ (see [And,01]).

Recall that if given a set U and a mapping $p \in {}^\alpha U$, then the set

$${}^\alpha U^{(p)} = \{x \in {}^\alpha U \; : \; x \text{ and } p \text{ are different only in finitely many places}\}$$

is called the *weak space* determined by p and U.

Definition 6.1.4 (Gw_α). It is a subclass of D_α such that the unit V is of the form $\bigcup_{k\in K} {}^\alpha U_k^{(p_k)}$ for some sets U_k, $k \in K$, where $p_k \in {}^\alpha U_k$.

The difference between the classical class Gs_α ([Hen-Mon-Tar,85], 3.1.1) and G_α is that the disjointness for U_k's in G_α is not assumed. The difference between the classes Gws_α ([Hen-Mon-Tar,85], 3.1.1) and Gw_α is analogous.

Definition 6.1.5. A cylindric-like algebra $\mathfrak{A}$ is *r-representable* if $\mathfrak{A} \in \mathbf{I}\mathsf{Crs}_\alpha$.

Lemma 6.1.6. *The following propositions* (i) *and* (ii) *hold for every* $i, j < \alpha$*:*

(i) *If* $\mathfrak{A} \in \mathsf{Crs}_\alpha$*, then* $\mathfrak{A} \in \mathsf{D}_\alpha$ *if and only if* $x \in V$ *implies* $x \circ [i/j] \in V$*.*

(ii) *If* $\mathfrak{B}$ *is a cylindric-like algebra such that* $\mathsf{s}_j^i 1 = 1$ *and* $\mathfrak{B}$ *is r-representable, then* $\mathfrak{B} \in \mathbf{I}\mathsf{D}_\alpha$*.*

Proof.

(i) The statement that $x \in V$ implies $x \circ [i/j] \in V$ means that $V \subseteq {}^V\mathsf{S}_j^i V$. But, ${}^V\mathsf{S}_j^i V \subseteq V$ is always satisfied, thus ${}^V\mathsf{S}_j^i V = V$. The latter together with $\mathfrak{A} \in \mathsf{Crs}_\alpha$ are equivalent to $\mathfrak{A} \in \mathsf{D}_\alpha$, by definition.

(ii) If h denotes an isomorphism between $\mathfrak{B}$ and an algebra in Crs_α, then $h\mathsf{s}^i_j 1 = \mathsf{S}^i_j h1$, where S^i_j is the abbreviation of ${}^V\mathsf{S}^i_j$. But, in the previous equality, $h\mathsf{s}^i_j 1 = h1 = V$, and $\mathsf{S}^i_j h1 = \mathsf{S}^i_j V$, i.e., $\mathsf{S}^i_j V = V$. ■

Now, we define some cylindric-like *abstract* classes connected with cylindric relativized set algebras.

Definition 6.1.7. The *merry-go-round properties* are:

$$\mathsf{s}^k_i \mathsf{s}^i_j \mathsf{s}^j_k\, \mathsf{c}_k x = \mathsf{s}^k_j \mathsf{s}^j_i \mathsf{s}^i_k\, \mathsf{c}_k x$$

$$\mathsf{s}^k_i \mathsf{s}^i_j \mathsf{s}^j_m \mathsf{s}^m_k\, \mathsf{c}_k x = \mathsf{s}^k_j \mathsf{s}^j_m \mathsf{s}^m_i \mathsf{s}^i_k\, \mathsf{c}_k x$$

for distinct ordinals i, j, k and n (see [Hen-Mon-Tar,85] 3.2.88). The two properties together are denoted by MGR (for an equivalent form of MGR, see also the note at the beginning of Section 2).

Definition 6.1.8 (CNA^+_α)**.** The axioms of this class are obtained from the cylindric axioms so that the axiom (C_4) is replaced by the property

$$(6.1.1) \qquad {}^-(C_4): \qquad \mathsf{s}^i_k \mathsf{s}^j_m x = \mathsf{s}^j_m \mathsf{s}^i_k x$$

$i, k \notin \{j, m\}$, and the set of axioms obtained is extended by the MGR property (see [Fer,07a]).

Other notations for CNA^+_α are: NA^+_α, CNA^M_α or F_α.

The following theorem is due to Resek, Thompson and Andréka, the published proof is due to Andréka ([And-Tho,88]), and the variant below is due to Ferenczi ([Fer,07a], Corollary 3.2):

Theorem 6.1.9 (Main r-representation theorem for cylindric-like algebras in CNA^+_α)**.**

$$\mathfrak{A} \in \mathsf{CNA}^+_\alpha \quad \textit{if and only if} \quad \mathfrak{A} \in \mathbf{I}\mathsf{D}_\alpha.$$

where $\alpha \geq 4$.

The theorem says that the class D_α is *axiomatizable* by the CNA^+_α axioms. We note that modifying ${}^-(C_4)$ and MGR a little, the theorem also remains true for $\alpha = 2$ and $\alpha = 3$, too.

Let CA^+_α denote the class of cylindric algebras satisfying the MGR property. D_α satisfies (C_6), thus the following holds:

Corollary 6.1.10. $\mathfrak{A} \in \mathsf{CA}_\alpha^+$ *if and only if* $\mathfrak{A} \in \mathbf{I}\big(\mathsf{D}_\alpha \cap \mathsf{Mod}\,(\mathrm{C}_4)\big)$, *where* $\alpha \geq 4$.

The lemma below lists some equivalents of the above weakening $^-(\mathrm{C}_4)$ of (C_4). Denote by Σ the set of cylindric axioms except for (C_4) and let us assume that $\alpha \geq 4$.

Lemma 6.1.11. *Under* Σ *the following properties are equivalent:*

(i) $\mathsf{s}_k^i \mathsf{s}_m^j x = \mathsf{s}_m^j \mathsf{s}_k^i x$ *(property* $^-(\mathrm{C}_4)$*)*

(ii) $\mathsf{c}_i \mathsf{s}_m^j x \leq \mathsf{s}_m^j \mathsf{c}_i x$

(iii) $\mathsf{d}_{ik} \cdot \mathsf{d}_{jm} \cdot \mathsf{c}_i \mathsf{c}_j x = \mathsf{d}_{jm} \cdot \mathsf{d}_{ik} \cdot \mathsf{c}_j \mathsf{c}_i x$

(iv) $\mathsf{d}_{ik} \cdot \mathsf{c}_i \mathsf{c}_j x \leq \mathsf{c}_j \mathsf{c}_i x$ *(property* $(\mathrm{C}_4)^*$*)*

where i, j, k *and* m *are different, except, possibly, for* $k = m$ (see [Fer,07a] and [Tho,90]).

If the axiom (C_4) is replaced by $(\mathrm{C}_4)^*$ in the axioms of CA_α, we get the class CNA_α (NA_α, for short, i.e., the class of the non-commutative cylindric algebras). The original Resek–Thompson–Andréka theorem is stated for CNA_α supplemented by MGR, instead of CNA_α^+ (see [And-Tho,88]).

In the Andréka's proof of the Resek–Thompson–Andréka theorem, the so-called step-by-step method (or iteration method, see [Hir-Hod,97a]) is applied to construct the suitable representation. We outline the proof below.

We decompose the proof of the non-trivial part of the theorem into parts (Parts 1–4) so that the beginning of the original proof is cited almost word by word (Parts 1–3), while the remainder of the proof is only sketched (Part 4).

The sketch of the proof of the non-trivial part of Theorem 6.1.9:

Part 1. About the framework of the proof:

$\mathfrak{A}$ can be assumed to be atomic. Namely, by [Hen-Mon-Tar,85], 2.7.5, 2.7.13, every Boolean algebra $\mathfrak{B}$ with operators can be embedded into an atomic one such that all the equations valid in $\mathfrak{B}$, and in which "$-$" does not occur, continue to hold in the atomic one. This latter condition is satisfied in $\mathfrak{B}$ because it is easy to eliminate the "$-$" from the axioms. As a consequence, from now on $\mathfrak{A}$ is assumed to be atomic, satisfying the axioms.

Let $At\,\mathfrak{A}$ denote the set of all atoms of $\mathfrak{A}$. An isomorphism $\mathrm{rep} : \mathfrak{A} \twoheadrightarrow \mathfrak{B}$ will be defined for some $\mathfrak{B} \in \mathsf{Crs}_\alpha$ such that the equality below holds:

$$\text{(6.1.2)} \qquad \mathrm{rep}\,(x) = \bigcup \{\, \mathrm{rep}\,(a) \,:\, a \in At\,\mathfrak{A}, a \leq x \,\} \quad \text{for every} \quad x \in A.$$

Let V be a set of α-sequences and for every $X \subseteq V$ and $i, j < \alpha$ let $\mathsf{C}_i X \stackrel{d}{=} \{\, f \in V \,:\, f(i/u) \in X \text{ for some } u \,\}$, $\mathsf{D}_{ij} \stackrel{d}{=} \{ f \in V \,:\, f_i = f_j \}$. Assume that $\mathrm{rep} : A \to \{X : X \subseteq V\}$ is a function for which (6.1.2) holds. Then it is easy to check that rep is an isomorphism onto a $\mathfrak{B} \in \mathsf{Crs}_\alpha$ with $1^{\mathfrak{B}} \subseteq V$ if and only if

(6.1.3) the conditions (i)–(v) below

hold for every $a, b \in At\,\mathfrak{A}$ and $i, j < \alpha$:

(i) $\mathrm{rep}\,(a) \cap \mathrm{rep}\,(b) = 0$ if $a \neq b$,

(ii) $\mathrm{rep}\,(a) \subseteq \mathsf{D}_{ij}$ if $a \leq \mathsf{d}^{\mathfrak{A}}_{ij}$ and $\mathrm{rep}\,(a) \cap \mathsf{D}_{ij} = \emptyset$ if $a \cdot \mathsf{d}^{\mathfrak{A}}_{ij} = 0$,

(iii) $\mathrm{rep}\,(a) \subseteq \mathsf{C}_i\,\mathrm{rep}\,(b)$ if $a \leq \mathsf{c}^{\mathfrak{A}}_i b$,

(iv) $\mathrm{rep}\,(a) \cap \mathsf{C}_i\,\mathrm{rep}\,(b) = \emptyset$ if $a \cdot \mathsf{c}^{\mathfrak{A}}_i b = 0$,

(v) $\mathrm{rep}\,(a) \neq \emptyset$.

The function rep (and a set V of α-sequences) will be constructed with the above properties, step-by-step.

Part 2. About the 0th step:

For the α-sequence f let $\ker\,(f) \stackrel{d}{=} \{\, (i, j) \in {}^2\alpha \,:\, f_i = f_j \,\}$.

For every $a \in At\,\mathfrak{A}$ let $\mathrm{Ker}\,(a) \stackrel{d}{=} \{\, (i, j) \in {}^2\alpha \,:\, a \leq \mathsf{d}^{\mathfrak{A}}_{ij} \,\}$.

Then $\mathrm{Ker}(a)$ is an equivalence relation on α by our axioms (C_5)–(C_7). For every $a \in At\,\mathfrak{A}$ let f_a be an α-sequence such that for every $a, b \in At\,\mathfrak{A}$ we have

a) $\ker\,(f_a) = \mathrm{Ker}\,(a)$, b) $\mathrm{Rg}\,(f_a) \cap \mathrm{Rg}\,(f_b) = \emptyset$ if $a \neq b$.

Such a system $\{f_a : a \in At\,\mathfrak{A}\}$ of α-sequences does exist.

Define $\mathrm{rep}_0(a) \stackrel{d}{=} \{f_a\}$, for every $a \in At\,\mathfrak{A}$.

Then the function rep_0 satisfies the *basic* conditions (i), (ii), (iv) and (v) but it does not satisfy condition (iii). Below, we shall make condition (iii) become true step-by-step, and later we shall check that conditions (i), (ii), (iv), (v) remain true in each step.

Part 3. About the successor, (n+1)th step, i.e., about the definition of the function rep_{n+1}, and about the definition of rep:

Let $R = At\,\mathfrak{A} \times At\,\mathfrak{A} \times \alpha$, ρ be an ordinal and let $r : \rho \to R$ be an enumeration of R such that for all $n \in \rho$ and $\langle a, b, i\rangle \in R$ there is $m \in \rho$, $m > n$ such that $r(m) = \langle a, b, i\rangle$. Such ρ and r clearly exists.

Assume that $n \in \rho$ and $\mathrm{rep}_n : At\,\mathfrak{A} \to \{X : X \subseteq V'\}$ is already defined where V' is a set of α-sequences. We define $\mathrm{rep}_{n+1} : At\,\mathfrak{A} \to \{X : X \subseteq V''\}$, where V'' is a set of α-sequences. Let $r(n) = \langle a, b, i\rangle$. If $a \not\leq \mathsf{c}_i b$, then

$$\mathrm{rep}_{n+1} \stackrel{d}{=} \mathrm{rep}_n\,.$$

Assume $a \leq \mathsf{c}_i b$. Then $\mathrm{rep}_{n+1}(e) \stackrel{d}{=} \mathrm{rep}_n(e)$ for all $e \in At\,\mathfrak{A}$, $e \neq b$.

Further,

case 1. $b \leq \mathsf{d}_{ij}$ for some $j < \alpha$, $j \neq i$. Then

$$\mathrm{rep}_{n+1}(b) = \mathrm{rep}_n(b) \cup \{\, f(i/f_j) \,:\, f \in \mathrm{rep}_n(a)\}$$

case 2. $b \not\leq \mathsf{d}_{ij}$ for all $j < \alpha$, $j \neq i$. For every $f \in \mathrm{rep}_n(a)$ let u_f be such that

(1) $$u_f \notin \bigcup \Big\{\mathrm{Rg}\,(h) \,:\, h \in \bigcup \{\,\mathrm{rep}_n(e) \,:\, e \in At\,\mathfrak{A}\}\Big\}$$

(2) $$u_f \neq u_h \quad \text{if} \quad f \neq h, \quad f, h \in \mathrm{rep}_n(a).$$

Now,

$$\mathrm{rep}_{n+1}(b) = \mathrm{rep}_n(b) \cup \{\, f(i/u_f) \,:\, f \in \mathrm{rep}_n(a)\}\,.$$

Let $n \in \rho$ be a limit ordinal and assume that rep_m is defined for all $m < n$. Then

$$\mathrm{rep}_n(e) \stackrel{d}{=} \bigcup \{\,\mathrm{rep}_m(e) \,:\, m < n\}$$

for all $e \in At\,\mathfrak{A}$.

By this, $\langle \mathrm{rep}_n \,:\, n \in \rho\rangle$ is defined. Now we define

$$\mathrm{rep}\,(a) \stackrel{d}{=} \bigcup \{\,\mathrm{rep}_n(a) \,:\, n \in \rho\}$$

for all $a \in At\,\mathfrak{A}$. Let

$$V \stackrel{d}{=} \bigcup \{\operatorname{rep}(a) : a \in At\,\mathfrak{A}\}.$$

It will be shown that conditions (i)–(v) hold for the above rep function and V.

Part 4. On the proof of the properties (i)–(v).

They are proven by induction. The proofs of the properties (ii), (iii) and (v) are relatively easy. Instead of (i) and (iv) a stronger property, denoted by (iv)$'$ is proven such that it implies both (i) and (iv). In the proof of (iv)$'$ Jónsson's famous theorem plays a key role ([Hen-Mon-Tar,85], 3.2.17, p. 68). It concerns the extension of a mapping, having certain fixed properties, from the elementary transformations $[i/j]$ and $[i,j]$, to arbitrary finite transformations. ■

Notice, that the above proof yields a complete representation. A similar proof can be given for the Resek–Thompson theorem by the technique "games", see Sayed Ahmed [Say,g] (similarly to the proof of the analogous theorem for relation algebras, see [Hir-Hod,02a]).

We note that the original result of Resek (in [Res,75]) is a bit stronger in a sense, than the Resek–Thompson theorem. Maddux published a result in [Mad,89b] which is equivalent to Resek's original theorem, but it contains the simplifications due to Thompson. Maddux proved the following:

If $\mathfrak{A} \in \mathsf{NA}_\alpha$ atomic, complete and $\mathfrak{A}$ satisfies the MGR *property, then* $\mathfrak{A} \in \mathbf{I}\,\mathsf{Rl}\,\mathsf{Cs}_\alpha$ (here Rl denotes relativization).

The following result of Andréka concerns the class G_α (see [And,01]). A simple axiom system is given for G_α if α is finite. In the proof, the step-by-step method is used, again.

Denote LS_α the subclass of CNA^+_α satisfying the axioms

$$\mathrm{Ax}_{ij}\ :\ x \leq \mathsf{c}_i\mathsf{c}_j\left(\mathsf{s}^i_j\mathsf{c}_j x \cdot \mathsf{s}^j_i\mathsf{c}_i x \cdot \prod_{k<\alpha, k\neq i,j} \mathsf{s}^k_i\mathsf{s}^i_j\mathsf{s}^j_k\mathsf{c}_k x\right)$$

$i, j < \alpha$.

Theorem 6.1.12. $\mathfrak{A} \in \mathsf{LS}_\alpha$ *if and only if* $\mathfrak{A} \in \mathbf{I}\mathsf{G}_\alpha$, *where α is finite* ([And,01], Theorem 1(i), [Hir-Hod,02a], Fact. 5.38).

A similar claim can be formulated for the class $\mathbf{H}\mathsf{G}_\alpha$ ($\mathbf{H}\mathsf{G}_\alpha$ denotes the class of the homomorphic images of the algebras in G_α, [And,01], Theorem 1(ii)).

2. Transposition Equality Algebras, Cylindric Quasi-polyadic Algebras

Let us consider the operation ${}_k\mathsf{s}(i,j)$ in CA_α, where ${}_k\mathsf{s}(i,j)y = \mathsf{s}_i^k\mathsf{s}_j^i\mathsf{s}_k^j y$ and i, j, k are different. The properties of ${}_k\mathsf{s}(i,j)$ are investigated in detail in [Hen-Mon-Tar,85] 1.5. Andréka and Thompson proved that the following property is equivalent to the two merry-go-round properties:

$$(6.2.1) \qquad {}_k\mathsf{s}(i,j)\,{}_k\mathsf{s}(j,m)\mathsf{c}_k x = {}_k\mathsf{s}(j,m)\,{}_k\mathsf{s}(m,i)\mathsf{c}_k x$$

under the other CNA_α axioms if $k \notin \{i,j,m\}$, $j \notin \{m,i\}$ (Proposition 3 in [And-Tho,88]). (6.2.1) is closely related to the known property of the transposition operation $[i,j]$ or abstract transposition p_{ij}, formulated as axiom (F_8) in [Sai-Tho,91] (we return to the background of the MGR property at the end of this Section). It is well-known that the transposition operation (or equivalently, the substitution operator s_τ for finite τ) cannot be introduced in arbitrary cylindric algebra.

The above observations, together with several related observations, have led to the idea of investigating r-representations for transposition algebras, quasi-polyadic and polyadic algebras.

Definition 6.2.1 (Trs_α). A *transposition relativized set algebra*

$$\langle A, \cup, \cap, \sim_V, \emptyset, V, \mathsf{C}_i^V, {}^V\mathsf{S}_j^i, [i,j]^V, \mathsf{D}_{ij}^V \rangle_{i,j<\alpha}$$

is a set algebra such that its cylindric reduct is in Crs_α, ${}^V\mathsf{S}_j^i$ is defined by ${}^V\mathsf{D}_{ij}$ and C_i^V in the usual way, and $\mathfrak{A}$ is closed under $[i,j]^V$, where

$$[i,j]^V X = \{ y \in V : y \circ [i,j] \in X \},$$

and $[i,j]$ denotes the ordinary transposition defined on α.

The upper index V is often omitted from $[i,j]^V$, in this case we can disambiguate $[i,j]$ and $[i,j]^V$ taking the context into consideration.

Notice that $[i,j]^V V = V$ in Trs_α. To see this, recall that $[i,j]^V V \subseteq V$, by definition. Now, let us apply $[i,j]^V$ to this inclusion. Then, the equality $y \circ [i,j] \circ [i,j] = y$ implies that, for the left hand side, $[i,j]^V\,[i,j]^V V = V$, thus we get the opposite inclusion $V \subseteq [i,j]^V V$.

Definition 6.2.2 (Gwt_α). A set algebra in Trs_α is called a *generalized weak transposition relativized set algebra* if there are sets U_k, $k \in K$ and sequences $p_k \in {}^\alpha U_k$ such that $V = \bigcup_{k \in K} {}^\alpha U_k^{(p_k)}$, where V is the unit.

With the class Gwt_α we can associate the cylindric set algebra class Gws_α (see [Hen-Mon-Tar,85] 3.1.1). Besides their different types, a further difference between these classes is that the disjointness of the sets U_k's is not assumed in Gwt_α. The subclass of Gwt_α in which the disjointness of the U_k's is assumed is denoted by $\overset{\bullet}{\mathsf{Gwt}}_\alpha$.

Now, let us define some abstract classes.

Definition 6.2.3 (TA_α). A *transposition algebra* of dimension α ($\alpha \geq 3$) is an algebra

$$\mathfrak{A} = \left\langle A, +, \cdot, -, 0, 1, \mathsf{c}_i, \mathsf{s}^i_j, \mathsf{p}_{ij}, \mathsf{d}_{ij} \right\rangle_{i,j<\alpha},$$

where c_i, s^i_j, p_{ij} are unary operations, d_{ij} are constants, and the axioms below are assumed for every $i, j, k < \alpha$:

(F$_0$) $\langle A, +, \cdot, -, 0, 1\rangle$ is a Boolean algebra, $\mathsf{s}^i_i = \mathsf{p}_{ii} = \mathsf{d}_{ii} = Id \restriction A$ and $\mathsf{p}_{ij} = \mathsf{p}_{ji}$,

(F$_1$) $x \leq \mathsf{c}_i x$,

(F$_2$) $\mathsf{c}_i(x+y) = \mathsf{c}_i x + \mathsf{c}_i y$,

(F$_3$) $\mathsf{s}^i_j \mathsf{c}_i x = \mathsf{c}_i x$,

(F$_4$) $\mathsf{c}_i \mathsf{s}^i_j x = \mathsf{s}^i_j x \quad i \neq j$,

(F$_5$)* $\mathsf{s}^i_j \mathsf{s}^k_m x = \mathsf{s}^k_m \mathsf{s}^i_j x$ if $i, j \notin \{k, m\}$,

(F$_6$) s^i_j and p_{ij} are Boolean endomorphisms (i.e., $\mathsf{s}^i_j(-x) = -\mathsf{s}^i_j x$, etc.),

(F$_7$) $\mathsf{p}_{ij}\mathsf{p}_{ij} x = x$,

(F$_8$) $\mathsf{p}_{ij}\mathsf{p}_{ik} x = \mathsf{p}_{jk}\mathsf{p}_{ij} x$ if i, j, k are distinct,

(F$_9$) $\mathsf{p}_{ij}\mathsf{s}^i_j x = \mathsf{s}^j_i x$,

(F$_{10}$) $\mathsf{s}^i_j \mathsf{d}_{ij} = 1$,

(F$_{11}$) $x \cdot \mathsf{d}_{ij} \leq \mathsf{s}^i_j x$.

Definition 6.2.4 (TAS_α). A *strong transposition algebra* TAS_α can be obtained from a transposition algebra TA_α by replacing $(\mathrm{F}_5)^*$ by the following, stronger axiom:

$$(\mathrm{F}5)\ :\ \mathsf{s}^i_j\mathsf{c}_k x = \mathsf{c}_k\mathsf{s}^i_j x\ \ k \notin \{i,j\}.$$

The class TAS_α is the same as the class of *finitary polyadic equality algebras* (class FPEA_α) introduced in [Sai-Tho,91]. Here, the notations of the axioms in [Sai-Tho,91] are preserved. But, it seems expedient to change the terminology of FPEA_α, especially in the case of the weaker version TA_α.

A transformation τ defined on α is called *finite* if $\tau i = i$ with finitely many exceptions ($i \in \alpha$).

As is known, *quasi-polyadic equality algebras* (QPEA_α) are polyadic algebras such that the transformations occurring in them are finite ones ([Hen-Mon-Tar,85], p 266). Sain and Thompson proved in [Sai-Tho,91] that FPEA_α (thus TAS_α, as well) and the class of quasi-polyadic equality algebras are definitionally equivalent ([Sag,thisVol], Section 1).

The following question arises: which subclass of quasi-polyadic equality algebras corresponds to TA_α under the above definitional equivalence of QPEA_α and FPEA_α?

It is easy to check that cylindrification is not commutative in TA_α, therefore simultaneous cylindrification does not exist in the expected class, only single cylindrification. We introduce the concept of *cylindric quasi-polyadic equality algebras:*

Definition 6.2.5. By a *cylindric quasi-polyadic equality algebra* of dimension α, briefly CQE_α, we mean an algebra $\mathfrak{A} = \langle \mathfrak{B}, \mathsf{c}_i, \mathsf{s}_\tau, \mathsf{d}_{ij}\rangle_{i,j<\alpha}$, where $\tau \in {}^\alpha\alpha$ is finite and the following equations (CQ_0)–(CQ_6), furthermore (E_1)–(E_3) are valid in $\mathfrak{A}$ for every finite $\tau, \sigma \in {}^\alpha\alpha$ and $i, j < \alpha$:

(CQ_0) $\mathfrak{B} = \langle B; \cdot, -, 0, 1\rangle$ *is a Boolean Algebra,*

(CQ_1) $x \leq \mathsf{c}_i x$,

(CQ_2) $\mathsf{c}_i(x \cdot \mathsf{c}_i y) = \mathsf{c}_i x \cdot \mathsf{c}_i y$,

(CQ_3) $\mathsf{s}_{Id} x = x$,

(CQ_4) $\mathsf{s}_{\sigma\circ\tau} x = \mathsf{s}_\sigma \mathsf{s}_\tau x$,

(CQ_5) $\mathsf{s}_\sigma(x + y) = \mathsf{s}_\sigma x + \mathsf{s}_\sigma y$ and $\mathsf{s}_\sigma(-x) = -\mathsf{s}_\sigma x$,

(CQ_6) $\mathsf{c}_i \mathsf{s}_\tau x \leq \mathsf{s}_\tau \mathsf{c}_j x$, if $\sigma^{-1*}\{i\}$ equals $\{j\}$ or the empty set (in the latter case c_j is the identity operator),

(E$_1$) $\mathsf{d}_{ii} = 1$,

(E$_2$) $x \cdot \mathsf{d}_{ij} \leq \mathsf{s}_{[i/j]} x$,

(E$_3$) $\mathsf{s}_\tau \mathsf{d}_{ij} = \mathsf{d}_{\tau i \tau j}$.

Here σ^{-1*} denotes complete inverse image. In [Fer] it is proven that the varieties CQE_α and TA_α are definitionally equivalent.

Definition 6.2.6. An algebra $\mathfrak{A} \in \mathsf{TA}_\alpha$ is *r-representable* if $\mathfrak{A} \in \mathbf{I}\,\mathsf{Trs}_\alpha$.

Lemma 6.2.7. *The following propositions* (i) *and* (ii) *hold:*

(i) *If* $\mathfrak{A} \in \mathsf{Trs}_\alpha$, *then* $\mathfrak{A} \in \mathsf{Gwt}_\alpha$ *if and only if* $x \in V$ *implies both* $x \circ [i, j] \in V$ *and* $x \circ [i/j] \in V$, *for every* $i, j < \alpha$.

(ii) *If* $\mathfrak{B} \in \mathsf{TA}_\alpha$ *and* $\mathfrak{B}$ *is r-representable, then* $\mathfrak{B} \in \mathbf{I}\,\mathsf{Gwt}_\alpha$.

Proof.

(i) If $\mathfrak{A} \in \mathsf{Gwt}_\alpha$, then, by the definition of V, V is closed under the operations $[i, j]$ and $[i/j]$. Conversely, we have to prove that V is of the form $\bigcup_{k \in K} {}^\alpha U_k^{(p_k)}$. The condition in (i) implies that V is closed under the finite transformations of α, i.e., $x \in V$ implies $x \circ \tau \in V$ if τ is finite. This is because, as is known, finite transformations can be composed by finitely many applications of elementary transpositions and replacements. We show now that V is of the form $\bigcup_{x \in V} {}^\alpha(\mathrm{Rg}\, x)^{(x)}$ (this latter is really a Gwt_α unit). $V \subseteq \bigcup_{x \in V} {}^\alpha(\mathrm{Rg}\, x)^{(x)}$ obviously holds by definition. Conversely, if $y \in \bigcup_{x \in V} {}^\alpha(\mathrm{Rg}\, x)^{(x)}$, then $y = x \circ \tau$ for some $x \in V$ and finite τ, by the definition of the weak space ${}^\alpha(\mathrm{Rg}\, x)^{(x)}$. But, $x \circ \tau \in V$, by assumption. Thus $\bigcup_{x \in V} {}^\alpha(\mathrm{Rg}\, x)^{(x)} \subseteq V$, and, consequently $V = \bigcup_{x \in V} {}^\alpha(\mathrm{Rg}\, x)^{(x)}$, as we claimed.

(ii) The proof is similar to that of Lemma 6.1.6, making use of the above part (i) and the fact that the foregoing isomorphism preserves the operations s_j^i and p_{ij}. ■

The following main r-representation theorem holds for algebras in TA_α ([Fer,12a], Theorem 3.1):

Theorem 6.2.8 (Ferenczi).

$$\mathfrak{A} \in \mathsf{TA}_\alpha \quad \textit{if and only if} \quad \mathfrak{A} \in \mathbf{I}\,\mathsf{Gwt}_\alpha$$

where $\alpha \geq 3$.

If we set out from the problem of the axiomatizability of the class Gwt_α of set algebras, then the reformulation of the theorem is the following one: *The class* Gwt_α *is axiomatizable and the* TA_α *axioms axiomatize this class.*

By Definition 6.2.4, the class TAS_α is obtained from TA_α so that axiom $(\mathrm{F}_5)^*$ is replaced by (F_5). Thus, we obtain the following:

Corollary 6.2.9. $\mathfrak{A} \in \mathsf{TAS}_\alpha$ *if and only if* $\mathfrak{A} \in \mathbf{I}\big(\mathsf{Gwt}_\alpha \cap \mathsf{Mod}\,(\mathrm{F}_5)\big)$ $(\alpha \geq 3)$.

As is known, TAS_α is not representable in the classical sense (see [Sai-Tho,91]), thus Gwt_α cannot replaced by $\overset{\bullet}{\mathsf{Gwt}}_\alpha$ in the Corollary and in Theorem 6.2.8. Taking into consideration the definitional equivalence of the class TAS_α and quasi-polyadic equality algebras (class QPEA_α, see [Sai-Tho,91]), Corollary 6.2.9 implies that the class QPEA_α is also r-representable in a sense (introducing the substitution operation S_τ in Gwt_α for finite τ's). Similarly, the definitional equivalence of TA_α and CQE_α implies that CQE_α is also r-representable (see [Fer]).

The proof of Theorem 6.2.8 follows Andréka's proof for the Resek–Thompson theorem (from now, AP or *cylindric case*), with some modifications in accordance with the transposition type of the algebras. The proof is a non-trivial modification of Andréka's proof. Among others, a difference between the cylindric and transposition cases is that the definition of the function rep_0 is more complex in the transposition case. We emphasize only the differences between the two proofs discussing the proof with respect to parts 1–4 of the AP. We refer to *the sketch of the proof* of Theorem 6.1.9 (in Section 1) and to the complete proof of the theorem (proof of Theorem 3.1 in [Fer,12a]).

The sketch of the proof of the non-trivial part of Theorem 6.2.8:

Let us consider Part 1 in AP. The only necessary change is that we need to prove the preservation of the operation p_{ij}. By [Sai-Tho,91], i.e., by the term definitional equivalence of TAS_α and quasi-polyadic equality algebras, p_{ij} may be considered as the transformation $\mathsf{s}_{[i,j]}$. We are going to use $\mathsf{s}_{[i,j]}$ rather than p_{ij}. So we have to prove:

(vi) $\mathrm{rep}\left(\mathsf{s}_{[i,j]}a\right) = [i,j]\,\mathrm{rep}\,(a)$.

We prove the following more general property

$$\text{(6.2.2)} \qquad \text{(vi)}' \qquad \mathrm{rep}\,(\mathsf{s}_\sigma a) = \mathsf{S}_\sigma\,\mathrm{rep}\,(a),$$

where σ is an arbitrary *finite* permutation of α.

We note that the original cylindric representation is *complete* and this will be transmitted to our construction too.

The next part, i.e., Part 2 in AP, is the *definition of the* 0*th step,* i.e., the definition of the function rep_0. *We have to change essentially the definition* of rep_0 to handle property (vi)′. First, as a preparation, we introduce two equivalence relations:

Let a be an arbitrary *fixed* atom. The definition of the relation $\equiv_a$ ($\equiv$, for short) on α is:

$$i \equiv j \quad \text{if and only if} \quad \mathsf{s}_{[i,j]}a = a.$$

Notice that

$$(i,j) \in \mathrm{Ker}\,(a) \quad \text{implies that} \quad i \equiv j.$$

Let us consider the following equivalence relation $\sim$ on $At\,\mathfrak{A}$:

$$a \sim b \quad \text{if and only if} \quad b = \mathsf{s}_\tau a \quad \text{for some finite permutation} \quad \tau$$

$a, b \in At\,\mathfrak{A}$.

It is not difficult to check that $\equiv$ and $\sim$ are equivalence relations. Let us choose and fix representative points for the equivalence classes concerning $\sim$ and let Rp denote this fixed set of representative points.

The definition of the function rep_0 *is:*

If $c \in Rp$, then let

$$\mathrm{rep}_0(c) = \{\mathsf{S}_\tau f_c \ : \ \mathsf{s}_\tau c = c\}$$

where f_c is the sequence defined in the original proof and τ is a finite permutation of α. If $b = \mathsf{s}_\sigma c$, then let

$$\mathrm{rep}_0(b) = \mathsf{S}_\sigma\, \mathrm{rep}_0(c).$$

Lemma 4 in [Fer,12a] states that this definition is unique, i.e., the definition of $\mathrm{rep}_0(b)$ does not depend on the permutation σ.

Lemma 5 in [Fer,12a] states the following: $\mathrm{rep}_0(\mathsf{s}_\sigma a) = \mathsf{S}_\sigma\, \mathrm{rep}_0(a)$, where σ is an arbitrary finite permutation of α and a is an arbitrary atom, i.e., the property (vi)′ in (6.2.2) is satisfied for rep_0.

Then (similarly to the Part 4 in AT), we can shown that rep_0 satisfies the conditions (i), (ii) and (iv) in (6.1.3) (see the Lemmas 6–8 in [Fer,12a]). The proof here requires a bit more complex considerations than in the cylindric case.

Let us consider Part 3 in AP. As regards the successor $(n+1)th$ *step* of the proof, i.e., the definition of the function rep_{n+1}, *the modified construction is the following:*

In order to ensure the validity of the property (vi)′ in (6.2.2), we modify the original cylindric construction. We consider equivalence classes of triples instead of single triples. From the point of view of the cylindric case this means that we classify the single triples according to an equivalence relation to be introduced below.

The cylindric construction uses an *arbitrary fixed* free transfinite enumeration of the $\langle a, b, i \rangle$ triples, where $a, b \in At\,\mathfrak{A}$, $i < \alpha$. In contrast to this, we assume certain restrictions for this enumeration, by which the triples are classified in a sense. The function rep_n is defined in accordance with this classification. Beyond this little modification we *do not change* the original cylindric procedure, so the cylindric proof works here as well. In Lemma 9 in [Fer,12a] it is proven that property (vi)′ in (6.2.2) is *preserved* in every step.

The *completion* of the proof:

In the cylindric case it is proven that there is an isomorphism, denoted by rep′, between the cylindric reduct $\mathfrak{Rd}_{ca}\mathfrak{A}$ of $\mathfrak{A}$ and some algebra $\mathfrak{B}' \in \mathsf{D}_\alpha$. In [Fer,12a] it is shown that this isomorphism preserves the operation s_σ for any finite permutation σ on α. Now, Lemma 6.2.7 implies that $\mathfrak{B}$' can be considered as an algebra in Gwt_α. So rep′ is an isomorphism between $\mathfrak{A}$ and a $\mathfrak{B} \in \mathsf{Gwt}_\alpha$. ■

As we mentioned at the beginning of this Section, we return to the analysis of the connection of the merry-go-round properties and the r-representability of certain *cylindric algebras.* We show that a non-commutative cylindric algebra (algebra in CNA_α) is r-representable if and only if it can be regarded as a "weak" transposition algebra, i.e., as a "partial transposition algebra".

Assume that $\mathfrak{B}$ is an algebra with the type of TA_α.

Definition 6.2.10 (PTA_α). $\mathfrak{B}$ is a *partial transposition algebra* if the TA_α axioms are satisfied in $\mathfrak{B}$ except for the axioms including the operation p_{ij}. These axioms must be satisfied only for the dimension complemented elements (i.e., for the elements x with $\Delta x \neq \alpha$).

The following theorem states that exactly those non-commutative cylindric algebras are r-representable which are cylindric reducts of certain partial transposition algebras.

Theorem 6.2.11. *Assume that* $\mathfrak{A} \in \mathsf{CNA}_\alpha$ $(\alpha \geq 4)$. *Then* $\mathfrak{A}$ *is* r*-representable if and only if* $\mathfrak{A}$ *is a cylindric reduct of some partial transposition algebra* $\mathfrak{B}$. (Theorem 3.5 in [Fer,11]).

It is easy to formulate the similar theorem for algebras in CA_α instead of CNA_α.

The following corollary of Theorem 6.2.8 is analogous to [Hen-Mon-Tar,85], 5.4.18.

Corollary 6.2.12. *If* $\mathfrak{B}$ *is a* cylindric reduct *of some* $\mathfrak{A} \in \mathsf{TA}_\alpha$*, then* $\mathfrak{B}$ *is representable, i.e.,* $\mathfrak{B} \in \mathbf{I}\,\mathsf{Gws}_\alpha$.

3. Cylindric Polyadic Equality Algebras

The algebras introduced in this Section have infinite substitution operations (S_τ, or s_τ, $\tau \in Q$, $Q \subseteq {}^\alpha\alpha$), and single cylindrifications (c_i, $i < \alpha$), instead of general cylindrifications $\mathsf{c}_{(\Gamma)}$ $(\Gamma \subseteq \alpha)$. These algebras, in the rigorous sense, can no longer be considered as cylindric-like algebras. These algebras a mixture of cylindric and polyadic algebras, therefore these algebras are called *cylindric polyadic equality algebras.* Throughout the Section the dimensions α are assumed to be infinite (the finite dimensional case is connected with the quasi-polyadic case).

The following investigations focus on the analysis of the substitution operations with *infinite* transformations and the equalities (in another terminology, *transformation systems* (see [Dai-Mon,63]) with equalities). The techniques needed for these investigations are different from the case of finite transformations. It is important to emphasize that we investigate structures *with equalities,* because the behaviour of the foregoing structures *without* equalities is completely different. For example, polyadic algebras without equality are representable in the classical sense (see [Dai-Mon,63]), whereas polyadic equality algebras are not (see [Sag,thisVol], Section 2).

As a consequence of the representation theorem for TA_α we obtained that quasi-polyadic equality algebras, i.e., polyadic equality algebras with finite substitutions, are r-representable. The question arises: what about the r- representability of cylindric polyadic equality algebras?

First, some classes of *set algebras* are introduced: the classes Cprs_α, Gp_α, $\mathsf{Gp}_\alpha^{\mathrm{reg}}$, Gpw_α and $\mathsf{Gpw}_\alpha^{\mathrm{reg}}$.

Definition 6.3.1 (Cprs_α). An algebra $\mathfrak{A}$ is a *cylindric polyadic relativized set algebra* of dimension α, with a unit set V of α-termed sequences if $V \subseteq {}^\alpha U$ (U is called the base) and $\mathfrak{A}$ is of the form:

$$\langle A, \cup, \cap, \sim_V, \emptyset, V, \mathsf{C}_i^V, \mathsf{S}_\tau^V, \mathsf{D}_{ij}^V \rangle_{\tau \in Q,\ i,j \in \alpha} \tag{6.3.1}$$

closed under the Boolean operations $\cup$, $\cap$, $\sim_V$, also closed under the cylindrifications

$$\mathsf{C}_i^V X = \{y \in V \,:\, y_u^i \in X \text{ for some } u \in U\}$$

where $\left(y_u^i\right)_j = y_j$ if $j \neq i$, and $\left(y_u^i\right)_j = u$ if $j = i$, and closed under the transformations

$$\mathsf{S}_\tau^V X = \{x \in {}^\alpha U \,:\, x \circ \tau \in X\}$$

and A contains $\emptyset$, V the diagonals

$$\mathsf{D}_{ij}^V = \{y \in V \,:\, y_i = y_j\}$$

(see also [Hen-Mon-Tar,85], Definition 5.4.22).

If the transformation S_τ^V is dropped in (6.3.1), and $Q = {}^\alpha\alpha$, we get a cylindric relativized set algebra, i.e., the cylindric reduct of a Cprs_α is a Crs_α.

Definition 6.3.2 (Gp_α and $\mathsf{Gp}_\alpha^{\mathrm{reg}}$). A set algebra $\mathfrak{A}$ in Cprs_α is called a *generalized polyadic relativized set algebra* ($\mathfrak{A} \in \mathsf{Gp}_\alpha$) if there are sets U_k such that $V = \bigcup_{k \in K} {}^\alpha U_k$, where V is the unit, and $Q = {}^\alpha\alpha$. The subclass of Gp_α such that the pairwise disjointness of the U_k's is assumed is denoted by $\dot{\mathsf{Gp}}_\alpha$. An algebra $\mathfrak{A}$ in Gp_α is called regular ($\mathfrak{A} \in \mathsf{Gp}_\alpha^{\mathrm{reg}}$) if for each $X \in A$, $x \in X$ and $y \in V$, the condition $(\Delta X \cup 1) \upharpoonright x \subseteq y$ implies $y \in X$.

The concept of *regularity* (see [Hen-Mon-Tar,85], Definition 3.1.1(viii)) compensates, in a sense, for the lack of general cylindrification $\mathsf{C}_{(\Gamma)}$ because if such a cylindrification exists, then $(\Delta X \cup 1) \upharpoonright x \subseteq y$ implies that $y \in \mathsf{C}_{(\alpha \sim (\Delta X \cup 1))} X = X$.

Definition 6.3.3. Assume that $m < \alpha$ and m is infinite. Given a set U and a fixed sequence $p \in {}^{\alpha}U$, the set

$$ {}^{\alpha}_{m}U^{(p)} = \{x \in {}^{\alpha}U \,:\, x \text{ and } p \text{ are different at most in } m\text{-many places}\} $$

is called the *m-weak space* (or m-weak Cartesian space) determined by p and U. Here p is called a *support* of the m-weak space and U is called the *base.*

Recall that the finite version (m is finite) of the above definition is the definition of the *weak space*, in notation ${}^{\alpha}U^{(p)}$ ([Hen-Mon-Tar,85], 3.1.2).

Definition 6.3.4. A transformation τ defined on α is said to be an *m-transformation* if $\tau i = i$ except for m-many $i \in \alpha$. The class of m-transformations is denoted by ${}_{m}\mathsf{T}_{\alpha}$.

Definition 6.3.5 (${}_{m}\mathsf{Gwp}_{\alpha}$ and ${}_{m}\mathsf{Gwp}_{\alpha}^{\mathrm{reg}}$). A set algebra $\mathfrak{A}$ in Cprs_{α} is called a *generalized m-weak polyadic relativized set algebra* ($m < \alpha$) ($\mathfrak{A} \in {}_{m}\mathsf{Gwp}_{\alpha}$) if there are sets U_k and $p_k \in {}^{\alpha}U_k$ such that $V = \bigcup_{k \in K} {}^{\alpha}_{m}U_k^{(p_k)}$, where V is the unit, and $Q = {}_{m}\mathsf{T}_{\alpha}$. The subclass of ${}_{m}\mathsf{Gwp}_{\alpha}$ such that the pairwise disjointness of the U_k's is assumed is denoted by ${}_{m}\dot{\mathsf{G}}\mathsf{wp}_{\alpha}$. An algebra $\mathfrak{A}$ in ${}_{m}\mathsf{Gwp}_{\alpha}$ is called *regular* ($\mathfrak{A} \in {}_{m}\mathsf{Gwp}_{\alpha}^{\mathrm{reg}}$) if for each $X \in A$, $x \in X$ and $y \in V$, such that x and y are different at most in m-many places, condition $(\Delta X \cup 1) \upharpoonright x \subseteq y$ implies $y \in X$.

Remarks.

a) One of the differences between the classical cylindric class Gs_{α} (generalized cylindric set algebras, see [Hen-Mon-Tar,85], Definition 3.1.2) and Gp_{α} is that in Gp_{α} the pairwise disjointness of the U_k's is not required.

b) The cylindric reduct of a Gp_{α} is obviously a G_{α}.

The known characterizations of the classes ${}_{m}\mathsf{Gwp}_{\alpha}$ and Gp_{α} are as follows:

Lemma 6.3.6. *If V is the unit of an $\mathfrak{A} \in \mathsf{Cprs}_{\alpha}$, then*

(i) *$\mathfrak{A} \in {}_{m}\mathsf{Gwp}_{\alpha}$ if and only if $y \in V$ implies $y \circ \tau \in V$ for every transformation τ, $\tau \in {}_{m}\mathsf{T}_{\alpha}$. Another equivalent condition for $\mathfrak{A} \in {}_{m}\mathsf{Gwp}_{\alpha}$ is: $\mathsf{S}_{\tau}V = V$ for every transformation τ, $\tau \in {}_{m}\mathsf{T}_{\alpha}$.*

(ii) *$\mathfrak{A} \in \mathsf{Gp}_{\alpha}$ if and only if $y \in V$ implies $y \circ \tau \in V$ for every transformation τ, $\tau \in {}^{\alpha}\alpha$. Another equivalent condition for $\mathfrak{A} \in \mathsf{Gp}_{\alpha}$ is: $\mathsf{S}_{\tau}V = V$ for every τ, $\tau \in {}^{\alpha}\alpha$* (see [And,01]).

This lemma is analogous to the Lemmas 6.1.6 and 6.2.7. As regards the equivalency of the property $\mathsf{S}_\tau V = V$ in (i), for example, the condition *$y \in V$ implies $y \circ \tau \in V$ for every transformation τ*, $\tau \in {}_m\mathsf{T}_\alpha$ means that $V \subseteq \mathsf{S}_\tau V$. Conversely, $\mathsf{S}_\tau V \subseteq V$ is always holds in ${}_m\mathsf{Gwp}_\alpha$.

Now, we turn to some concepts of classes of *abstract algebras*. Three classes will be defined: CPE_α, CPES_α and ${}_m\mathsf{CPE}_\alpha$.

Definition 6.3.7 (CPE_α). A *cylindric polyadic equality algebra* of dimension α is an algebraic structure

$$\mathfrak{A} = \left\langle A, +, \cdot, -, 0, 1, \mathsf{c}_i, \mathsf{s}_\tau, \mathsf{d}_{ij} \right\rangle_{\tau \in {}^\alpha\alpha,\ i,j \in \alpha} \tag{6.3.2}$$

where $+$ and $\cdot$ are binary operations on A, $-$, c_i and s_τ are unary operations on A, 0, 1 and d_{ij} are elements of A such that for every $i, j \in \alpha$, $x, y \in A$, $\sigma, \tau \in {}^\alpha\alpha$ the following postulates are satisfied:

(CP_0) $\langle A, +, \cdot, -, 0, 1 \rangle$ *is a* BA,

(CP_1) $\mathsf{c}_i 0 = 0$,

(CP_2) $x \leq \mathsf{c}_i x$,

(CP_3) $\mathsf{c}_i(x \cdot \mathsf{c}_i y) = \mathsf{c}_i x \cdot \mathsf{c}_i y$,

(CP_4) $\mathsf{s}_{Id} x = x$,

(CP_5) $\mathsf{s}_{\sigma \circ \tau} x = \mathsf{s}_\sigma \mathsf{s}_\tau x$,

(CP_6) $\mathsf{s}_\sigma(x + y) = \mathsf{s}_\sigma x + \mathsf{s}_\sigma y$,

(CP_7) $\mathsf{s}_\sigma(-x) = \sim \mathsf{s}_\sigma x$,

(CP_8) $\mathsf{d} \cdot \mathsf{s}_\sigma x = \mathsf{d} \cdot \mathsf{s}_\tau x$ if the product d of the elements $\mathsf{d}_{\tau i \sigma i}$ ($i \in \Delta x$) exists.

$(CP_9)^*$ $\mathsf{c}_i \mathsf{s}_\sigma x \leq \mathsf{s}_\sigma \mathsf{c}_j x$ if $\sigma^{-1*}\{i\}$ equals $\{j\}$ or the empty set (in the latter case c_j is the identity operator), and the equality holds instead of $\leq$ if σ is a permutation of α,

(E_1) $\mathsf{d}_{ii} = 1$,

(E_2) $x \cdot \mathsf{d}_{ij} \leq \mathsf{s}_{[i/j]} x$,

(E_3) $\mathsf{s}_\tau \mathsf{d}_{ij} = \mathsf{d}_{\tau i \tau j}$.

Definition 6.3.8 (CPES_α). A *strong cylindric polyadic equality algebra* of dimension α is an algebra in CPE_α such that instead of $(\mathrm{CP}_9)^*$ the axiom

$$(\mathrm{CP}_9)\ :\ \mathsf{c}_i\mathsf{s}_\sigma x = \mathsf{s}_\sigma \mathsf{c}_j x$$

is required if $\sigma^{-1*}\{i\}$ equals $\{j\}$ or the empty set (in the latter case c_j is the identity operator), and, in addition, the axiom

$$(\mathrm{C}_4)\ :\ \mathsf{c}_i\mathsf{c}_j x = \mathsf{c}_j\mathsf{c}_i x$$

must also be present, where $i, j \in \alpha$, $\sigma \in {}^\alpha\alpha$.

Definition 6.3.9 (${}_m\mathsf{CPE}_\alpha$). If the transformations τ and σ are assumed to be m-transformations in the definition of CPE_α, i.e. $\tau, \sigma \in {}_m\mathsf{T}_\alpha$, then the concept of *cylindric m-quasi-polyadic equality algebra* of dimension α (${}_m\mathsf{CPE}_\alpha$) is obtained.

Lemma 6.3.10. ${}_m\mathsf{Gwp}_\alpha^{\mathrm{reg}} \cup \mathsf{Gp}_\alpha^{\mathrm{reg}} \subset \mathsf{CPE}_\alpha$.

Proof. As examples, we check the validity of (CP_8) and $(\mathrm{CP}_9)^*$ for an algebra $\mathfrak{A} \in \mathsf{Gp}_\alpha^{\mathrm{reg}}$.

Axiom (CP_8). Assume that $z \in \mathsf{d} \cap \mathsf{S}_\sigma X$, where $X \in A$. Then, $\mathsf{S}_\sigma z \in X$, by definition. $z \in \mathsf{d}$ implies $z_{\tau i} = z_{\sigma i}$ if $i \in \Delta X$, i.e. $(\mathsf{S}_\sigma z)_i = (\mathsf{S}_\tau z)_i$ if $i \in \Delta X$. The regularity of $\mathfrak{A}$ implies that $\mathsf{S}_\tau z \in X$, as well. Thus, $z \in \mathsf{S}_\tau X$. Therefore, $z \in \mathsf{S}_\sigma X$ implies $z \in \mathsf{S}_\tau X$, i.e., $\mathsf{S}_\sigma X \subseteq \mathsf{S}_\tau X$. By symmetry, $\mathsf{S}_\sigma X = \mathsf{S}_\tau X$.

Axiom $(\mathrm{CP}_9)^*$. Assume that $z \in \mathsf{C}_i\mathsf{S}_\sigma X$. Then, $z_u^i \in \mathsf{S}_\sigma X$ for some u. By definition, $\mathsf{S}_\sigma z_u^i \in X$. Notice that $\mathsf{S}_\sigma z_u^i = (\mathsf{S}_\sigma z)_u^j$, where $\{j\} = \sigma^{-1*}\{i\}$ and $\mathsf{S}_\sigma z \in V$ (the latter follows from the fact that $\mathrm{Rg}\,\mathsf{S}_\sigma z = \mathrm{Rg}\, z$ and the definition of a Gp_α unit). Thus, $(\mathsf{S}_\sigma z)_u^j \in X$, as well. $(\mathsf{S}_\sigma z)_u^j \in X$ means that $z \in \mathsf{S}_\sigma\mathsf{C}_j X$. Therefore, $\mathsf{C}_i\mathsf{S}_\sigma X \subseteq \mathsf{S}_\sigma\mathsf{C}_j X$.

We check the converse inclusion, assuming that σ is a permutation of α. Assume that $z \in \mathsf{S}_\sigma\mathsf{C}_j X$, where $\{j\} = \sigma^{-1*}\{i\}$. This means that $(\mathsf{S}_\sigma z)_u^j \in X$ for some u. If σ is a permutation, then the $\mathrm{Rg}\ (\mathsf{S}_\sigma z)_u^j = \mathrm{Rg}\, z_u^i$, therefore, by definition of a Gp_α unit, $z_u^i \in V$. In this case, the argument above can be repeated, i.e., $(\mathsf{S}_\sigma z)_u^j = \mathsf{S}_\sigma z_u^i$ implies $z \in \mathsf{C}_i\mathsf{S}_\sigma X$. Thus, $\mathsf{S}_\sigma\mathsf{C}_j X \subseteq C_i\mathsf{S}_\sigma X$.
■

Remarks.

a) In *polyadic equality algebras* (in PEA_α), the cylindrification is defined for any subset Γ of α (in notation, $\exists(\Gamma)$ or $\mathsf{c}_{(\Gamma)}$ and it is called *quantification,* see [Hal,56] or [Hen-Mon-Tar,85], Definition 5.4.1). If the sets Γ are restricted to finite sets, then the previous definition is equivalent to the definition of CPES_α (because if only single cylindrifications are defined in an algebra, and their commutativity is assumed, this kind of definition of the cylindrification is equivalent to the definition of the cylindrification $\mathsf{c}_{(\Gamma)}$ defined for *finite* sets Γ, $\Gamma \subset \alpha$).

b) An algebra in Cprs_α with $Q = {}^\alpha\alpha$ satisfies all the CPES_α axioms, with the possible exceptions of the axioms (C_4), (CP_5), (CP_7), (CP_8), (CP_9) and (E_3) (see [Hen-Mon-Tar,85], Theorem 5.4.15). ${}_m\mathsf{Gwp}^{\mathrm{reg}}_\alpha \cup \mathsf{Gp}^{\mathrm{reg}}_\alpha \not\subseteq \mathsf{CPES}_\alpha$, that is, the CPES_α axioms (C_4) and (CP_9) fail to hold for the union on the left hand side. But, ${}_m\dot{\mathsf{G}}\mathsf{wp}^{\mathrm{reg}}_\alpha \cup \dot{\mathsf{G}}\mathsf{p}^{\mathrm{reg}}_\alpha \subseteq \mathsf{CPES}_\alpha$. Notice that ${}_m\mathsf{Gwp}_\alpha \cup \mathsf{Gp}_\alpha$ satisfies all the CPE_α axioms except for (CP_8).

c) We note that CPE_α and CPES_α can be conceived of as so-called *transformation systems* equipped by diagonals and cylindrifications (see [Dai-Mon,63], 3§ and 4§).

Definition 6.3.11. An algebra $\mathfrak{A}$ of the type of Cprs_α is *r-representable* if it is isomorphic to an algebra in Cprs_α.

The next theorem *motivates* the representation theorems following it. For r-representable algebras it gives *necessary* conditions for the representants.

Theorem 6.3.12. *Suppose that $\mathfrak{A} \in \mathbf{I}\mathfrak{B}$, where $\mathfrak{B} \in \mathsf{Cprs}_\alpha$, i.e., $\mathfrak{A}$ is r-representable. Then the following propositions* (i) *and* (ii) *hold:*

(i) *If $\mathfrak{A} \in {}_m\mathsf{CPE}_\alpha$, then $\mathfrak{A} \in \mathbf{I}\,{}_m\mathsf{Gwp}_\alpha$.*

(ii) *If $\mathfrak{A} \in \mathsf{CPE}_\alpha \cup \mathsf{CPES}_\alpha$, then $\mathfrak{A} \in \mathbf{I}\,\mathsf{Gp}_\alpha$.*

Proof.

(i) $\mathfrak{A} \in \mathbf{I}\mathfrak{B}$ implies that $f(\mathsf{s}_\lambda 1) = \mathsf{S}_\lambda f1$, where f is an isomorphism between $\mathfrak{A}$ and $\mathfrak{B}$, and λ is an arbitrary m-transformation (i.e., $\lambda \in {}_m\mathsf{T}_\alpha$). But, $\mathsf{s}_\lambda 1 = 1$ and $f1 = V$, therefore $f1 = \mathsf{S}_\lambda V$, i.e., $V = \mathsf{S}_\lambda V$. By Lemma 6.3.6(i), $\mathfrak{B} \in \mathbf{I}\,{}_m\mathsf{Gwp}_\alpha$.

(ii) The proof is similar to the previous one, but we need to use Lemma 6.3.6(ii) instead of (i). ∎

Assume that $m < \alpha$ and m is infinite. An algebra $\mathfrak{A} \in {}_m\mathsf{CPE}_\alpha$ is locally-m dimensional (locally-m, for short) if $|\Delta b| \leq m$ for each $b \in A$. The α-dimensional class of locally-m algebras is denoted by $\mathsf{L}m_\alpha$.

Remark.

The main r-representation theorems concerning *cylindric polyadic equality algebras* are the following ones (see [Fer,12b]):

Representation theorem for ${}_m\mathsf{CPE}_\alpha \cap \mathsf{L}m_\alpha$:

$\mathfrak{A} \in {}_m\mathsf{CPE}_\alpha \cap \mathsf{L}m_\alpha$ if and only if $\mathfrak{A} \in \mathbf{I}({}_m\mathsf{Gwp}_\alpha^{\mathrm{reg}} \cap \mathsf{L}m_\alpha)$, where m is infinite, $m < \alpha$.

The theorem generalizes Halmos's classical theorem that locally finite, infinite dimensional, quasi-polyadic algebras are representable (see [Hal,56], Sági [Sag,thisVol]).

Representation theorem for CPE_α and CPES_α:

$\mathfrak{A} \in \mathsf{CPE}_\alpha$ if and only if $\mathfrak{A} \in \mathbf{I}\,\mathsf{Gp}_\alpha^{\mathrm{reg}}$. $\mathfrak{A} \in \mathsf{CPES}_\alpha$ if and only if

$$\mathfrak{A} \in \mathbf{I}\big(\mathsf{Gp}_\alpha^{\mathrm{reg}} \cap \mathsf{Mod}\,\{\,(\mathrm{C}_4), (\mathrm{CP}_9)\}\,\big).$$

With the second proposition the following cylindric algebraic theorem can be associated: cylindric algebras satisfying the merry-go-round axioms are representable by set algebras in $\mathsf{Crs}_\alpha \cap \mathsf{Mod}\,\{\,(\mathrm{C}_4), (\mathrm{C}_6)\}$ (see [Hen-Mon-Tar,85], 3.2.88). The theorem above answers the problem raised in [And,01] and [And-Gol-Nem,98], whether $\mathsf{Gp}_\alpha^{\mathrm{reg}}$ ($\mathsf{G}_\alpha^{\mathrm{reg}}$) is a variety. Moreover, the formulae of the axioms in CPE_α imply that $\mathsf{Gp}_\alpha^{\mathrm{reg}}$ is a canonical variety.

As regards the proofs, see some hints in the next Section.

4. On the Connection of r-representability and Neat Embeddability

In the theory of *cylindric algebras*, neat embeddability is an important abstract algebraic equivalent of representability. The equivalence of the two concepts is stated by the so-called *neat embedding theorem* (see [Hen-Mon-Tar,85] 3.2.10). Representability means here that the algebra in question, is a subdirect product of cylindric set algebras or, what is equivalent to this, the algebra is isomorphic to some generalized cylindric set algebra (to an algebra in Gs_α).

In this Section, first, we investigate if there is some neat embedding theorem for the concept of r-representability, for the class CNA_α^+.

We recall a bit more general definition of neat embeddability, than that of neat embeddability for cylindric algebras.

Denote by K_β the class of algebras with the type of algebras in CPE_β.

Definition. Let $\alpha < \beta$ and $\mathfrak{B} \in \mathsf{K}_\beta$. Let

$$\mathfrak{Nr}_\alpha \mathfrak{B} = \left\langle \mathfrak{B}^0, +, \cdot, -, 0, 1, \mathsf{c}_i, \mathsf{s}'_\tau, \mathsf{d}_{ij} \right\rangle_{\tau \in {}^\alpha\alpha,\ i,j \in \alpha}$$

where $\mathfrak{B}^0 = \{b \in B : \mathsf{c}_i b = b \text{ for every } i \in \beta \sim \alpha\}$, and $\mathsf{s}'_\tau = \mathsf{s}_\sigma$ with $\sigma = \tau \cup \{i : i \in \beta \sim \alpha\}$ for each $\tau \in {}^\alpha\alpha$. An algebra $\mathfrak{A} \in \mathsf{CPE}_\alpha$ can be *neatly embedded* into some $\mathfrak{B}$, $\mathfrak{B} \in \mathsf{K}_\beta$ if $\mathfrak{A}$ is isomorphic to a subalgebra of $\mathfrak{Nr}_\alpha \mathfrak{B}$, in notation, $\mathfrak{A} \in \mathsf{SNr}_\alpha \mathsf{K}_\beta$ (see also [Hen-Mon-Tar,85], Definition 5.4.16).

Now, an *unusual class* of cylindric-like algebras is introduced, denoted by $\mathsf{F}_{\alpha+\varepsilon}^\alpha$, which can be regarded as a class of *many sorted* algebras. This class is obtained from $\mathsf{CA}_{\alpha+\varepsilon}$ such a way that instead of the axioms (C_4) and (C_6) certain *consequences* of them are postulated, and, furthermore, the schemes of these consequences are *restricted* to certain ordinals depending on α and ε.

Definition 6.4.1. The axioms of $\mathsf{F}_{\alpha+\varepsilon}^\alpha$ are obtained from those of $\mathsf{CA}_{\alpha+\varepsilon}$ so that the axioms (C_4) and (C_6) are replaced by the axioms $(\mathrm{C}_4)^-$, $(\mathrm{C}_6)^-$ below, where $\alpha \geq 4$ and $\alpha + \varepsilon$ is denoted by β:

$(\mathrm{C}_4)^-$ is the set of the following properties:

a) $\mathsf{s}_m^i \mathsf{s}_n^j x = \mathsf{s}_n^j \mathsf{s}_m^i x$, if $i, j, m, n \in \beta$ $i \neq j$ except for two cases $i, j \in \alpha$, $m \notin \alpha$ and $i, j \in \alpha$, $n \notin \alpha$,

b) $\mathsf{s}_m^i \mathsf{s}_n^j x \leq \mathsf{s}_n^j \mathsf{s}_m^i x$, if $i, j, n \in \alpha$, $m \notin \alpha$ (i, j, n, m are different),

c) $\mathsf{d}_{ik} \cdot \mathsf{s}_m^i \mathsf{s}_n^j x \leq \mathsf{s}_n^j \mathsf{s}_m^i x$, if $i, j, k \in \alpha$, $n \notin \alpha$ (i, j, k, n, m are different),

d) $\mathsf{c}_i \mathsf{c}_j x = \mathsf{c}_j \mathsf{c}_i x$, $m \notin \alpha$,

$(C_6)^-$ is the set of the following properties:

a) $\mathsf{d}_{ij} = \mathsf{d}_{ji}, \quad i, j \in \beta$,

b) $\mathsf{d}_{ij} \cdot \mathsf{d}_{jk} \leq \mathsf{d}_{ik}, \quad i, j, k \in \beta$,

c) $\mathsf{c}_k \mathsf{d}_{ij} = \mathsf{d}_{ij}$, $k \notin \{i, j\}$, $i, j, k \in \beta$,

d) $\mathsf{c}_i \mathsf{d}_{ij} = 1$ if $i, j \in \beta$ except for the case $i \in \alpha$, $j \notin \alpha$

(see [Fer,10b], Definition 4.1).

Theorem 6.4.2 (Ferenczi). *Let ε be any fixed infinite ordinal. Then*

$$\mathfrak{A} \in \mathbf{S}\mathsf{Nr}_\alpha \mathsf{F}^\alpha_{\alpha+\varepsilon} \text{ if and only if } \mathfrak{A} \in \mathbf{ID}_\alpha$$

where $\alpha \geq 4$ ([Fer,10b], Corollary 3.2).

An immediate consequence of this theorem is that if ε is any fixed infinite ordinal, then $\mathfrak{A}$ is *r-representable* if and only if $\mathfrak{A} \in \mathbf{S}\mathsf{Nr}_\alpha \mathsf{F}^\alpha_{\alpha+\varepsilon}$.

We note that the classical neat embedding theorem for cylindric algebras can be obtained from this one. The results above are immediate generalizations of the results included in Sayed Ahmed [Say,thisVol,a], Section 4.

Similar neat embedding theorems can be formulated for the r-representability of the classes TA_α, CQE_α, as well.

The topic has several applications to logic. A proof theoretical application can be found in [Fer,09]. A model theoretical application of the representation theory of cylindric and quasi-polyadic algebras has found its place in the research of the Vaught's conjecture (see Sági, Sziráki [Sag-Szi,12]).

The following neat embedding theorem plays an important role in the proofs of the representations theorems formulated in the previous Section.

Let us recall the definitions of polyadic and polyadic equality algebras (PA_α and PEA_α, see [Sag,thisVol] or [Hen-Mon-Tar,85], 5.4.1) and the following important result, closely related to our subject:

Theorem (Daigneault–Monk–Keisler). *If $\mathfrak{A} \in \mathsf{PA}_\alpha$, then $\mathfrak{A} \in \mathbf{S}\mathsf{Nr}_\alpha \mathsf{PA}_{\alpha+\varepsilon}$, where α is a fixed infinite ordinal and $\varepsilon > 1$.*

This form of the theorem (apart from terminology) is due to Daigneault and Monk ([Dai-Mon,63], Theorem 4.3). Keisler published the proof theoretical variant of the theorem in the same issue ([Kei,63]). The theorem holds for equality algebras, too (see [Hen-Mon-Tar,85], 5.4.17).

The classes CPE_α, CPES_α and ${}_m\mathsf{CPE}_\alpha$ are neatly embeddable in the classical sense (as a consequence of the variants of the Daigneault–Monk–Keisler theorem), but they are not representable in the classical sense (see e.g., [Dai-Mon,63]), i.e. classical neat embedding theorems do not exist for these classes. Nevertheless, as the theorem below illustrates, there are *unusual* neat embedding theorems for these classes, which, in turn, proves that *neat embeddability* is equivalent to a kind of r-representability.

Remark.

The following theorem implies that a kind of neat embeddability of an algebra in ${}_m\mathsf{CPE}_\alpha \cap \mathsf{L}m_\alpha$ is equivalent to r-representability (see [Fer,12b]).

Neat embedding theorem for ${}_m\mathsf{CPE}_\alpha \cap \mathsf{L}m_\alpha$:

Assume that $\mathfrak{A} \in {}_m\mathsf{CPE}_\alpha \cap \mathsf{L}m_\alpha$, *where* m *is infinite and* $m < \alpha$. *Then* $\mathfrak{A} \in \mathbf{S}\mathsf{Nr}_\alpha({}_m\mathsf{CPE}^-_{\alpha+\varepsilon})$, *if and only if* $\mathfrak{A} \in \mathbf{I}\,{}_m\mathsf{Gwp}_\alpha$, *where* ε *is infinite and* ${}_m\mathsf{CPE}^-_{\alpha+\varepsilon}$ *is a given superclass of* $\mathsf{CPE}_{\alpha+\varepsilon}$ (see [Fer,12b]).

Problem 6.4.3. Find an elementary proof for the following theorem: If an $\mathfrak{A} \in \mathsf{CNA}_\alpha$ satisfies the MGR property (i.e., $\mathfrak{A} \in \mathsf{CNA}^+_\alpha$), then $\mathfrak{A} \in \mathbf{S}\mathsf{Nr}_\alpha\mathsf{F}^\alpha_{\alpha+\varepsilon}$ (together with Theorem 6.4.2 this yields a new proof for the Resek–Thompson theorem).

Problem 6.4.4. Does any r-representation theorem exist for algebras in PEA_α?

BOLYAI SOCIETY
MATHEMATICAL STUDIES, 22

Cylindric-like Algebras
and Algebraic Logic
pp. 163–182.

Representing all Cylindric Algebras by Twisting
On a Problem of Henkin

ANDRÁS SIMON*

Had the class of cylindric set algebras turned out to be finitely (and/or nicely) axiomatizable, algebraic logic would have evolved along a markedly different path than it did in the past 40 years. Among other things, it would have probably spelled the end of the "abstract" class CA_α as a separate subject of research; after all, why bother with abstract algebras, if a few nice extra equations can get us from there to concrete algebras of relations. As it is, Monk's 1969 result (and its various improvements by algebraic logicians from Andréka to Venema), stating that for $\alpha > 2$, RCA_α is not axiomatizable finitely, meant, among other things, that CA_α was here to stay, and its "distance" from RCA_α became an important research topic. To mention just one example of how this distance can be measured, we refer to the representation theory of CAs, where sufficient conditions for a CA to be representable are sought. This line of research is typical: the question one asks here is "what is missing from CAs to be representable?". But Henkin (himself a prolific contributor to representation theory) turned around this question and instead of asking how much CAs needed to be representable, he asked how much set algebras needed to be "distorted" to provide representations for all CAs. Now, if the answer is "very much", then we haven't got closer to understanding either CAs or the possible causes of their non-representability. On the other hand, if the distortions that are needed are such that they do not destroy the geometric nature

*Dedicated to the memory of Leon Henkin.

The results in this chapter first appeared in my Ph.D. thesis [Sim,99]. I would like to thank my supervisor, Ildikó Sain, and Hajnal Andréka and István Németi for their support over the years. I'm also grateful to Leon Henkin and Steven Givant for the comments and suggestions I received from them at around the time my thesis was written, and to Ian Hodkinson, for his insights.

of representable algebras, it means that CAs themselves have geometric character, even if in a less standard way than the representables do.

In fact, Henkin did have some distortions in mind: relativization, dilation and twisting (the last two are his own inventions; all will be defined in due course); he used these to get non-representable algebras from representable ones for various purposes (such as showing the independence of certain equations from the CA-axioms; we will see one of his constructions in 1.2.15 in subsection 2.2 below). The problem he posed in [Hen,67, p. 48] was whether, for $\alpha > 2$, all CA_αs were isomorphic to algebras obtained from RCA_αs by means of these (and the usual algebraic, such as taking subalgebras) operations[1].

The aim of this paper is to outline the steps leading to the proof of a positive answer (Theorem 1.3.1) to Henkin's problem for $\alpha = 3$, and to discuss some ramifications of this result. Details omitted here can be found in [Sim,99].

1. A Few Facts about Cylindric Algebras

In this section we briefly recall the main facts from CA-theory which we need in stating our results.

RCA_α is a variety (a result of Tarski and Henkin [Tar,55], [Hen-Tar,61]), and, in case α is finite, a discriminator variety, whose simple members are, up to isomorphism, the Cs_αs with non-empty base set. It is easy to check that RCA_α validates all axioms of CA_α, and it is natural to ask, whether the reverse inclusion holds, that is, whether every CA_α is representable. The answer is "yes" for $\alpha \leq 1$[2] (cf. 3.2.54, 55 in [Hen-Mon-Tar,85]), and "no" for $\alpha > 1$. The real cutting point, however, is 2, as shown by the following results of Henkin and Monk:

Theorem 1.1.1 (Henkin). RCA_2 *is axiomatizable by finitely many equations, cf.* [Hen-Mon-Tar,85, 3.2.65].

[1]Open Problem 5.30 in [Res,75] is a variant of this question, also answered affirmatively for $\alpha = 3$ by Theorem 1.3.1 below.

[2]For $\alpha = 0$, this is just a restatement of Stone's representation theorem for Boolean algebras.

The equations that characterize RCA_2 are the CA_2 axioms together with the equation (known as the *Henkin equation*) that appears in Example 1.2.15 in subsection 2.2 below.

Theorem 1.1.2 (Monk). RCA_α *is not axiomatizable by finitely many equations if* $\alpha > 2$ ([Mon,69], cf. [Hen-Mon-Tar,85, 4.1.3, 4.1.7]).

This result, together with its subsequent improvements by Andréka [And,97a], Bíró [Bir,92], Hodkinson [Hod,97c], Maddux [Mad,89a], Venema [Ven,97a] and others, shows that narrowing the class of CA_αs by adding finitely many equations, or even reasonably simple schemes of equations, will not be sufficient to obtain a characterization of the representable algebras.

Definition 1.1.3. For $i, j, k < \alpha$, we set

$$\mathsf{s}^i_j x \stackrel{\text{def}}{=} \begin{cases} x & \text{if } i = j \\ \mathsf{c}_i(\mathsf{d}_{ij} \cdot x) & \text{if } i \neq j \end{cases} \qquad \text{and} \qquad {}_k\mathsf{s}(i,j)x \stackrel{\text{def}}{=} \mathsf{s}^k_i \mathsf{s}^i_j \mathsf{s}^j_k x$$

([Hen-Mon-Tar,85, 1.5.1,1.5.12]).

The operations s^i_j are called *substitutions*. In representable algebras, s^i_j copies the ith argument of a c_j-closed relation to its jth coordinate. E.g., if x is an element of a Cs_3 with base U, then $\mathsf{s}^0_1 x = \{\langle u, v, w\rangle \in {}^3U : \langle v, v, w\rangle \in x\}$; hence if x is c_1-closed, i.e., if it is a binary relation between the zeroth and second coordinates, then $\mathsf{s}^0_1 x$ is the same binary relation, but between the first and second coordinates. Consequently, in representable algebras, ${}_k\mathsf{s}(i,j)$ interchanges the the ith and jth coordinate of a c_k-closed relation, using the kth coordinate as a "place holder".

Section 1.5 of [Hen-Mon-Tar,85] contains a thorough discussion of these operations.

Next we define a set of CA_3 equations (known as the *merry-go-round equations*). We restrict ourselves to the 3-dimensional case (but see Remark 1.1.6 below) because this is what we need in later sections.

Definition 1.1.4. Let $\{k, l, m\} = 3$. MGR_k is the equation

$${}_k\mathsf{s}(l,m)\mathsf{c}_k x = {}_k\mathsf{s}(m,l)\mathsf{c}_k x$$

([Hen-Mon-Tar,85, 3.2.88]).

By Lemma 1.1.5 below, MGR_k says that interchanging the ith and jth coordinate of a c_k-closed element twice (using the kth coordinate as a place holder) leaves it fixed. While obviously valid in representable CAs, the MGR equations do not follow from the CA axioms; this was first shown by Henkin. The importance of these equations was recognized when Resek [Res,75] proved that every CA in which MGR holds is representable as a relativized set algebra. We note that the present form of MGR is due to Thompson [Tho,79].

Lemma 1.1.5. *Let* $\mathfrak{A} \in \mathsf{CA}_3$ *and* $\{k, l, m\} = 3$. *Then*

(i) $\mathfrak{A} \models \mathrm{MGR}_k$ *iff* $\mathfrak{A} \models {}_k\mathsf{s}(l,m){}_k\mathsf{s}(l,m)\mathsf{c}_k x = \mathsf{c}_k x$

(ii) $\mathfrak{A} \models \mathrm{MGR}_k$ *iff* $\mathfrak{A} \models \mathrm{MGR}_l$.

By Lemma 1.1.5(ii), we may write MGR for any of the equations MGR_k ($k < 3$).

Remark 1.1.6. If $\alpha > 3$, then MGR consists of two schemes of equations,

($_3$MGR) $\qquad {}_k\mathsf{s}(l,m)\mathsf{c}_k x = {}_k\mathsf{s}(m,l)\mathsf{c}_k x \quad \text{when } \big|\{k,l,m\}\big| = 3$, and

($_4$MGR)
$$ {}_k\mathsf{s}(l,m){}_k\mathsf{s}(m,n)\mathsf{c}_k x = {}_k\mathsf{s}(n,l){}_k\mathsf{s}(l,m)\mathsf{c}_k x \quad \text{when } \big|\{k,l,m,n\}\big| = 4.$$

It's not hard to show that all instances of $_3$MGR involving a fixed three-element subset of α are equivalent.

Just as it is natural to look for CAs of smaller dimension in a CA (see Sayed Ahmed [Say,thisVol,a]), for $\alpha \geq 3$ we can find algebras of the same types as relation algebras (RAs) in CA_αs.

Definition 1.1.7. Let $\mathfrak{C} \in \mathsf{CA}_\alpha$, $\alpha \geq 3$. $\mathfrak{Ra}\,\mathfrak{C}$, the RA-*reduct of* $\mathfrak{C}$ is the algebra
$$\langle \mathfrak{Ra}\,\mathfrak{C}, +, \cdot, -, 0, 1, ;, {}^\smile, 1\text{'} \rangle,$$
where $\mathfrak{Ra}\,\mathfrak{C} = \{x \in C \colon \mathsf{c}_2 x = x\}$, $x \,;\, y = \mathsf{c}_2\big(\mathsf{s}_2^1 x \cdot \mathsf{s}_2^0 y\big)$, $x^\smile = {}_2\mathsf{s}(0,1)x$ and $1\text{'} = \mathsf{d}_{01}$. For a class K of CA_αs we define $\mathsf{Ra}\,\mathsf{K} \stackrel{\text{def}}{=} \mathbf{I}\{\mathfrak{Ra}\,\mathfrak{C} : \mathfrak{C} \in \mathsf{K}\}$ ([Hen-Mon-Tar,85, 5.3.7]).

From now on, whenever convenient, we will pretend that $^\smile$ and ; are operations of every CA_α (if $\alpha > 2$). We note that the RA-reduct of a CA_3 is not necessarily an RA: in fact, it is known that none of the axioms (R1),

(R4), (R6) and (R7) holds in $\mathsf{Ra}\,\mathsf{CA}_3$. For example, (R4) holds in the RA-reduct of a $\mathfrak{C} \in \mathsf{CA}_3$ iff MGR holds in $\mathfrak{C}$.

The next section introduces the tools one can use to construct cylindric algebras with such RA-reducts.

2. Non-Representable Algebras

The purpose of this section is to recall methods for constructing non-representable cylindric algebras: *dilation, twisting* and *relativization,* and to establish their basic properties.

Relativization, historically the first (and perhaps the simplest) of these contructions, is a straightforward generalization of the corresponding Boolean algebraic notion. Dilation and twisting, on the other hand, do not have BA-theoretic counterparts; they were devised to construct algebras of relations of higher ranks. We note that dilation and twisting, as well as the use of relativization in constructing non-representable CAs, originate with Henkin.

2.1. Atom structures

In both dilation and twisting one starts out with a complete and atomic CA, adjoins new elements and/or changes the operations to get a new, complete and atomic CA with certain prescribed properties. In CAs of this kind, and actually in all complete and atomic normal Boolean algebras with operators (BAOs) where the extra-Boolean operations distribute over arbitrary joins, the operators are determined by their behaviour on the atoms. Hence it is possible, and often simpler, to work with the *atom structure* of such an algebra instead of the algebra itself.

In the first part of this subsection we recall the notion of atom structure of an algebra and of the *complex algebra* of a relational structure. (More on these notions can be found in Section 2.7 of [Hen-Mon-Tar,85] and Venema's chapter in the present volume.) Since there is nothing in these constructions that would require the strong axioms of CAs, and because we will also be interested in atom structures of certain definitional extensions of CAs, we present these notions in the context of BAOs.

Definition 1.2.1. An algebra $\mathfrak{A} = \langle \mathfrak{A}_0, f_i \rangle_{i \in I}$ is a *Boolean algebra with operators,* or BAO, if $\mathfrak{A}_0 \in \mathsf{BA}$, and the extra-Boolean operations f_i $(i \in I)$ distribute over the Boolean join in each of their arguments. The operations f_i are called *operators.* An operator is *normal* if its value is 0 whenever one of its arguments is 0. A BAO is normal if all its operators are normal. A BAO is said to be *complete* if its Boolean reduct is complete and its operators distribute over arbitrary joins ([Jon-Tar,51]).

Examples of normal BAOs include almost all known algebras of relations. E.g., CAs are normal BAOs, and complete CAs are complete BAOs by (C1) and [Hen-Mon-Tar,85, 1.2.6].

Definition 1.2.2. Let $\mathfrak{A} = \langle \mathfrak{A}_0, f_i \rangle_{i \in I}$ be a normal, atomic BAO. The *atom structure* $\mathfrak{At}\,\mathfrak{A}$ of $\mathfrak{A}$ is the structure $\langle At\,\mathfrak{A}, R_i \rangle_{i \in I}$, where $At\,\mathfrak{A}$ is the set of atoms of $\mathfrak{A}$, and if f_i is an n-ary operator, then R_i is the $n+1$-ary relation $\{ \langle x_1, \ldots, x_n, y \rangle \,:\, y \leq f_i(x_1, \ldots, x_n) \}$ ([Jon-Tar,51]).

The relation R_i is sometimes referred to as the *accessibility relation* of f_i.

When speaking of atom structures of CAs, the relations corresponding to cylindrifications and diagonals will always be denoted by (variants of) T_i and E_{ij}, respectively. Note that by the definition above, in a complete and atomic CA, E_{ij} is the set of atoms below d_{ij}.

Definition 1.2.3. Let $\mathfrak{B} = \langle B, R_i \rangle_{i \in I}$ be a relational structure. The *complex algebra* $\mathfrak{Cm}\,\mathfrak{B}$ of $\mathfrak{B}$ is the algebra $\left\langle \mathcal{P}(B), \cup, \cap, \sim, \emptyset, B, R_i^* \right\rangle_{i \in I}$ ([Hen-Mon-Tar,85, 2.7.33]).

In particular, if $\mathfrak{B}$ is a structure $\langle B, T_i, E_{ij} \rangle_{i,j<\alpha}$, the relations T_i are binary and E_{ij} are unary, then $\mathfrak{Cm}\,\mathfrak{B}$ is the algebra $\left\langle \mathcal{P}(B), \cup, \cap, \sim, \emptyset, B, T_i^*, E_{ij} \right\rangle_{i,j<\alpha}$, similar to CA_αs.

Theorem 1.2.4. *If $\mathfrak{A}$ is a normal, complete and atomic* BAO, *then the map $h : x \mapsto \{a : a \leq x,\, a \in At\,\mathfrak{A}\}$ is an isomorphism from $\mathfrak{A}$ to $\mathfrak{Cm}\,\mathfrak{At}\,\mathfrak{A}$. If $\mathfrak{B}$ is a relational structure, then $\mathfrak{Cm}\,\mathfrak{B}$ is a normal, complete and atomic* BAO, *and the map $g : b \mapsto \{b\}$ is an isomorphism from $\mathfrak{B}$ to $\mathfrak{At}\,\mathfrak{Cm}\,\mathfrak{B}$* ([Jon-Tar,51], [Hen-Mon-Tar,85, 2.7.34, 2.7.35]).

Remark 1.2.5. Because of Theorem 1.2.4, we may identify normal, complete and atomic BAOs with the complex algebra of their atom structures, and relational structures with the atom structure of their complex algebras. We will do so without explicitly referring to 1.2.4.

Now we recall the first-order characterization of atom structures of CAs.

Definition 1.2.6. Let α be an ordinal. A structure

$$\mathfrak{B} = \langle B, T_i, E_{ij} \rangle_{i,j<\alpha}$$

with binary relations T_i and unary relations E_{ij} is a *cylindric atom structure of dimension* α iff the following conditions hold for all $i, j, k < \alpha$:

(i) T_i is an equivalence relation on B

(ii) $T_i \mid T_j = T_j \mid T_i$

(iii) $E_{ii} = B$

(iv) $E_{ij} = T_k{}^*(E_{ik} \cap E_{kj})$ if $k \notin \{i, j\}$

(v) $T_j \cap (E_{ij} \times E_{ij}) \subseteq Id$ if $i \neq j$.

Ca_α is the class of cylindric atom structures of dimension α (cf. [Hen-Mon-Tar,85, 2.7.40]).

The commuting equivalence relations T_i in a cylindric atom structure correspond to cylindrifications, and the unary relations E_{ij} to the diagonals. We note that e.g. axiom (v) says that each block of T_j has at most one representative that belongs to E_{ij}, provided $i \neq j$. (Using the rest of the axioms, one can actually show that in each block of T_j there is exactly one such representative, cf. Lemma 1.2.10(i).)

Theorem 1.2.7. Ca_α *is the class of atom structures of complete and atomic* CA_α*s, and a structure* $\mathfrak{B}$ *of the appropriate type is in* Ca_α *iff its complex algebra* $\mathfrak{Cm}\,\mathfrak{B} \in \mathsf{CA}_\alpha$ ([Hen-Mon-Tar,85, 2.7.38, 2.7.40]).

Definition 1.2.8. Let $\alpha \geq 3$ be an ordinal. For any structure $\mathfrak{B} = \langle B, T_i, E_{ij} \rangle_{i,j<\alpha}$ with binary relations T_i and unary relations E_{ij} we introduce the binary relations ${}_kK_{ij}^{\mathfrak{B}}$ (for $i, j, k < \alpha$) and the ternary relation $C^{\mathfrak{B}}$ defined by the following formulas (we will always omit the superscript when this cannot lead to confusion):

$$a \; {}_kK_{ij} \; b \stackrel{\text{def}}{\Longleftrightarrow} a \in E_{jk} \wedge (\exists c \in E_{ij})(\exists d \in E_{ik}) a \; T_j \; c \; T_i \; d \; T_k \; b \tag{1.2.1}$$

$$C(a, b, c) \stackrel{\text{def}}{\Longleftrightarrow} a \in E_{12} \wedge b \in E_{02} \wedge \exists d (a \; T_1 \; d \; T_0 \; b \wedge d \; T_2 \; c), \tag{1.2.2}$$

and $K_{ij} \stackrel{\text{def}}{=} {}_kK_{ij}$, where k is the smallest ordinal distinct from i and j.

Lemma 1.2.9. *Let $\alpha \geq 3$, $\mathfrak{B} \in \mathsf{Ca}_\alpha$ and let $x, y \in C$ for some $\mathfrak{C} \subseteq \mathfrak{Cm}\,\mathfrak{B}$. Then ${}_kK^*_{ij}x = {}_k\mathsf{s}(i,j)x$ for distinct $i, j, k < \alpha$, and $k < \alpha$, and $C^*(x,y) = \mathsf{c}_2(\mathsf{s}^1_2 x \cdot \mathsf{s}^0_2 y)$.*

So the relations $K_{01} = {}_2K_{01}$ and C reflect the derived operations $\smallsmile$ and ; (cf. Definition 1.1.7) just as T_i and E_{ij} reflect the operations c_i and d_{ij} on the level of atoms. In other words, K_{01} and C are the accessibility relations of $\smallsmile$ and ;.

The next lemma contains some basic facts about Ca_αs.

Lemma 1.2.10. *Let $\mathfrak{B} \in \mathsf{Ca}_\alpha$, $\mathfrak{C} = \mathfrak{Cm}\,\mathfrak{B}$, and let $i, j, k < \alpha$ be distinct. Then*

(i) *for every $a \in B$, $\mathsf{d}_{ij} \cdot \mathsf{c}_i a$ is the unique element in E_{ij} that is T_i-related to a*

(ii) $T_k{}^* E_{ij} = E_{ij}$

(iii) *${}_kK_{ij} \cap {}^2E_{jk}$ is a permutation of E_{jk}.*

The following lemma illustrates how certain properties of a complete and atomic algebra are reflected in its atom structure. The MGR equations were defined in 1.1.4 in Section 1.

Lemma 1.2.11. *Let $\mathfrak{B} \in \mathsf{Ca}_3$. Then*

$$\mathfrak{Cm}\,\mathfrak{B} \models \mathrm{MGR}_k \quad \textit{iff} \quad \mathfrak{B} \models T_k \mid {}_kK_{ij} = T_k \mid {}_kK_{ji}$$

provided $\{i, j, k\} = 3$.

Remark 1.2.12. An advantage of working with atom structures is that they can be drawn. One possibility is to draw T_i-classes as lines (or segments) and elements of a class as points on them. As an illustration, consider once again the MGR-equations. Keeping in mind that each T_i-class has exactly one member that belongs to E_{ij} (when $j \neq i$), we obtain the following picture of the equation ${}_k\mathsf{s}(i,j)\mathsf{c}_k x = {}_k\mathsf{s}(j,i)\mathsf{c}_k x$ (see Fig. 1.2.12). The interesting thing about this picture is that one can practically read off from it both parts of Lemma 1.1.5.

We close our brief survey of atom structures by recalling from [Hen-Mon-Tar,85, 3.2.68] a simple method (used in the proof of 1.3.3 below) for constructing atomic Css. Let α be an ordinal, and let $\mathfrak{A}$ be the full Cs_α

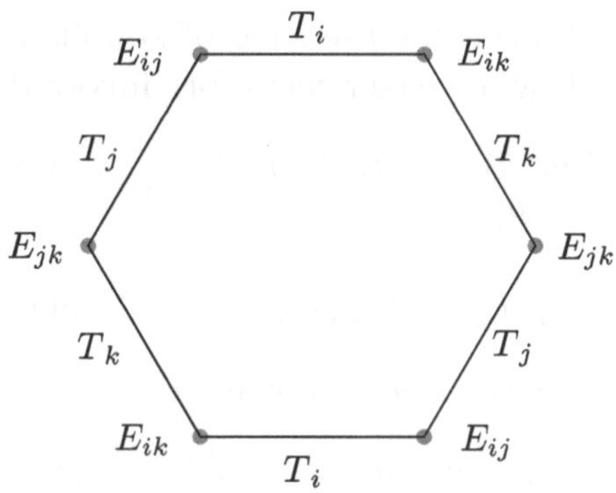

Fig. 1. merry-go-round

on a set U. If G is a group of permutations of U, then each element g of G induces a permutation $s \mapsto g \circ s$ of U-sequences of a fixed length. Call two U-sequences s and z G-equivalent (written $s \equiv_G z$) if $z = g \circ s$ for some $g \in G$. If s is a U-sequence, $s^- \stackrel{\text{def}}{=} s/\equiv_G = \{g \circ s : g \in G\}$ denotes the orbit of s under the permutations induced by elements of G. It is easy to see that the set[3] $Fx_G \mathfrak{A} \stackrel{\text{def}}{=} \{X \subseteq {}^\alpha U : (\forall g \in G) X = \{g \circ s : s \in X\}\}$ is a subuniverse of $\mathfrak{A}$. (This is true by virtue of the fact that the operations of a Cs are *permutation invariant* in the sense of Tarski–Givant [Tar-Giv,87] or Németi [Nem,91].) We denote by $\mathfrak{Fr}_G \mathfrak{A}$ the subalgebra of $\mathfrak{A}$ with universe $Fx_G \mathfrak{A}$. Clearly, $\mathfrak{Fr}_G \mathfrak{A}$ is atomic, and its atoms are exactly the relations s^-, where $s \in {}^\alpha U$. Further, two atoms s^- and z^- are T_i-related in the atom structure of $\mathfrak{Fr}_G \mathfrak{A}$ iff $s \restriction (\alpha \setminus \{i\}) \equiv_G z \restriction (\alpha \setminus \{i\})$, and $s^- \in E_{ij}$ iff $s_i = s_j$.

2.2. Dilation

The idea behind dilation may be expressed vaguely as follows: in a CA, if it is not outright impossible to have an atom in a certain position, then insert a new atom there. Here "impossible" means that the existence of the new atom would contradict the CA-axioms; and the "position" of an atom is how it is related to other atoms (and itself) via the relations T_i and E_{ij}. The definition (a slightly modified version of [Hen-Mon-Tar,85, 3.2.69] which allows inserting a set of new atoms simultaneously) below is an attempt to formalize this idea. It is, however, a compromise, in that it

[3] This notation is slightly different from that used in [Hen-Mon-Tar,85].

sacrifices certain possibilities for the sake of simplicity: namely, condition (D1) could be replaced by a weaker but more involved condition.

Definition 1.2.13. Let $\mathfrak{B} = \langle B, T_i, E_{ij}\rangle_{i,j<\alpha} \in \mathsf{Ca}_\alpha$, let $\Psi \subseteq {}^\alpha B$ and suppose that for all $\psi \in \Psi$,

$$\text{(D1)} \qquad \psi_i \notin E_{jk} \quad \text{if} \quad \big|\{i,j,k\}\big| = 3, \quad \text{and}$$

$$\text{(D2)} \qquad \psi_i \; T_i \mid T_j \; \psi_j \quad \text{for all} \quad i,j < \alpha.$$

Then the result of dilating $\mathfrak{B}$ with Ψ is $\mathfrak{B}^\Psi = \big\langle B^\Psi, T_i^\Psi, E_{ij}^\Psi \big\rangle_{i,j<\alpha}$, where $B^\Psi \stackrel{\text{def}}{=} B \cup \{\nu_\psi : \psi \in \Psi\}$ (here the ν_ψs are assumed to be distinct and not in B), $E_{ii}^\Psi \stackrel{\text{def}}{=} \mathfrak{B}^\Psi$ and $E_{ij}^\Psi \stackrel{\text{def}}{=} E_{ij}$ if $i \neq j$. To define T_i^Ψ it is convenient to introduce the following notation: for $a \in B^\Psi$ and $i < \alpha$,

$$|a|_i = \begin{cases} a & \text{if } a \in B \\ \psi_i & \text{if } a = \nu_\psi. \end{cases}$$

Then we let $a \; T_i^\Psi \; b \stackrel{\text{def}}{\iff} |a|_i \; T_i \; |b|_i$.

When Ψ is the singleton $\{\psi\}$, we write $\mathfrak{B}^\psi$ instead of $\mathfrak{B}^{\{\psi\}}$.

So the new atoms are all outside the diagonals, and each of them is "coordinatized" by an α-sequence of old atoms: the ith element of this sequence determines how the new atom behaves with respect to T_i.

The definition of dilation in [Hen-Mon-Tar,85, 3.2.69] allows only dilating with one sequence. Our definition is not a true generalization, however. For it is not hard to show that dilating a Ca_α $\mathfrak{B}$ with the set $\Psi = \{\psi_\beta : \beta < \gamma\}$ (where γ is an ordinal) can be simulated by taking the union of $\mathfrak{B}_\beta$, where $\mathfrak{B}_\beta$ is $\mathfrak{B}_\delta^{\psi_\delta}$ if β is the successor of δ, or $\bigcup\{\mathfrak{B}_\delta : \delta < \beta\}$ if β is a limit ordinal.

The next theorem says that dilation does not lead out of CA.

Theorem 1.2.14. *Let $\mathfrak{B} \in \mathsf{Ca}_\alpha$ and suppose that $\Psi \subseteq {}^\alpha B$ satisfies conditions* (D1) *and* (D2) *of 1.2.13. Then $\mathfrak{B}^\Psi$, the result of dilating $\mathfrak{B}$ with Ψ, is a Ca_α* (cf. [Hen-Mon-Tar,85, 3.2.69]).

Example 1.2.15. The drawing in Fig. 1.2.15 shows a simple (but already useful) example of dilation. The atom structure on the left hand side shows the atoms of the full Cs_2 on a 2-element set; the dashed lines represent

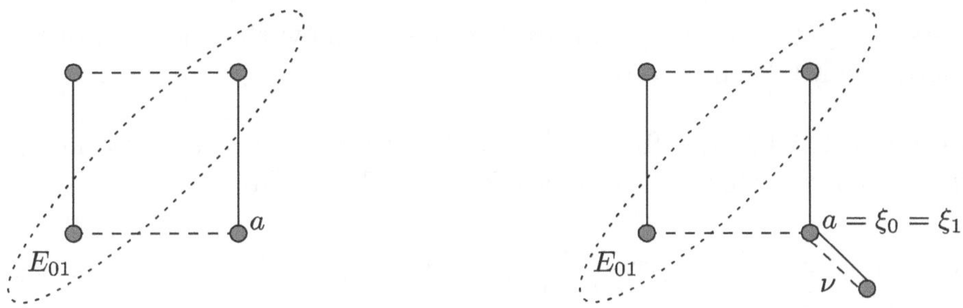

Fig. 2. Dilation

T_0 and the solid lines represent T_1. It is obvious that the sequence $\langle a, a\rangle$ satisfies conditions (D1) and (D2). The atom structure on the right hand side is what we get by dilating with $\langle a, a\rangle$. This example was used by Henkin to show that the equation[4] $\mathsf{d}_{01} \cdot \mathsf{c}_1(-x \cdot \mathsf{c}_0 x) \cdot -\mathsf{c}_0(\mathsf{c}_1 x \cdot -\mathsf{d}_{01}) = 0$, true in RCA_2, is not valid in CA_2 (it fails when $x = a$).

The following lemma relates the accessibility relations corresponding to converse and composition in $\mathfrak{B}^\Psi$ to those of $\mathfrak{B}$.

Lemma 1.2.16. *Let $\mathfrak{B} \in \mathsf{Ca}_\alpha$, $\alpha > 2$, and suppose that $\Psi \subseteq {}^\alpha B$ satisfies the dilating conditions* (D1) *and* (D2). *If $i, j < \alpha$ are distinct, and k is the smallest element of $\alpha \setminus \{i, j\}$, then*

(i) $K^\Psi_{ij} = K_{ij} \mid T_k{}^\Psi = \{\, \langle a, b\rangle : a\ K_{ij}\ |b|_k\}$

(ii) $(\mathsf{c}^\Psi_2 a)^{\smile\Psi} = \mathsf{c}^\Psi_2\big((\mathsf{c}_2 a)^\smile\big)$ *for all $a \in B$*

(iii) $C^\Psi = \big\{\langle a, b, c\rangle : C\big(a, b, |c|_2\big)\big\} \cup \{\, \langle a, b, c\rangle : (\exists \psi \in \Psi)\psi_1\ T_1\ a \in E_{12}, \psi_0\ T_0\ b \in E_{02}$ *and* $\psi_2\ T_2\ |c|_2\}$

(iv)

$$\mathsf{c}^\Psi_2 a \,;^\Psi \mathsf{c}^\Psi_2 b = \mathsf{c}^\Psi_2(\mathsf{c}_2 a \,; \mathsf{c}_2 b) + \sum \{\mathsf{c}^\Psi_2 \psi_2 : \psi \in \Psi, \mathsf{d}_{12} \cdot \mathsf{c}_1\psi_1 \le \mathsf{c}_2 a, \mathsf{d}_{02} \cdot \mathsf{c}_0\psi_0 \le \mathsf{c}_2 b\}$$

for all $a, b \in B$.

In (ii) *and* (iv) *above the operations without superscripts are understood to be computed in $\mathfrak{Cm}\,\mathfrak{B}$, the dilated operations (the ones with superscripts) in $\mathfrak{Cm}\,\mathfrak{B}^\Psi$.*

[4]This is actually a simplified (but equivalent) form of Henkin's equation, due to de Rijke–Venema [Rij-Ven,95].

As an application of the above lemmas we show that dilation does not destroy the MGR equations.

Theorem 1.2.17. *Let* $\mathfrak{B} \in \mathsf{Ca}_3$ *and let* $\mathfrak{B}^\Psi$ *be the result of dilating* $\mathfrak{B}$ *with* Ψ. *Then* $\mathfrak{Cm}\,\mathfrak{B}^\Psi \models \mathrm{MGR}$ *provided* $\mathfrak{Cm}\,\mathfrak{B} \models \mathrm{MGR}$.

Proof. Let $\mathfrak{B} = \langle B, T_i, E_{ij}\rangle_{i,j<3}$ and $\mathfrak{B}^\Psi = \left\langle B^\Psi, T_i^\Psi, E_{ij}^\Psi\right\rangle_{i,j<3}$ as before. By Lemma 1.2.11 it is enough to check that

$$T_k \mid {}_kK_{ij} \subseteq T_k \mid {}_kK_{ji} \quad \text{implies} \quad T_k^\Psi \mid {}_kK_{ij}^\Psi \subseteq T_k^\Psi \mid {}_kK_{ji}^\Psi$$

when $\{i,j,k\} = 3$. So assume that $T_k \mid {}_kK_{ij} \subseteq T_k \mid {}_kK_{ji}$. Then

$$\begin{aligned}
T_k^\Psi \mid {}_kK_{ij}^\Psi &= T_k^\Psi \mid T_k \mid {}_kK_{ij}^\Psi && \text{by } T_k \subseteq T_k^\Psi \\
&= T_k^\Psi \mid T_k \mid {}_kK_{ij} \mid T_k^\Psi && \text{by 1.2.16(i)} \\
&\subseteq T_k^\Psi \mid T_k \mid {}_kK_{ji} \mid T_k^\Psi && \text{by assumption} \\
&= T_k^\Psi \mid T_k \mid {}_kK_{ji}^\Psi && \text{by 1.2.16(ii)} \\
&\subseteq T_k^\Psi \mid {}_kK_{ji}^\Psi && \text{by } T_k \subseteq T_k^\Psi. \quad \blacksquare
\end{aligned}$$

2.3. Twisting

The second method of constructing non-representable cylindric algebras from representable ones is called *twisting*. It is due to Henkin, but here we will extend his construction (described in [Hen-Mon-Tar,85, 3.2.71]) in order to be able to produce CAs that do not arise as the result of Henkin's method. For example, by using our extended version of twisting we will be able to construct a CA_3 in which the quasi-equation $x^{\smile\smile} \leq x \rightarrow x^{\smile\smile} = x$ fails (see Theorem 1.3.3 below), and we will show that this is impossible using Henkin's construction[5].

Just like dilation, twisting is used to "distort" atomstructures. First, we describe the construction and show that it indeed produces Ca_αs from Ca_αs.

[5]finitely many times

Definition 1.2.18. Let $\mathfrak{B} = \langle B, T_i, E_{ij} \rangle_{i,j<\alpha} \in \mathsf{Ca}_\alpha$, $t \in \alpha$ and $\xi \in {}^I B$ for some set I, and suppose that

(T1) $$\langle \xi_i, \xi_j \rangle \notin T_t \quad \text{for all distinct } i, j \in I$$

(T2) $$\xi_i \notin E_{jk} \quad \text{for all } i \in I \text{ and all distinct } j, k < \alpha \text{ such that } t \notin \{j, k\}.$$

For $i \in I$, let Ξ_i denote the T_t-class of ξ_i, let π be a permutation of I, and for all $i \in I$, let Ξ_i be partitioned into Ξ'_i and Ξ''_i. Assume that for all $i \in I$ and $j < \alpha$, $j \neq t$,

(T3) $$\begin{aligned} &\mathrm{dom}\big(T_j \cap (\Xi'_i \times \Xi'_{\pi i})\big) \supseteq \Xi'_i, && \mathrm{ran}\big(T_j \cap (\Xi'_i \times \Xi'_{\pi i})\big) \supseteq \Xi'_{\pi i}, \\ &\mathrm{dom}\big(T_j \cap (\Xi''_i \times \Xi''_{\pi i})\big) \supseteq \Xi''_i, && \mathrm{ran}\big(T_j \cap (\Xi''_i \times \Xi''_{\pi i})\big) \supseteq \Xi''_{\pi i}. \end{aligned}$$

Then we form a new relational structure $\mathfrak{B}' = \big\langle B, T'_i, E_{ij} \big\rangle_{i,j<\alpha}$ by letting T'_t be the equivalence relation on B with equivalence classes x/T_t for $x \in B \setminus \bigcup_{i \in I} \Xi_i$, together with the classes $\Xi'_i \cup \Xi''_{\pi i}$ $(i \in I)$, and $T'_i \stackrel{\text{def}}{=} T_i$ if $i \neq t$. We say that $\mathfrak{B}'$ ($\mathfrak{Cm}\,\mathfrak{B}'$) is a twisted version of $\mathfrak{B}$ ($\mathfrak{Cm}\,\mathfrak{B}$). For a class K of CA_αs,

$$\mathsf{Tw}\,\mathsf{K} \stackrel{\text{def}}{=} \mathbf{I}\mathsf{K} \cup \mathbf{I}\{\mathfrak{A} : \mathfrak{A} \text{ is a twisted version of some } \mathfrak{C} \in \mathsf{K}\}.$$

We note that by Lemma 1.2.10(ii), (T2) is equivalent with the condition

$$E_{jk} \cap \bigcup_{i \in I} \Xi_i = \emptyset \quad \text{for all distinct } j, k < \alpha \text{ such that } t \notin \{j, k\}.$$

The difference between Henkin's version of twisting and ours is that he only allowed two T_t-classes to be "twisted" (i.e., in his construction, $|I| = 2$ always), so it's essentially the restriction of our version to finite I[6]. Fig. 2 gives a schematic picture of twisting. The solid vertical lines represent T_t-classes before twisting, and the dashed lines illustrate the T'_t classes. The arrows on the top represent the action of the permutation π.

Theorem 1.2.19. *Let $\alpha > 2$ and let $\mathfrak{B}'$ be the result of twisting $\mathfrak{B} \in \mathsf{Ca}_\alpha$. Then $\mathfrak{B}' \in \mathsf{Ca}_\alpha$.*

[6] The possibility of twisting with an infinite set seems to have been first considered by Resek in [Res,75].

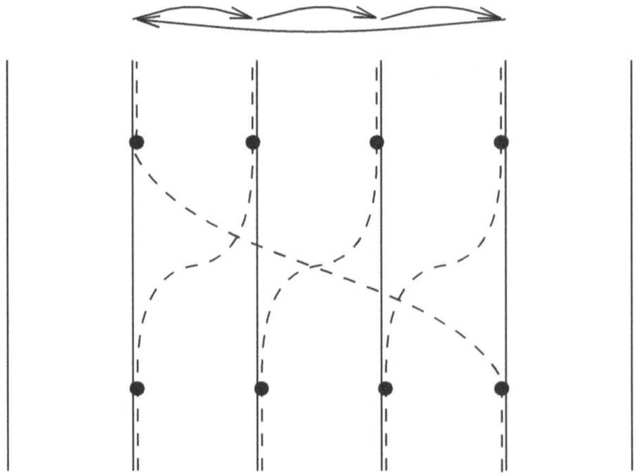

Fig. 3. Twisting

The next lemma is used to simplify computations in a twisted algebra. It says that the effect of twisting the tth cylinder of an algebra is invisible if c_t is buried deep enough in a term. (That cylindrifications other than c_t and the diagonals are not affected by twisting dimension t is clear from the definition.) We use the notation of Definition 1.2.18.

Lemma 1.2.20. *Let $\alpha > 2$ and let $\mathfrak{B}'$ be the result of twisting dimension t of $\mathfrak{B} \in \mathsf{Ca}_\alpha$. Then the term functions $\mathsf{c}_i\mathsf{c}_t$, $\mathsf{c}_t\mathsf{c}_i$, and $\mathsf{s}_i^t\mathsf{c}_j$ are the same in $\mathfrak{Cm}\,\mathfrak{B}'$ as in $\mathfrak{Cm}\,\mathfrak{B}$, provided $\big|\{t, i, j\}\big| = 3$.*

The following lemma describes the behaviour of $^\smile$ in a twisted algebra.

Lemma 1.2.21. *Let $\mathfrak{B} \in \mathsf{Ca}_\alpha$, $\alpha > 2$, and let $\mathfrak{B}'$ be the result of twisting T_1 in $\mathfrak{B}$. Assume that $\xi_i \in E_{12} \cap \Xi_i'$ for all $i \in I$. Then*

(i)

$$\xi_i\ K_{01}'\ b \iff \begin{cases} \xi_i\ K_{01}\ b \ \text{ and } \ \Xi_i'' \cap E_{01} = \emptyset, \ \text{ or} \\ \xi_{\pi i}\ K_{01}\ b \ \text{ and } \ \Xi_i'' \cap E_{01} \neq \emptyset \end{cases}$$

for all $i \in I$ and $b \in B$

(ii)

$$(\mathsf{c}_2\xi_i)^{\smile\prime} = \begin{cases} (\mathsf{c}_2\xi_i)^\smile & \text{if } \Xi_i'' \cap E_{01} = \emptyset \\ (\mathsf{c}_2\xi_{\pi i})^\smile & \text{if } \Xi_i'' \cap E_{01} \neq \emptyset \end{cases}$$

for all $i \in I$ in the complex algebra $\mathfrak{Cm}\,\mathfrak{B}'$ of $\mathfrak{B}'$.

We will see twisting in practice in the proof of Theorem 1.3.3. That proof will use our infinite version of twisting, and we will also show (in Theorem 1.3.4) that this is necessarily so.

In passing we note that twisting can be used to answer questions about amalgamation; see [Mad-Say,thisVol] on amalgamation and related topics.

2.4. Relativization

Of the constructions mentioned in the introduction to this section, relativization is both the most well-known, and the least used in the proof of Theorem 1.3.1. For this reason we will confine ourselves to giving only the basic definitions.

Definition 1.2.22. Let $\mathfrak{A} = \langle A, +, \cdot, -, 0, 1, \mathsf{c}_i, \mathsf{d}_{ij}\rangle_{i,j<\alpha} \in \mathsf{CA}_\alpha$ and let $w \in A$. The *relativization of* $\mathfrak{A}$ *to* w is the algebra

$$\mathfrak{Rl}_w\,\mathfrak{A} \stackrel{\text{def}}{=} \left\langle \mathfrak{Rl}_w\,\mathfrak{A}, +^w, \cdot^w, -^w, 0^w, 1^w, \mathsf{c}_i^w, \mathsf{d}_{ij}^w \right\rangle_{i,j<\alpha},$$

where

$$\mathfrak{Rl}_w\,\mathfrak{A} \stackrel{\text{def}}{=} \{x \in A : x \le w\},$$

and $x +^w y = x + y$, $x \cdot^w y = x \cdot y$, $-^w x = w \cdot -x$, $0^w = 0$, $1^w = w$, $\mathsf{c}_i^w x = w \cdot \mathsf{c}_i x$ and $\mathsf{d}_{ij}^w = w \cdot \mathsf{d}_{ij}$ for all x, $y \in \mathfrak{Rl}_w\,\mathfrak{A}$ and i, $j < \alpha$. For a class K of CA_αs, $\mathsf{Rl}\,\mathsf{K} \stackrel{\text{def}}{=} \{\mathfrak{Rl}_w\,\mathfrak{A} : \mathfrak{A} \in \mathsf{K},\ w \in A\}$ ([Hen-Mon-Tar,85, 2.2.1]).

When $\alpha > 1$, $\mathsf{CA}_\alpha \subset \mathsf{Rl}\,\mathsf{CA}_\alpha$, that is, not every algebra obtained by relativization of a CA_α is a CA_α itself (see [Hen-Mon-Tar,85, 2.2.4]). To distinguish relativizations that do not lead out of CA_α, we introduce the notation $\mathsf{Rl}^{ca}\,\mathsf{K} \stackrel{\text{def}}{=} \mathsf{CA}_\alpha \cap \mathsf{Rl}\,\mathsf{K}$, where $\mathsf{K} \subseteq \mathsf{CA}_\alpha$. Finally, $\mathsf{Srl}^{ca}\,\mathsf{K} \stackrel{\text{def}}{=} \mathsf{CA}_\alpha \cap \mathbf{S}\,\mathsf{Rl}\,\mathsf{K}$.

The following theorem of Resek and Thompson will be an important building block in our proof of the "twisted-relativized" representation theorem for CA_3s (Theorem 1.3.1 in the next section).

Theorem 1.2.23 (Resek–Thompson). $\mathsf{CA}_3 \cap \mathsf{Mod}\,(\text{MGR}) \subseteq \mathsf{Srl}^{ca}\,\mathsf{Cs}_3$.

Proof. The most accessible proof of this theorem is in Andréka–Thompson [And-Tho,88]. ■

Actually, the Resek–Thompson theorem is more general than the form quoted in Theorem 1.2.23, in two respects. First, it is about algebras of arbitrary dimensions ≥ 2; for dimensions greater than 3, MGR stands for the conjunction of ${}_3$MGR and ${}_4$MGR (cf. Remark 1.1.6). Second, it gives representation as concrete, set algebras (i.e., as subalgebras of relativized Cs_αs) of members of a slightly bigger class than $\mathsf{CA}_\alpha \cap \mathsf{Mod}\,$(MGR).

3. Untwisting Cylindric Algebras

By now we have met all the notions and methods that are needed to formulate and prove the main result of this chapter, which says that every CA_3 can be obtained from a representable CA_3 (in fact, a Cs_3) by relativization and twisting. After a brief outline of the proof of this theorem, we will mention some questions that it suggests, and, along the way, will also show that our extension of the definition of twisting is essential, by exhibiting a CA_3 which cannot be obtained by relativization and finite twisting.

Theorem 1.3.1. $\mathsf{CA}_3 \subseteq \mathsf{S}\,\mathsf{Rl}^{ca}\,\mathsf{Tw}\,\mathsf{Srl}^{ca}\,\mathsf{Cs}_3$. *That is, for every* $\mathfrak{A} \in \mathsf{CA}_3$ *there are* $\mathfrak{A}_1, \mathfrak{A}_2, \mathfrak{A}_3 \in \mathsf{CA}_3$ *and* $\mathfrak{A}_4 \in \mathsf{Cs}_3$ *such that* $\mathfrak{A}_3 \subseteq \mathfrak{Rl}\,\mathfrak{A}_4$*,* $\mathfrak{A}_2$ *is a twisted version of* $\mathfrak{A}_3$*,* $\mathfrak{A}_1 = \mathfrak{Rl}\,\mathfrak{A}_2$*, and* $\mathfrak{A} \subseteq \mathfrak{A}_1$.

The proof of this theorem is long and somewhat involved, but its plan is simple enough: One starts with a CA_3 $\mathfrak{A}$, and checks if MGR holds in it. If yes, then $\mathfrak{A}$ is a subalgebra of a relativized Cs_3 by the Resek–Thompson theorem 1.2.23. Otherwise $\mathfrak{A}$ is embedded in its canonical embedding algebra $\mathfrak{A}^\sigma$ ($\mathfrak{A}_1$ in the formulation of the theorem) in order to be able to repair the failure of MGR by twisting its atomstructure. ($\mathfrak{A}_1$ is complete and atomic, and is in CA_3 by [Hen-Mon-Tar,85] 2.7.5 and 2.7.15.) At this point it is clear how the parameters in twisting should be chosen, but one has to apply dilation first to be able to satisfy the conditions of Definition 1.2.18. This is where the first "Rl^{ca}" in the theorem comes in: $\mathfrak{A}^\sigma$ can be gotten back by relativizing the dilated algebra $\mathfrak{A}_2$ with the top element (i.e., the sum of the atoms) of $\mathfrak{A}^\sigma$. The next step is to apply twisting to the dilated algebra and get a CA_3 $\mathfrak{A}_3$ in which MGR holds, and use the Resek–Thompson theorem to represent the latter as an $\mathsf{Srl}^{ca}\,\mathsf{Cs}_3$. Since the effect of twisting can always be undone by twisting the twisted algebra, the procedure we have described now will indeed show that $\mathfrak{A}$ can be gotten from a subalgebra of a relativized Cs_3 by applying twisting, relativization and the operation of taking subalgebras.

Let $FB(\mathsf{Cs}_3)$ denote the class of 3-dimensional cylindric set algebras with finite base (this is a strictly smaller class than the class of finite Cs_3s: the Cs_3 on the rationals generated by the usual ordering is finite but is not representable on a finite set, see e.g. Maddux [Mad,91a, p. 371]), let $\mathsf{Fin}\,(\mathsf{CA}_3)$ stand for the class of finite CA_3s, and if K is a class of cylindric algebras, then let $\mathsf{Tw}_f\,\mathsf{K}$ denote the class of algebras that can be gotten from members of K by an application of *finite* twisting (i.e., by choosing a finite set I in 1.2.18). The following corollary of the proof of 1.3.1 was pointed out by Leon Henkin and Hajnal Andréka.

Corollary 1.3.2. $\mathsf{Fin}\,(\mathsf{CA}_3) \subseteq \mathsf{Rl}^{ca}\,\mathsf{Tw}_f\,\mathsf{Srl}^{ca}\,FB(\mathsf{Cs}_3)$.

Proof. If the algebra $\mathfrak{A}$ we start out with in the proof of 1.3.1 is finite, then there is no need to take its perfect extension since it is already complete and atomic. This is why the first "S" in the formulation of the theorem can be omitted. Next, what we get after dilating and twisting the atom structure of $\mathfrak{A}$ ($\mathfrak{A}_3$ in the formulation of 1.3.1) is also finite, because we put in only finitely many new atoms in the dilation phase, and twisting doesn't change the number of elements. By Theorem 1 in [And-Tho,88] it is isomorphic to a finite $\mathsf{Srl}^{ca}\,\mathsf{Cs}_3$. But then, by a result of Andréka–Hodkinson–Németi [And-Hod-Nem,99] saying that every finite algebra in $\mathsf{S}\,\mathsf{Rl}\,\mathsf{Cs}_3$ is representable on a finite set, there is a finite set U such that $\mathfrak{A}_3$ is isomorphic to a subalgebra of a relativized Cs_3 with base U. ■

One might ask whether the extension of twisting to infinite sets it necessary to obtain the "non-standard" representation theorem 1.3.1. The next results indicate that it probably is. Consider the quasi-equation

(e) $$ {}_2\mathsf{s}(0,1)_2\mathsf{s}(0,1)\mathsf{c}_2 x \le \mathsf{c}_2 x \to {}_2\mathsf{s}(0,1)_2\mathsf{s}(0,1)\mathsf{c}_2 x = \mathsf{c}_2 x $$

valid in representable CA_3s, and in every finite CA_α.

Theorem 1.3.3. *There are a* CA_3 $\mathfrak{C}'$ *and an element* $x \in \mathfrak{Ra}\,\mathfrak{C}'$ *such that* $x^{\smile\smile} < x$ *in* $\mathfrak{Ra}\,\mathfrak{C}'$. *Hence* $\mathsf{CA}_3 \not\models e$.

Proof. Let U be the set $2 \times \mathbb{Z}$, suc the permutation $\langle i, z\rangle \mapsto \langle i, z+1\rangle$ and f the permutation $\langle i, z\rangle \mapsto \langle j, z\rangle$ of U, where $z \in \mathbb{Z}$ and $\{i, j\} = 2$. Let $\mathfrak{C}$ be the full Cs_3 with base U, G the group of permutations generated by

$\{\text{suc}, f\}$, and let $\mathfrak{C} = \mathfrak{Fr}_G\mathfrak{C}$. The proofs of some of the statements below make use of the simple fact that

$$(1) \qquad G \text{ is semiregular.}$$

Note that if $s = \big\langle \langle i_0, z_0\rangle, \langle i_1, z_1\rangle, \langle i_2, z_2\rangle\big\rangle$ then

$$\begin{aligned} s^- = \Big\{ & \big\langle \langle i_0, z_0 + z\rangle, \langle i_1, z_1 + z\rangle, \langle i_2, z_2 + z\rangle\big\rangle, \\ & \big\langle \langle j_0, z_0 + z\rangle, \langle j_1, z_1 + z\rangle, \langle j_2, z_2 + z\rangle\big\rangle \;:\; z \in \mathbb{Z}, \\ & \{i_0, j_0\} = \{i_1, j_1\} = \{i_2, j_2\} = 2\Big\}. \end{aligned}$$

Let $\mathfrak{B}$ be the atom structure of $\mathfrak{C}$. Again, we want to set up things as in Definition 1.2.18. Let $t \stackrel{\text{def}}{=} 1$, $I \stackrel{\text{def}}{=} \mathbb{Z}$ and let π be the permutation $z \mapsto -(z+1)$ of I (hence π is of order 2). For $k \in I$ let $\xi_k = \big\langle \langle 0,0\rangle, \langle 1,k\rangle, \langle 1,k\rangle\big\rangle^-$. Condition (T1) then follows from (1), and (T2) obviously holds. For future reference we note that

$$\begin{aligned} (\mathsf{c}_2\xi_k)^\smile &= \big(\mathsf{c}_2\big\langle \langle 0,0\rangle, \langle 1,k\rangle, \langle 1,k\rangle\big\rangle^-\big)^\smile = \mathsf{c}_2\big\langle \langle 1,k\rangle, \langle 0,0\rangle, \langle 0,0\rangle\big\rangle^- \\ &= \mathsf{c}_2\big\langle \langle 0,k\rangle, \langle 1,0\rangle, \langle 1,0\rangle\big\rangle^- = \mathsf{c}_2\big\langle \langle 0,0\rangle, \langle 1,-k\rangle, \langle 1,-k\rangle\big\rangle^- = \mathsf{c}_2\xi_{-k}. \end{aligned}$$

To complete the construction, we partition $\Xi_k = \xi_k/T_1$ into two parts and check that condition (T3) holds. Using (1), it is easy to see that

$$\Xi_k = \Big\{\big\langle \langle 0,0\rangle, u, \langle 1,k\rangle\big\rangle^- \;:\; u \in U\Big\}$$

and that the displayed atoms are pairwise distinct. Hence (again by (1)), $T_0 \cap (\Xi_k \times \Xi_{\pi k})$ is the bijection

$$\big\langle \langle 0,0\rangle, \langle i,z\rangle, \langle 1,k\rangle\big\rangle^- \mapsto \big\langle \langle 0,0\rangle, \langle i, z-2k-1\rangle, \langle 1, -(k+1)\rangle\big\rangle^-$$

from Ξ_k onto $\Xi_{\pi k}$ and $T_2 \cap (\Xi_k \times \Xi_{\pi k})$ is the bijection

$$\big\langle \langle 0,0\rangle, u, \langle 1,k\rangle\big\rangle^- \mapsto \big\langle \langle 0,0\rangle, u, \langle 1, -(k+1)\rangle\big\rangle^-$$

from Ξ_k onto $\Xi_{\pi k}$. From this it follows that the partition

$$\Xi'_k \stackrel{\text{def}}{=} \Big\{\big\langle \langle 0,0\rangle, \langle 1,z\rangle, \langle 1,k\rangle\big\rangle^- \;:\; z \in \mathbb{Z}\Big\},$$

$$\Xi''_k \stackrel{\mathrm{def}}{=} \{\langle \langle 0,0\rangle, \langle 0,z\rangle, \langle 1,k\rangle\rangle^- : z \in \mathbb{Z}\}$$

of Ξ_k satisfies condition (T3). We let $\mathfrak{B}'$ be the twisted atom structure and $\mathfrak{C}'$ its complex algebra.

Since $\xi_k \in E_{12} \cap \Xi'_k$ and $\Xi''_k \cap E_{01} \neq \emptyset$ for all $k \in I$, Lemma 1.2.21 can be applied to conclude that in $\mathfrak{Ra}\,\mathfrak{C}'$,

$$(\mathsf{c}_2\xi_k)^{\smile\prime} = (\mathsf{c}_2\xi_{\pi k})^{\smile} = \big(\mathsf{c}_2\xi_{-(k+1)}\big)^{\smile} = \mathsf{c}_2\xi_{k+1}$$

for all $k \in I$. Hence there is a $\mathbb{Z}$-type chain

$$\dots \stackrel{\smile}{\longrightarrow} \mathsf{c}_2\xi_{-1} \stackrel{\smile}{\longrightarrow} \mathsf{c}_2\xi_0 \stackrel{\smile}{\longrightarrow} \mathsf{c}_2\xi_1 \stackrel{\smile}{\longrightarrow} \dots$$

in $\mathfrak{Ra}\,\mathfrak{C}'$. Therefore, letting $x = \sum_{k\in\omega} \mathsf{c}_2\xi_k$, by [Hen-Mon-Tar,85, 1.5.3] we get $x^{\smile\prime\smile\prime} = \sum_{1<k\in\omega} \mathsf{c}_2\xi_k < x$ in $\mathfrak{Ra}\,\mathfrak{C}'$, as desired. ■

Theorem 1.3.4. *Let $\mathfrak{B} \in \mathsf{Ca}_3$, and let $\mathfrak{B}'$ be the result of twisting $\mathfrak{B}$ with a finite set I. If there is no $x \in \mathfrak{Ra}\,\mathfrak{Cm}\,\mathfrak{B}$ with $x^{\smile\smile} < x$, then the same is true for $\mathfrak{Ra}\,\mathfrak{Cm}\,\mathfrak{B}'$. In other words, Henkin's twisting does not lead out of* $\mathsf{CA}_3 \cap \mathsf{Mod}\,(e)$.

Theorem 1.3.5. *Let $\mathfrak{A}$, $\mathfrak{C} \in \mathsf{CA}_3$ with $\mathfrak{A} \subseteq \mathfrak{Rl}_w\,\mathfrak{C}$ for some $w \in C$. If $\mathfrak{C}$ satisfies the quasi-equation* (e)*, then so does $\mathfrak{A}$.*

We note that it is very easy to construct from an RCA_3 by finite twisting a CA_3 $\mathfrak{C}$ and an element $w \in C$ such that $\mathfrak{C} \models e$ but $(\mathsf{c}_2x)^{\smile\smile} = 0$ in $\mathsf{Rl}_w\,\mathfrak{C}$ for some nonzero $x \leq w$.

Corollary 1.3.6. *Let $\mathcal{S}$ be any finite string of the class operators $\{\mathbf{S}, \mathsf{Up}, \mathbf{P}, \mathsf{Tw}_f, \mathsf{Rl}^{ca}, \mathsf{Srl}^{ca}\}$. Then $\mathcal{S}\,\mathsf{RCA}_3 \subset \mathsf{CA}_3$.*

Proof. By 1.3.3, the quasi-equation (e) fails in CA_3; but it is valid in RCA_3, and it is obviously preserved by $\mathbf{S}$, Up and $\mathbf{P}$. By 1.3.4 and 1.3.5, it is also preserved by Tw_f and Srl^{ca} (and hence by Rl^{ca}, too). ■

We note that if we allow the operations of taking complex algebras and atom structures in $\mathcal{S}$, then our proof (and probably the corollary itself) breaks down.

The proof of Theorem 1.3.1 and the preceding discussion motivates the following

Problem 1.3.7. Is $\mathsf{CA}_3 \subseteq \mathbf{S}\,\mathsf{Tw}\,\mathsf{Srl}^{ca}\,\mathsf{Cs}_3$? Is

$$\mathsf{CA}_3 \subseteq \mathbf{S}\,\mathfrak{Cm}\,\mathfrak{At}\,\mathsf{Up}\,\mathsf{Tw}_f\,\mathsf{Srl}^{ca}\,\mathsf{Cs}_3?$$

More generally: what other "routes" *via* CA_3s are there to get to every CA_3 from representable ones?

In connection with the first question in 1.3.7 we note that the proof of Theorem 1.3.1 could probably be modified to give

$$(*) \qquad \mathsf{CA}_3 \subseteq \mathbf{S}\,\mathsf{Tw}\,\mathsf{S}\,\mathsf{Rl}\,\mathsf{Cs}_3$$

(after, of course, extending twisting to members of $\mathbf{S}\,\mathsf{Rl}\,\mathsf{Cs}_3$). But we believe that the presence of an algebra that is not in CA_3 in the chain leading from a CA_3 to a Cs_3 makes $(*)$ less attractive.

The most important problem left open by the results in this section is the following.

Problem 1.3.8 (Henkin). Does Theorem 1.3.1 extend in some form to dimensions > 3?

We note that if $\alpha > 3$, then, in order to be able to use the Resek–Thompson theorem, one has to cure the failures of *all* equations in Remark 1.1.6. Richard Thompson has some unpublished results in connection with this.

III. Notions of Logic Related to Cylindric-Like Algebras

BOLYAI SOCIETY
MATHEMATICAL STUDIES, 22

Cylindric-like Algebras
and Algebraic Logic
pp. 185–203.

Representable Cylindric Algebras and Many-Dimensional Modal Logics

AGI KURUCZ*

The equationally expressible properties of the cylindrifications and the diagonals in finite-dimensional representable cylindric algebras can be divided into two groups:

(i) 'One-dimensional' properties describing individual cylindrifications. These can be fully characterised by finitely many equations saying that each c_i, for $i < n$, is a normal ($\mathsf{c}_i 0 = 0$), additive ($\mathsf{c}_i(x+y) = \mathsf{c}_i x + \mathsf{c}_i y$) and complemented closure operator:

$$x \leq \mathsf{c}_i x \qquad \mathsf{c}_i \mathsf{c}_i x \leq \mathsf{c}_i x \qquad \mathsf{c}_i(-\mathsf{c}_i x) \leq -\mathsf{c}_i x. \tag{2.0.1}$$

(ii) 'Dimension-connecting' properties, that is, equations describing the diagonals and interaction between different cylindrifications and/or diagonals. These properties are much harder to describe completely, and there are many results in the literature on their complexity.

The main aim of this chapter is to study generalisations of (i) while keeping (ii) as unchanged as possible. In other words, we would like to analyse how much of the complexity of RCA_n is due to its 'many-dimensional' character and how much of it to the particular properties of the cylindrifications. Note that this direction is kind of orthogonal to the one taken by *relativised cylindric algebras* [Hen-Mon-Tar,85, Section 5.5], where (i) is kept unchanged, while generalisations of (ii) are considered.

---

*I am grateful to Ian Hodkinson for discussions and for his many comments on the preliminary version. Thanks are also due to Rob Goldblatt, Stanislav Kikot, András Simon and Misha Zakharyaschev for discussions.

In our investigations we view representable cylindric algebras from the perspective of *multimodal logic.* This approach is explained in detail in [Ven,thisVol] and [Mar-Ven,97]. In particular, we look at n-dimensional cylindric set algebras as subalgebras of the complex algebra of 'n-dimensional' *relational structures* of the form

$$\langle {}^nU, \equiv_i, Id_{ij} \rangle_{i,j<n}, \tag{2.0.2}$$

where, for all $i, j < n$, $\mathbf{u}, \mathbf{v} \in {}^nU$,

$$\mathbf{u} \equiv_i \mathbf{v} \iff u_k = v_k \text{ for all } k < n, \; k \neq i, \text{ and}$$

$$\mathbf{u} \in Id_{ij} \iff u_i = u_j.$$

Instead of equations in the algebraic language having operators d_{ij} and c_i, we use formulas of the corresponding propositional multimodal language having modal constants δ_{ij} and unary diamonds $\Diamond_i$ (and their duals $\Box_i$), for $i, j < n$. As the variety RCA_n of n-dimensional representable cylindric algebras is generated by cylindric set algebras, equations valid in RCA_n correspond to multimodal formulas valid in all structures described in (2.0.2). The above classification of equational properties now translates to the following classification of modally expressible properties:

(i) Modal formulas saying that each $\Diamond_i$ is normal and distributes over $\vee$, and axioms of modal logic **S5**, for each $i < n$:

$$\Box_i p \to p \qquad \Box_i p \to \Box_i \Box_i p \qquad \Diamond_i p \to \Box_i \Diamond_i p. \tag{2.0.3}$$

(ii) Multimodal formulas describing 'dimension-connecting' properties of the n-dimensional structures described in (2.0.2).

Our investigations can also be motivated from a purely modal logic point of view. The n-dimensional relational structures described in (2.0.2) are examples of *products of Kripke frames,* a notion introduced in [Seg,73, She,78]. Product frames have been widely used for modelling interactions between modal operators representing time, space, knowledge, actions, etc.; see [Gab-Kur-Wol-Zak,03, Kur,07] and references therein. One can also consider this chapter as a demonstration of how cylindric algebraic results and techniques can be used for studying combinations of modal logics.

1. Special Varieties of Complex Algebras

This chapter is not self-contained in the sense that we use without explicit reference standard notions and results from basic modal logic; such as *p-morphisms* (also known as *bounded morphisms* or *zigzag morphisms*), *inner substructures* (also known as *generated subframes*), Sahlqvist formulas and canonicity, and duality between relational structures and *Boolean algebras with operators* (BAOs). For notions and statements not defined or proved here, see other chapters of this volume (like that of [Hir-Hod,thisVol], [Ven,thisVol]) or [Cha-Zak,97, Bla-Rij-Ven,01, Gol,89].

We begin with introducing some notation and terminology. If x is a point in a relational structure $\mathfrak{F}$ then we denote by $\mathfrak{F}^x$ the smallest inner substructure of $\mathfrak{F}$ containing x. We call $\mathfrak{F}^x$ a *point-generated inner substructure* of $\mathfrak{F}$. If $\mathfrak{F} = \mathfrak{F}^x$ for some x, then $\mathfrak{F}$ is called *rooted.* Rooted structures are important in modal logic, as in any model over $\mathfrak{F}$, the truth-values of modal formulas at x depend only on how the model behaves at points in $\mathfrak{F}^x$.

Apart from the usual operators **H**, **S** and **P** on classes of algebras (see [Hen-Mon-Tar-And-Nem,81, Ch. 0]) we use the following operators on classes of relational structures of the same signature:

$$\mathbb{I}\,\mathcal{C} = \text{isomorphic copies of inner substructures of structures in } \mathcal{C},$$

$$\mathbb{I}_{\mathsf{p}}\,\mathcal{C} = \text{isomorphic copies of point-generated inner substructures of structures in } \mathcal{C},$$

$$\mathbb{U}\mathrm{p}\,\mathcal{C} = \text{isomorphic copies of ultraproducts of structures in } \mathcal{C}.$$

The *(full) complex algebra* of a relational structure $\mathfrak{F}$ is denoted by $\mathfrak{Cm}\,\mathfrak{F}$. We can describe properties of $\mathfrak{F}$ in the corresponding *(multi)modal language*, having a k-ary modal $\Diamond$ for each $k+1$-ary relation. *Validity* of a set Σ of such modal formulas in a relational structure $\mathfrak{F}$ (in symbols: $\mathfrak{F} \models \Sigma$) is defined as usual. Formulas of this modal language can also be considered as terms of an algebraic language, where each k-ary $\Diamond$ is regarded as a k-ary function symbol. The starting point of the duality between modal logic and BAOs is the following property: for every relational structure $\mathfrak{F}$ and every modal formula φ,

$$\text{(2.1.1)} \qquad \mathfrak{F} \models \varphi \quad \iff \quad \mathfrak{Cm}\,\mathfrak{F} \models (\varphi = 1).$$

Given a class $\mathcal{C}$ of relational structures of the same signature, we denote by $\mathsf{Cm}\,\mathcal{C}$ the class of complex algebras of structures in $\mathcal{C}$, and by $\mathsf{Log}\,(\mathcal{C})$ the set of all modal formulas that are valid in every structure in $\mathcal{C}$. We then have the following consequence of (2.1.1): for any relational structure $\mathfrak{F}$,

$$(2.1.2) \qquad \mathfrak{F} \models \mathsf{Log}\,(\mathcal{C}) \quad \iff \quad \mathfrak{Cm}\,\mathfrak{F} \in \mathbf{HSP}\,\mathsf{Cm}\,\mathcal{C}.$$

We are interested in varieties of BAOs generated by complex algebras of (special) structures (these are called *complex varieties* in [Gol,89]). The following general result will be used throughout this chapter:

Theorem 2.1.1 (Goldblatt). *If $\mathcal{C}$ is a class of relational structures that is closed under* $\mathbb{U}\mathrm{p}$, *then* $\mathbf{SP}\,\mathsf{Cm}\,\mathbb{I}\,\mathcal{C}$ *is a canonical variety* ([Gol,95]).

Let us have a closer look at the subdirectly irreducible algebras of these varieties.

Lemma 2.1.2. *For any class $\mathcal{C}$ of relational structures, the subdirectly irreducible members of $\mathbf{SP}\,\mathsf{Cm}\,\mathbb{I}\,\mathcal{C}$ belong to $\mathbf{S}\,\mathsf{Cm}\,\mathbb{I}_{\mathsf{p}}\,\mathcal{C}$* ([Kur,10]).

Proof. Let $\mathfrak{A} \in \mathbf{SP}\,\mathsf{Cm}\,\mathbb{I}\,\mathcal{C}$ and let $\mathfrak{A} \rightarrowtail \prod_{i\in I} \mathfrak{A}_i$ be a subdirect embedding, for some $\mathfrak{A}_i \in \mathbf{S}\,\mathsf{Cm}\,\mathbb{I}\,\mathcal{C}$, $i \in I$. If $\mathfrak{A}$ is subdirectly irreducible then there is an $i \in I$ such that $\mathfrak{A} \cong \mathfrak{A}_i$, and so $\mathfrak{A}$ is isomorphic to a subalgebra of $\mathfrak{Cm}\,\mathfrak{F}$ for some $\mathfrak{F} \in \mathbb{I}\,\mathcal{C}$. Then for each point x in $\mathfrak{F}$, $\mathfrak{F}^x \in \mathbb{I}_{\mathsf{p}}\,\mathbb{I}\,\mathcal{C} \subseteq \mathbb{I}_{\mathsf{p}}\,\mathcal{C}$. It is not hard to show (see e.g. [Gol,89, 3.3]) that $\mathfrak{Cm}\,\mathfrak{F} \rightarrowtail \prod_{x\in\mathfrak{F}} \mathfrak{Cm}\,\mathfrak{F}^x$ is a (subdirect) embedding. So there exist subalgebras $\mathfrak{B}_x$ of $\mathfrak{Cm}\,\mathfrak{F}^x$ such that $\mathfrak{A} \rightarrowtail \prod_{x\in\mathfrak{F}} \mathfrak{B}_x$ is a subdirect embedding as well. As $\mathfrak{A}$ is subdirectly irreducible, there is some x in $\mathfrak{F}$ such that $\mathfrak{A} \cong \mathfrak{B}_x$, and so $\mathfrak{A} \in \mathbf{S}\,\mathsf{Cm}\,\mathbb{I}_{\mathsf{p}}\,\mathcal{C}$. ■

Now Theorem 2.1.1 and Lemma 2.1.2 imply the following characterisation of varieties generated by certain classes of complex algebras.

Theorem 2.1.3. *If $\mathcal{C}$ is a class of relational structures that is closed under* $\mathbb{U}\mathrm{p}$ *and* $\mathbb{I}_{\mathsf{p}}$, *then* $\mathbf{SP}\,\mathsf{Cm}\,\mathcal{C} = \mathbf{HSP}\,\mathsf{Cm}\,\mathcal{C}$ *is a canonical variety* ([Kur,10]).

We can also have a 'dual' structural characterisation of subdirectly irreducible algebras of these varieties. We denote by $\mathfrak{Uf}\,\mathfrak{A}$ the *ultrafilter frame* of a BAO $\mathfrak{A}$, and by $\mathfrak{Ue}\,\mathfrak{F} = \mathfrak{Uf}\,\mathfrak{Cm}\,\mathfrak{F}$ the *ultrafilter extension* of a relational structure $\mathfrak{F}$.

Theorem 2.1.4. *Let $\mathcal{C}$ be a class of relational structures that is closed under $\mathbb{U}\mathrm{p}$. Then for every subdirectly irreducible algebra $\mathfrak{A}$,*

$$\mathfrak{A} \in \mathbf{SP}\,\mathsf{Cm}\,\mathcal{C} \iff \mathfrak{A} \in \mathbf{S}\,\mathsf{Cm}\,\mathcal{C}$$

$$\iff \mathfrak{Uf}\,\mathfrak{A} \textit{ is a p-morphic image of some } \mathfrak{G} \in \mathcal{C}.$$

Proof. $\Leftarrow$: By Jónsson and Tarski's [Jon-Tar,51] theorem, $\mathfrak{A}$ is embeddable into $\mathfrak{Cm}\,\mathfrak{Uf}\,\mathfrak{A}$. And by duality, $\mathfrak{Cm}\,\mathfrak{Uf}\,\mathfrak{A}$ is embeddable into $\mathfrak{Cm}\,\mathfrak{G} \in \mathsf{Cm}\,\mathcal{C}$.

$\Rightarrow$: If $\mathfrak{A} \in \mathbf{SP}\,\mathsf{Cm}\,\mathcal{C}$ then there is a subdirect embedding $\mathfrak{A} \rightarrowtail \prod_{i\in I} \mathfrak{A}_i$, for some $\mathfrak{A}_i \in \mathbf{S}\,\mathsf{Cm}\,\mathcal{C}$, $i \in I$. As $\mathfrak{A}$ is subdirectly irreducible, there is an $i \in I$ such that $\mathfrak{A} \cong \mathfrak{A}_i$, that is, $\mathfrak{A}$ is isomorphic to a subalgebra of $\mathfrak{Cm}\,\mathfrak{F}$ for some $\mathfrak{F} \in \mathcal{C}$. By duality, $\mathfrak{Uf}\,\mathfrak{A}$ is a p-morphic image of $\mathfrak{Ue}\,\mathfrak{F}$. As $\mathfrak{Ue}\,\mathfrak{F}$ is a p-morphic image of an ultrapower of $\mathfrak{F}$ (see [Fin,75] and van Benthem [Ben,79, Ben,80]) and $\mathcal{C}$ is closed under taking ultraproducts, the proof is completed. ■

Remark 2.1.5. We can have a similar characterisation of arbitrary, not necessarily subdirectly irreducible, algebras in these kinds of varieties. If $\mathcal{C}$ is closed under $\mathbb{U}\mathrm{p}$ then

$$\mathfrak{A} \in \mathbf{SP}\,\mathsf{Cm}\,\mathcal{C} \tag{2.1.3}$$

$$\iff \mathfrak{Uf}\,\mathfrak{A} \text{ is a p-morphic image of } \dot{\bigcup_{i\in I}} \mathfrak{G}_i \text{ for some } \mathfrak{G}_i \in \mathcal{C},$$

where $\dot{\bigcup}_{i\in I} \mathfrak{G}_i$ is the *disjoint union* of $\mathfrak{G}_i$, for $i \in I$. The proof of (2.1.3) is similar to that of Theorem 2.1.4, but we need to use some additional properties of the various operators such as:

- $\prod_{i\in I} \mathfrak{Cm}\,\mathfrak{G}_i \cong \mathfrak{Cm}\,\dot{\bigcup}_{i\in I} \mathfrak{G}_i$.
- [Gol,91] An ultrapower of a disjoint union of structures is a p-morphic image of a disjoint union of some ultraproducts formed from the same structures.

An example of a class $\mathcal{C}$ of relational structures that is closed under taking ultraproducts and point-generated inner substructures is the class of n-dimensional *full cylindric set algebra atom structures* (described in (2.0.2)). Disjoint unions of such structures are atom structures of *generalised cylindric set algebras.* By duality, a surjective p-morphism from such a structure onto $\mathfrak{Uf}\,\mathfrak{A}$ corresponds to a *complete representation* (see [Hir-Hod,thisVol], [Ven,thisVol]) of the *canonical embedding algebra* $\mathfrak{Cm}\,\mathfrak{Uf}\,\mathfrak{A}$ of $\mathfrak{A}$. So, as a special case of (2.1.3) one can obtain the following result of Monk. For every cylindric-type algebra $\mathfrak{A}$,

$$\mathfrak{A} \in \mathsf{RCA}_n \quad \iff \quad \mathfrak{Cm}\,\mathfrak{Uf}\,\mathfrak{A} \text{ has a complete representation.}$$

Corollary 2.1.6. *Let $\mathcal{C}$ be a class of relational structures that is closed under $\mathbb{U}\mathrm{p}$ and $\mathbb{I}_{\mathsf{p}}$. Then for every rooted structure $\mathfrak{F}$,*

$$\mathfrak{F} \models \mathsf{Log}\,(\mathcal{C}) \quad \iff \quad \mathfrak{Ue}\,\mathfrak{F} \textit{ is a p-morphic image of some } \mathfrak{G} \in \mathcal{C}.$$

Proof. By (2.1.2) and Theorem 2.1.3,

$$\mathfrak{F} \models \mathsf{Log}\,(\mathcal{C}) \quad \iff \quad \mathfrak{Cm}\,\mathfrak{F} \in \mathbf{SP}\,\mathsf{Cm}\,\mathcal{C}.$$

As the complex algebra of a rooted structure is subdirectly irreducible [Gol,89], the statement follows from Theorem 2.1.4. ∎

As the ultrafilter extension of a finite relational structure is isomorphic to the structure itself, we obtain:

Corollary 2.1.7. *Let $\mathcal{C}$ be a class of relational structures that is closed under $\mathbb{U}\mathrm{p}$ and $\mathbb{I}_{\mathsf{p}}$. Then for every rooted finite structure $\mathfrak{F}$,*

$$\mathfrak{F} \models \mathsf{Log}\,(\mathcal{C}) \quad \iff \quad \mathfrak{F} \textit{ is a p-morphic image of some } \mathfrak{G} \in \mathcal{C}.$$

2. The Diagonal-Free Case

We would like to apply the general results of the previous section to special classes of 'n-dimensional' relational structures. To this end, for any $0 < n < \omega$, we define an *n-frame* to be a structure of the form $\langle W, T_i \rangle_{i<n}$, where W is a non-empty set and T_i is a binary relation on W, for each $i < n$. Multimodal formulas matching n-frames are called *n-modal formulas* (that is, n-modal formulas are built up from propositional variables using the Booleans and unary modal operators $\Diamond_i$ and $\Box_i$, $i < n$).

The following notion is a generalisation of atom structures of n-dimensional full diagonal-free cylindric set algebras (cf. [Hen-Mon-Tar-And-Nem,81, 2.7.38]).

Definition 2.2.1. Given 1-frames $\mathfrak{F}_i = \langle W_i, R_i \rangle$, $i < n$, their *product* is the n-frame

$$\mathfrak{F}_0 \times \cdots \times \mathfrak{F}_{n-1} = \big\langle W_0 \times \cdots \times W_{n-1}, \bar{R}_i \big\rangle_{i<n},$$

where $W_0 \times \cdots \times W_{n-1}$ is the Cartesian product of the W_i and for all $\mathbf{u}, \mathbf{v} \in W_0 \times \cdots \times W_{n-1}$ and $i < n$,

$$\mathbf{u}\bar{R}_i\mathbf{v} \quad \text{iff} \quad u_i R_i v_i \text{ and } u_j = v_j \text{ for } j \neq i, \;\; j < n.$$

Such n-frames we call *n-dimensional product frames.*

It is not hard to see that the product operation commutes with taking ultraproducts and point-generated inner substructures:

Proposition 2.2.2. *Let U be an ultrafilter over some index set I, and let $\mathfrak{F}_k^i$ be a 1-frame, for $i \in I$, $k < n$. Then:*

$$\prod_{i \in I} \left(\mathfrak{F}_0^i \times \cdots \times \mathfrak{F}_{n-1}^i \right) / U \cong \Big(\prod_{i \in I} \mathfrak{F}_0^i / U \Big) \times \cdots \times \Big(\prod_{i \in I} \mathfrak{F}_{n-1}^i / U \Big).$$

Proposition 2.2.3. *Let $\mathfrak{F} = \mathfrak{F}_0 \times \cdots \times \mathfrak{F}_{n-1}$ and $\mathbf{x}$ be a point in $\mathfrak{F}$. Then:*

$$\mathfrak{F}^{\mathbf{x}} = \mathfrak{F}_0^{x_0} \times \cdots \times \mathfrak{F}_{n-1}^{x_{n-1}}.$$

Remark 2.2.4. Given classes $\mathcal{C}_i$ of 1-frames, for $i < n$, we can define a class $\mathcal{C}$ of n-dimensional product frames by taking

$$\mathcal{C} = \mathcal{C}_0 \times \cdots \times \mathcal{C}_{n-1} = \{ \mathfrak{F}_0 \times \cdots \times \mathfrak{F}_{n-1} \, : \, \mathfrak{F}_i \in \mathcal{C}_i, \;\; i < n \}.$$

As a consequence of Propositions 2.2.2 and 2.2.3, we obtain that if each class $\mathcal{C}_i$ is defined by a set of *universal* formulas in the first-order language having one binary predicate symbol and possibly equality, then $\mathcal{C}$ is closed under taking ultraproducts and point-generated inner substructures. Here are some examples of this kind:

$$\begin{aligned}
\mathcal{C}^n_{all} &= \text{the class of all } n\text{-dimensional product frames,} \\
\mathcal{C}^n_{trans} &= \text{the class of all } n\text{-dimensional products of transitive frames,} \\
\mathcal{C}^n_{equiv} &= \text{the class of all } n\text{-dimensional products of equivalence frames,} \\
\mathcal{C}^n_{univ} &= \text{the class of all } n\text{-dimensional products of universal frames} \\
&= \text{the class of all } n\text{-dimensional full diagonal-free cylindric} \\
&\quad\; \text{set algebra atom structures.}
\end{aligned}$$

By Theorem 2.1.3, in each of these cases $\mathbf{SP}\,\mathsf{Cm}\,\mathcal{C}$ is a canonical variety. In particular, we can obtain the class RDf_n of *representable diagonal-free cylindric algebras of dimension* n: As $\mathbb{I}_\mathsf{p}\,\mathcal{C}^n_{equiv} = \mathcal{C}^n_{univ}$, $\mathsf{RDf}_n = \mathbf{SP}\,\mathsf{Cm}\,\mathcal{C}^n_{univ} = \mathbf{SP}\,\mathsf{Cm}\,\mathcal{C}^n_{equiv}$ holds. Moreover, by Johnson [Joh,69] (see also Halmos [Hal,57] and [Hen-Mon-Tar,85, Section 5.1]), we also have $\mathsf{RDf}_n = \mathbf{SP}\,\mathsf{Cm}\,\mathcal{C}^n_{cube}$, where

$$\mathcal{C}^n_{cube} = \{\, \underbrace{\mathfrak{F} \times \cdots \times \mathfrak{F}}_{n} : \mathfrak{F} = \langle U, U \times U \rangle \text{ for some non-empty set } U \}$$

is yet another class of n-dimensional product frames that is closed under $\mathbb{U}\mathrm{p}$ and $\mathbb{I}_\mathsf{p}$.

Let us introduce notation for the corresponding n-modal logics:

$$\begin{aligned}
\mathbf{K}^n &= \mathsf{Log}\left(\mathcal{C}^n_{all}\right),\\
\mathbf{K4}^n &= \mathsf{Log}\left(\mathcal{C}^n_{trans}\right),\\
\mathbf{S5}^n &= \mathsf{Log}\left(\mathcal{C}^n_{equiv}\right) = \mathsf{Log}\left(\mathcal{C}^n_{univ}\right) = \mathsf{Log}\left(\mathcal{C}^n_{cube}\right).
\end{aligned}$$

The following theorem shows that any n-frame having n equivalence relations and being a p-morphic image of an arbitrary n-dimensional product frame is also a p-morphic image of a product of n equivalence frames.

Theorem 2.2.5. *Let $\mathfrak{F} = \langle W, T_i \rangle_{i<n}$ be an n-frame such that every T_i is an equivalence relation, for $i < n$. Suppose that $f : \mathfrak{G}_0 \times \cdots \times \mathfrak{G}_{n-1} \to \mathfrak{F}$ is a surjective p-morphism, for some 1-frames $\mathfrak{G}_i = \langle U_i, R_i \rangle$, $i < n$. Then there exist 1-frames $\mathfrak{G}^*_i = \langle U_i, R^*_i \rangle$, $i < n$, such that*

- *each R^*_i is an equivalence relation extending R_i, and*
- *$f : \mathfrak{G}^*_0 \times \cdots \times \mathfrak{G}^*_{n-1} \to \mathfrak{F}$ is still a surjective p-morphism*

([Kur,10]).

Proof. Let κ be an infinite cardinal $\geq \max_{i<n} |U_i|$. Then we can 'fix' the domain of f by playing the following 2-player κ-long game over f. The players $\forall$ and $\exists$ are building an increasing sequence $\langle R^\alpha_i : \alpha < \kappa \rangle$ of binary relations on U_i, for each $i < n$. At round 0, R^0_i is the reflexive closure of R_i, $i < n$. Then clearly f is still a surjective p-morphism from $\langle U_0, R^0_0 \rangle \times \cdots \times \langle U_{n-1}, R^0_{n-1} \rangle$ to $\mathfrak{F}$, as each relation T_i in $\mathfrak{F}$ is reflexive.

At round $\alpha + 1 < \kappa$, $\forall$ picks

(i) either a tuple $\langle i, x, y\rangle$ such that $i < n$, $x, y \in U_i$, and $xR_i^\alpha y$;

(ii) or a tuple $\langle i, x, y, z\rangle$ such that $i < n$, $x, y, z \in U_i$, and $xR_i^\alpha yR_i^\alpha z$.

$\exists$ has to respond with $R_i^{\alpha+1} \supseteq R_i^\alpha$ such that f is still a surjective p-morphism from $\langle U_0, R_0^{\alpha+1}\rangle \times \cdots \times \langle U_{n-1}, R_{n-1}^{\alpha+1}\rangle$ to $\mathfrak{F}$, and either $yR_i^{\alpha+1}x$ (in case (i)), or $xR_i^{\alpha+1}z$ (in case (ii)).

At round β for limit ordinals $\beta < \kappa$, they take $R_i^\beta = \bigcup_{\alpha<\beta} R_i^\alpha$, for $i < n$. If at each round $\alpha < \kappa$ $\exists$ can respond, then $R_i^* = \bigcup_{\alpha<\kappa} R_i^\alpha$, $i < n$, would clearly be an equivalence relation as required.

Let us define a winning strategy for $\exists$. Suppose that in round α $\forall$ chooses a tuple like in (ii) (the case of (i) is similar). Then let $R_i^{\alpha+1} = R_i^\alpha \cup \{\langle x, z\rangle\}$ and $R_j^{\alpha+1} = R_j^\alpha$ for all $j < n$, $j \neq i$. We claim that f is still a surjective p-morphism from $\langle U_0, R_0^{\alpha+1}\rangle \times \cdots \times \langle U_{n-1}, R_{n-1}^{\alpha+1}\rangle$ to $\mathfrak{F}$. Indeed, the 'backward condition' clearly holds, as we added pairs only to the domain of f. As concerns f being a homomorphism, take a 'new' pair (if there is such) $\langle \mathbf{u}, \mathbf{v}\rangle$ from $\bar{R}_i^{\alpha+1}$. Then $u_i = x$ and $v_i = z$, and $u_j = v_j$ for $j < n$, $j \neq i$. Let $\mathbf{w} = \langle u_0, \ldots, u_{i-1}, y, u_{i+1}, \ldots, u_{n-1}\rangle$. Then $\mathbf{u}\bar{R}_i^\alpha\mathbf{w}\bar{R}_i^\alpha\mathbf{v}$ and, as f is p-morphism from $\langle U_0, R_0^\alpha\rangle \times \cdots \times \langle U_{n-1}, R_{n-1}^\alpha\rangle$ to $\mathfrak{F}$, $f(\mathbf{u})T_if(\mathbf{w})T_if(\mathbf{v})$. As T_i is transitive, we have $f(\mathbf{u})T_if(\mathbf{v})$ as required. ■

Remark 2.2.6. Note that a similar proof would prove a stronger statement. The property of each T_i being an equivalence relation can be replaced with any property of T_i that can be defined by a set of *universal Horn* formulas in the first-order language having a binary predicate symbol and possibly equality (and there can be different such properties for different i).

Theorem 2.2.7. *Let L be any canonical n-modal logic*[1] *such that* $\mathbf{K}^n \subseteq L \subseteq \mathbf{S5}^n$. *Then* $\mathbf{S5}^n$ *is finitely axiomatizable over L:* $\mathbf{S5}^n$ *is the smallest n-modal logic containing L and the* **S5***-axioms* (2.0.3), *for $i < n$* ([Kur,10]).

Proof. One inclusion is clear, let us prove the other. The **S5**-axioms are well-known examples of Sahlqvist formulas, and their first-order correspondent is the property of being an equivalence relation. So, by Sahlqvist's

[1] By an *n-modal logic* we mean any set of n-modal formulas that contains all propositional tautologies, the formulas (K) for each $\Box_i$, and is closed under the derivation rules of Substitution, Modus Ponens and Necessitation, for $i < n$.

completeness theorem, the smallest n-modal logic containing L and the **S5**-axioms is canonical, and so Kripke complete. So it is enough to show that every rooted n-frame $\mathfrak{F}$ for this logic is a frame for $\mathbf{S5}^n$.

Take such an n-frame $\mathfrak{F}$. As $\mathfrak{F}$ is a frame for $\mathbf{K}^n = \mathsf{Log}\left(\mathcal{C}^n_{all}\right)$, by Corollary 2.1.6, $\mathfrak{Ue}\,\mathfrak{F}$ is a p-morphic image of some n-dimensional product frame $\mathfrak{G}$. As $\mathfrak{F}$ validates the canonical **S5**-axioms, they also hold in $\mathfrak{Ue}\,\mathfrak{F}$, and so all the relations in $\mathfrak{Ue}\,\mathfrak{F}$ are equivalence relations. Now by Theorem 2.2.5, $\mathfrak{Ue}\,\mathfrak{F}$ is a p-morphic image of some $\mathfrak{G}^* \in \mathcal{C}^n_{equiv}$, and so by Corollary 2.1.6 again, $\mathfrak{F}$ is a frame for $\mathbf{S5}^n = \mathsf{Log}\left(\mathcal{C}^n_{equiv}\right)$. ∎

Let us formulate a consequence of Theorems 2.1.3 and 2.2.7 in an algebraic form:

Theorem 2.2.8. *Let $\mathcal{C}$ be any class of n-dimensional product frames that is closed under $\mathbb{U}\mathrm{p}$ and $\mathbb{I}_{\mathsf{p}}$. Then the equational theory of RDf_n is finitely axiomatizable over the equational theory of $\mathbf{SP}\,\mathsf{Cm}\,\mathcal{C}$: one only has to add the equations* (2.0.1), *for $i < n$.*

Remark 2.2.9. By Remarks 2.2.4 and 2.2.6, we can have similar statements for any $\mathbf{SP}\,\mathsf{Cm}\,\mathcal{K}$ in place of RDf_n, whenever $\mathcal{K} = \mathcal{C}_0 \times \cdots \times \mathcal{C}_{n-1}$ for come classes $\mathcal{C}_i$ of 1-frames, each of which is definable by Sahlqvist formulas having universal Horn first-order correspondents.

Theorems 2.2.7 and 2.2.8 show that any negative result on the equational axiomatization of RDf_n (such as its non-finiteness [Joh,69], for $n \geq 3$) transfers to many varieties generated by complex algebras of n-dimensional product frames (or, to many-dimensional modal logics like $\mathbf{K}^n$). In other words, these theorems also mean that all the complexity of RDf_n (or its logic counterpart $\mathbf{S5}^n$) comes from the many-dimensional structure and is already present in $\mathbf{K}^n$. Though, by a general result of [Gab-She,98], $\mathbf{K}^n$ is known to be recursively enumerable, an axiomatization of $\mathbf{K}^n$ should be quite complex, whenever $n \geq 3$: any such axiomatization should contain n-modal formulas of arbitrary modal depth for each modality [Kur,00], n-modal formulas without first-order correspondents [Kur,10], and infinitely many propositional variables [Kur,08]. At the moment we cannot use Theorem 2.2.7 to infer the latter, as it is not known whether $\mathbf{S5}^n$ (or RDf_n) can be axiomatized using finitely many variables, whenever $n \geq 3$.

For $n = 2$, the following generalisation of $\mathsf{RDf}_2 = \mathsf{Df}_2$ (see e.g. [Hen-Mon-Tar,85, 5.1.47]) holds:

Theorem 2.2.10 (Gabbay and Shehtman). *Let Σ_0 and Σ_1 be sets of 1-modal formulas having universal Horn first-order correspondents, and let*

$$\mathcal{C} = \{\mathfrak{F}_0 \times \mathfrak{F}_1 : \mathfrak{F}_0 \models \Sigma_0, \ \mathfrak{F}_1 \models \Sigma_1\}.$$

Then $\mathsf{Log}\,(\mathcal{C})$ *is the smallest 2-modal logic containing Σ_0 for $\Diamond_0$, Σ_1 for $\Diamond_1$, and the interaction axioms*

$$\Diamond_0\Diamond_1 p \leftrightarrow \Diamond_1\Diamond_0 p \qquad \text{and} \qquad \Diamond_0\Box_1 p \to \Box_1\Diamond_0 p$$

([Gab-She,98]).

So, in particular, $\mathbf{K}^2$, $\mathbf{K4}^2$ and $\mathbf{S5}^2$ (the logic counterpart of RDf_2) are all finitely axiomatizable. Note that in case of $\mathbf{S5}^2$ the second interaction axiom (*confluence*) follows from the first (*commutativity*).

Next, we make use of a result of Hirsch and Hodkinson [Hir-Hod,01d] saying that representability of finite subdirectly irreducible *relation algebras* is undecidable. To begin with, this result implies the following:

Theorem 2.2.11. *If $n \geq 3$ then it is undecidable whether a finite subdirectly irreducible n-dimensional diagonal-free cylindric algebra is representable* ([Hir-Hod-Kur,02a]).

Proof. Monk [Mon,61] introduced a construction that turns any finite subdirectly irreducible relation algebra $\mathfrak{A}$ to a finite 3-dimensional cylindric algebra $\mathsf{Ca}_3\,\mathfrak{A}$ such that

- $\mathfrak{A}$ is representable as a relation algebra iff $\mathsf{Ca}_3\,\mathfrak{A} \in \mathsf{RCA}_3$;
- the diagonal-free reduct $\mathsf{Df}_3\,\mathfrak{A}$ of $\mathsf{Ca}_3\,\mathfrak{A}$ is subdirectly irreducible and generated by 2-dimensional elements.

Now using the results of Halmos [Hal,57] and Johnson [Joh,69] (see also [Hen-Mon-Tar,85, 5.1]), we obtain that $\mathsf{Ca}_3\,\mathfrak{A} \in \mathsf{RCA}_3$ iff $\mathsf{Df}_3\,\mathfrak{A} \in \mathsf{RDf}_3$. Next, for every $n > 3$, we can extend $\mathsf{Df}_3\,\mathfrak{A}$ to an n-dimensional diagonal-free cylindric algebra $\mathsf{Df}_n\,\mathfrak{A}$ by keeping the same domain and c_0, c_1 and c_2 as in $\mathsf{Df}_3\,\mathfrak{A}$, and defining c_i as the identity function, for each $3 \leq i < n$. Then it is straightforward to show that $\mathsf{Df}_3\,\mathfrak{A} \in \mathsf{RDf}_3$ iff $\mathsf{Df}_n\,\mathfrak{A} \in \mathsf{RDf}_n$. ∎

Now take any finite subdirectly irreducible n-dimensional diagonal-free cylindric algebra $\mathfrak{A}$, and consider its atom structure $\mathfrak{At}_{\mathfrak{A}}$. Then $\mathfrak{At}_{\mathfrak{A}}$ is an n-frame, so by (2.1.2),

$$(2.2.1) \qquad \mathfrak{At}_{\mathfrak{A}} \models \mathbf{S5}^n \quad \Longleftrightarrow \quad \mathfrak{A} \cong \mathfrak{Cm}\,\mathfrak{At}_{\mathfrak{A}} \in \mathsf{RDf}_n\,.$$

Theorem 2.2.12. *Let $n \geq 3$ and let L be any set of n-modal formulas with $\mathbf{K}^n \subseteq L \subseteq \mathbf{S5}^n$. Then the following hold:*

(i) *It is undecidable whether $\mathfrak{F} \models L$ for a finite rooted n-frame $\mathfrak{F}$.*

(ii) *L is not finitely axiomatizable*

([Hir-Hod-Kur,02a]*).*

Proof. Take any finite subdirectly irreducible n-dimensional diagonal-free cylindric algebra $\mathfrak{A}$. Then its atom structure $\mathfrak{At}_{\mathfrak{A}}$ is rooted and all its relations are equivalence relations, so by Corollary 2.1.7 and Theorem 2.2.5,

$$\mathfrak{At}_{\mathfrak{A}} \models L \quad \Longleftrightarrow \quad \mathfrak{At}_{\mathfrak{A}} \models \mathbf{S5}^n.$$

Therefore, item (i) follows from (2.2.1) and Theorem 2.2.11.

Item (ii) clearly follows from (i), as long as a *finite axiomatization* of L means a finitary proof system that is suitable for testing whether $\mathfrak{F} \models L$ for a finite n-frame $\mathfrak{F}$. This does not necessarily mean that only the so-called 'orthodox' derivation rules (Substitution, Modus Ponens and Necessitation) of modal logic are allowed. However, certain 'non-orthodox' rules such as some versions of the *irreflexivity rule* (see [Ven,thisVol]) are not suitable for this purpose. ■

Observe that the atom structure $\mathfrak{At}_{\mathfrak{A}} = \langle W, T_i \rangle_{i<n}$ of a finite subdirectly irreducible n-dimensional diagonal-free cylindric algebra $\mathfrak{A}$ is not only rooted but, having chosen any of its points r as root, has the following property:

(2.2.2)
$$\forall x \in W\ \exists\, y_0, \ldots, y_n\ \big(y_0 = r \wedge y_n = x \wedge \forall i < n\ (y_i = y_{i+1} \vee y_i T_i y_{i+1})\big)\,.$$

Now, for each $w \in W$, let us introduce a propositional variable p_w, and define an n-modal formula $\varphi_{\mathfrak{A}}$ by taking

$$(2.2.3)\quad \Box_0^+ \dots \Box_{n-1}^+ \Bigg(\bigvee_{w \in W} p_w \wedge \bigwedge_{w \neq w' \in W} \neg(p_w \wedge p_{w'}) \\ \wedge \bigwedge_{\substack{i<n,\ w,w' \in W \\ wT_iw'}} p_w \to \Diamond_i p_{w'} \wedge \bigwedge_{\substack{i<n,\ w,w' \in W \\ \neg(wR_iw')}} p_w \to \neg\Diamond_i p_{w'} \Bigg),$$

where $\Box_i^+\psi$ abbreviates $\psi \wedge \Box_i\psi$. This formula is the n-modal version of the *frame formula* (also known as *splitting formula,* see [Yan,68, Fin,74]) of the n-frame $\mathfrak{At}_\mathfrak{A}$. It is clearly satisfiable in $\mathfrak{At}_\mathfrak{A}$, and it is supposed to describe $\mathfrak{At}_\mathfrak{A}$ 'up to p-morphism' in n-frames with property (2.2.2). However, as the following lemma shows, it is quite powerful in arbitrary product frames as well:

Lemma 2.2.13. *If $\varphi_\mathfrak{A}$ is satisfied in any n-dimensional product frame, then there is some $\mathfrak{G} \in \mathcal{C}^n_{univ}$ such that $\mathfrak{At}_\mathfrak{A}$ is a p-morphic image of $\mathfrak{G}$* ([Hir-Hod-Kur,02a]).

Proof. Suppose $\mathfrak{M}, \mathbf{x} \models \varphi_\mathfrak{A}$ for some model $\mathfrak{M}$ over $\mathfrak{F} = \mathfrak{F}_0 \times \cdots \times \mathfrak{F}_{n-1}$, for some $\mathfrak{F}_i = \langle U_i, R_i \rangle$, $i < n$. Take $U_i^- = \{u \in U_i : u = x_i \text{ or } x_iR_iu\}$, and define a function f from $U_0^- \times \cdots U_{n-1}^-$ to $\mathfrak{At}_\mathfrak{A}$ by taking,

$$f(\mathbf{u}) = w \qquad \Longleftrightarrow \qquad \mathfrak{M}, \mathbf{u} \models p_w.$$

It is not hard to show that f is well-defined and a p-morphism from the n-dimensional product $\mathfrak{G}$ of universal frames over U_i^- onto $\mathfrak{At}_\mathfrak{A}$. ■

This lemma now implies that, for every L with $\mathbf{K}^n \subseteq L \subseteq \mathbf{S5}^n$,

$$\neg\varphi_\mathfrak{A} \notin L \qquad \Longleftrightarrow \qquad \mathfrak{At}_\mathfrak{A} \models \mathbf{S5}^n.$$

So, by (2.2.1) and Theorem 2.2.11, we obtain the following generalisation of Maddux's result [Mad,80] on the undecidability of the equational theory of RDf_n, for $n \geq 3$:

Theorem 2.2.14. *Let $n \geq 3$ and let L be any set of n-modal formulas with $\mathbf{K}^n \subseteq L \subseteq \mathbf{S5}^n$. Then L is undecidable* ([Hir-Hod-Kur,02a]).

Remark 2.2.15. The undecidability of $\mathbf{S5}^n$ can also be derived from Theorem 2.2.12 and (2.2.1) as follows. As observed by Tarski [Tar,54], embeddability of a finite algebra $\mathfrak{A}$ can be described by an existential first-order sentence in the language of $\mathfrak{A}$. As $\mathsf{RDf}_n = \mathbf{SP}\,\mathsf{Cm}\,\mathcal{C}^n_{univ}$ is a *discriminator variety*, non-embeddability of a finite diagonal-free cylindric algebra into a representable one can be described in RDf_n by an equation. Note, however, that other varieties of the form $\mathbf{SP}\,\mathsf{Cm}\,\mathcal{C}$ (such as, say, $\mathbf{SP}\,\mathsf{Cm}\,\mathcal{C}^n_{all}$) might not be discriminator varieties and we have to use something like Lemma 2.2.13.

Note that one can find undecidable many-dimensional modal logics already in dimension 2. Gabelaia *et al.* [Gab-Kur-Wol-Zak,05] provide a wide choice of these logics, in a sense the most surprising among them being $\mathbf{K4}^2$. This undecidable logic is finitely axiomatizable with the natural axioms by Theorem 2.2.10. In the algebraic setting, we obtain that the equational theory of two commuting and confluent closure operators is undecidable.

The reader might have the impression by now that metalogical properties of, say, $\mathbf{K}^n$ and $\mathbf{S5}^n$ always go hand in hand. We mention a property for which this is not the case: while $\mathbf{K}^n$ does have the *finite model property* [Gab-She,98], $\mathbf{S5}^n$ does not [Kur,02b], whenever $n \geq 3$.

3. With Diagonals

We define an *$n\delta$-frame* to be a structure of the form $\langle W, T_i, E_{ij}\rangle_{i,j<n}$, where $\langle W, T_i\rangle_{i<n}$ is an n-frame and E_{ij} is a subset of W, for $i, j < n$. Multimodal formulas matching $n\delta$-frames, *$n\delta$-modal formulas*, are built up from propositional variables using the Booleans, unary modal operators $\Diamond_i$ and $\Box_i$, and constants δ_{ij}, for $i, j < n$.

The following notion is a generalisation of atom structures of n-dimensional full cylindric set algebras.

Definition 2.3.1. Given 1-frames $\mathfrak{F}_i = \langle W_i, R_i\rangle$, $i < n$, their *δ-product* is the $n\delta$-frame

$$(\mathfrak{F}_0 \times \cdots \times \mathfrak{F}_{n-1})^\delta = \big\langle W_0 \times \cdots \times W_{n-1}, \bar{R}_i, Id_{ij}\big\rangle_{i,j<n},$$

where $\big\langle W_0 \times \cdots \times W_{n-1}, \bar{R}_i\big\rangle_{i<n} = \mathfrak{F}_0 \times \cdots \times \mathfrak{F}_{n-1}$ and, for $i, j < n$,

$$Id_{ij} = \{\mathbf{u} \in W_0 \times \cdots \times W_{n-1} \,:\, u_i = u_j\}.$$

Such n-frames we call *n-dimensional δ-product frames*.

We have the analogues of Propositions 2.2.2 and 2.2.3:

Proposition 2.3.2. *Let U be an ultrafilter over some index set I, and let $\mathfrak{F}_k^i$ be a 1-frame, for $i \in I$, $k < n$. Then:*

$$\prod_{i\in I} \left(\mathfrak{F}_0^i \times \cdots \times \mathfrak{F}_{n-1}^i\right)^\delta / U \cong \left(\left(\prod_{i\in I} \mathfrak{F}_0^i/U\right) \times \cdots \times \left(\prod_{i\in I} \mathfrak{F}_{n-1}^i/U\right)\right)^\delta .$$

Proposition 2.3.3. *Let $\mathfrak{F} = (\mathfrak{F}_0 \times \cdots \times \mathfrak{F}_{n-1})^\delta$ and $\mathbf{x}$ be a point in $\mathfrak{F}$. Then:*

$$\mathfrak{F}^{\mathbf{x}} = \left(\mathfrak{F}_0^{x_0} \times \cdots \times \mathfrak{F}_{n-1}^{x_{n-1}}\right)^\delta .$$

Given a class $\mathcal{C}$ of n-dimensional product frames, we denote by $\mathcal{C}^\delta$ the corresponding class of δ-product frames:

$$\mathcal{C}^\delta = \left\{ (\mathfrak{F}_0 \times \cdots \times \mathfrak{F}_{n-1})^\delta \;:\; \mathfrak{F}_0 \times \cdots \times \mathfrak{F}_{n-1} \in \mathcal{C} \right\}.$$

It is straightforward to see the following:

Proposition 2.3.4. *For any class $\mathcal{C}$ of n-dimensional product frames, $\mathsf{Log}\,(\mathcal{C}^\delta)$ is a conservative extension of $\mathsf{Log}\,(\mathcal{C})$.*

Remark 2.3.5. Just like in Remark 2.2.4, observe that as a consequence of Propositions 2.3.2 and 2.3.3 we obtain the following. If each class $\mathcal{C}_i$ of 1-frames, for $i < n$, is defined by *universal* formulas in the first-order language having one binary predicate symbol and possibly equality, then $(\mathcal{C}_0 \times \cdots \times \mathcal{C}_{n-1})^\delta$ is closed under taking ultraproducts and point-generated inner substructures. Here are some examples of this kind:

$$\mathcal{C}^{n\delta}_{all} = \text{the class of all } n\text{-dimensional } \delta\text{-product frames},$$

$$\mathcal{C}^{n\delta}_{equiv} = \text{the class of all } n\text{-dimensional } \delta\text{-products of equivalence frames},$$

$$\mathcal{C}^{n\delta}_{univ} = \text{the class of all } n\text{-dimensional } \delta\text{-products of universal frames}.$$

Since $\mathcal{C}$ is closed under $\mathbb{U}\mathrm{p}$ and $\mathbb{I}_\mathsf{p}$ in each of these cases, $\mathbf{SP}\,\mathsf{Cm}\,\mathcal{C}$ is a canonical variety by Theorem 2.1.3. As $\mathbb{I}_\mathsf{p}\,\mathcal{C}^{n\delta}_{equiv} = \mathcal{C}^{n\delta}_{univ}$, we have $\mathbf{SP}\,\mathsf{Cm}\,\mathcal{C}^{n\delta}_{univ} = \mathbf{SP}\,\mathsf{Cm}\,\mathcal{C}^{n\delta}_{equiv}$.

However, consider now the class

$$\mathcal{C}^{n\delta}_{cube} = \{ (\underbrace{\mathfrak{F} \times \cdots \times \mathfrak{F}}_{n})^\delta \;:\; \mathfrak{F} = \langle U, U \times U\rangle \text{ for some non-empty set } U\}$$

= the class of all n-dimensional full cylindric set algebra atom structures

that is also closed under $\mathbb{U}\mathrm{p}$ and $\mathbb{I}_{\mathsf{p}}$. Unlike in the diagonal-free case, $\mathbf{SP}\,\mathsf{Cm}\,\mathcal{C}^{n\delta}_{cube} = \mathsf{RCA}_n$ is properly contained in $\mathbf{SP}\,\mathsf{Cm}\,\mathcal{C}^{n\delta}_{equiv}$, as for instance the equations $\mathsf{c}_i d_{ij} = 1$ fail in the latter. (As we shall see below, in a sense they are the only missing ones.)

Let us introduce notation for some $n\delta$-modal logics:

$$\mathbf{K}^{n\delta} = \mathsf{Log}\left(\mathcal{C}^{n\delta}_{all}\right)$$

$$\mathbf{S5}^{n\delta} = \mathsf{Log}\left(\mathcal{C}^{n\delta}_{equiv}\right) = \mathsf{Log}\left(\mathcal{C}^{n\delta}_{univ}\right).$$

The proof of the following two theorems are completely analogous to the respective proofs of Theorems 2.2.5 and 2.2.7:

Theorem 2.3.6. *Let $\mathfrak{F} = \langle W, T_i, E_{ij}\rangle_{i,j<n}$ be an $n\delta$-frame such that every T_i is an equivalence relation, for $i < n$. Suppose that $f : (\mathfrak{G}_0 \times \cdots \times \mathfrak{G}_{n-1})^{\delta} \to \mathfrak{F}$ is a surjective p-morphism, for some 1-frames $\mathfrak{G}_i = \langle U_i, R_i\rangle$, $i < n$. Then there exist 1-frames $\mathfrak{G}^*_i = \langle U_i, R^*_i\rangle$, $i < n$, such that*

- *each R^*_i is an equivalence relation extending R_i, and*
- *$f : \left(\mathfrak{G}^*_0 \times \cdots \times \mathfrak{G}^*_{n-1}\right)^{\delta} \to \mathfrak{F}$ is still a surjective p-morphism.*

Theorem 2.3.7. *Let L be any canonical $n\delta$-modal logic with $\mathbf{K}^{n\delta} \subseteq L \subseteq \mathbf{S5}^{n\delta}$. Then $\mathbf{S5}^{n\delta}$ is finitely axiomatizable over L: $\mathbf{S5}^{n\delta}$ is the smallest $n\delta$-modal logic containing L and the* **S5**-*axioms* (2.0.3), *for $i < n$.*

It turns out that the equations $\mathsf{c}_i d_{ij} = 1$ (or, the $n\delta$-modal formulas $\Diamond_i \delta_{ij}$) are quite strong in the sense that they can 'force' the **S5**-properties in the presence of 'many-dimensionality', as the following surprising theorems show:

Theorem 2.3.8. *Let $\mathfrak{F} = \langle W, T_i, E_{ij}\rangle_{i,j<n}$ be an $n\delta$-frame such that, for all $i, j < n$,*

(2.3.1) *for all $w \in W$ there is some $w' \in E_{ij}$ with wT_iw'.*

Suppose that $f : (\mathfrak{G}_0 \times \cdots \times \mathfrak{G}_{n-1})^{\delta} \to \mathfrak{F}$ is a surjective p-morphism, for some 1-frames $\mathfrak{G}_i = \langle U_i, R_i\rangle$, $i < n$. Then $U_i = U_j$ and R_i is the universal relation on U_i, for all $i, j < n$.

Proof. We show first that $U_i \subseteq U_j$, for any $i, j < n$, $i \neq j$. To this end, let $u \in U_i$ and take any $\mathbf{x} \in U_0 \times \cdots \times U_{n-1}$ such that $x_i = u$. By (2.3.1), there is some $w \in E_{ij}$ such that $f(\mathbf{x})T_j w$. As f is a p-morphism, there is $\mathbf{y} \in Id_{ij}$ such that $\mathbf{x}\bar{R}_j\mathbf{y}$, so $u = x_i = y_i = y_j \in U_j$ as required.

Next, we show that uR_iu' hold, for all $i < n$, $u, u' \in U_i$. To this end, take some $j \neq i$ and $\mathbf{x} \in {}^nU_i$ such that $x_i = u$ and $x_j = u'$. As f is a p-morphism, there is $\mathbf{y} \in Id_{ij}$ such that $\mathbf{x}\bar{R}_i\mathbf{y}$, so $u = x_iR_iy_i = y_j = x_j = u'$. ■

Theorem 2.3.9. *Let L be any canonical $n\delta$-modal logic with $\mathbf{K}^{n\delta} \subseteq L \subseteq \mathsf{Log}\left(\mathcal{C}^{n\delta}_{cube}\right)$. Then $\mathsf{Log}\left(\mathcal{C}^{n\delta}_{cube}\right)$ is finitely axiomatizable over L: $\mathsf{Log}\left(\mathcal{C}^{n\delta}_{cube}\right)$ is the smallest $n\delta$-modal logic containing L and the $n\delta$-formulas $\Diamond_i\delta_{ij}$, for $i, j < n$.*

Proof. Like that of Theorem 2.2.7, using Theorem 2.3.8 and that each $\Diamond_i\delta_{ij}$ is a Sahlqvist formula, with property (2.3.1) being its first-order correspondent. ■

A consequence of Theorems 2.1.3 and 2.3.9 formulated in an algebraic setting is as follows:

Theorem 2.3.10. *Let $\mathcal{C}$ be any class of n-dimensional δ-product frames that is closed under $\mathbb{U}\mathrm{p}$ and $\mathbb{I}_{\mathsf{p}}$. Then the equational theory of RCA_n is finitely axiomatizable over the equational theory of $\mathbf{SP}\,\mathsf{Cm}\,\mathcal{C}$: one only has to add the equations $\mathsf{c}_i\mathsf{d}_{ij} = 1$, for $i, j < n$.*

Theorems 2.3.9 and 2.3.10 show that any negative result on the equational axiomatization of RCA_n transfers to many varieties generated by complex algebras of n-dimensional δ-product frames (or, to many-dimensional modal logics like $\mathbf{K}^{n\delta}$). There are many such, whenever $n \geq 3$: $\mathsf{RCA}_n = \mathbf{SP}\,\mathsf{Cm}\,\mathcal{C}^{n\delta}_{cube}$ is not only not finitely axiomatizable [Mon,69], but cannot be axiomatized using finitely many variables and finitely many occurrences of the diagonals [And,97a]. RCA_n cannot be axiomatized using only Sahlqvist equations [Hod,97c, Ven,97a]. Moreover, it is also known [Hir-Hod,09] that the class of all $n\delta$-frames $\mathfrak{F}$ such that $\mathfrak{F} \models \mathsf{Log}\left(\mathcal{C}^{n\delta}_{cube}\right)$ (*strongly representable cylindric atom structures*) is not closed under ultraproducts, so $\mathsf{Log}\left(\mathcal{C}^{n\delta}_{cube}\right)$ cannot be axiomatized by any set of $n\delta$-modal formulas having first-order correspondents. The even more general result of [Hod-Ven,05] saying that *representable relation algebras* do not have a canonical axiomatization might also hold for RCA_n[2]. So all these are true for, say,

[2] It does hold, see [Bul-Hod].

$\mathsf{Log}\left(\mathcal{C}^{n\delta}_{all}\right) = \mathbf{K}^{n\delta}$. Though it is not hard to see that $\mathbf{K}^{n\delta}$ is recursively enumerable, there is a further difficulty in finding an explicit infinite axiomatization for it. The known explicit (infinite) equational axiomatizations for RCA_n (for $n \geq 3$) [Mon,69, Hen-Mon-Tar-And-Nem,81, Hir-Hod,97a] (see also [Hen-Mon-Tar,85, 4.1] and [Hir-Hod,02a, 8.3]) all make use of RCA_n being a discriminator variety. But the algebraic counterpart $\mathbf{SP}\,\mathsf{Cm}\,\mathcal{C}^{n\delta}_{all}$ of $\mathbf{K}^{n\delta}$ is not such.

Remark 2.3.11. Moreover, when we have diagonals, finding an axiomatization can be tricky even for $n = 2$. Though RCA_2 is known to be finitely axiomatizable (see e.g. [Hen-Mon-Tar,85, 3.2.65]), and it is also finitely axiomatizable over $\mathbf{K}^{2\delta}$ by Theorem 2.3.10, somewhat surprisingly $\mathbf{K}^{2\delta}$ is not even axiomatizable using finitely many variables, as shown by Kikot [Kik,10].

Let us next turn to decision problems. Hodkinson [Hod,12] shows that it is undecidable whether a finite subdirectly irreducible n-dimensional cylindric algebra is representable, for any finite $n \geq 3$. Using this, with diagonals we can have a bit better than Theorem 2.2.12:

Theorem 2.3.12. *Let $n \geq 3$ and let L be any set of $n\delta$-modal formulas with $\mathbf{K}^{n\delta} \subseteq L \subseteq \mathsf{Log}\left(\mathcal{C}^{n\delta}_{cube}\right)$. Then the following hold:*

(i) *It is undecidable whether $\mathfrak{F} \models L$ for a finite rooted $n\delta$-frame $\mathfrak{F}$.*

(ii) *L is not finitely axiomatizable.*

Proof. Like that of Theorem 2.2.12, using Theorem 2.3.8. ■

Observe that the atom structure $\mathfrak{At}_{\mathfrak{A}} = \langle W, T_i, E_{ij}\rangle_{i,j<n}$ of a finite subdirectly irreducible n-dimensional cylindric algebra $\mathfrak{A}$ not only has property (2.2.2), but it also has (2.3.1). Now define an $n\delta$-modal formula $\psi_{\mathfrak{A}}$ by adding the following conjunct to $\varphi_{\mathfrak{A}}$ in (2.2.3):

$$\Box_0^+ \ldots \Box_{n-1}^+ \bigwedge_{i,j<n} \left(\delta_{ij} \leftrightarrow \bigvee_{w \in E_{ij}} p_w\right). \tag{2.3.2}$$

Then $\psi_{\mathfrak{A}}$ is satisfied in $\mathfrak{At}_{\mathfrak{A}}$, and we have the following analogue of Lemma 2.2.13:

Lemma 2.3.13. *If $\psi_{\mathfrak{A}}$ is satisfied in any n-dimensional δ-product frame, then there is some $\mathfrak{G}^{\delta} \in \mathcal{C}^{n\delta}_{cube}$ such that $\mathfrak{At}_{\mathfrak{A}}$ is a p-morphic image of $\mathfrak{G}^{\delta}$.*

Proof. Suppose that $\psi_{\mathfrak{A}}$ is satisfied in $\mathfrak{F}^\delta$ for some n-dimensional product frame $\mathfrak{F}$. Then $\varphi_{\mathfrak{A}}$ is satisfied in $\mathfrak{F}$. Now define $\mathfrak{G} \in \mathcal{C}^n_{univ}$ and f as in the proof of Lemma 2.2.13. As $\varphi_{\mathfrak{A}}$ is satisfied in $\mathfrak{F}$, f is a p-morphism from $\mathfrak{G}$ onto the diagonal-free reduct of $\mathfrak{At}_{\mathfrak{A}}$. However, by (2.3.2), f is in fact a p-morphism from $\mathfrak{G}^\delta$ onto $\mathfrak{At}_{\mathfrak{A}}$. As $\mathfrak{At}_{\mathfrak{A}}$ has property (2.3.1), Theorem 2.3.8 implies that $\mathfrak{G}^\delta \in \mathcal{C}^{n\delta}_{cube}$, as required. ■

Now we can have the analogue of Theorem 2.2.14:

Theorem 2.3.14. *Let $n \geq 3$ and let L be any set of $n\delta$-modal formulas with $\mathbf{K}^{n\delta} \subseteq L \subseteq \mathsf{Log}\left(\mathcal{C}^{n\delta}_{cube}\right)$. Then L is undecidable.*

Proof. Like that of Theorem 2.2.14, using Lemma 2.3.13. Note that if $\mathbf{K}^{n\delta} \subseteq \mathsf{Log}(\mathcal{C}) \subseteq \mathbf{S5}^{n\delta}$ for some class $\mathcal{C}$ of n-dimensional δ-product frames, then the undecidability of $\mathsf{Log}(\mathcal{C})$ already follows from Theorem 2.2.14 and Proposition 2.3.4. ■

Remark 2.3.15. There are cases when adding the diagonal does matter in the decision problem. An example is $\mathbf{K}^2$ that is decidable [Gab-She,98], while $\mathbf{K}^{2\delta}$ is not [Kik-Kur,11].

BOLYAI SOCIETY
MATHEMATICAL STUDIES, 22

Cylindric-like Algebras and Algebraic Logic
pp. 205–221.

Completions, Complete Representations and Omitting Types

TAREK SAYED AHMED

Algebraic logic arose as a subdiscipline of algebra mirroring constructions and theorems of mathematical logic. It is similar in this respect to such fields as algebraic geometry and algebraic topology, where the main constructions and theorems are algebraic in nature, but the main intuitions underlying them are respectively geometric and topological. The main intuitions underlying algebraic logic are, of course, those of formal logic. Investigations in algebraic logic can proceed in two conceptually different, but often (and unexpectedly) closely related ways. First one tries to investigate the algebraic essence of constructions and results in logic, in the hope of gaining more insight that one could add to his understanding, thus to his knowledge. Second, one can study certain "particular" algebraic structures (or simply algebras) that arise in the course of his first kind of investigations as objects of interest in their own right and go on to discuss questions which naturally arise independently of any connection with logic. But often such purely algebraic results have impact on the logic side. Examples are the undecidability of the representation problem for finite relation algebras [Hir-Hod,01d] that led to deep results concerning undecidability of product modal logics [Hir-Hod-Kur,02a] answering problems of Gabbay and Shehtman. Another example is the interconnection of the metalogical notion of omitting types and algebraic notions of atom canonicity and complete representations, first presented in [Say,05a] and elaborated upon in [And-Nem-Say,08]. In this paper we study such connections in some depth. We investigate the classical Orey Henkin omitting types theorem for various modifications of first order logic, be it reducts or expansions. Our investigation will be algebraic.

First order logic (*FOL*) posseses some desirable properties, for example the completeness theorem, the omitting types theorem, the Craig interpolation theorem, and Beth's theorem. Daniele Mundici initiated the following type of investigations for *FOL*. Concerning various positive properties like Craig's interpolation theorem, Mundici suggested to investigate how resourse sensitive the positive result is. For example, Craig's theorem says that to an implication $\phi \to \psi$ there exists an interpolant θ with $\phi \to \theta$ and $\theta \to \psi$. Now the question is, how much does θ depend on ϕ and ψ, or how complicated is θ relative to ϕ and ψ? Recent work measures expensiveness with the number of variables needed for θ. For example if both ψ and θ are built up of k variables, do we guarantee that the number of variables in the interpolant does not exceed k? Another example for such investigations is Monk's classical result that for any bound $k \in \omega$ there is a valid 3 variable formula which cannot be proved using only k variables. Note that ϕ can be proved using m variables for some finite $m > k$, for formulas contain finitely many variables and proofs are finite strings of formulas. This result was refined by Hirsch, Hodkinson and Maddux, showing that given any such k we can find a 3 variable formula as above, but also subject to the condition that it can be proved using $k + 1$ variables. Such results concerning proof theory for finite variable fragments of first order logic were first proved using algebraic logic. In this paper we apply this "resource-oriented" kind of the investigation to the classical Henkin–Orey omitting types theorem.

Let $\mathfrak{L}_n$ denote first order logic restricted to the first n variables. A systematic study of the fragments $\mathfrak{L}_n$ via cylindric algebras was initiated by Leon Henkin via cylindric algebras of dimension n. The issue of "resource-sensitivity" is often addressed in the following form. We ask ourselves if certain distinguished properties of *FOL* are inherited by $\mathfrak{L}_n$. Examples of such distinguished properties studied in the literature for $\mathfrak{L}_n$ include interpolation, Beth definability [And-Com-Mad-Nem-Say,09], submodel preservation and completeness theorems [Gra-Ros,99], [And-Ben-Nem,95b]. A general first impression might be that, usually positive properties for *FOL* turn resourse sensitive in such a strong way that a goal formulatable in $\mathfrak{L}_n$ cannot be soved in $\mathfrak{L}_n$. One might go further, by stipulating that a goal formulatable in $\mathfrak{L}_n$ cannot be solved, even in $\mathfrak{L}_{n+k}$ for every finite (fixed in advance) k. (Like Monk's result stated above). However, this is not true in such generality, some natural properties of substitutions in $\mathfrak{L}_n$ which are not provable in $\mathfrak{L}_n$ can be proved in $\mathfrak{L}_{n+2}$. A further counterexample is provided by the guarded fragment of *FOL* introduced by Andréka, Németi and van Benthem. Negative results (for finite variable fragments of first order logic) mentioned above do not occur for the guarded fragment of first order logic introduced in [And-Ben-Nem,98]. The guarded fragment (*GF*) was introduced as a fragment of first order logic which combines a great expressive power with nice modal behavior. It consists of relational first order formulas whose quantifiers are relativized by atoms in a certain way. *GF* has been established as a particularly well-behaved fragment of first order logic in many respects. The main point of the *GF* (and its variants e.g. the packed fragments) is that (inside the *GF*) we are safe of the above negative results for $\mathfrak{L}_n$, like essential incompleteness [Ben,thisVol]. The omitting types theorem has not been investigated for the *GF*. However, omitting types was investigated algebraically for other modifications of first order logic, the so called finitary logics of infinitary relations.

The layout of this paper is as follows. In Section 1 we study omitting types for $\mathfrak{L}_n$ and in Section 2 we study omitting types for certain extensions of *FOL* that have an infinitary flavor.

1. Omitting Types Fails in $\mathfrak{L}_n$

We work in usual *FOL*. For a formula ϕ and a first order structure $\mathfrak{M}$ in the language of ϕ we write ϕ^M to denote the set of all assignments that satisfy ϕ in M, i.e.

$$\phi^{\mathfrak{M}} = \left\{ \mathsf{s} \in {}^{\omega}M \,:\, \mathfrak{M} \models \phi[\mathsf{s}] \right\}.$$

For example, if $\mathfrak{M} = (\mathbb{N}, <)$ and ϕ is the formula $x_1 < x_2$ then a sequence $\mathsf{s} \in {}^{\omega}\mathbb{N}$ is in $\phi^{\mathfrak{M}}$ iff $\mathsf{s}_1 < \mathsf{s}_2$. Let Γ be a set of formulas (Γ may contain free variables). We say that Γ is realized in $\mathfrak{M}$ if $\bigcap_{\phi\in\Gamma} \phi^{\mathfrak{M}} \neq \emptyset$. Let ϕ be a formula and T be a theory. We say that ϕ ensures Γ in T if $T \models \phi \to \mu$ for all $\mu \in \Gamma$ and $T \models \exists \bar{x}\phi$. The classical Henkin–Orey omitting types theorem, *OTT* for short, states that if T is a consistent theory in a countable language $\mathfrak{L}$ and $\Gamma(x_1 \dots x_n) \subseteq \mathfrak{L}$ is realized in every model of T, then there is a formula $\phi \in \mathfrak{L}$ such that ϕ ensures Γ in T. The formula ϕ is called a T-witness for Γ. Now the problem of resource sensitivity can be applied to *OTT* in the following sense. Can we always guarantee that the witness uses the same number of variables as T and Γ, or do we need extra variables? If we do need extra variables, is there perhaps an upper bound on the number of extra variables needed. In other words, let $\mathfrak{L}_n$ denote the set of formulas of $\mathfrak{L}$ which are built up using only n variables. The question is: If $T\cup\Gamma \subseteq \mathfrak{L}_n$, is there any guarantee that the witness stays in $\mathfrak{L}_n$, or do we occasionally have to step outside $\mathfrak{L}_n$?

Assume that $T \subseteq \mathfrak{L}_n$. We say that T is n complete iff for all sentences $\phi \in \mathfrak{L}_n$ we have either $T \models \phi$ or $T \models \neg\phi$. We say that T is n atomic iff for all $\phi \in \mathfrak{L}_n$, there is $\psi \in \mathfrak{L}_n$ such that $T \models \psi \to \phi$ and for all $\eta \in \mathfrak{L}_n$ either $T \models \psi \to \eta$ or $T \models \psi \to \neg\eta$.

Theorem 3.1.1. *Assume that $\mathfrak{L}$ is a countable first order language containing a binary relation symbol. For $n > 2$ and $k \geq 0$, there are a consistent n complete and n atomic theory T using only n variables, and a set $\Gamma(x_1)$ using only 3 variables (and only one free variable) such that Γ is realized in all models of T but each T-witness for T uses more that $n + k$ variables.*

Theorem 3.1.1 is proved using algebraic logic in [And-Nem-Say,08]. For undefined terminology in the coming key Lemma the reader is referred to [Hir-Hod,thisVol].

Theorem 3.1.2. *Suppose that n is a finite ordinal with $n > 2$ and $k \geq 0$. There is a countable symmetric integral representable relation algebra $\mathfrak{R}$ such that*

(i) *Its completion, i.e. the complex algebra of its atom structure is not representable, so $\mathfrak{R}$ is representable but not completely representable.*

(ii) *$\mathfrak{R}$ is generated by a single element.*

(iii) *The (countable) set $\mathfrak{B}_n\mathfrak{R}$ of all n by n basic matrices over $\mathfrak{R}$ constitutes an n-dimensional cylindric basis. Thus $\mathfrak{B}_n\mathfrak{R}$ is a cylindric atom structure and the full complex algebra $\mathfrak{Cm}(\mathfrak{B}_n\mathfrak{R})$ with universe the power set of $\mathfrak{B}_n\mathfrak{R}$ is an n-dimensional cylindric algebra.*

(iv) *The term algebra over the atom structure $\mathfrak{B}_n\mathfrak{R}$, which is the countable subalgebra of $\mathfrak{Cm}(\mathfrak{B}_n\mathfrak{R})$ generated by the countable set of n by n basic matrices, $\mathfrak{Tm}(B_n\mathfrak{R})$ for short, is a countable representable CA_n, but $\mathfrak{Cm}(\mathfrak{B}_n)$ is not representable.*

(v) *Hence $\mathfrak{C}$ is a simple, atomic representable but not completely representable CA_n.*

(vi) *$\mathfrak{C}$ is generated by a single 2 dimensional element g, the relation algebraic reduct of $\mathfrak{C}$ does not have a complete representation and is also generated by g as a relation algebra, and $\mathfrak{C}$ is a sub-neat reduct of some simple representable $\mathfrak{D} \in \mathsf{CA}_{n+k}$ such that the relation algebraic reducts of $\mathfrak{C}$ and $\mathfrak{D}$ coincide.*

Sketch of proof. We prove everything except that $\mathfrak{R}$ can be generated by a single element, to which we refer to [And-Nem-Say,08]. Let k be a cardinal. Let $\mathfrak{E}_k = \mathfrak{E}_k(2,3)$ denote the relation algebra which has k non-identity atoms, in which $a_i \leq a_j ; a_l$ if $\left|\{i,j,l\}\right| \in \{2,3\}$ for all non-identity atoms a_i, a_j, a_k. (This means that all triangles are allowed except the monochromatic ones.) These algebras were defined by Maddux, see also [Hir-Hod,thisVol]. Let k be finite, let I be the set of non-identity atoms of $\mathfrak{E}_k(2,3)$ and let $P_0, P_1 \dots P_{k-1}$ be an enumeration of the elements of I. Let $l \in \omega$, $l \geq 2$ and let J_l denote the set of all subsets of I of cardinality l. Define the symmetric ternary relation on ω by $E(i,j,k)$ if and only if i, j, k are evenly distributed, that is

$$(\exists p,q,r)\{p,q,r\} = \{i,j,k\}, \quad r-q = q-p.$$

Now assume that $n > 2$, $l \geq 2n-1$, $k \geq (2n-1)l$, $k \in \omega$. Let $\mathfrak{M} = \mathfrak{E}_k(2,3)$. Then $\mathfrak{M}$ is a simple, symmetric finite atomic relation algebra. Also,

$$(\forall V_2 \dots, V_n, W_2 \dots W_n \in J_l)(\exists T \in J_l)(\forall 2 \leq i \leq n)$$

$$(\forall a \in V_i)(\forall b \in W_i)(\forall c \in T_i)(a \leq b;c).$$

That is $(J4)_n$ formulated in [And-Nem-Say,08] p. 72 is satisfied. Therefore, as proved in [And-Nem-Say,08] p. 77, B_n the set of all n by n basic matrices is a cylindric basis of dimension n. But we also have

$$(\forall P_2, \dots, P_n, Q_2 \dots Q_n \in I)(\forall W \in J_l)(W \cap P_2; Q_2 \cap \dots \cap P_n \ : \ Q_n \neq 0)$$

That is $(J5)_n$ formulated on p. 79 of [And-Nem-Say,08] holds. According to Definition 3.1(ii) (J,E) is an n blur for $\mathfrak{M}$, and clearly E is definable in $(\omega, <)$. Let $\mathfrak{C}$ be as defined in Lemma 4.3 in [And-Nem-Say,08]. Then, by Lemma 4.3, $\mathfrak{C}$ is a subalgebra of $\mathfrak{Cm}\,\mathfrak{B}_n$, hence it contains the term algebra $\mathfrak{Tm}\,\mathfrak{B}_n$. Denote $\mathfrak{C}$ by $\mathfrak{Bb}_n(\mathfrak{M},J,E)$. Then by Theorem 4.6 in [And-Nem-Say,08] $\mathfrak{C}$ is representable, and by Theorem 4.4 in [And-Nem-Say,08] for $m < n$ $\mathfrak{Bb}_m(\mathfrak{M},J,E) = \mathfrak{Nr}_m\,\mathfrak{Bb}_n(\mathfrak{M},J,E)$. However $\mathfrak{Cm}\,\mathfrak{B}_n$ is not representable. In [And-Nem-Say,08] $\mathfrak{R} = \mathfrak{Bb}(\mathfrak{M},J,E)$ is proved to be generated by a single element. ■

Now we give a proof of Theorem 3.1.1 modulo Theorem 3.1.2.

Proof of Theorem 3.1.1. Let g, $\mathfrak{C}$ and $\mathfrak{D}$ be as in Theorem 3.1.2(vi). Then g generates $\mathfrak{C}$ and g is 2 dimensional in $\mathfrak{C}$. We can write up a theory $T \subseteq \mathfrak{L}_n$ such that for any model $\mathfrak{M}$ we have

$$\mathfrak{M} = (M,G) \models T \quad \text{iff} \quad \mathfrak{C}_n(\mathfrak{M}) \cong \mathfrak{C}$$

and G corresponds to g via this isomorphism.

Now $T \subseteq L_n$, T is consistent and n complete and n atomic because C is simple and atomic. We now specify $\Gamma(x,y)$. For $a \in At$, let τ_a be a relation algebraic term such that $\tau_a(g) = a$ in R, the relation algebra reduct of $\mathfrak{C}$. For each τ_a there is a formula $\mu_a(x,y)$ such that $\tau_a(g) = \mu_a^{\mathfrak{M}}$. Define $\Gamma(x,y) = \{\neg\mu_a : a \in At\}$. We will show that Γ is as required. First we show that Γ is realized in every model of T. Let $\mathfrak{M} \models T$. Then $\mathfrak{C}_n(\mathfrak{M}) \cong \mathfrak{C}$, hence $\mathfrak{M}$ gives a representation of $\mathfrak{R}$ because $\mathfrak{R}$ is the relation algebraic reduct of $\mathfrak{C}_n(\mathfrak{M})$. But $\mathfrak{R}$ has no complete representation, which means that $X = \bigcup\{\mu_a^{\mathfrak{M}} : a \in At\} \subset M \times M$, i.e. proper subset, so let $(u,v) \in M \times M \sim X$. This means that Γ is realized by (u,v) in $\mathfrak{M}$. We have seen that Γ is realized in each model of T. Assume that that $\phi \in \mathfrak{L}_{n+k}$ such that $T \models \exists\bar{x}\phi$. We may assume that ϕ has only two free variables, say x, y. Take the representable $\mathfrak{D} \in \mathsf{CA}_{n+k}$ from Theorem 3.1.2(iv). Recall that $g \in C \subseteq D$ and $\mathfrak{D}$ is simple. Let $\mathfrak{M} = (M,g)$ where M is the base set of $\mathfrak{D}$. Then $\mathfrak{M} \models T$ because $\mathfrak{C}$ is a subreduct of $\mathfrak{D}$ generated by g. By $T \models \exists\bar{x}\phi$, we have $\phi^{\mathfrak{M}} \neq \emptyset$. Also $\phi^{\mathfrak{M}} \in \mathfrak{D}$ and is 2 dimensional, hence $\phi^{\mathfrak{M}} \in R$, since $\mathfrak{R}$ is the relation algebraic reduct of $\mathfrak{D}$, as well. But $\mathfrak{R}$ is atomic hence $\phi^{\mathfrak{M}} \cap \mu_a \neq \emptyset$ for some $a \in At$. This shows that it is not the case that $\mathfrak{M} \models \phi \to \neg\mu_a$ where $\neg\mu_a \in \Gamma$, thus ϕ is not a T-witness for Γ. Now we modify T, Γ so that Γ uses only one free variable. We use the technique of so-called partial pairing functions. Let g, $\mathfrak{C}$, $\mathfrak{D}$ be as in Theorem 3.1.2(iv) with $\mathfrak{D} \in \mathsf{CA}_{2n+2k}$. We may assume that g is disjoint from the identity $1'$ because $1'$ is an atom in the relation algebraic reduct of $\mathfrak{C}$. Let U be the base set of $\mathfrak{C}$. We may assume that U and $U \times U$ are disjoint. Let $M = U \cup (U \times U)$, let $G = g \cup \big\{(u,(u,v)) : u,v \in U\big\} \cup \big\{((u,v),v) : u,v \in U\big\} \cup \big\{((u,v),(u,v)) : u,v \in U\big\}$ and let $\mathfrak{M} = (M,G)$. From G we can define $U \times U$ as $\big\{x : G(x,x)\big\}$ and from $U \times U$ and G we can define the projection functions between $U \times U$ and U, and g. All these definitions use only 3 variables. Thus for all $t \geq 3$ for all $\phi(x,y) \in \mathfrak{L}_t$ there is a $\psi(x) \in \mathfrak{L}_t$ such that $\psi^{\mathfrak{M}} = \big\{(u,v) \in U \times U : \phi^{(U,g)}(u,v)\big\}$. For any $a \in At$ let $\psi_a(x)$ be the formula corresponding to $\mu_a(x,y)$ this way. Conversely for any $\psi \in \mathfrak{L}_t$ there is a $\phi \in \mathfrak{L}_{2t}$ such that the projection of $\psi^{\mathfrak{M}}$ to U is $\phi^{(U,g)}$. Now define T as the $\mathfrak{L}_n$ theory of $\mathfrak{M}$, and set $\Gamma(x) = \big\{\neg\psi_a(x) : a \in At\big\}$. Then it can be easily checked that Γ and T are as required. ■

Let us call an atom structure $\mathfrak{M}$ strongly representable if $\mathfrak{Cm}\,\mathfrak{M}$ is representable, and weakly representable if $\mathfrak{Tm}\,\mathfrak{M}$ is representable. As shown above, the construction of weakly representable atom structures that are not strongly representable [And-Nem-Say,08], [Say,05a] leads to atomic algebras with no complete representations and proves that RCA_n is not closed under completions i.e. is not atom-canonical, for $\mathfrak{Cm}\,At\mathfrak{A}$ is the completion of $\mathfrak{Tm}\,At\mathfrak{C}$. Such algebras were first constructed by Hirsch and Hodkinson [Hir-Hod,02a]. Several variations on such constructions can be found in [Kha-Say,09b], [Kha-Say,09a]. It also shows that OTT fails in L_n, the first order logic restricted to the first n variables, as long as $n > 2$. For $n = 2$ the Omitting types theorem fails for L_n in a more subtle way. A moment's reflection reveals that what we actually showed above is that Vaught's famous theorem on existence of atomic models for atomic theories fails for L_n when $n > 2$ [Say,09c]. For L_2 the analogue of Vaught's theorem holds [Kha-Say,10], but the omitting types fails (an unpublished result of Andréka and Németi, cf. concluding remarks [And-Nem-Say,08] p. 87.) In passing we note that for usual FOL Vaught's theorem follows from the omitting types theorem, so this method cannot be used for L_2. Hirsch and Hodkinson show that the class of strongly representable atom structures of relation algebras (and cylindric algebras) is not elementary [Hir-Hod,09]. The construction makes use of the probabilistic method of Erdős to show that there are finite graphs with arbitrarily large chromatic number and girth. In his pioneering paper of 1959, Erdős took a radically new approach to constructing such graphs: for each n he defined a probability space on the set of graphs with n vertices, and showed that, for some carefully chosen probability measures, the probability that an n vertex graph has these properties is positive for all large enough n. This approach, now called the *probabilistic method* has since unfolded into a sophisticated and versatile proof technique, in graph theory and in other branches of discrete mathematics. This method was used first in algebraic logic by Hirsch and Hodkinson to show that the class of strongly representable atom structures of cylindric and relation algebras is not elementary and that varieties of representable relation algebras are barely canonical. But yet again, using these methods of Erdős in [Gol-Hod-Ven,04] it is shown that there exist continuum-many canonical equational classes of Boolean algebras with operators that are not generated by the complex algebras of any first-order definable class of relational structures. Using a variant of this construction the authors resolve the long-standing question of Fine, by exhibiting a bimodal logic that is valid in its canonical frames, but is not sound and complete for any first-order definable class of Kripke frames.

2. Omitting Types in Other Contexts

The algebraic counterpart of first order logic is the class Lf_ω. That is, if Λ is an ordinary first order language and Σ is a set of formulas, then $\mathfrak{Fm}_\Sigma$ as defined in [Hen-Mon-Tar,85] Section 4.3 is locally finite. Monk [Mon,78] proved algebraically the omitting types theorem by proving a strong representation theorem for Lf_ω. See also [Hen-Mon-Tar,85] Remark 4.3.68 (11). In this paper, Monk also algebraised the notion of individual constants, and gave an algebraic treatment of Henkin constructions using individual constants [Hen-Mon-Tar,85]. Here we show that a version of OTT holds in certain extensions of first order logic. We follow the (more or less) standard notation of [Hen-Mon-Tar,85, §4.3]. We start by defining our languages. We consider certain infinitary languages that are extensions of first order languages. A language Λ is a triple $(\alpha, \mathbf{R}, \rho)$ where α is a countable infinite ordinal. ρ and $\mathbf{R}$ are functions with common domain β, β a cardinal. That is $Do\mathbf{R} = Do\rho = \beta$. α specifies the number of variables available and $\mathbf{R}$ is the sequence of relation symbols with domain β. ρ is a function from β to α. For $i < \beta$, $\rho(i)$ is the rank of $\mathbf{R}_i$. The cardinality of this language is defined to be β. All languages considered in this paper are countable, i.e. $\beta \leq \omega$. We do not allow function symbols. When $\rho(i)$ is finite for all i then we are dealing with an ordinary first order language. But we shall deal with more general languages, namely languages such that $\alpha \smallsetminus \rho(i)$ is infinite. Such languages are called *variable rich* in [Nem,90]. In particular, for such languages the arity of atomic formulas may be infinite. However, formulas are built up from atomic formulas the usual way, using $=$, $\neg$, $\wedge$, $\vee$, and $\exists x$. In other words, only quantification on finitely many variables is allowed. Since only finite conjunctions and disjunctions are allowed, only finitely many relation symbols can occur in formulas. But let us specify our atomic formulas. An atomic formula is either an equation or one of the form $\mathbf{R}_\eta(v_0, v_1, \ldots v_i \ldots)_{i<\rho(\eta)}$ (where $\eta < \beta$.) Such atomic formulas are called restricted atomic formulas, meaning that the variables occur only in their natural order. We allow only these. A restricted formula, or a formula for short, is one whose atomic subformulas are restricted (for first order logic this is not a restriction because any (ordinary) formula is equivalent to a restricted one). We let Fm_r^Λ stand for the set of all Λ (restricted) formulas. We use the system $\vdash_{r,\Lambda}$ for our language Λ introduced in [Hen-Mon-Tar,85, §4.3]. Such a proof system is complete for first order languages. Let $Ax \subseteq Fm_r^\Lambda$ and $\phi \in Fm_r^\Lambda$. We write $Ax \vdash_{r,\Lambda} \phi$ or even simply $Ax \vdash \phi$ if ϕ can be derived from Ax by the above mentioned proof system. A theory T is a set of formulas. T is consistent if not $T \vdash \bot$.

We have specified our syntax. We now turn to semantical notions.

Definition 3.2.1.

(i) Let α be an ordinal and M be a set. A weak space of dimension α and base M is a set of the form

$$\left\{\mathsf{s} \in {}^{\alpha}M \,:\, \left|\{i \in \alpha \,:\, \mathsf{s}_i \neq p_i\}\right| < \omega\right\}$$

for some $p \in {}^{\alpha}M$. We denote this set by ${}^{\alpha}M^{(p)}$.

(ii) Let $\Lambda = (\alpha, \mathbf{R}, \rho)$ be a language with $Do\mathbf{R} = \beta$. A weak structure for Λ is a triple $\mathfrak{M} = (M, R, p)$ where M is a non empty set $p \in {}^{\alpha}M$ and R is a function with domain β assigning to each $i < \beta$ a subset $R_i^{\mathfrak{M}}$ of the weak space ${}^{\rho i}M^{(p\restriction\rho i)}$. A sequence $\mathsf{s} \in {}^{\alpha}M^{(p)}$ satisfies an atomic formula $R_i(v_0, \dots v_i \dots)_{i<\rho(i)}$ if $\mathsf{s} \restriction \rho(i) \in R_i^{\mathfrak{M}}$. s satisfies $v_i = v_j$ if $\mathsf{s}(i) = \mathsf{s}(j)$.

(iii) A standard structure, or simply a structure, for Λ is defined the usual way. That is, a Λ structure is a pair $\mathfrak{M} = (M, R)$ where M is a non-empty set and R is a function with domain β assigning to each $i < \beta$, a subset $R_i^{\mathfrak{M}}$ of ${}^{\rho(i)}M$. Satisfiability for atomic formulas are defined as in the previous item.

We can extend the notion of satisfiability to all formulas in the usual way. In the following, notions formulated for weak structures apply equally well to structures. We will not explicitly mention this.

For a weak structure $\mathfrak{M} = (M, R, p)$ and a sequence $\mathsf{s} \in {}^{\alpha}M^{(p)}$, we write $\mathfrak{M} \models \phi[\mathsf{s}]$ if s satisfies ϕ in $\mathfrak{M}$. We write $\mathfrak{M} = (M, R, p) \models \phi$ if every $\mathsf{s} \in {}^{\alpha}M^{(p)}$ satisfies ϕ in $\mathfrak{M}$, in which case $\mathfrak{M}$ is called a weak model of ϕ.

If K is a class of weak structures for Λ, then $K \models \phi$ if $\mathfrak{M}$ is a weak model of ϕ for all $\mathfrak{M} \in K$. $Md\Sigma$ for a set of formulas Σ is the class of all weak models of all formulas in Σ.

We write $\Sigma \models \phi$ if $Md\Sigma \models \phi$. Finally $\models \phi$ means $K \models \phi$ where K is the class of all weak structures for Λ. We note that for such languages $\Gamma \models \phi$ if and only if $\Gamma \vdash \phi$, so that we have a completeness Theorem [Hen-Mon-Tar,85, 4.3.23]. Here, it is not trivial to show that $\Gamma \models \phi$ with respect to (standard) models is equivalent to $\Gamma \models \phi$ with respect to weak models. For more on these and similar languages see [Say,04a], [Mad,91a], [Nem,91], [Nem,90] and [And-Nem-Sai,01]. The reference [Nem,90], together with [Sam-Say,07b], can be considered as a follow-up of [Hen-Mon-Tar,85, 4.3] for such extended languages.

For formulating an omitting types theorem, we recall some notions:

Definition 3.2.2.

(i) Let Λ be a language and Σ be a set of Λ formulas. A theory T in Λ is said to *isolate* Σ if there is a formula ϕ in Λ such that $\{\phi\} \cup T$ is consistent and for all $\sigma \in \Sigma$, $T \vdash \phi \implies \sigma$. Implication is defined the usual way.

(ii) T *locally omits* Σ iff T does not isolate Σ. Thus T locally omits Σ iff for every formula ϕ which is consistent with T, there exists $\sigma \in \Sigma$ such that $\phi \wedge \neg\sigma$ is consistent with T.

(iii) A *weak model* $\mathfrak{M} = (M, R, p)$ *realizes* Σ if there is an $\mathsf{s} \in {}^{\alpha}M^{(p)}$ that satisfies every formula in Σ. $\mathfrak{M}$ omits Σ if $\mathfrak{M}$ does not realize Σ. That is for every assignment $\mathsf{s} \in {}^{\alpha}M^{(p)}$ there is a formula $\phi \in \Sigma$ such that s does not satisfy ϕ in $\mathfrak{M}$.

We introduce a certain cardinal that plays an essential role in Omitting Types Theorems [New,87]. In the process we review certain notions from descriptive Set Theory [Azr,79].

Definition 3.2.3.

(i) A *Polish space* is a complete separable metric space.

(ii) For a Polish space X, $K(X)$ denotes the ideal of meager subsets of X. Set

$$\operatorname{cov} K(X) = \min \big\{ |C| \ : \ C \subseteq K(X),\ \cup C = X \big\}.$$

If X is the real line, or the Baire space ${}^{\omega}\omega$, or the Cantor set ${}^{\omega}2$, which are the prime examples of Polish spaces, we write K instead of $K(X)$.

The above three spaces are sometimes referred to as *real* spaces since they are all Baire isomophic to the real line, [Azr,79] p. 223. It is straightforward to see that $\omega < \operatorname{cov} K \leq 2^{\aleph_0}$. $\operatorname{cov} K$ is the least cardinal κ such that the real space can be covered by κ many closed nowhere dense sets, that is the least cardinal such that the Baire Category Theorem fails [Mil,80]. This cardinal has been studied quite intensely, [Mil,82], [She,97]. Martin's axiom, and for that matter, the continuum hypothesis imply that $\operatorname{cov} K = 2^{\aleph_0}$ but it is consistent that $\operatorname{cov} K = \omega_1 < 2^{\aleph_0}$. $\operatorname{cov} K < 2^{\aleph_0}$ is true in the random real model and is mentioned on p. 170 of [Mar-Sol,70]. It also holds in models constructed by forcings which do not add Cohen reals [Bar-Jud,95]. The following Theorem is proved in [Sam-Say,07b]:

Theorem 3.2.4. *Let κ be a cardinal $< \operatorname{cov} K$. Let $\Lambda = (\alpha, \mathbf{R}, \rho)$ be a countable language and Γ a consistent Λ theory. Let ϕ be a formula consistent with Γ. For each $i < \kappa$, let Σ_i be a set of formulas. If Γ locally omits each Σ_i, then Γ has a weak model $\mathfrak{M}$ which omits each Σ_i. Furthermore, ϕ is satisfiable in $\mathfrak{M}$.*

Proof. Let $\mathfrak{A} = \mathfrak{Fm}_\Gamma$. Then $\mathfrak{A}$ is countable and $\mathfrak{A} \in \mathsf{Dc}_\alpha$. Let $X_i = \Sigma_i / \equiv_\Gamma$. Let $a = \phi / \equiv_\Gamma$. Then $a \in \mathfrak{A}$ is non zero. Also $\prod X_i = 0$. We have by [Hen-Mon-Tar,85, 1.11.6] that

$$(3.2.1) \qquad (\forall j < \alpha)(\forall x \in B)\Big(\mathsf{c}_j x = \sum_{i \in \alpha \smallsetminus \Delta x} \mathsf{s}_i^j x\Big).$$

Here $\sum$ denotes supremum and for distinct $i, j < \beta$, $\mathsf{s}_i^j x$ is defined by $\mathsf{c}_j(x \cdot \mathsf{d}_{ij})$. $\mathsf{s}_i^i x$ is defined to be x. If x is a formula, then $\mathsf{s}_i^j x$ is the operation of replacing the free occurrences of variable v_j by v_i such that the substitution is free. Now let V be the weak space ${}^\omega\omega^{(Id)} = \Big\{\mathsf{s} \in {}^\omega\omega : \big|\{i \in \omega : \mathsf{s}_i \neq i\}\big| < \omega\Big\}$. For each $\tau \in V$ for each $i \in \kappa$, let

$$X_{i,\tau} = \{\mathsf{s}_\tau x : x \in X_i\}.$$

Here s_τ is the unary operation as defined in [Hen-Mon-Tar,85, 1.11.9]. s_τ is the algebraic counterpart of the metalogical operation of the simultaneous substitution of variables (indexed by the range of τ) for variables (indexed by its domain) [Hen-Mon-Tar,85, 1.11.8]. For each $\tau \in V$, s_τ is a complete boolean endomorphism on $\mathfrak{B}$ by [Hen-Mon-Tar,85, 1.11.12(iii)]. It thus follows that

$$(3.2.2) \qquad (\forall \tau \in V)(\forall i \in \kappa) \prod{}^{\mathfrak{A}} X_{i,\tau} = 0$$

Let S be the Stone space of the boolean part of $\mathfrak{A}$, and for $x \in \mathfrak{A}$, let N_x denote the clopen set consisting of all boolean ultrafilters that contain x.

Then from (3.2.1), (3.2.2), it follows that for $x \in \mathfrak{A}$, $j < \beta$, $i < \kappa$ and $\tau \in V$, the sets

$$\mathbf{G}_{j,x} = N_{\mathsf{c}_j x} \setminus \bigcup_{i \notin \Delta x} N_{\mathsf{s}_i^j x} \quad \text{and} \quad \mathbf{H}_{i,\tau} = \bigcap_{x \in X_i} N_{\mathsf{s}_{\bar{\tau}} x}$$

are closed nowhere dense sets in S. Let

$$\mathbf{G} = \bigcup_{j \in \beta} \bigcup_{x \in B} \mathbf{G}_{j,x} \quad \text{and} \quad \mathbf{H} = \bigcup_{i \in \kappa} \bigcup_{\tau \in V} \mathbf{H}_{i,\tau}.$$

By properties of cov K, it can be shown that $\mathbf{H}$ is a countable collection of nowhere dense sets. By the Baire Category theorem for compact Hausdorff spaces, we get that $X = S \sim \mathbf{H} \cup \mathbf{G}$ is dense in S. Accordingly let F be an ultrafilter in $N_a \cap X$. By the very choice of F, it follows that $a \in F$ and we have the following

$$(3.2.3) \qquad (\forall j < \beta)(\forall x \in B)\big(\mathsf{c}_j x \in F \implies (\exists j \notin \Delta x)\mathsf{s}_j^i x \in F\big)$$

and

$$(3.2.4) \qquad (\forall i < \kappa)(\forall \tau \in V)(\exists x \in X_i)\mathsf{s}_\tau x \notin F.$$

Next we form the canonical representation corresponding to F in which satisfaction coincides with genericity. To handle equality we define

$$E = \big\{ (i,j) \in {}^2\alpha : \mathsf{d}_{ij} \in F \big\}.$$

E is an equivalence relation on α. E is reflexive because $\mathsf{d}_{ii} = 1$ and symmetric because $\mathsf{d}_{ij} = \mathsf{d}_{ji}$. E is transitive because F is a filter and for all $k, l, u < \alpha$, with $l \notin \{k, u\}$, we have

$$\mathsf{d}_{kl} \cdot \mathsf{d}_{lu} \leq \mathsf{c}_l(\mathsf{d}_{kl} \cdot \mathsf{d}_{lu}) = \mathsf{d}_{ku}.$$

Let $M = \alpha/E$ and for $i \in \omega$, let $q(i) = i/E$. Let W be the weak space ${}^\alpha M^{(q)}$. For $h \in W$, we write $h = \bar{\tau}$ if $\tau \in V$ is such that $\tau(i)/E = h(i)$ for all $i \in \omega$. τ of course may not be unique. Define f from $\mathfrak{A}$ to the full weak set algebra with unit W as follows:

$$f(x) = \{\bar{\tau} \in W : \mathsf{s}_\tau x \in F\}, \quad \text{for} \quad x \in \mathfrak{A}.$$

Then it can be checked that f is a homomorphism such that $f(a) \neq 0$ and $\bigcap f(X_i) = \emptyset$ for all $i \in \kappa$, hence the desired. ■

An important difference from first order logic is that, in our present context, we do not assume an upper bound on the number of (free) variables occurring in the types omitted, i.e. these types need not be finitary, they can have infinitely many free variables. Indeed for first order logic the (classical) omitting types theorem fails when infinitary types are considered as the following (simple) example illustrates:

Let T be the theory of dense linear order without endpoints. Then T is complete. Let $\Gamma(x_0, x_1 \ldots)$ be the set

$$\{x_1 < x_0, x_2 < x_1, x_3 < x_2 \ldots\}.$$

(Here there is no bound on free variables.) A model $\mathfrak{M}$ omits Γ if and only if $\mathfrak{M}$ is a well ordering. But T has no well ordered models, so no model of T omits Γ. However T locally omits Γ because if $\phi(x_0, \ldots x_{n-1})$ is consistent with T, then $\phi \wedge \neg x_{n+2} < x_{n+1}$ is consistent with T. Note that Γ can be omitted in a weak model.

But there is a price we pay for this improvement. The model omitting this types is not a *standard model.* For cardinality considerations this is expected. Let us explain why. The classical proof (by forcing) of omitting types for first order logic breaks down when types consisting of formulas having infinitely many free variables are considered, because there are uncountably many assignments to free variables, but only countably many stages of the forcing construction to consider them in. When we consider only those assignments that are eventually constant, this problem of cardinality disappears.

Remark. The above proof depended on the following topological property. If X is a second countable compact Hausdorff space and $(A_i : i < \kappa)$ is a family of nowhere dense sets then $X \sim \bigcup_{i<\kappa} A_i$ is dense. The idea of proof is that the (possibly uncountable union) $\bigcup_{i<\kappa} A_i$ can be written as a *countable* union of nowhere dense sets and then a direct application of the Baire category theorem for compact Hausdorff spaces becomes possible, so we can find the desired filter F. The question arises as to what happens if we replace $\operatorname{cov} K$ by ${}^{\omega}2$. (Recall that it is consistent that they are not equal). In this case the theorem cannot be proved in ZFC. We would need extra (independent) axioms. One possible axiom is Martin's axiom (MA). This follows from the fact that MA implies that if we have a union of nowhere dense sets over an indexing set I with $|I| < {}^{\omega}2$ then it is a countable union. But MA is too strong. For any ordinal α let P_α be the statement: Given a collection $< 2^{\omega_\alpha}$ subsets of ω_α such that the intersection of any $< \omega_\alpha$ has cardinality ω_α, then there is $B \subseteq \omega_\alpha$ of cardinality ω_α such that for each element A of the collection $|B - A| < \omega_\alpha$. P_0 is the statement: Whenever $\mathcal{A}$ is a family of subsets of ω such that $|\mathcal{A}| < {}^\omega 2$ and $A_0 \cap A_1 \cap \ldots A_n$ is infinite, whenever $A_0, A_1 \ldots A_n \in \mathcal{A}$, then there is a subset I of ω such that $I \sim A$ is finite for every $A \in I$. It can be shown that $MA \implies P_0$ and that MA is strictly stronger than P_0. The condition P_0 is essentially the combinatorial part of MA. Under P_0 the following can be proved: If X is a topological space with countable base, then the family of nowhere dense sets J has the property that whenever $J_1 \subseteq J$ and $|J_1| < {}^\omega 2$, there is a countable $J_0 \subseteq J$ such that every member of J_1 is included in a member of J_0. And thats all we need. P_0 is equivalent to Martin's axiom restricted to the so called σ-centered partially ordered sets, so it is a restricted form of Martin's axiom. But actually what we need is even, yet, a weaker assumption, and that is Martin's axiom restricted to *countable* partially ordered sets, called MA(countable). In passing we note that $\operatorname{cov} K$ is the largest cardinal such that MA(countable) is true, so that in some exact sense the cardinal $\operatorname{cov} K$ is the best possible. In short when we loosen the statement $\kappa < \operatorname{cov} K$ to $\kappa < {}^\omega 2$ we are led to an independent statement in set theory. In fact such a statement is a consequence of MA, and like MA, it is independent from $ZFC + \neg CH$. For more on such connections we refer the reader to [Fre,84].

Now what happens if we consider finitary types, as the case with first order logic. A finitary type is a type such that the free variables occurring in its formulas are uniformy bounded. That is if we let $\mathfrak{Fr}(\phi)$ be the set of free variables in ϕ, then Γ is finitary if there exists $n \in \omega$ such that $\mathfrak{Fr}(\phi) \subseteq \{x_0, \ldots x_{n-1}\}$ for all $\phi \in \Gamma$. Now the standard version of the omitting types theorem holds. And indeed we have:

Theorem 3.2.5. *Let κ be a cardinal $< \operatorname{cov} K$. Let $\Lambda = (\alpha, \mathbf{R}, \rho)$ be a countable language and Γ a consistent Λ theory. Let ϕ be a formula consistent with Γ. For each $i < \kappa$, let Σ_i be a finitary type. If Γ locally omits each Σ_i, then Γ has a standard model $\mathfrak{M}$ which omits each Σ_i. Furthermore, ϕ is satisfiable in $\mathfrak{M}$.*

Proof. The proof uses ideas of Andréka and Németi, reported in [Hen-Mon-Tar,85], on how one can "square" weak spaces using ultraproducts. So using the previous theorem, one first finds a weak set algebra and then using ultraproducts one shows that this weak set algebra is isomorphic to a set algebra. Because types considered are finitary, this isomorphism still preserves the required meets ([Kha-Say,09c]). ■

Corollary 3.2.6.

(i) *Let $\mathfrak{A} \in \mathsf{Nr}_n \mathsf{CA}_\omega$ be countable. Let $(X_i : i \in \omega)$ be a countable family of subsets of $\mathfrak{A}$ such that $\prod X_i = 0$ for all $i \in \kappa$. Then for every non-zero $a \in A$, there exists a representation $f : \mathfrak{A} \to \wp({}^n M)$ such that $f(a) \neq 0$ and $\bigcap_{x \in X_i} f(x) = \emptyset$ for all $i \in \kappa$.*

(ii) *Let $\mathfrak{A} \in \mathsf{Nr}_n \mathsf{CA}_\omega$ be countable and atomic. Then there exists a representation $f : \mathfrak{A} \to \wp({}^n X)$ such that $f\big(\prod X\big) = \bigcap_{x \in X} f(x)$ whenever $\prod X$ exists.*

Proof. (i) follows from the proof of Theorem 3.2.4, by noting that the n neat reduct of a weak set algebra is a set algebra of dimension n. Now consider (ii). Let $Y = \{-x : x \in At\mathfrak{A}\}$. Then $\prod X = 0$. Hence there exists a representation $f : \mathfrak{A} \to \wp({}^n X)$ such that $\bigcap f(x) = \emptyset$. In fact for every $a \in A$, a non-zero, there exists a set algebra $\mathfrak{C}_a$ with base set M and $h : \mathfrak{A} \to \mathfrak{C}$ such that $h(a) \neq 0$, $\bigcap_{x \in X} h(x) = \emptyset$. Let $\mathfrak{D} = \mathbf{P}_{a \in A} \mathfrak{C}_a$, then $\mathfrak{D} \in \mathsf{RCA}_n$ and there exists an isomorphism $f : \mathfrak{A} \to \mathfrak{D}$ such that

$$f(a) = \bigcup \big\{ f(b) : b \text{ is an atom } \leq a \big\}.$$

Let X be a subset of $\mathfrak{A}$ such that $\sum X$ exists. Then $\mathsf{s} \in f(\sum X)$ iff $\mathsf{s} \in f(b)$ for some atom $b \leq \sum X$ iff $\mathsf{s} \in f(b)$ for some atom b with some x with $b \leq x \in X$, iff $\mathsf{s} \in f(x)$ iff $\mathsf{s} \in \bigcup f(X)$. ∎

Let $n < \omega$. A classical theorem of Henkin's states that the class of representable cylindric algebras of dimension n coincides with the class of algebras having the neat embedding property, in symbols $\mathsf{RCA}_n = S \mathsf{Nr}_n \mathsf{CA}_\omega$. In particular, every $\mathfrak{A} \in \mathsf{Nr}_n \mathsf{CA}_\omega$ is representable. While RCA_n is a variety, the class $\mathsf{Nr}_n \mathsf{CA}_\omega$ is a pseudo elementary class, that is not elementary; furthermore; its elementary closure, $UpUr \mathsf{Nr}_n \mathsf{CA}_\omega$ is not finitely axiomatizable. The class of neat reducts is treated at length in [Say,thisVol,a]. In [Say,f] the following question is investigated. When does $\mathfrak{A} \in \mathsf{Nr}_n \mathsf{CA}_\omega$ posses a cylindric representation preserving a given set of (infinite) meets carrying them to set theoretic intersection? Then if $\mathfrak{A}$ has a representation preserving arbitrary meets, then $\mathfrak{A}$ is atomic. Conversely, when $\mathfrak{A}$ is countable and atomic then $\mathfrak{A}$ has such a representation. The example used above to violate OTT for $\mathfrak{L}_n$ together with an unpublished example of the author, can be used to show that countability is essential and we cannot replace $\mathsf{Nr}_n \mathsf{CA}_\omega$ by $\mathsf{Nr}_n \mathsf{CA}_{n+k} \cap \mathsf{RCA}_n$ for any finite k. We also investigate the question of when representations preserve a given (possibly infinite) set of meets. More concretely, if $\mathfrak{A} \in \mathsf{Nr}_n \mathsf{CA}_\omega$ is countable, κ is a cardinal and $(X_i : i < \kappa)$ is a family of subsets of $\mathfrak{A}$ such that $\prod X_i = 0$, when does there exist a (generalized) set algebra $\mathfrak{B}$ and isomorphism $f : \mathfrak{A} \to \mathfrak{B}$ such that for all $i \in \kappa$, $\bigcap_{x \in X_i} f(x) = \emptyset$. (This *is* an algebraic version of omitting κ many types in n variable logics). Let $2^{\aleph_0}$ denote the power of the continuum. We show that when the meets are ultrafilters then preservation of $< 2^{\aleph_0}$ many meets is possible (in ZFC), while if they are not then we are led to a statement that is independent of ZFC. The consistency of such a statement is proved by showing that the statement is a consequence of a combinatorial consequence of Martin's axiom, namely P_0 stated before. The independence is proved using iterated forcing. Let us be even more explicit and formulate and state the results of [Say,f].

Definition 3.2.7.

(i) Let κ be a cardinal. Let $OTT(\kappa)$ be the following statement. $\mathfrak{A} \in \mathsf{Nr}_n \mathsf{CA}_\omega$ is countable and for $i \in \kappa$, $X_i \subseteq A$ are such that $\prod X_i = 0$, then for all $a \neq 0$, there exists a set algebra $\mathfrak{C}$ with countable base, $f : \mathfrak{A} \to \mathfrak{C}$ such that $f(a) \neq 0$ and for all $i \in \kappa$, $\bigcap_{x \in X_i} f(x) = 0$.

(ii) Let OTT be the statement that

$$(\forall \kappa < 2^{\aleph_0})\, OTT(\kappa)$$

(iv) Let $OTT_m(\kappa)$ be the statement obtained from $OTT(\kappa)$ by replacing X_i with "nonprincipal ultrafilter F_i" and OTT_m be the statement

$$(\forall \kappa < 2^{\aleph_0})\, OTT_m(\kappa).$$

The proofs of the following theorems can be found in [Say,f].

Theorem 3.2.8.

(i) *OTT is independent from $ZFC + \neg CH$. In fact, for any regular cardinal $\kappa > \omega_1$, there is a model of ZFC in which $\kappa = 2^{\aleph_0}$ and OTT holds. Conversely, there is a model of ZFC in which $\omega_3 = 2^{\aleph_0}$ and $OTT(\omega_2)$ is false.*

(ii) *OTT_m is provable in ZFC.*

Using Shelah's techniques from stability theory, we also investigate preservation of $< 2^\lambda$ many (maximal) meets, where λ is a regular uncountable cardinal, for uncountable algebras in $\mathsf{Nr}_n \mathsf{CA}_\omega$. In more detail:

Theorem 3.2.9. *Let $\mathfrak{A} \in \mathsf{Nr}_n \mathsf{CA}_\omega$ be infinite such that $|A| = \lambda$, λ is a regular cardinal. Let $\kappa < 2^\lambda$. Let $(X_i : i \in \kappa)$ be a family of non-principal ultrafilters of $\mathfrak{A}$. Then there exists a representation $f : \mathfrak{A} \to \wp({}^n X)$ such that $\bigcap_{x \in X_i} f(x) = \emptyset$ for all $i \in \kappa$.*

The last theorem is a new omitting types theorem addressing the uncountable case for $\mathfrak{L}_n$. Before we give a logical counterpart of the above theorems, we review briefly the notion of quantifier elimination. Quantifier elimination is a concept that occurs in mathematical logic, model theory, and theoretical computer science. One way of classifying formulas is by the amount of quantification. Formulae with less depth of quantifier alternation are thought of as simpler and the quantifier free formulae as the simplest. *A theory has quantifier elimination if for every formula α there exists a formula α_{QF} without quantifiers which is equivalent to it (modulo the theory).* Quantifier elimination is particularly useful in proving that a given theory is decidable. Examples of theories that have been shown decidable using quantifier elimination are Presburger arithmetic, real closed fields, atomless Boolean algebras, term algebras, dense linear orders, and random graphs.

Now one metalogical reading of the last two theorems is

Theorem 3.2.10. *Let T be an $\mathfrak{L}_n$ consistent theory that admits elimination of quantifiers. Assume that $|T| = \lambda$ is a regular cardinal. Let $\kappa < 2^\lambda$. Let $(\Gamma_i : i \in \kappa)$ be a set of non-principal maximal types in T. Then there is a model $\mathfrak{M}$ of T that omits all the Γ_i's.*

Proof. If $\mathfrak{A} = \mathfrak{Fm}_T$ denotes the cylindric algebra corresponding to T, then since T admits elimination of quantifiers, then $\mathfrak{A} \in \mathsf{Nr}_n \mathsf{CA}_\omega$. This follows from the following reasoning. Let $\mathfrak{B} = \mathfrak{Fm}_{T_\omega}$ be the locally finite cylindric algebra based on T but now allowing ω many variables. Consider the map $\phi/T \mapsto \phi/T_\omega$. Then this map is from $\mathfrak{A}$ into $\mathfrak{Nr}_n \mathfrak{B}$. But since T admits elimination of quantifiers the map is onto. The theorem now follows. ■

We feel that some clarification is in order. We mentioned above that we have an uncountable atomic simple algebra $\mathfrak{A} \in \mathsf{Nr}_n \mathsf{CA}_\omega$ which is not completely representable. First impression might be that this is incompatible with Theorem 3.2.9. To construct a complete representation of this $\mathfrak{A}$ one has to construct a representation that preserves $X = \{-a : a \in At\mathfrak{A}\}$. But this is not an ultrafilter so the last theorem does not apply. Even more the atoms of algebra $\mathfrak{A}$ are mutually disjoint and uncountable, so even Martin's axiom cannot offer solace in this context, for the algebra in question does not satisfy the countable chain condition.

BOLYAI SOCIETY
MATHEMATICAL STUDIES, 22

Cylindric-like Algebras and Algebraic Logic
pp. 223–248.

ELEMENTS OF CYLINDRIC ALGEBRAIC MODEL THEORY

GYÖRGY SERÉNY

According to J. Donald Monk, one of the authors of 'Cylindric Algebras', the basic monograph on algebraic logic, the fact that the sets of all $\varphi^{\mathfrak{M}}$'s consisting of sequences satisfying the first order formula φ in the model $\mathfrak{M}$ constitutes the universe of a cylindric set algebra is 'the main motivating force for [...] the whole topic of algebraic logic.' (cf. [Mon,00] p. 453). Therefore, the investigation of cylindric set algebras from the point of view of their close links to first order models has a distinguished role in algebraic logic. In the course of this investigation, the specific properties of models (e.g. universality, homogeneity, saturatedness) become algebraic ones, and the various connections between models correspond to different kinds of isomorphisms between the cylindric set algebras concerned (see e.g. [Hen-Mon-Tar,85] 4.3.68(7) and (10), [Hen-Mon-Tar,85] pp. 37 and 45, [Mon,00] Sections 5 and 6).

The algebraic versions of ordinary first order models only correspond to elements of a specific segment of the class of all cylindric set algebras (Cs_α's). Consequently, the algebraic versions of model theoretical results cannot always (or rather cannot at all) be generalized to the whole class. The answers given to the questions concerning both the conditions under which the generalization of the algebraic versions of ordinary model theoretical results is possible and the scope of this generalization is the subject matter of cylindric algebraic model theory.

This paper is an overview of the results of István Németi and the present author in cylindric algebraic model theory. It is an updated and extended version of that part of [Ser,97] which examines the cylindric set algebraic versions of large models. The results presented here are connected to the work of Hajnal Andréka, Leon Henkin, J. Donald Monk, István Németi, and Alfred Tarski (see [Hen-Mon-Tar,85] 4.3, especially 4.3.68, [Hen-Mon-Tar,85], [Mon,00], and [Nem,90]). In the first part of the paper, we describe the initial state of the algebraic generalization process, that is, the main facts that characterize cylindric set algebras from an ordinary model theoretic point of view. Then, in the second part, choosing saturatedness as the central notion governing the generalization process, we give the negative results showing that the ordinary model theoretical results concerning saturatedness and some other basic notions related to it cannot be extended in an obvious way to the natural generalizations of the algebraic versions of ordinary models. On the other hand, these investigations reveal interesting connections between apparently unrelated structural properties of Cs_α's (see Theorem 1.2.8 below), and lead to a new class of algebras, that of the so-called pf-regular Cs_α's, which can be considered a possible proper algebraic generalization of ordinary models.

Pf-regular Cs_α's are examined in the third part of the paper. These algebras retain many important properties of algebras corresponding to ordinary models and the way this new class deviates from the original one seems to be in accord with the spirit of the Lindström theorem. Further, the elements of the algebras in this class can be looked at from a topological point of view. We present both a simple non-constructive topological proof and an elementary constructive proof based on König lemma of the interesting fact that all pf-regular finite models are ordinary models, which adds one more trait to the unique character of ordinary logic (cf. Theorem 1.3.5).

It should be noted that István Németi has also pursued an alternative way of doing algebraic model theory. He has formulated some generalizations of ordinary model theoretical results in a model theoretical setting. For example, he proved the generalization of the Łoš lemma to some infinitary models (cf. [Nem,90] Proposition 21, p. 63). But here we confine ourselves to the cylindric algebraic context.

As far as the basic algebraic and set theoretical notions are concerned, throughout the paper, we shall use exclusively the terminology and notation of the basic monograph [Hen-Mon-Tar,85] (among others, $f^*G = \{fx : x \in G\}$ for any subset $G \subseteq H$ and function f defined on H, see [Hen-Mon-Tar,85] p. 28) with the exception that the restriction of a function f to a set H will be denoted by $f \restriction H$. Throughout the paper, α is always an (arbitrary but fixed) *infinite ordinal.* We shall frequently use the following notation: $\overline{u} = \langle u : \lambda \in \alpha \rangle$ for any u. '$\subseteq_\omega$' means '*finite* subset of'. For any set U, $\Gamma \subseteq \alpha$, and $p, q \in {}^\alpha U$, $p[\Gamma/q] = p \restriction (\alpha \sim \Gamma) \cup q \restriction \Gamma$. Finally, we denote by $\mathsf{S}^\kappa_\lambda$ the λ-for-κ substitution operator for Cs_α's, i.e., $\mathsf{S}^\kappa_\lambda X = \mathsf{C}_\kappa(\mathsf{D}_{\kappa\lambda} \cap X)$ for any $\kappa, \lambda \in \alpha$ and $X \in A$, $\mathfrak{A} \in \mathsf{Cs}_\alpha$.

1. Connections Between Ordinary Models and Cylindric Set Algebras

From the point of view of algebraic model theory, the relevance of the class of cylindric set algebras stems from the fact that, among them, we can find the structures corresponding to first order models. Actually, let $\mathcal{L}$ be a first order language with a sequence of variables $v = \langle v_i : i \in \alpha \rangle$ and let $\mathfrak{M}$ be a model for $\mathcal{L}$. For any formula φ of $\mathcal{L}$, let $\varphi^{\mathfrak{M}} = \{q \in {}^{\alpha}M : \mathfrak{M} \models \varphi[q]\}$, the set of all sequences with elements in M of length α that satisfy φ in $\mathfrak{M}$. Now, the set

$$\mathsf{Cs}_{\alpha}^{\mathfrak{M}} = \{\varphi^{\mathfrak{M}} : \varphi \text{ is a formula of } \mathcal{L}\}$$

is a universe of a Cs_{α}, and we denote it by $\mathsf{Cs}_{\alpha}^{\mathfrak{M}}$. We can assign a Cs_{α} to any first order model in a natural way (for example, in [Fer,thisVol,a], Section 2, Ferenczi investigates probabilities defined on $\mathsf{Cs}_{\alpha}^{\mathfrak{M}}$ instead of being defined on the model $\mathfrak{M}$). The Cs_{α}'s obtained in this way constitute a specific subclass of Cs_{α}. Indeed, whether or not a sequence satisfies a formula only depends on those elements of the sequence that correspond to the free variables of the formula. This property of ordinary models is reflected in the fact that $\mathsf{Cs}_{\alpha}^{\mathfrak{M}}$ is always regular. Further, $\mathsf{Cs}_{\alpha}^{\mathfrak{M}}$ is also locally finite-dimensional, since ordinary first order formulas have only finitely many free variables. Consequently, for any first order model $\mathfrak{M}$, $\mathsf{Cs}_{\alpha}^{\mathfrak{M}} \in \mathsf{Cs}_{\alpha}^{\mathrm{reg}} \cap \mathsf{Lf}_{\alpha}$. On the other hand, for any $\mathfrak{A} \in \mathsf{Cs}_{\alpha}^{\mathrm{reg}} \cap \mathsf{Lf}_{\alpha}$, there are a first order language $\mathcal{L}$ with a sequence of variables of length α and a model $\mathfrak{M}$ for $\mathcal{L}$ such that $\mathfrak{A} = \mathsf{Cs}_{\alpha}^{\mathfrak{M}}$ (cf. [Hen-Mon-Tar,85] 4.3.10).

It is obvious from these facts that we can take elements of $\mathsf{Cs}_{\alpha}^{\mathrm{reg}} \cap \mathsf{Lf}_{\alpha}$ as the algebraic versions of ordinary first order models. The two most natural ways to improve the results concerning algebras in this class is to examine the algebras in the classes $\mathsf{Cs}_{\alpha} \cap \mathsf{Lf}_{\alpha}$ and $\mathsf{Cs}_{\alpha}^{\mathrm{reg}} \cap \mathsf{Dc}_{\alpha}$ (obtained by the natural weakening[1] of the restrictions defining the class $\mathsf{Cs}_{\alpha}^{\mathrm{reg}} \cap \mathsf{Lf}_{\alpha}$). Actually, these algebras are the most frequently investigated ones in the cylindrical algebraic model theory, since they are considered to be the most natural algebraic extensions of ordinary models. It should be added that Németi has intensively investigated more general classes as well (see e.g. [Nem,90] pp. 48–49), but here we shall only examine the two most basic classes.

[1] Generally, the omission of *all* restrictions concerning the dimension set of elements is too rough a generalization, excluding the application of the algebraic version of substitutions (cf. [Hen-Mon-Tar,85] 1.11.9 and the remarks preceding it).

Now, let us recall from [Hen-Mon-Tar,85] 3.1.41 the algebraic notions corresponding to the basic *relations* between models (cf. also [Mon,00] pp. 458, 460). Let $\mathfrak{A}$ be a Cs_α with base U.

(a) Let g be a one-one function mapping U onto a set W. For any $X \in A$, let $\widetilde{g}X = \{y \in {}^\alpha W : g^{-1} \circ y \in X\}$. Then, obviously, $\widetilde{g}$ is an isomorphism from $\mathfrak{A}$ onto a Cs_α $\mathfrak{B}$ (cf. [Hen-Mon-Tar,85] 3.1.36). $\widetilde{g}$ is called the **base-isomorphism** induced by g. In the case of two Cs_α's $\mathfrak{A}$ and $\mathfrak{B}$, $\mathfrak{A}$ and $\mathfrak{B}$ are called **base-isomorphic** if there is a base-isomorphism from $\mathfrak{A}$ onto $\mathfrak{B}$.

(b) Let $W \subseteq U$ be any set, and let $\mathfrak{B}$ be a Cs_α with unit element $V = {}^\alpha W$. If the function $rl_V^{\mathfrak{A}}$ $(= rl_V) = \langle X \cap V : X \in A\rangle$ is an isomorphism from $\mathfrak{A}$ onto $\mathfrak{B}$, then it is called an **ext-isomorphism.** The inverse of an ext-isomorphism is said to be a **sub-isomorphism.**

(c) A function f is called an **ext-base-isomorphism** if f is the composition of an ext-isomorphism and a base-isomorphism (in any order, cf. [Hen-Mon-Tar,85] 3.1.42). Again, in this case f^{-1} is called a **sub-base-isomorphism.**

The motivation for introducing these notions is that they are the set algebraic counterparts of the fundamental relations between models (cf. [Mon,00] pp. 453, 457, 460–461):

Proposition 1.1.1. *Let $\mathfrak{M}$ and $\mathfrak{N}$ be models of $\mathcal{L}$, and let $f = \{\langle \varphi^{\mathfrak{M}}, \varphi^{\mathfrak{N}}\rangle :$ φ is a formula of $\mathcal{L}\}$.*

(a) *The following conditions are equivalent:*

  (i) *$\mathfrak{M}$ is elementary equivalent to $\mathfrak{N}$*

  (ii) *f is an isomorphism from $\mathsf{Cs}_\alpha^{\mathfrak{M}}$ onto $\mathsf{Cs}_\alpha^{\mathfrak{N}}$.*

(b) *The following conditions are equivalent:*

  (i) *g is an isomorphism of $\mathfrak{M}$ onto $\mathfrak{N}$*

  (ii) *$f = \widetilde{g}$ is a base-isomorphism from $\mathsf{Cs}_\alpha^{\mathfrak{M}}$ onto $\mathsf{Cs}_\alpha^{\mathfrak{N}}$.*

(c) *The following conditions are equivalent:*

  (i) *$\mathfrak{N}$ is elementary substructure of $\mathfrak{M}$*

  (ii) *$f = rl_{{}^\alpha N}$ is an ext-isomorphism from $\mathsf{Cs}_\alpha^{\mathfrak{M}}$ onto $\mathsf{Cs}_\alpha^{\mathfrak{N}}$.*

(d) *The following conditions are equivalent:*

(i) *g is an elementary embedding of $\mathfrak{N}$ into $\mathfrak{M}$.*

(ii) *$f = \widetilde{g}^{-1} \circ rl_{{}^\alpha g^* N}$ is an ext-base-isomorphism from $\mathsf{Cs}_\alpha^{\mathfrak{M}}$ onto $\mathsf{Cs}_\alpha^{\mathfrak{N}}$.*

This proposition shows the way we can turn from model theoretic language concerning the relations between models to the set algebraic one concerning the relations between the corresponding set algebras. In order to formulate the set algebraic versions (and possible generalizations) of elementary model theoretical propositions describing the way the structure of a model determines its relation to other ones, we need the set algebraic counterparts of the basic properties of models. Since, in this paper, we shall never consider Cs_α's as models (for the language of cylindric algebras) and thus we shall never apply model theoretical notions to them, there will not be any possibility to confuse the model theoretical properties of a Cs_α as a model with the cylindric set algebraic properties introduced below. Therefore, we generally use the model theoretical terms to denote their set algebraic counterparts (recall from [Hen-Mon-Tar,85] 2.6.28 the notion of the neat κ-reduct of a Cs_α $\mathfrak{A}$ and our convention that α is always an infinite ordinal):

Definition 1.1.2. Let $\mathfrak{A}$ be an arbitrary Cs_α with base U.

(i) $\mathfrak{A}$ is said to be *compact* if the intersection of any subset of A with the finite intersection property is not empty, that is, $\cap H \neq 0$ for any $H \subseteq A$ with the property that $\cap H' \neq 0$ whenever $H' \subseteq H$ is finite. For any cardinal κ, $\mathfrak{A}$ is *κ-compact* if $\mathfrak{Nr}_\kappa \mathfrak{A}$ is compact.

(ii) (a) For any $u \in U$, we set $k_u = \{x \in {}^\alpha U : x_0 = u\}$, say that k_u is a *hyperplane,* and set $\mathfrak{A}_X = \mathfrak{Sg}^{\mathfrak{Sb}^\alpha U}\big(A \cup \{k_u : u \in X\}\big)$ for any $X \subseteq U$.

(b) Let κ be an arbitrary cardinal. We say that a $\mathfrak{A}$ is *ordinarily κ-saturated* if $\mathfrak{Nr}_1 \mathfrak{A}_X$ is compact for any $X \subseteq U$ whenever $|X| < \kappa$. $\mathfrak{A}$ is called *ordinarily saturated* if $\mathfrak{A}$ is ordinarily $|U|$-saturated.

(iii) Let κ be a cardinal and $K \subseteq \mathsf{Cs}_\alpha$. $\mathfrak{A}$ is *κ-universal over K* if every isomorphism from $\mathfrak{A}$ onto an element $\mathfrak{B}$ of K with base of power $< \kappa$ is an ext-base-isomorphism.

(iv) Let κ be a cardinal. A Cs_α with base U is *κ-homogeneous* if for any $W \subseteq U$, $u \in U$, $f \in {}^W U$ and $g \in Is(\mathfrak{A}_W, \mathfrak{A}_{f^*W})$ such that $|W| < \kappa$,

$g \restriction A = Id \restriction A$ and $gk_w = k_{fw}$ for any $w \in W$, there are a $t \in U$ and a $h \in Is(\mathfrak{A}_{W\cup\{u\}}, \mathfrak{A}_{f^*W\cup\{t\}})$ such that $h \restriction A_W = g$ and $hk_u = k_t$. $\mathfrak{A}$ is called *homogeneous* if $\mathfrak{A}$ is $|U|$-homogeneous.

Now, it is easy to see that the algebraic versions of the model theoretical notions are adequate in the following sense:

Proposition 1.1.3. *Let $\mathfrak{M}$ be an arbitrary first order model and let κ be a cardinal.*

(i) *If $\alpha \geq \kappa$, then $\mathfrak{M}$ is κ-compact*[2] *iff $\mathsf{Cs}_\alpha^{\mathfrak{M}}$ is κ-compact; in particular, $\mathfrak{M}$ is α-compact iff $\mathsf{Cs}_\alpha^{\mathfrak{M}}$ is compact.*

(ii) *$\mathfrak{M}$ is κ-saturated iff $\mathsf{Cs}_\alpha^{\mathfrak{M}}$ is ordinarily κ-saturated.*

(iii) *$\mathfrak{M}$ is κ-universal iff $\mathsf{Cs}_\alpha^{\mathfrak{M}}$ is κ-universal over $\mathsf{Cs}_\alpha^{\mathrm{reg}}$.*

(iv) *$\mathfrak{M}$ is κ-homogeneous iff $\mathsf{Cs}_\alpha^{\mathfrak{M}}$ is κ-homogeneous.*

Now, the relation between first order models and the elements of $\mathsf{Cs}_\alpha^{\mathrm{reg}} \cap \mathsf{Lf}_\alpha$ makes it possible to formulate the algebraic versions of some elementary model theoretic results:

Theorem 1.1.4.

(a) *Let $\mathfrak{A}$ be an arbitrary $\mathsf{Cs}_\alpha^{\mathrm{reg}} \cap \mathsf{Lf}_\alpha$, and let κ be a cardinal.*

  (i) *If $\omega \leq \kappa \leq \alpha$, then $\mathfrak{A}$ is ordinarily κ-saturated iff for any subset X of the base of $\mathfrak{A}$, $\mathfrak{A}_X$ is κ-compact whenever $|X| < \kappa$.*

  (ii) *(1) In case $\kappa \leq \alpha$, $\mathfrak{A}$ is ordinarily κ-saturated iff $\mathfrak{A}$ is κ-compact and κ-homogeneous. (2) In case $\kappa \geq \alpha$, if $\mathfrak{A}$ is ordinarily κ-saturated, then $\mathfrak{A}$ is compact and κ-homogeneous.*

  (iii) *If $\mathfrak{A}$ is ordinarily κ-saturated, then $\mathfrak{A}$ is κ^+-universal over $\mathsf{Cs}_\alpha^{\mathrm{reg}}$.*

(b) (i) *Every isomorphism between two ordinarily saturated $\mathsf{Cs}_\alpha^{\mathrm{reg}} \cap \mathsf{Lf}_\alpha$'s is a base-isomorphism if the algebras concerned have bases of the same cardinality.*

---

[2]Since the compactness of models is not a widely used notion, we define it as follows: Let κ be an arbitrary ordinal. A model $\mathfrak{M}$ is **κ-compact** if considering it as a model for a language with a sequence of variables of length $\alpha \geq \kappa \cup \omega$, any set of formulas with free variables among the first κ ones is satisfiable in $\mathfrak{M}$ whenever it is finitely satisfiable in $\mathfrak{M}$. (Any set Σ of formulas is, of course, finitely satisfiable in a model if every finite subset of Σ is satisfiable in this model.)

(ii) *Every* $\mathsf{Cs}_\alpha^{\mathrm{reg}} \cap \mathsf{Lf}_\alpha$ *with a finite base is compact and ordinarily* κ*-saturated for any cardinal* κ.

(iii) *Every isomorphism from a* $\mathsf{Cs}_\alpha^{\mathrm{reg}} \cap \mathsf{Lf}_\alpha$ *with a finite base onto a* $\mathsf{Cs}_\alpha^{\mathrm{reg}}$ *is a base-isomorphism.*

(c) *Let* $\mathfrak{A}$ *be an arbitrary* $\mathsf{Cs}_\alpha^{\mathrm{reg}} \cap \mathsf{Lf}_\alpha$.

(i) *If* $\mathfrak{A}$ $|\alpha|^+$*-universal over* $\mathsf{Cs}_\alpha^{\mathrm{reg}}$, *then* $\mathfrak{A}$ *is compact, provided* $|A| \leq |\alpha|$.

(ii) *If* $\mathfrak{A}$ *is compact, then* $\mathfrak{A}$ *is* $|\alpha|^+$*-universal over* $\mathsf{Cs}_\alpha^{\mathrm{reg}}$.

(d) *Every* $\mathsf{Cs}_\alpha^{\mathrm{reg}} \cap \mathsf{Lf}_\alpha$ *is isomorphic to a compact one.*

(e) *Let* $\mathfrak{A}$ *be an arbitrary* $\mathsf{Cs}_\alpha^{\mathrm{reg}} \cap \mathsf{Lf}_\alpha$ *with base* U. $\mathfrak{A}$ *is homogeneous iff for every* $W, T \subseteq U$ *such that* $|W| = |T| < |U|$, *every* $f \in Is(\mathfrak{A}_W, \mathfrak{A}_T)$ *can be extended into a base-automorphism of* $\mathfrak{A}_U$.

(f) *Each of a pair of two isomorphic* $\mathsf{Cs}_\alpha^{\mathrm{reg}} \cap \mathsf{Lf}_\alpha$*'s is sub-base-isomorphic to a common* $\mathsf{Cs}_\alpha^{\mathrm{reg}} \cap \mathsf{Lf}_\alpha$.

There are two ways to obtain the proof of this theorem. One of them is simply translating the proof of the well-known elementary model theoretical results using the 'dictionary' provided by the Proposition 1.1.1 and Proposition 1.1.3 above, which describe the relations between the model theoretical and set algebraic notions. Nevertheless, the algebraic results can be obtained by purely algebraic means (that is, without having recourse to logical notions such as e.g. that of the first order language). We illustrate this latter method by a short proof that shows that every $\mathfrak{A} \in \mathsf{Cs}_\alpha^{\mathrm{reg}} \cap \mathsf{Lf}_\alpha$ with a finite base is κ-saturated for any κ:

Let the base of $\mathfrak{A}$ be U, and let $W \subseteq U$ be arbitrary. Suppose that $\cap H = 0$ for some $H \subseteq Nr_1 \mathfrak{A}_W$. Now, for any $u \in U$ there are a $y \in {}^\alpha U$ and a $Y_u \in H$ such that $y_0 = u$ and $y \notin Y_u$. Let $H' = \{Y_u : u \in U\}$. Then the finiteness of U implies that $H' \subseteq H$ is finite. On the other hand, by [Hen-Mon-Tar,85] 3.1.64, $\mathfrak{A}_W$ is regular. Therefore, for any $x \in {}^\alpha U$, $x \notin Y_{x_0}$. Consequently, $\cap H' = 0$, that is, H does *not* have the finite intersection property. This concludes the proof.

On the other hand, as we shall see, generally, the statements in Theorem 1.1.4 cannot straightforwardly be generalized to the distinguished subclasses of Cs_α properly containing $\mathsf{Cs}_\alpha^{\mathrm{reg}} \cap \mathsf{Lf}_\alpha$. We illustrate this fact by the result of Németi that solves Problem 3.2 of [Hen-Mon-Tar,85] as well as Problem II.3.11 of [Hen-Mon-Tar,85] negatively, showing that Theorem 1.1.4(f) cannot be generalized to $\mathsf{Cs}_\alpha^{\mathrm{reg}} \cap \mathsf{Dc}_\alpha$:

Theorem 1.1.5. *Let $\kappa > 1$. Then there are isomorphic $\mathfrak{A}, \mathfrak{B} \in \mathsf{Cs}_\alpha^{\mathrm{reg}} \cap \mathsf{Dc}_\alpha$ with the same base κ such that no $\mathsf{Cs}_\alpha^{\mathrm{reg}}$ is ext-base-isomorphic to both $\mathfrak{A}$ and $\mathfrak{B}$. Moreover, $\mathfrak{A}$ can be chosen to be locally countable dimensional, i.e. $(\forall x \in A)|\Delta x| \le \omega$.*

Proof. Let $\kappa > 1$, $H, L \subseteq \alpha$ with $|H| = |L| = \omega$, $\big|\alpha \smallsetminus (H \cup L)\big| \ge \omega$, $H \cap L = 0$. Let us introduce the following notation (recall that $\overline{u} = \langle u : \lambda \in \alpha \rangle$ for any u): $p = \overline{1}$, $q = (\overline{1} \upharpoonright L) \cup (\overline{0} \upharpoonright H)$, $t = (\overline{0} \upharpoonright L) \cup (\overline{1} \upharpoonright H)$, $V = {}^\alpha\kappa$, $K = L \cup H$, $P = \{f \in V : p \upharpoonright K \subseteq f\}$, $Q = \{f \in V : q \subseteq f\}$, $T = \{f \in V : t \subseteq f\}$, $E = \{f \in V : f_0 < 2\}$, $\mathfrak{C} = \mathfrak{Sb}\, V$, $\mathfrak{A} = \mathfrak{Sg}^{\mathfrak{C}}\{E, P, Q\}$ and $\mathfrak{B} = \mathfrak{Sg}^{\mathfrak{C}}\{E, P, T\}$. By [Hen-Mon-Tar,85] 3.1.63, $\mathfrak{A}$ and $\mathfrak{B}$ are regular, hence $\mathfrak{A}, \mathfrak{B} \in {}_\kappa\,\mathsf{Cs}_\alpha^{\mathrm{reg}} \cap \mathsf{Dc}_\alpha$.

First, we show that $\mathfrak{A} \cong \mathfrak{B}$. Let $V_q = \bigcup\big\{{}^\alpha\kappa^{(s)} : s \in Q\big\}$, $V_t = \bigcup\big\{{}^\alpha\kappa^{(s)} : s \in T\big\}$, $V_r = V \smallsetminus (V_q \cup V_t)$, $\mathfrak{C}_q = \mathfrak{Sb}\, V_q$, $\mathfrak{C}_t = \mathfrak{Sb}\, V_t$, $\mathfrak{C}_r = \mathfrak{Sb}\, V_r$. Then $\mathfrak{C} \cong \mathfrak{C}_q \times \mathfrak{C}_t \times \mathfrak{C}_r$ and indeed, by [Hen-Mon-Tar,85] 3.1.76, $h = \big\langle (x \cap V_q, x \cap V_t, x \cap V_r) : x \in C \big\rangle \in \mathcal{I}s(\mathfrak{C}_q \times \mathfrak{C}_t \times \mathfrak{C}_r)$. Clearly, $\mathfrak{C}_q$ is base-isomorphic to $\mathfrak{C}_t$; let $i \in \mathcal{I}s(\mathfrak{C}_q, \mathfrak{C}_t)$ be this base-isomorphism. Then $f = \big\langle (i^{-1}(b), i(a), c) : (a, b, c) \in C_q \times C_t \times C_r \big\rangle$ is an automorphism of $\mathfrak{C}_q \times \mathfrak{C}_t \times \mathfrak{C}_r$, thus $k = h^{-1} \circ f \circ h$ is an automorphism of $\mathfrak{C}$. It can be checked that $k(E) = E$, $k(P) = P$, and $k(Q) = T$, showing that $\mathfrak{A} = \mathfrak{Sg}^{(\mathfrak{C})}\{E, P, Q\} \cong \mathfrak{Sg}^{(\mathfrak{C})}\{E, P, T\} = \mathfrak{B}$.

Assume now that $\mathfrak{N} \in \mathsf{Cs}_\alpha^{\mathrm{reg}}$ is ext-base-isomorphic to both $\mathfrak{A}$ and $\mathfrak{B}$. Say, $f \in \mathcal{I}s(\mathfrak{N}, \mathfrak{A})$ and $h \in \mathcal{I}s(\mathfrak{N}, \mathfrak{B})$ are these two ext-base-isomorphisms. By the choice of f, there are W and $\mathfrak{M} \in \mathsf{Cs}_\alpha^{\mathrm{reg}}$ with $rl_{{}^\alpha W} \in \mathcal{I}s(\mathfrak{N}, \mathfrak{M})$ and $\mathfrak{M}$ base-isomorphic to $\mathfrak{A}$. We may assume that $\mathfrak{M} = \mathfrak{A}$. Let U be the base of $\mathfrak{N}$. Then $\kappa \subseteq U$ and $f = rl_{{}^\alpha\kappa} \in \mathcal{I}s(\mathfrak{N}, \mathfrak{A})$. There are $E^+, P^+, Q^+ \in N$ with $fE^+ = E$, $fP^+ = P$ and $fQ^+ = Q$. By the regularity of $\mathfrak{N}$ and by $\Delta(E^+) = \Delta(E) = \{0\}$, there is an $H \subseteq U$ such that $E^+ = \{g \in {}^\alpha U : g_0 \in H\}$. $E \cdot c_0(d_{01} \cdot E) \cdot c_0(d_{02} \cdot E) \cdot \overline{d}_{\{0,1,2\}} = 0$ implies that $|H| \le 2$, and it follows from $E = E^+ \cap {}^\alpha\kappa$ that $2 \subseteq H$. Thus $H = 2$. Therefore $E^+ = \{g \in {}^\alpha U : g_0 \in 2\}$. Further, $(\forall g \in P^+) g \upharpoonright K \in {}^K 2$ is implied by $(\forall i \in K) c_0(d_{0i} \cdot E) \supseteq P$ and $P = P^+ \cap {}^\alpha\kappa$, $(\forall i, j \in K) P \le dij$ implies that $P^+ = \{g \in {}^\alpha U : g \upharpoonright K \subseteq p\}$. Similarly, $Q^+ = \{g \in {}^\alpha U : q \subseteq g\}$. The structure of P^+ and $Q^+ \in N$ is similar to that of P and $Q \in A$ respectively. The crucial question is 'what are $hP^+, hQ^+ \in B$ like?' Can they be different from $P, T \in B$? Let $x = hP^+$. Then $x \in B$, $\Delta x = K$ and $x \le d_{ij}$ for all $i, j \in K$. Assume that $s \in x$ with $s \upharpoonright K \ne p \upharpoonright K$. Let $Q = {}^\alpha\kappa^{(s)}$. Then rl_Q is a homomorphism on $\mathfrak{B}$, since $\Delta Q = 0$. Further, $rl_Q P = rl_Q T = 0$, therefore $rl_Q^* \mathfrak{B} \in \mathsf{Lf}_\alpha$. Thus $\Delta(x \cap Q)$ is finite, contradicting $0 \ne x \cap Q \le d_{ij}$

for all $i, j \in K$. Therefore $(\forall s \in x) s \restriction K = p \restriction K$ and thus $x = P$ because $\mathfrak{B}$ is regular. The proof of $hQ^+ = T$ is completely analogous, we omit it. But then $hQ^+ = T$ and $hP^+ = P$ prove that h cannot be an ext-base-isomorphism since $(\forall i \in L)[q_i = p_i$ but $t_i \neq p_i]$. This proves that no $\mathsf{Cs}_\alpha^{\mathrm{reg}}$ is ext-base-isomorphic to both $\mathfrak{A}$ and $\mathfrak{B}$, which is what was to be proved. ■

Theorem 1.1.4 is a kind of a reference statement in the sense that the investigations described below are governed by the aim to find a class $K \subseteq \mathsf{Cs}_\alpha$ properly containing $\mathsf{Cs}_\alpha^{\mathrm{reg}} \cap \mathsf{Lf}_\alpha$ in such a way that the generalized version of Theorem 1.1.4 holds for the algebras in K.

2. Algebraic Generalizations of Saturatedness

Németi has introduced a notion corresponding to the model theoretical notion of saturatedness (cf. [Nem,90] Definitions 4 and 7, pp. 41 and 45, resp.) that is more of an algebraic flavour than that of the ordinary saturatedness (which is simply the literal translation of its model theoretical counterpart):

Definition 1.2.1. Let $\kappa \leq |\alpha|^+$ be a cardinal and $\mathfrak{A} \in \mathsf{Cs}_\alpha$.

(i) $\mathfrak{A}$ is *κ-saturating* if, for all $\beta < \kappa$, for every filter F of $\mathfrak{Nr}_\beta \mathfrak{A}$, and for all $i \in \alpha$,

$$C_i \left(\bigcap F \right) = \bigcap C_i^* F.$$

(ii) $\mathfrak{A}$ is *saturating* if $\mathfrak{A}$ is $|\alpha|^+$-saturating.

The proposition below shows that saturatingness is indeed a possible algebraic version of the notion of saturatedness, see [Nem,90] Theorems 1 and 3 (pp. 43 and 45, resp.):

Proposition 1.2.2. *Let $\mathfrak{M}$ be a model of an ordinary language and $\kappa \leq |\alpha|^+$. Then*

$\mathfrak{M}$ is κ-saturated iff $\mathsf{Cs}_\alpha^{\mathfrak{M}}$ is κ-saturating. In particular,

$\mathfrak{M}$ is $|\alpha|^+$-saturated iff $\mathsf{Cs}_\alpha^{(\mathfrak{M})}$ is saturating.

It follows from Proposition 1.2.2 that the statements in Theorem 1.1.4 concerning ordinarily $|\alpha|^+$-saturated algebras remain true if ordinarily $|\alpha|^+$-saturatedness is replaced by saturatingness in them. The facts to this effect are formulated in [Nem,90] Corollary 9 (p. 46). Among others, for example, every $\mathsf{Cs}_\alpha^{\text{reg}} \cap \mathsf{Lf}_\alpha$ with a finite base is saturating, and every saturating $\mathsf{Cs}_\alpha^{\text{reg}} \cap \mathsf{Lf}_\alpha$ is compact. On the other hand, the saturating algebras fail to retain their most important properties within the classes $\mathsf{Cs}_\alpha \cap \mathsf{Lf}_\alpha$ and $\mathsf{Cs}_\alpha^{\text{reg}} \cap \mathsf{Dc}_\alpha$ (cf. [Nem,90] Theorem 4, p. 48):

Proposition 1.2.3. *Let $1 < \kappa < \omega$ or $\kappa > \alpha$. Then*

(i) *there are saturating but not compact $\mathsf{Cs}_\alpha^{\text{reg}} \cap \mathsf{Dc}_\alpha$'s and $\mathsf{Cs}_\alpha \cap \mathsf{Lf}_\alpha$'s with base κ.*

(ii) (a) *there are two isomorphic but not base-isomorphic saturating $\mathsf{Cs}_\alpha^{\text{reg}} \cap \mathsf{Dc}_\alpha$'s with the same base κ.*

(b) *there are two isomorphic but not base-isomorphic saturating $\mathsf{Cs}_\alpha \cap \mathsf{Lf}_\alpha$'s with the same base κ.*

As a matter of fact, we are not better off if we examine the behaviour of ordinary saturatedness, since it is easy to find examples to show that *none* of the results of Theorem 1.1.4 concerning the properties of ordinary saturated algebras in $\mathsf{Cs}_\alpha^{\text{reg}} \cap \mathsf{Lf}_\alpha$ remains true if we change the class under consideration to $\mathsf{Cs}_\alpha^{\text{reg}} \cap \mathsf{Dc}_\alpha$ or $\mathsf{Cs}_\alpha \cap \mathsf{Lf}_\alpha$. This fact can be explained easily in an informal way. In fact, regularity and locally finiteness together guarantee that many properties of the whole algebra is determined by the behaviour of its neat 1-reduct, while in the absence of any of them, this connection vanishes; loosely speaking, generally, non-regular elements and those with infinite dimension sets have nothing to do with regular elements having finite dimension sets. These informal considerations explain that, among others, one of the most basic property of ordinary saturatedness, namely that it implies compactness (cf. Theorem 1.1.4(a)(ii)) does remain true within neither $\mathsf{Cs}_\alpha^{\text{reg}} \cap \mathsf{Dc}_\alpha$ nor $\mathsf{Cs}_\alpha \cap \mathsf{Lf}_\alpha$:

Proposition 1.2.4. *There are ordinarily saturated but non-compact algebras with bases of any power > 1 in both $\mathsf{Cs}_\alpha^{\text{reg}} \cap \mathsf{Dc}_\alpha$ and $\mathsf{Cs}_\alpha \cap \mathsf{Lf}_\alpha$.*

Proof. We shall first show that

$$\mathfrak{C} = \mathfrak{Sg}^{\mathfrak{BlSb}\,{}^{\alpha}U}\{k_w : w \in W\} \text{ is compact for any sets } U \text{ and } W \subseteq U \text{ whenever } U \sim W \neq 0. \tag{1.2.1}$$

Obviously, by ultrafilter theorem, it is enough to prove that $\cap F \neq 0$ for any ultrafilter F of $\mathfrak{C}$, so let F be such an ultrafilter. There are two cases. If

$k_a \in F$ for some $a \in W$, then, obviously, $x \in \cap F$ for any $x \in {}^\alpha U$ such that $x_0 = a$. Let us suppose that $-k_w \in F$ for all $w \in W$. Since $U \sim W \neq 0$, we can choose a $b \in U \sim W$. Let $q \in {}^\alpha U$ be arbitrary such that $q_0 = b$. We shall show that $q \in \cap F$. Let us introduce the following abbreviations: $K = \{k_w : w \in W\}$, $K' = \{-Z : Z \in K\}$ and let $X \in F$ be arbitrary. Then there is a finite system $\{Q_{ij}\}_{i\in n,\ j\in m(i)}$ of elements of $K \cup K'$ such that $X = \sum_{i\in n}\left(\prod_{j\in m(i)} Q_{ij}\right)$. F is an ultrafilter thus $X \in F$ implies that $\prod_{j\in m(k)} Q_{kj} \in F$ for some $k \in n$. Since $k_w \notin F$ for any $w \in W$, $Q_{kj} \in K'$ for every $j \in m(k)$. Consequently, $q \in \prod_{j\in m(k)} Q_{kj} \subseteq X$, since $q_0 = b \in U \sim W$.

Now, let $\kappa > 1$ be an arbitrary cardinal. Recall that $\overline{u} = \langle u : \lambda \in \alpha\rangle$ for any u.

1) First we consider the class $\mathsf{Cs}^{\mathrm{reg}}_\alpha \cap \mathsf{Dc}_\alpha$.

Let $\Gamma, \Delta \subseteq \alpha$ be such that $\Gamma\cup\Delta = \alpha$, $\Gamma\cap\Delta = 0$, $|\Gamma|\cap|\Delta| \geq \omega$, and $0 \in \Gamma$. Further, let $X_0 = \{x \in {}^\alpha\kappa : x \restriction \Gamma = \overline{0} \restriction \Gamma\}$ and $\mathfrak{A} = \mathfrak{Sg}^{\mathfrak{Sb}\,{}^\alpha\kappa}\{X_0, k_0\}$. Then $\mathfrak{A} \in \mathsf{Cs}^{\mathrm{reg}}_\alpha \cap \mathsf{Dc}_\alpha$ (cf. [Hen-Mon-Tar,85] 2.1.5(ii), 3.1.56, 3.1.63) and $\mathfrak{A}$ is not compact, since $H = \{k_0 \cdot \mathsf{D}_{0\lambda} \cdot -X_0 : \lambda \in \Gamma\}$ has the finite intersection property but $\cap H = 0$. In order to prove that $\mathfrak{A}$ is ordinarily saturated, first we show that

$$(1.2.2)\qquad Nr_1\mathfrak{A}_W \subseteq Sg^{\mathfrak{Bl}\mathfrak{Sb}\,{}^\alpha\kappa}\{k_w : w \in W\cup\{0\}\} \text{ for any } W \subseteq \kappa.$$

Let $X \in Nr_1\mathfrak{A}_W$. Then there is a finite $W' \subseteq W\cup\{0\}$ such that $0 \in W'$ and $X \in Nr_1\mathfrak{A}_{W'}$. Since $\mathfrak{A}$ is regular, there is a $T \subseteq \kappa$ such that $X = \{x \in {}^\alpha\kappa : x_0 \in T\}$. Let $\mathfrak{C}$ is the Boolean set algebra with universe $C = \mathfrak{Sb}\,\kappa = \{c : c \subseteq \kappa\}$. To prove (1.2.2), it is obviously enough to show that $T \in Sg^{\mathfrak{C}}\{\{w\} : w \in W'\}$. To this end, we first show that either $T \subseteq W'$ or $\kappa \sim W' \subseteq T$. Now, let us suppose that $T \nsubseteq W'$. Then there is a $u \in T$ such that $u \in \kappa \sim W'$. Let $v \in \kappa \sim W'$ be arbitrary and let φ be a one-one mapping of κ onto oneself such that $\varphi u = v$, $\varphi v = u$ and $\varphi w = w$ for any $w \in \kappa \sim \{u, v\}$. Now, $0 \notin \{u, v\}$ since $0 \in W'$. Let $G = \{X_0\}\cup\{k_w : w \in W'\}$. Since $\mathfrak{A}_{W'}$ is generated by G and $\widetilde{\varphi} \restriction G = Id \restriction G$, by [Hen-Mon-Tar,85] 0.2.14(iii), we get $\widetilde{\varphi} \restriction A_{W'} = Id \restriction A_{W'}$, that is, $\widetilde{\varphi}X = X$. Since $u \in T$, there is an $x \in X$ such that $x_0 = u$. Then $\varphi\circ x \in \widetilde{\varphi}X = X$, consequently $(\varphi\circ x)_0 = \varphi u = v \in T$. As $v \in \kappa \sim W'$ was arbitrary, we have proved that $\kappa \sim W' \subseteq T$ provided $T \nsubseteq W'$. This implies $T \in Sg^{\mathfrak{C}}\{\{w\} : w \in W'\}$ by the finiteness of W', which concludes the proof of (1.2.2).

Now, let $W \subseteq \kappa$ such that $|W| < \kappa$. Then $\mathfrak{Sg}^{\mathfrak{Bl}\mathfrak{Sb}\,{}^\alpha\kappa}\{k_w : w \in W \cup \{0\}\}$ is compact either by (1.2.1), if κ is infinite, or by its finiteness in case κ is finite. Thus (1.2.2) finishes the proof of 1).

2) Let us consider the class $\mathsf{Cs}_\alpha \cap \mathsf{Lf}_\alpha$.

Let $X_0 = {}^\alpha\kappa^{(\overline{0})}$ and $\mathfrak{B} = \mathfrak{Sg}^{\mathfrak{Sb}\,{}^\alpha\kappa} X_0$. Then $\mathfrak{B} \in \mathsf{Cs}_\alpha \cap \mathsf{Lf}_\alpha$ by [Hen-Mon-Tar,85] 2.1.5(i), since clearly $\Delta X_0 = 0$. Further, $\mathfrak{B}$ is not compact. In fact, $H = \{X_0 \cdot -\mathsf{D}_{01}\} \cup \{\mathsf{D}_{0\kappa} : \kappa < \alpha,\ \kappa \text{ is even}\} \cup \{\mathsf{D}_{1\lambda} : \lambda < \alpha,\ \lambda \text{ is odd}\}$ has the finite intersection property but $\cap H = 0$. In order to prove that $\mathfrak{B}$ is ordinarily saturated, first we note that, for any $W \subseteq \kappa$, $Nr_1\mathfrak{B}_W \subseteq Sg^{\mathfrak{Bl}\mathfrak{Sb}\,{}^\alpha\kappa}(\{k_w : w \in W\} \cup \{X_0\})$ by [Hen-Mon-Tar,85] 2.2.24, since $\mathfrak{B}$ is monadic generated. Therefore, using the notation $\mathfrak{D} = \mathfrak{Sg}^{\mathfrak{Bl}\mathfrak{Sb}\,{}^\alpha\kappa}\{k_w : w \in W\}$, every $X \in Nr_1\mathfrak{B}_W$ has the following form: $X = P \cdot X_0 + Q \cdot -X_0$ for some $P, Q \in D$. Consequently, for every $H \subseteq Nr_1\mathfrak{B}_W$, there are systems $\{P_i\}_{i\in I}$, $\{Q_i\}_{i\in I} \subseteq D$ such that $H = \{P_i \cdot X_0 + Q_i \cdot -X_0 : i \in I\}$. Now, let $W \subseteq \kappa$ be such that $|W| < \kappa$. If H has the finite intersection property, so does at least one of $\{P_i\}_{i\in I}$ and $\{Q_i\}_{i\in I}$. By symmetry, we may suppose that $\{P_i\}_{i\in I}$ has the finite intersection property. In this case, by (1.2.1), there is a $p \in {}^\alpha\kappa$ such that $p \in \bigcap_{i\in I} P_i$. Then $x \in \bigcap_{i\in I} P_i$ for every $x \in {}^\alpha\kappa$ with $x_0 = p_0$. Consequently, $\overline{0}^0_{p_0} \in \left(\bigcap_{i\in I} P_i\right) \cap X_0 \subseteq \cap H$, which concludes the second part of the proof. ■

If we would like as many results in Theorem 1.1.4 as possible to remain true within the classes $\mathsf{Cs}_\alpha \cap \mathsf{Lf}_\alpha$ and $\mathsf{Cs}^{\mathrm{reg}}_\alpha \cap \mathsf{Dc}_\alpha$, we should choose a stronger notion than both saturatingness and ordinary saturatedness as the algebraic version of the original model theoretical notion. Of course, it has to be a proper generalization giving back the original notion within $\mathsf{Cs}^{\mathrm{reg}}_\alpha \cap \mathsf{Lf}_\alpha$. Ordinary saturatedness lends itself readily to such a strengthening. Actually, Németi has introduced the following natural generalization of the original notion ([Nem,90] Def. 31, p. 72):

Definition 1.2.5. For any cardinality κ, $\mathfrak{A} \in \mathsf{Cs}_\alpha$ with base U is *κ-hereditarily compact* if $\mathfrak{A}_T$ is β-compact for any $\beta < \kappa$ and $T \subseteq U$ such that $|T| < \kappa$.

In view of Proposition 1.2.2, the proposition below show that this notion is also a possible algebraic generalization of saturatedness of ordinary models (cf. [Nem,90] Fact 34, p. 72):

Proposition 1.2.6. *Let $\omega \le \kappa \le |\alpha|^+$ and $\mathfrak{A} \in \mathsf{Cs}^{\mathrm{reg}}_\alpha \cap \mathsf{Lf}_\alpha$. Then $\mathfrak{A}$ is κ-saturating iff $\mathfrak{A}$ is κ-hereditarily compact.*

On the other hand, in the notion of κ-hereditarily compactness, κ plays a double role. Actually, it sets a limit for both the cardinality of extension and for the dimension sets of elements of the algebra whose compactness is considered. Therefore, using this notion, we cannot examine separately the role played by two completely unrelated components constituting this property. Moreover, clearly, if $\kappa > |\alpha|$, then the limitation concerning the size of dimension sets become vacuous and, as the results below will show, if $\kappa \leq |\alpha|$, then *only* the limit for the size of the extension is relevant. Consequently, it is better to split the notion in question into two parts at the cardinal $|\alpha|$. (It will turn out that $|\alpha|$ is indeed the crucial cardinal in those respects that are relevant for our investigations.)

Definition 1.2.7.

(i) For some cardinality κ, $\mathfrak{A} \in \mathsf{Cs}_\alpha$ with base U is *weakly κ-saturated* if $\mathfrak{A}_T$ is β-compact for any $\beta < |\alpha|$ and $T \subseteq U$ such that $|T| < \kappa$. $\mathfrak{A}$ is *weakly saturated* if it is weakly $|U|$-saturated.

(ii) For some cardinality κ, $\mathfrak{A} \in \mathsf{Cs}_\alpha$ with base U is *κ-saturated* if $\mathfrak{A}_T$ is compact for any $T \subseteq U$ such that $|T| < \kappa$. $\mathfrak{A}$ is *saturated* if it is $|U|$-saturated.

Obviously, by definition, κ-hereditarily compactness is implied by weakly κ-saturatedness for any $\kappa \leq |\alpha|$ and by κ-saturatedness for any κ. So, for every $\kappa \leq |\alpha|$, any of these two newly defined notions at least is not weaker than Németi's original notion. Still, as we shall see below, even these notions are not strong enough to make true any part of our reference statement (Theorem 1.1.4) with $\mathsf{Cs}_\alpha^{\mathrm{reg}} \cap \mathsf{Dc}_\alpha$ or $\mathsf{Cs}_\alpha \cap \mathsf{Lf}_\alpha$ in place of $\mathsf{Cs}_\alpha^{\mathrm{reg}} \cap \mathsf{Lf}_\alpha$. On the other hand, unlike the notion of κ-hereditarily compactness, that of κ-saturatedness makes it possible to reveal interesting connections between apparently unrelated structural properties of Cs_α's. Actually, according to the nice picture exhibited by the theorem below, every essential change in the cardinality of the parameter of saturatedness brings about substantial structural consequences. We present the simple proofs of the universal claims, the existential ones are witnessed by the constructions that Theorems 1.2.9, 1.2.10, and 1.2.12 below are based on. We say that $\mathfrak{A} \in \mathsf{Cs}_\alpha$ is *non-trivial,* if $|A| > 2$.

Theorem 1.2.8.

(i) (a) *There is a non-trivial* 1*-saturated non-simple* $\mathsf{Cs}_\alpha \cap \mathsf{Lf}_\alpha$.

(b) *Every non-trivial* 2*-saturated* $\mathsf{Cs}_\alpha \cap \mathsf{Lf}_\alpha$ *is simple.*

(ii) (a) *There is a k-saturated non-regular* $\mathsf{Cs}_\alpha \cap \mathsf{Lf}_\alpha$ *for any* $k \in \omega$.

(b) *Every* ω*-saturated* $\mathsf{Cs}_\alpha \cap \mathsf{Lf}_\alpha$ *is regular.*

(iii) (a) *There is a non-trivial* $|\alpha|$*-saturated non-simple* $\mathsf{Cs}_\alpha^{\mathrm{reg}} \cap \mathsf{Dc}_\alpha$.

(b) *Every non-trivial* $|\alpha|^+$*-saturated* Cs_α *is simple and regular.*

Proofs of the second parts. First of all, if $\mathfrak{A} \in \mathsf{Cs}_\alpha$ is non-trivial, then its base λ is of cardinality > 1. Actually, by [Hen-Mon-Tar,85] I.1.5(i) and (iii), if $\lambda = 0$, then $A = \{0\}$, and if $\lambda = 1 = \{0\}$, then $1^{\mathfrak{A}} = \{\overline{0}\}$ (recall that $\overline{0} = \langle 0 : \lambda \in \alpha \rangle$), and hence $A = \{0, 1^{\mathfrak{A}}\}$. We shall need the following simple fact:

Fact. Let $\mathfrak{A} \in \mathsf{Cs}_\alpha$. If there is an $X \in A$ such that $0 < X < 1$, $\Delta X = 0$, then $\mathfrak{A}$ is not 2-saturated.

Proof (of the fact). We may suppose that the base of $\mathfrak{A}$ is λ. Then $\lambda > 1$, otherwise no $X \in A$ satisfies the given conditions. Further,

$$(1.2.3) \qquad {}^{\alpha}\lambda^{(x)} \subseteq -X \quad \text{for any} \quad x \in -X.$$

For, let $x \in -X$, $y \in {}^{\alpha}\lambda^{(x)}$ be arbitrary, and $\Gamma = \{\kappa < \alpha : y_\kappa \neq x_\kappa\}$. Then $|\Gamma| < \omega$ and, by [Hen-Mon-Tar,85] 1.2.12(i), $\mathsf{C}_{(\Gamma)} - X = -X$. Thus $\{x\} \subseteq -X$ implies $y \in \mathsf{C}_{(\Gamma)}\{x\} \subseteq \mathsf{C}_{(\Gamma)} - X = -X$.

Clearly, either $\overline{0} \in X$ or $\overline{0} \in -X$. The other case being completely analogous, we may suppose that $\overline{0} \in X$. Now, let $H = \{-X\} \cup \{\mathsf{D}_{0\kappa} : \kappa < \alpha\} \cup \{k_0\}$. By (1.2.3), H has the finite intersection property: for any $x \in -X \neq 0$ and any finite $\Gamma \subseteq \alpha$, $x(\Gamma \cup \{0\}/\overline{0}) \in {}^{\alpha}\lambda^{(x)} \cap \{\mathsf{D}_{0\kappa} : \kappa \in \Gamma\} \cap \{k_0\}$. On the other hand, $\cap H = 0$, since $\cap\{\mathsf{D}_{0\kappa} : \kappa < \alpha\} \cap \{k_0\} = \{\overline{0}\}$ and $\overline{0} \notin -X$. This means that $\mathfrak{A}_{\{0\}}$ is not compact, i.e. $\mathfrak{A}$ is not 2-saturated, which completes the proof of the fact.

Now, the proof of (i)(b) is immediate. Actually, if $\mathfrak{A} \in \mathsf{Cs}_\alpha \cap \mathsf{Lf}_\alpha$ is an arbitrary non-trivial non-simple algebra, then, by [Hen-Mon-Tar,85] 2.3.14, there is an $X \in A$ such that $0 < X < 1$, $\Delta X = 0$. Therefore, by the above fact, $\mathfrak{A}$ is not 2-saturated.

As far as the proof of (ii)(b) is concerned, we argue by contradiction. Let $\mathfrak{A} \in \mathsf{Cs}_\alpha \cap \mathsf{Lf}_\alpha$ be an arbitrary non-regular algebra. We shall show that there is a finite $T \subseteq \lambda$ such that $\mathfrak{A}_T$ is not 2-saturated. By the above fact, it is enough to show that there is an $Y \in A_T$ such that $0 < Y < 1$, $\Delta Y = 0$.

It follows from the non-regularity of $\mathfrak{A}$ that there is an $X \in A$, $x \in X$, $y \notin X$ such that $x \restriction \Delta X = y \restriction \Delta X$. Now, $|\Delta X| < \omega$, since $\mathfrak{A} \in \mathsf{Lf}_\alpha$. Let

$Z = X \cdot \bigcap \{ \mathsf{S}^0_\kappa k_{x_\kappa} : \kappa \in \Delta X \}$ and $Y = \mathsf{C}_{(\Delta X)} Z$. On the one hand, $Y > 0$, since $x \in X$ implies that $x \in Z$, i.e. $Z \neq 0$. On the other hand, $Y \neq 1$. To see this, we show that $y \notin Y$. In fact, $y \in Y$ implies the existence of a $z \in Z$ such that $z \upharpoonright (\alpha \sim \Delta X) = y \upharpoonright (\alpha \sim \Delta X)$, from which, in turn, we can infer that $y = z$, since, as a consequence of $z \in Z$, we have $z \upharpoonright \Delta X = x \upharpoonright \Delta X = y \upharpoonright \Delta X$. Thus $y \in Z \subseteq X$ contradicting the original assumption concerning y. Consequently, $0 < Y < 1$. Moreover, $\Delta Y = 0$, since, by [Hen-Mon-Tar,85] 1.6.6, $\Delta Z \subseteq \Delta X$, thus we are done.

Finally, in order to prove (iii)(b), let us suppose that $\mathfrak{A}$ is a non-trivial $|\alpha|^+$-saturated Cs_α with base U. First we show that

(1.2.4) for any $X \in A$ and $x \in X$, there is a finite $\Gamma \subseteq \alpha$ such that

$$y \upharpoonright \Gamma = x \upharpoonright \Gamma \text{ implies } y \in X \text{ for any } y \in {}^\alpha U.$$

Let us suppose that, on the contrary, there are $X \in A$ and $x \in X$ such that, for any $\Gamma \subseteq_\omega \alpha$, there is a $y_\Gamma \notin X$ such that $\Gamma \upharpoonright y_\Gamma = \Gamma \upharpoonright x$. Let $H = \{-X\} \cup \{ \mathsf{S}^0_\kappa k_{x_\kappa} : \kappa < \alpha \}$. Then H has the finite intersection property, since $y_{\Gamma'} \in -X \cap \bigcap_{\kappa \in \Gamma'} \mathsf{S}^0_\kappa k_{x_\kappa}$ for any $\Gamma' \subseteq_\omega \alpha$. On the other hand, $\cap H = -X \cap \{x\}$, and $x \notin -X$ implies $\cap H = 0$.

All that remains to do is to show that every Cs_α satisfying the condition (1.2.4) is simple and regular. First, we shall use [Hen-Mon-Tar,85] 2.3.14, according to which, in the case of non-trivial algebras, simplicity can be inferred from the fact that, for any $X \in A$, $X \neq 0$, there is a $\Gamma \subseteq_\omega \alpha$ such that $\mathsf{C}_{(\Gamma)} X = 1$. This condition will be shown to be a consequence of (1.2.4). Actually, let $x \in X$, $q \in {}^\alpha U$ be arbitrary. Then there is a $\Gamma \subseteq_\omega \alpha$ satisfying (1.2.4), i.e., $y \upharpoonright \Gamma = x \upharpoonright \Gamma$ implies $y \in X$ for any $y \in {}^\alpha U$. But $q[\Gamma/x] \upharpoonright \Gamma = x \upharpoonright \Gamma$, so $q[\Gamma/x] \in X$, from which, in turn, we can infer that $q \in \mathsf{C}_{(\Gamma)} X$. As far as regularity concerned, let $X \in A$, $x \in X$, and $q \in {}^\alpha U$ be arbitrary such that $q \upharpoonright \Delta X = x \upharpoonright \Delta X$, and let again $\Gamma \subseteq_\omega \alpha$ be satisfying (1.2.4), i.e., for any $y \in {}^\alpha U$, if $y \upharpoonright \Gamma = x \upharpoonright \Gamma$, then $y \in X$. Let $p = q[\Gamma \sim \Delta X/x]$. Then $p \upharpoonright \Gamma = x \upharpoonright \Gamma$, therefore $p \in X$, so $q = p[\Gamma \sim \Delta X/q] \in \mathsf{C}_{(\Gamma \sim \Delta X)} X = X$, which completes the proof. ■

Now, let us turn to the question as to whether saturated elements of the classes $\mathsf{Cs}_\alpha \cap \mathsf{Lf}_\alpha$ and $\mathsf{Cs}^{\text{reg}}_\alpha \cap \mathsf{Dc}_\alpha$ preserve the basic properties of ordinary saturated models. The theorems below show that, as far as $\mathsf{Cs}_\alpha \cap \mathsf{Lf}_\alpha$ is concerned, even supplemented either by compactness or simplicity, weak-saturatedness is not enough:

Theorem 1.2.9.

(i) *Let $1 < \kappa < \omega$ or $|\alpha| \leq \kappa$. Then there are two isomorphic but not base-isomorphic weakly saturated and compact $\mathsf{Cs}_\alpha \cap \mathsf{Lf}_\alpha$'s with the same base κ.*

(ii) *Let $1 < \kappa < \omega$ and $\mu = \kappa$ or else $|\alpha| \leq \kappa$ and $\omega \leq \mu \leq \kappa$. Then there are a weakly saturated and compact $\mathfrak{A} \in \mathsf{Cs}_\alpha \cap \mathsf{Lf}_\alpha$ with base κ and a $\mathfrak{B} \in \mathsf{Cs}_\alpha$ with base μ such that $\mathfrak{A}$ is isomorphic but not ext-base-isomorphic to $\mathfrak{B}$.*

(iii) *Let $2 < \kappa < \omega$ or $|\alpha| \leq \kappa$. Then there is a weakly saturated but non-compact and not homogeneous $\mathsf{Cs}_\alpha \cap \mathsf{Lf}_\alpha$ with base κ.*

(iv) *For any finite $n > 1$, there is a not weakly saturated and non-compact $\mathsf{Cs}_\alpha \cap \mathsf{Lf}_\alpha$ with base n.*

Theorem 1.2.10. *Let $1 < \kappa < \omega$ or $|\alpha| \leq \kappa$ and let $k \in \omega$, $1 < k < \kappa$. Then there are two isomorphic but not base-isomorphic weakly saturated and k-saturated simple $\mathsf{Cs}_\alpha \cap \mathsf{Lf}_\alpha$'s with the same base κ.*

In the light of Proposition 1.2.8(ii)(b), the construction this theorem is based on cannot straightforwardly be improved, since ω-saturatedness leads us back to $\mathsf{Cs}_\alpha^{\mathrm{reg}} \cap \mathsf{Lf}_\alpha$. Informally, this means that, within Lf_α, only algebras corresponding to ordinary models can be 'large'. In order to formulate this fact in model theoretical terms, let us recall the notion of infinitary models from [Hen-Mon-Tar,85] Part II p. 152 and [Nem,90] Section 3.3. These are models of languages that, along ordinary relation symbols, may contain others whose rank (the length of the sequences of variables to which they are applied) is infinite. Now, Proposition 1.2.8(ii)(b) means that, among infinitary models with the property that – from the point of view of satisfiability in the model concerned – every formula is equivalent to its closure with respect to any but finitely many variables, only the ordinary models can be saturated. Consequently, for infinite κ's, the effects exerted by κ-saturatedness can only be examined in our other basic class, $\mathsf{Cs}_\alpha^{\mathrm{reg}} \cap \mathsf{Dc}_\alpha$:

Theorem 1.2.11.

(i) *Let $\kappa \geq |\alpha|$. Then there are two isomorphic but not base-isomorphic weakly saturated and $|\alpha|$-saturated $\mathsf{Cs}_\alpha^{\mathrm{reg}} \cap \mathsf{Dc}_\alpha$'s with the same base κ.*

(ii) *Let $\kappa \leq |\alpha|$. Then there is a weakly saturated and κ-saturated $\mathfrak{A} \in \mathsf{Cs}_\alpha^{\mathrm{reg}} \cap \mathsf{Dc}_\alpha$ which is not κ^+-universal over $\mathsf{Cs}_\alpha^{\mathrm{reg}}$.*

(iii) *Let $2 < \kappa < \omega$ or $|\alpha| \leq \kappa$. Then there is a weakly saturated but non-compact and not homogeneous $\mathsf{Cs}_\alpha^{\mathrm{reg}} \cap \mathsf{Dc}_\alpha$ with base κ.*

(iv) *For any finite $n > 1$, there is a not weakly saturated $\mathsf{Cs}_\alpha^{\mathrm{reg}} \cap \mathsf{Dc}_\alpha$ with base n.*

Together with the results in the next section, which, among others, show that $|\alpha|^+$-saturated Cs_α's retain some important properties of algebras corresponding to ordinal models, the first two items of Theorem 1.2.11 demonstrates that, as far as κ-saturatedness is concerned, the salient cardinal is $|\alpha|$.

It is worth noting the previous three theorems add new facts to the group of already known results concerning the possibility of generalizing the algebraic version of uniqueness of finite models (see [Bir,89], [Hen-Mon-Tar,85] 3.1.38(1), [Hen-Mon-Tar,85] I.3.11 (p. 43) and 3.5(iv) (p. 162), [Nem,90] Theorem 4(iv) and (v) (p. 48), and [Ser,93]).

Finally, we would like to supplement the negative results in this section concerning universality, homogeneity and uniqueness of saturated algebras by those related to compact representability within the classes under investigation, making the 'negative version' of our reference statement (Theorem 1.1.4) more or less complete. Németi has proved that every Cs_α with an infinite base is isomorphic to a compact one (cf. [Nem,90] Theorem 5(ii) (p. 49, the proof is on p. 55.). The theorem below shows that we cannot expect more than that. Moreover, interestingly enough, both the statements (ii) and (iii) below and their proofs exhibit some kind of duality between the two classes concerned.

Theorem 1.2.12.

(i) *Let $\kappa \geq 2$. Then there is an $\mathfrak{A} \in \mathsf{Cs}_\alpha^{\mathrm{reg}} \cap \mathsf{Dc}_\alpha$ with base κ such that no compact $\mathsf{Cs}_\alpha^{\mathrm{reg}}$ is isomorphic to $\mathfrak{A}$.*

(ii) *Let $n \geq 2$ be finite. Then there is an $\mathfrak{A} \in \mathsf{Cs}_\alpha \cap \mathsf{Lf}_\alpha$ with base n such that no compact Cs_α is isomorphic to $\mathfrak{A}$.*

(iii) *Let $n \geq 2$ be finite. Then there is an $\mathfrak{A} \in \mathsf{Cs}_\alpha^{\mathrm{reg}} \cap \mathsf{Dc}_\alpha$ with base n such that no compact Cs_α is isomorphic to $\mathfrak{A}$.*

(iv) *There is a simple* $\mathfrak{A} \in \mathsf{Cs}_\alpha^{\mathrm{reg}} \cap \mathsf{Dc}_\alpha$ *with base* 2 *such that no compact* Cs_α *is isomorphic to* $\mathfrak{A}$.

Proof of (ii) **and** (iii)**.** Let $n \in \omega$, $n > 1$ be arbitrary. Recall the notation $\overline{u} = \langle u : \lambda \in \alpha \rangle$.

1) First we prove (ii). Let $Z_k = {}^\alpha n^{(\overline{k})}$ for any $k \in n$, $Z_n = {}^\alpha n \sim \bigcup_{k \in n} Z_k$ and $\mathfrak{A} = \mathfrak{Sg}^{\mathfrak{Sb}\,{}^\alpha n}\{Z_k : k \le n\}$. Then $\mathfrak{A} \in \mathsf{Cs}_\alpha \cap \mathsf{Lf}_\alpha$ since $\Delta Z_k = 0$ for all $k \le n$. We shall prove that $\mathfrak{A}$ is an algebra satisfying the conditions of (ii). Let $\mathfrak{B} \in \mathsf{Cs}_\alpha$ and $f \in Is(\mathfrak{A}, \mathfrak{B})$ be arbitrary. We can suppose that the base of $\mathfrak{B}$ is n (cf. [Hen-Mon-Tar,85] I.3.3). Let $H_k = \{f Z_k \cap \mathsf{D}_{0\kappa} : \kappa < \alpha\} \subseteq B$ for any $k \le n$. Clearly, every H_k has the finite intersection property, since $\Delta f Z_k = \Delta Z_k = 0$. On the other hand, $\cap H_k \subseteq \cap \mathsf{D}_{0\kappa} = \{\overline{k} : k < n\}$, $\big|\{\overline{k} : k < n\}\big| = n < n + 1 = \big|\{H_k : k \le n\}\big|$ together with the fact that the sets H_k are pairwise disjoint imply that there must be a $k \le n$ such that $\cap H_k = 0$.

2) To prove (iii), let $\Gamma, \Delta \subseteq \alpha$ be such that $\Gamma \cup \Delta = \alpha$, $\Gamma \cap \Delta = 0$, $|\Gamma| \cap |\Delta| \ge \omega$, and $0 \in \Gamma$. Further, for any $k \in n$, let $Z_k = \{x \in {}^\alpha n : x \restriction \Gamma = \overline{k} \restriction \Gamma\}$ and $Z_n = {}^\alpha n \sim \bigcup_{k \in n} Z_k$. Now we show that $\mathfrak{A} = \mathfrak{Sg}^{\mathfrak{Sb}\,{}^\alpha n}\{Z_k : k \le n\}$ is an algebra witnessing the truth of (iii). In fact, $\mathfrak{A} \in \mathsf{Cs}_\alpha^{\mathrm{reg}} \cap \mathsf{Dc}_\alpha$ by [Hen-Mon-Tar,85] 3.1.63 and 2.1.5(ii), while the proof of the fact that $\mathfrak{A}$ is not isomorphic to any compact Cs_α is completely analogous to that appearing in 1). Indeed, let $\mathfrak{B} \in \mathsf{Cs}_\alpha$ and $f \in Is(\mathfrak{A}, \mathfrak{B})$. The base of $\mathfrak{B}$ can be chosen to be n. Let $H_k = \{f Z_k \cap \mathsf{D}_{0\kappa} : \kappa < \alpha\} \subseteq B$ for any $k \le n$. Clearly, every H_k has the finite intersection property since it inherits this property from $f^{-1*} H_k$. Further, $\bigcap H_k \subseteq \bigcap_{\kappa \in \alpha} \mathsf{D}_{0\kappa} = \{\overline{k} : k < n\}$, $\big|\{\overline{k} : k < n\}\big| = n < n + 1 = \big|\{H_k : k \le n\}\big|$. Since the sets H_k are pairwise disjoint, there must be a H_k with an empty intersection.

In the proof of Theorem 1.2.8(iii)(b), we saw that $|\alpha|^+$-saturatedness implies a structural property (cf. (1.2.4) in the proof of the theorem) showing close resemblance to regularity. This property defines a new and important subclass of Cs_α that will turn out to be a fairly faithful algebraic generalization of ordinary models. ■

3. The class $\mathsf{Cs}_\alpha^{\mathrm{pfr}}$

3.1. Basic properties

The two properties characterizing those cylindric set algebras that correspond to ordinary models (locally finite-dimensionality and regularity) are related closely:

Fact 1.3.1. *Let $\mathfrak{A}$ be a Cs_α with unit element V, and let $X \in A$. Then the following conditions are equivalent:*

(i) *X is regular and its dimension set is finite*

(ii) *there is a finite $\Gamma \subseteq \alpha$ such that, for every $p \in X$,*

$$q \restriction \Gamma = p \restriction \Gamma \;\; \textit{implies} \;\; q \in X \;\; \textit{for every} \;\; q \in V.$$

The second condition in this proposition can be considered as a global variant of the following (local or pointwise) property, which leads to a natural extension of the class of algebraic versions of ordinary models:

Definition 1.3.2. Let $\mathfrak{A}$ be a Cs_α with unit element V, and let $X \in A$. We say that X is *pointwise finitely regular* if, for every $p \in X$, there is a finite $\Gamma \subseteq \alpha$ such that

$$q \restriction \Gamma = p \restriction \Gamma \;\; \text{implies} \;\; q \in X \;\; \text{for every} \;\; q \in V.$$

An $\mathfrak{A} \in \mathsf{Cs}_\alpha$ is *pointwise finitely regular* (or shortly *pf-regular*) if every $X \in A$ is pointwise finitely regular. The class of all pointwise regular Cs_α's is denoted by $\mathsf{Cs}_\alpha^{\mathrm{pfr}}$.

Let U be any set, and let us call an $X \in \mathfrak{Sb}\,{}^\alpha U$ to be *finite dimensional* if its dimension set is finite. It is obvious from the definition that an $X \in \mathfrak{Sb}\,{}^\alpha U$ is pf-regular iff it is a union of regular and finite dimensional elements of $\mathfrak{Sb}\,{}^\alpha U$. Now, based on its proof, the last item of Theorem 1.2.8 can be improved:

Theorem 1.3.3. *Every $|\alpha|^+$-saturated Cs_α is pf-regular.*

Let us look at some basic properties of $\mathsf{Cs}_\alpha^{\mathrm{pfr}}$. The first item of the proposition below is trivial, the last two were established in the proof of Theorem 1.2.8(iii)(b):

Proposition 1.3.4.

(i) $\mathsf{Cs}_\alpha^{\mathrm{reg}} \cap \mathsf{Lf}_\alpha \subseteq \mathsf{Cs}_\alpha^{\mathrm{pfr}}$

(ii) $\mathsf{Cs}_\alpha^{\mathrm{pfr}} \subseteq \mathsf{Cs}_\alpha^{\mathrm{reg}}$

(iii) *Every non-trivial pf-regular* Cs_α *is simple.*

In the case of algebras with finite base, the inclusion in the first item of this proposition can be replaced by equality. We give two different proofs of the fact that pf-regular Cs_α's with finite base are locally finite-dimensional. The first proof, which is due to András Simon (cf. [Ser-Sim,98]), can be carried out for any dimension and sheds light on the interesting possibility to look at algebras in $\mathsf{Cs}_\alpha^{\mathrm{pfr}}$ from a topological point of view. Nevertheless, this general proof, being built on Tychonoff's theorem, makes use of the Axiom of Choice and, consequently, is highly non-constructive. The second proof, on the other hand, is an elementary constructive one, which, however, being based on König's lemma, works only for $\alpha = \omega$.

Theorem 1.3.5. *Every* $\mathsf{Cs}_\alpha^{\mathrm{pfr}}$ *with a finite base is locally finite-dimensional.*

Proof 1. Let $\mathfrak{A} \in \mathsf{Cs}_\alpha^{\mathrm{pfr}}$ with a non-empty finite base U and $S = \{\mathsf{S}_\kappa^0 k_u : \kappa < \alpha,\ u \in U\}$, i.e., the set of all hyperplanes and 'translated' hyperplanes. Then S is a sub-base for the product topology of the αth topological power ${}^\alpha\mathbf{U}$ of the topological space $\mathbf{U} = \langle U, \tau\rangle$, where τ is the discrete topology. But the set of all unions of finite intersections of the elements of S, which is the product topology of ${}^\alpha\mathbf{U}$, is just the set of all pf-regular elements of $\mathfrak{Sb}\,{}^\alpha U$. Since the complement of every element of some pf-regular Cs_α is also pf-regular, the elements of A are all clopen.

Now, being finite, $\mathbf{U}$ is compact, therefore, by Tychonoff's theorem, ${}^\alpha\mathbf{U}$ is also compact. Consequently, being a closed set of a compact space, any element of A is compact, therefore, as it is a union of the elements of the base defined by S (i.e., that of finite intersections of elements of S), it should be the union of *finitely many* base elements, or in our algebraic terms, any element of a pf-regular algebra as the union of regular and finite dimensional elements can be obtained as a *finite* union of such elements, so it is itself regular and finite dimensional (cf. [Hen-Mon-Tar,85] 1.6.5 and 3.1.64), which was to be proved.[3] ■

[3]Since the elements of a $\mathsf{Cs}_\alpha^{\mathrm{pfr}}$ are all closed, it also follows from the compactness of ${}^\alpha\mathbf{U}$ that the intersection of any subset of the universe of a $\mathsf{Cs}_\alpha^{\mathrm{pfr}}$ with the finite intersection property is not empty. This simple argument immediately yields that every $\mathsf{Cs}_\alpha^{\mathrm{pfr}}$ with a finite base (and hence every $\mathsf{Cs}_\alpha^{\mathrm{reg}} \cap \mathsf{Lf}_\alpha$ with a finite base) is compact.

Proof 2 for $\alpha = \omega$. First of all, let us introduce the following notation for any set U, $X \in \mathfrak{Sb}\,{}^{\alpha}U$ and $x \in X$:

$$\Gamma(x, X) = \big\{ H \subseteq_\omega \alpha : \text{ for every } \ q \in {}^{\alpha}U,\ q \restriction H = x \restriction H \ \text{ implies } \ q \in X \big\}.$$

Let $\mathfrak{A} \in \mathsf{Cs}^{\mathrm{pfr}}_{\alpha}$ with a non-empty base $\kappa < \omega$. Assume that, contrary to the claim, there is an $X \in A$ such that $|\Delta X| \geq \omega$. Further, let $Y = \langle Y_n : n \in \omega \rangle$ and $G = \langle G_n : n \in \omega \rangle$ be defined as follows: $Y_0 = 0$, $Y_n = \big\{ x \in X \sim \bigcup_{k \in n} Y_k : n \in \Gamma(x, X) \big\}$ for any $n > 0$, $G_n = \{ x \restriction n : x \in Y_n \}$ for any $n \geq 0$, and let $G_\omega = \bigcup_{n \in \omega} G_n$. Since X is pf-regular,

$$X = \bigcup_{n \in \omega} Y_n. \tag{1.3.1}$$

It directly follows from these definitions that

(1.3.2) for any $n, m \in \omega$, $n \leq m$ and $g \in G_n$, $h \in G_m$,

$h \restriction n = g$ implies $n = m$, and hence, $h = g$.

Let us observe that

(1.3.3) there is a $\lambda \in \kappa$ such that $\big|\{ h \in G_\omega : h_0 = \lambda \}\big| \geq \omega$.

In fact, suppose that, on the contrary, $\{ h \in G_\omega : h_0 = \lambda \}$ is finite for all $\lambda \in \kappa$. But κ is itself finite, thus there are only finitely many λ's, consequently G_ω is also finite. This implies the existence of a longest element in it, which in turn means that all Y_n's with greater indices are empty. But, by (1.3.1), this latter means that X is the finite union of Y_n's, that is, the union of finitely many finite dimensional elements, thus itself is finite dimensional, contradicting the original assumption. Using (1.3.3), by induction on ω, we can define a

(1.3.4)
$z \in {}^{\omega}\kappa$ such that $\big|\{ h \in G_\omega : z \restriction n = h \restriction n \}\big| \geq \omega$ for any $n \in \omega,\ n > 0$.

Actually, by (1.3.3), there is a $\lambda \in \kappa$ such that $\big|\{ h \in G_\omega : h_0 = \lambda \}\big| \geq \omega$. Let $z_0 = \lambda$. Then $\big|\{ h \in G_\omega : h \restriction 1 = z \restriction 1 \}\big| \geq \omega$. Now, let us suppose that $n \in \omega$, $n > 0$ and $z \in {}^{n}\kappa$ has already been defined in such a way that $\big|\{ h \in G_\omega : h \restriction n = z \restriction n \}\big| \geq \omega$. Now, there is a $\lambda \in \kappa$ such that $\big|\{ h \in G_\omega : h \restriction n = z \restriction n \text{ and } h_n = \lambda \}\big| \geq \omega$. For, clearly, $\{ h \in G_\omega : h \restriction n = z \restriction n \} = \bigcup_{i \in \kappa} \{ h \in G_\omega : h \restriction n = z \restriction n \text{ and } h_n = i \}$. Let $z_n = \lambda$.

Then $z \in {}^{n+1}\kappa$ is such that $\left|\{h \in G_\omega : h \restriction (n+1) = z \restriction (n+1)\}\right| \geq \omega$ concluding the induction step.

Now, if there existed an $n \in \omega$ such that $z \restriction n \in G_n$, then, by (1.3.4), there would exist an $m > n$ and an $h \in G_m$ such that $h \restriction n = z \restriction n$, since $\bigcup_{i \in k} G_i$ is finite for any $k \in \omega$. But then $n = m$ by (1.3.2), which is a contradiction. Thus $z \restriction n \notin G_n$ for any $n \in \omega$. Therefore, using (1.3.1) and the definition of G_n,

$$z \notin X. \tag{1.3.5}$$

We conclude the proof by showing that $-X$ is not pf-regular. Indeed, $z \in -X$. Let us suppose that there is a $\Delta \in \Gamma(z, -X)$. Let $n = \max \Delta + 1$. Again, as in the argument leading to (1.3.5), it follows from (1.3.4) and (1.3.2) that there is an $m > n$ and $h \in G_m$ such that $h \restriction n = z \restriction n$. Then, by the definition of G_m, there is an $x \in Y_m$ such that $h = x \restriction m$. Thus $z \restriction n = h \restriction n = (x \restriction m) \restriction n = x \restriction n$, that is, $z \restriction \Delta = x \restriction \Delta$ since $\Delta \subseteq n$. Now, with $\Delta \in \Gamma(z, -X)$, this implies that $x \in -X$ contradicting the fact that $x \in Y_m \subseteq X$. ■

In informal model theoretical terms, Theorem 1.3.3 and 1.3.5 together imply that *infinitary ω_1-saturated finite models are ordinary models* expressing a unique characteristic of ordinary first order logic regarding saturatedness. Generally, however, $\mathsf{Cs}_\alpha^{\mathrm{pfr}} \cap \mathsf{Dc}_\alpha$ does have elements outside $\mathsf{Cs}_\alpha^{\mathrm{reg}} \cap \mathsf{Lf}_\alpha$. In fact, saturatedness in $\mathsf{Cs}_\alpha^{\mathrm{pfr}} \cap \mathsf{Dc}_\alpha$ is a proper generalization of ordinary saturatedness within $\mathsf{Cs}_\alpha^{\mathrm{reg}} \cap \mathsf{Lf}_\alpha$ (cf. [Ser,91b]):

Theorem 1.3.6.

(i) *Let $\mathfrak{A} \in \mathsf{Cs}_\alpha^{\mathrm{reg}} \cap \mathsf{Lf}_\alpha$ and $\kappa \geq |\alpha|$. $\mathfrak{A}$ is κ-saturated iff $\mathfrak{A}$ is ordinarily κ-saturated.*

(ii) *For every cardinal $\kappa \geq \alpha$, there is a saturated $\mathsf{Cs}_\alpha^{\mathrm{pfr}} \cap \mathsf{Dc}_\alpha$ with base κ, which is not locally finite-dimensional.*

On the other hand, there are even simple $\mathsf{Cs}_\alpha^{\mathrm{reg}} \cap \mathsf{Dc}_\alpha$'s that are *not* pf-regular. What is more,

Proposition 1.3.7. *There exist a simple $\mathfrak{A} \in \mathsf{Cs}_\alpha^{\mathrm{reg}} \cap \mathsf{Dc}_\alpha$ such that no $\mathsf{Cs}_\alpha^{\mathrm{pfr}}$ is isomorphic to $\mathfrak{A}$.*

Now, denoting by $\mathbf{Sm}K$ the class of simple elements of any class $K \subseteq \mathsf{Cs}_\alpha$, let us summarize what we know about the status of the class $\mathsf{Cs}_\alpha^{\mathrm{pfr}} \cap \mathsf{Dc}_\alpha$ within Cs_α. On the one hand, it is a proper extension of the class corresponding to the ordinary models, on the other, it is strictly narrower than the narrowest proper extension of the class of algebraic counterparts of ordinary models that has been investigated so far:

Theorem 1.3.8. $\mathsf{Cs}_\alpha^{\mathrm{reg}} \cap \mathsf{Lf}_\alpha \subsetneqq \mathsf{Cs}_\alpha^{\mathrm{pfr}} \cap \mathsf{Dc}_\alpha \subsetneqq \mathbf{Sm}\,\mathsf{Cs}_\alpha^{\mathrm{reg}} \cap \mathsf{Dc}_\alpha$.

3.2. Saturatedness in the class of pf-regular Cs_α's

The comparison of the following theorem to our reference statement (Theorem 1.1.4) shows that the class $\mathsf{Cs}_\alpha^{\mathrm{pfr}} \cap \mathsf{Dc}_\alpha$ can, in many respects, be considered a proper algebraic generalization of ordinary models, since it retains the properties of 'large' ordinary models we consider to be basic. What is more, as we shall see, there are arguments pointing in the direction that it may turn out to be the best possible generalization. The next theorem – together with the succeeding one – summarizes our most important positive results concerning the algebraic generalization of ordinary models.

Theorem 1.3.9.

(a) *Let $\mathfrak{A}$ be an arbitrary $\mathsf{Cs}_\alpha^{\mathrm{pfr}} \cap \mathsf{Dc}_\alpha$, and let κ be a cardinal.*

- (i) *If $\omega \leq \kappa \leq \alpha$, then $\mathfrak{A}$ is κ-saturated iff for any subset X of the base of $\mathfrak{A}$, $\mathfrak{A}_X$ is κ-compact whenever $|X| < \kappa$.*
- (ii) *If $\mathfrak{A}$ is κ-saturated, then $\mathfrak{A}$ is compact and $\mathfrak{A}$ is κ-homogeneous; in particular, if $\mathfrak{A}$ is saturated, then $\mathfrak{A}$ is compact and homogeneous.*
- (iii) *If $\mathfrak{A}$ is κ-saturated, then it is κ^+-universal over $\mathsf{Cs}_\alpha^{\mathrm{pfr}}$.*

(b)
- (i) *Every isomorphism between two saturated $\mathsf{Cs}_\alpha^{\mathrm{pfr}} \cap \mathsf{Dc}_\alpha$'s is a base-isomorphism if the algebras concerned have bases of the same cardinality.*
- (ii) *Every isomorphism between two saturated $\mathsf{Cs}_\alpha \cap \mathsf{Dc}_\alpha$'s is a base-isomorphism if the algebras concerned have bases of the same cardinality $> |\alpha|$.*
- (iii) *Every $\mathsf{Cs}_\alpha^{\mathrm{pfr}}$ with a finite base is compact and κ-saturated for any cardinal κ.*

(iv) *Every isomorphism from a* $\mathsf{Cs}_\alpha^{\mathrm{pfr}}$ *with a finite base onto a* $\mathsf{Cs}_\alpha^{\mathrm{reg}}$ *is a base-isomorphism.*

(c) *If* $\mathfrak{A} \in \mathsf{Cs}_\alpha^{\mathrm{pfr}} \cap \mathsf{Dc}_\alpha$ *is compact, then it is* $|\alpha|^+$*-universal over* $\mathsf{Cs}_\alpha^{\mathrm{pfr}}$.

(d) *Let* $\mathfrak{A}$ *be an arbitrary* $\mathsf{Cs}_\alpha^{\mathrm{pfr}}$ *with base* U*.* $\mathfrak{A}$ *is homogeneous iff for every* $W, T \subseteq U$ *such that* $|W| = |T| < |U|$*, every* $f \in Is(\mathfrak{A}_W, \mathfrak{A}_T)$ *can be extended into a base-automorphism of* $\mathfrak{A}_U$*.*

The algebraic version of downward Löwenheim–Skolem theorem ([Hen-Mon-Tar,85] 3.1.45) also remains true within $\mathsf{Cs}_\alpha^{\mathrm{pfr}}$, that is, [Hen-Mon-Tar,85] 3.1.45 can be supplemented by a fourth unit:

Theorem 1.3.10. *Let* $\mathfrak{A}$ *be a* $\mathsf{Cs}_\alpha^{\mathrm{pfr}}$ *with base* U *and let* κ *be an infinite cardinal such that* $|A| \leq \kappa \leq |U|$*. Let us suppose that* $S \subseteq U$ *and* $|S| \leq \kappa$*. Then there is a set* W *such that* $S \subseteq W \subseteq U$*,* $|W| = \kappa$ *and* $\mathfrak{A}$ *is ext-isomorphic to a* $\mathsf{Cs}_\alpha^{\mathrm{pfr}}$ *with base* W*.*

If we call the model theoretical objects corresponding to the elements of $\mathsf{Cs}_\alpha^{\mathrm{pfr}} \cap \mathsf{Dc}_\alpha$ pf-regular models, we can describe informally the essence of the theorems above as follows:

> The main properties of large pf-regular models with respect to the elements of their class are, *mutatis mutandis,* the same as those of their ordinary counterparts with respect to the elements of their own class, and the class of pf-regular models obeys the downward Löwenheim–Skolem theorem.

On the other hand, the similarity ends at this point. Namely, in the class of ordinary models, one can find a certain symmetry not only having the upward Löwenheim–Skolem theorem along with the downward one, but even in the description of large models in the sense that their main properties do characterize them: they and *only* they have these properties. In contrast to this class, that of the pf-regular models do not have this symmetrical character. In fact, as we shall see below, the upward Löwenheim–Skolem theorem does not hold here, and, for a pf-regular model, neither universality is sufficient to be compact nor universality + homogeneity to be saturated.

An informal remark is in order here. Bearing the Lindström theorem in mind, one cannot help expecting that, with a true variant of the downward Löwenheim–Skolem theorem at hand for a strictly wider class than that of the ordinary models, those properties of this class that are connected to compactness theorem must fail, even if, as in our case, in absence of a syntax, this connection could only be implicit. Now, let us consider the following theorems for ordinary logic: upward Löwenheim–Skolem theorem, the characterizability of compact models by their universality, that of the saturated models by their universality and homogeneity, and, finally, the possibility to represent models by compact ones modulo elementary equivalence. Since the proof of every one of these theorems is based on the compactness theorem of ordinary logic, it is not too surprising that all these facts fail to remain true for pf-regular models. This is witnessed by the existence of a single $\mathsf{Cs}_\omega^{\mathrm{pfr}}$. For any $\mathfrak{A} \in \mathsf{Cs}_\alpha$, $K \subseteq \mathsf{Cs}_\alpha$, and cardinal κ, we say that $\mathfrak{A}$ is **κ-categorical over** K if every isomorphism from $\mathfrak{A}$ onto an element of K with base of power κ is a base-isomorphism.

Theorem 1.3.11. *There exists a countably infinite, homogeneous and non-compact $\mathfrak{A} \in \mathsf{Cs}_\omega^{\mathrm{pfr}} \cap \mathsf{Dc}_\omega$ with a countably infinite base such that $\mathfrak{A}$ is κ-categorical over $\mathsf{Cs}_\omega^{\mathrm{pfr}}$ for every cardinal κ.*

It follows from this theorem that none of the algebraic versions of the four theorems we listed above as having proofs based on the compactness theorem holds for $\mathsf{Cs}_\omega^{\mathrm{pfr}} \cap \mathsf{Dc}_\omega$:

Corollary 1.3.12.

(i) *There is a countable $\mathsf{Cs}_\omega^{\mathrm{pfr}} \cap \mathsf{Dc}_\omega$ which is homogeneous and ω_1-universal over $\mathsf{Cs}_\omega^{\mathrm{pfr}}$, but not ω-saturated.*

(ii) *There is an $\mathfrak{A} \in \mathsf{Cs}_\omega^{\mathrm{pfr}} \cap \mathsf{Dc}_\omega$ with a countably infinite base such that every $\mathsf{Cs}_\omega^{\mathrm{pfr}}$ which is isomorphic to $\mathfrak{A}$ has a countable base.*

(iii) *There is a countable non-compact $\mathsf{Cs}_\omega^{\mathrm{pfr}} \cap \mathsf{Dc}_\omega$ which is ω_1-universal over $\mathsf{Cs}_\omega^{\mathrm{pfr}}$.*

(iv) *There is an $\mathfrak{A} \in \mathsf{Cs}_\omega^{\mathrm{pfr}} \cap \mathsf{Dc}_\omega$ such that no compact $\mathsf{Cs}_\omega^{\mathrm{pfr}}$ is isomorphic to $\mathfrak{A}$.*

What we said about the class of pf-regular models in connection with the compactness theorem, can be said about any other class of models properly containing that of the ordinary ones and satisfying the downward Löwenheim–Skolem theorem. Therefore, the positive results formulated in Theorem 1.3.9 supports our informal conjecture that, *from the point of view of similarity to ordinary models with respect to saturatedness and compactness, pf-regularity is close to being the best possible generalization of the notion of ordinary models.* Finally, it is worth noting that these considerations concerning Lindström theorem seem to be related to the problem formulated by W. Craig at the 1987 Algebraic Logic Conference (Asilomar, Calif.), asking to find a counterpart of Lindström theorem in algebraic logic (cf. [And-Mon-Nem,91a] Problem 65, p. 744).

BOLYAI SOCIETY
MATHEMATICAL STUDIES, 22

**Cylindric-like Algebras
and Algebraic Logic**
pp. 249–269.

Cylindric Modal Logic

YDE VENEMA*

The formalism of *cylindric modal logic* can be motivated from two directions. In its own right, it forms an interesting bridge over the gap between propositional formalisms and first-order logic, in that it formalizes first-order logic as if it were a modal formalism: The assignments of first-order variables can be seen as states or possible worlds of the modal formalism, and the quantifiers $\exists$ and $\forall$ may be studied as special cases of the modal operators $\Diamond$ and $\Box$, respectively. Elaborating this idea, we find that from this modal viewpoint, the standard semantics of first-order logic corresponds to just one of many possible classes of Kripke frames, and that other classes might be of interest as well.

From the algebraic logic perspective of this volume, cylindric modal logic provides a channel for the application of tools and ideas from modal logic in the theory of cylindric algebras. That the resulting formalism of cylindric modal logic fits in a volume on cylindric algebras, follows from the dualities between relational structures and Boolean algebras with operators. In this light, cylindric modal logic is nothing but cylindric algebra theory, studied from the dual perspective in which atom structures (of complete, atomic Boolean algebras with completely additive algebras) are the primary objects of study.

The leading question in this chapter is which facts about (representable) cylindric algebras can be explained from the general theory of modal logic and modal algebras. To mention two examples, we will see that the canonicity of the variety RCA_α follows from more general results in the duality theory of modal algebras, and that we can obtain a finite axiomatization of the equational theory of RCA_α by employing so-called non-orthodox derivation rules.

---

*The research of the author has been made possible by VICI grant 639.073.501 of the Netherlands Organization for Scientific Research (NWO).

1. First-Order Logic as Modal Logic

In this section we introduce the syntax and semantics of cylindric modal logic, we show in which sense it is a modal version of first-order logic, and we discuss some of its basic theory. Readers unfamiliar with the theory of modal logic may consult [Bla-Rij-Ven,01] for background information.

Syntax of *CML*. Starting with syntax: our purpose is to define a language that has two readings, both as (a restricted version of) first-order logic, and as a multi-modal language. Suppose that we consider a language of first-order logic with the constraints that we have a set $\{v_i \mid i < \alpha\}$ of first-order variables (with α a fixed but arbitrary ordinal), and a countable set of predicate symbols, each of arity α. The only admissible *atomic* formulas are of the form $v_i = v_j$ or $P_l(v_0 v_1 \dots v_i \dots)_{i<\alpha}$ (i.e., with a fixed order of the first-order variables). The motivation for adopting this particular restriction stems from the desire to stay close to the formalism of Cylindric Algebras [Hen-Mon-Tar,85, p. 152–153]; at the end of this section we will see how to handle more standard versions of first-order logic. For $\alpha < \omega$, we get a logic with finitely many variables. Such logics have been studied in the literature, for purely logical reasons [Hen,67, Tar-Giv,87, Hen-Mon-Tar,85] or because of their relation with temporal logics in computer science [Imm-Koz,87, Ott,97]; see also [And-Nem-Sai,01, Section 7] for an investigation from the algebraic logic point of view.

As their order is fixed, the variables in atomic relational formulas do not provide any information. Thus we may just as well leave them out, writing p_l for $P_l(v_0 \dots v_i \dots)_{i<\alpha}$, cf. [Hen-Mon-Tar,85, Remark 4.3.2]. This *restricted first-order logic* becomes *cylindric modal logic* if we replace the identity $v_i = v_j$ with the *modal constant* d_{ij}, and the existential quantification $\exists v_i$ with the *diamond* $\Diamond_i$. In order not to confuse the reader with too much notation, henceforth we will use modal notation and terminology mainly, occasionally referring to the first-order interpretation for motivation or clarification.

Definition 2.1.1. Let α be an arbitrary but fixed ordinal with $2 \leq \alpha$. CML_α is the modal language having constants d_{ij} for $i, j < \alpha$ and unary connectives ('diamonds') $\Diamond_i$ for $i < \alpha$. Given a set of propositional variables Q, the set of α-dimensional cylindric modal formulas in Q, or for short, *α-formulas (in Q)*, is built up as usual. The (modal or boolean) constants, and the variables from Q are the atomic formulas, and if φ and ψ are formulas, then so are $\varphi \wedge \psi$, $\neg\varphi$ and $\Diamond_i\varphi$. We will use standard abbreviations like $\wedge$, $\rightarrow$ and $\Box_i$.

SEMANTICS OF *CML*. Consider the basic declarative statement in first-order logic concerning the truth of a formula in a model under an assignment s:

$$\mathfrak{M} \models \varphi\,[s]. \tag{2.1.1}$$

The basic observation underlying our approach, is that we can read (2.1.1) from a modal perspective as: 'the formula φ is true in $\mathfrak{M}$ *at state* s.' But since we have exactly α variables at our disposal, we can identify assignments with maps from α $\big(= \{i \mid i < \alpha\}\big)$ to U, or equivalently, with α-tuples over the domain U of the structure $\mathfrak{M}$. We will denote the set of such α-tuples by ${}^\alpha U$. As a consequence, we find ourselves in the setting of multi-dimensional modal logic: the universe of our modal models will be of the form ${}^\alpha U$ for some base set U. More information on this branch of modal logic can be found in [Kur,thisVol] of this volume; for monographs, see [Mar-Ven,97, Gab-Kur-Wol-Zak,03].

Recall the truth definition of the existential quantifier:

$$\mathfrak{M} \models \exists v_i\, \varphi[s] \text{ iff there is a } u \in U \text{ such that } \mathfrak{M} \models \varphi\big[s^i_u\big],$$

where s^i_u is the assignment defined by $s^i_u(k) = u$ if $k = i$ and $s^i_u(k) = s(k)$ otherwise. We can replace the above truth definition with the more 'modal' equivalent,

$$\mathfrak{M} \Vdash \Diamond_i\, \varphi[s] \text{ iff there is an assignment } s' \text{ with } s \equiv_i s' \text{ and } \mathfrak{M} \Vdash \varphi\,[s'],$$

where the binary relation $\equiv_i$ is given by

$$s \equiv_i s' \text{ iff for all } j \neq i,\ s_j = s'_j. \tag{2.1.2}$$

In other words: existential quantification behaves like a modal *diamond*, having $\equiv_i$ as its *accessibility relation*.

Since the semantics of the boolean connectives in the predicate calculus is the same as in modal logic, this shows that the inductive clauses in the truth definition of first-order logic neatly fit a modal pattern. So let us now concentrate on the atomic formulas. To start with, *equality* formulas do not cause any problem: the formula $v_i = v_j$, with truth definition

$$\mathfrak{M} \models v_i = v_j[s] \text{ iff } s \in \mathit{Id}_{ij}\ \big(:= \{s \in {}^\alpha U \mid s_i = s_j\}\big). \tag{2.1.3}$$

is indeed interpreted as a modal *constant.* Concerning the other atomic formulas, recall that in first-order logic, an α-ary predicate symbol P_l is interpreted as an α-ary relation, that is, as a subset of ${}^{\alpha}U$. This is exactly how a modal *valuation* interprets the propositional variable p_l, given our multi-dimensional setting where ${}^{\alpha}U$ provides the set of states of the model.

The above shows that, indeed, we can present the semantics of first-order logic in a completely modal framework. In order to bring this presentation in line with standard modal terminology, recall that every modal language automatically comes equipped with a relational semantics of (Kripke) *frames*, i.e. abstract structures having an (arbitrary) $n+1$-ary accessibility relation for every n-ary modal operator. From this more abstract semantic perspective on CML_α, its interpretation as a first-order logic can be captured by restricting the relational semantics to a rather special class of so-called *cube* frames and models.

Definition 2.1.2. An *α-frame* is a structure $\mathfrak{F} = \langle W, T_i, E_{ij} \rangle_{i,j<\alpha}$ with every $T_i \subseteq W \times W$ and every $E_{ij} \subseteq W$. The first-order language used to describe these structures (having monadic predicates E_{ij} and dyadic predicates T_i, $i, j < \alpha$), is denoted by $\mathcal{L}_\alpha$. Given a set U, the α-frame $\mathfrak{C}_\alpha(U) = \langle {}^{\alpha}U, \equiv_i, Id_{ij} \rangle_{i,j<\alpha}$, with $\equiv_i$ and Id_{ij} as given by (2.1.2) and (2.1.3), respectively, is called the *α-cube over U*. The class of α-cubes is denoted by C_α.

An *α-model* is a pair $\mathfrak{M} = (\mathfrak{F}, V)$ with $\mathfrak{F}$ an α-frame and V a *valuation,* i.e. a map assigning a subset of the universe of $\mathfrak{F}$ to each propositional variable in the language. *Truth* of a formula φ at a world w in the model $\mathfrak{M}$ is defined by the usual induction, e.g.

$$\mathfrak{M}, w \Vdash p \iff w \in V(p),$$

$$\mathfrak{M}, w \Vdash \mathsf{d}_{ij} \iff w \in E_{ij},$$

$$\mathfrak{M}, w \Vdash \Diamond_i \psi \iff \text{there is a } v \text{ with } wT_iv \text{ and } \mathfrak{M}, v \Vdash \psi.$$

Validity of a formula or set of formulas in a model/frame/class of frames is defined and denoted as usual, e.g. $\mathsf{C}_\alpha \Vdash \varphi$ iff $\mathfrak{F}, V, w \Vdash \varphi$ for all frames $\mathfrak{F}$ in C_α, all valuations V on $\mathfrak{F}$ and all worlds w in $\mathfrak{F}$.

RELATIVIZED CUBES. In the general semantics of α-frames, states are no longer assignments but rather abstractions thereof. It is interesting to see what happens to familiar laws of the predicate calculus in this new set-up. The abstract modal perspective on the semantics of first-order logic imposes a certain *degree of validity* on familiar theorems of the predicate calculus. Some theorems are valid in all α-frames, such as distribution: $\forall v_i(\varphi \to \psi) \to (\forall v_i\varphi \to \forall v_i\psi)$. Others, such as the axiom schema $\varphi \to \exists\, v_i\varphi$ will only be valid in α-frames where T_i is a reflexive relation (below we will see many more of such correspondences). Clearly, narrowing down the class of frames increases the set of valid formulas, and vice versa.

Of particular interest are some classes of frames that differ only slightly from the cube structures, but have much nicer computational and/or logical properties. A *relativized cube* is a structure in which the states are still α-assignments on some set U (and the accessibility relations are as in $\mathfrak{C}_\alpha(U)$), but not all α-assignments on U are *available* as states. Formally, for $W \subseteq {}^\alpha U$, define $\mathfrak{C}^W_\alpha(U) := \left\langle W, \equiv_i \cap ({}^\alpha U \times {}^\alpha U), Id_{ij} \cap {}^\alpha U \right\rangle_{i,j<\alpha}$. A nice and natural intermediate class consists of multi-dimensional frames that are *locally cube*, which intuitively means that if $s \in {}^\alpha U$ is an available tuple, then any tuple 'drawing its coordinates from the set $\{s_i \mid i < \alpha\}$' should be available as well. Formally, a local cube is a relativized cube $\mathfrak{C}^W_\alpha(U)$ such that $\left\langle s_{\sigma(i)} \right\rangle_{i<\alpha} \in W$, for every $s \in W$ and every map $\sigma : \alpha \to \alpha$. Clearly, widening the semantics to such frame classes we lose some familiar first-order validities, such as $\exists\, v_i \exists\, v_j \varphi \to \exists\, v_j \exists\, v_i \varphi$; but others remain valid.

Many results are known about the CML_α-logic of these relativized cube frame classes (see [Mar-Ven,97] for an overview). We just mention Németi's seminal result [Nem,95] that the classes of relativized and of local cubes have a *decidable* logic. It was this result which led Andréka, van Benthem and Németi [And-Ben-Nem,98] to the discovery of the *guarded fragment*, an inductively defined, *decidable* fragment of first-order logic. More on this can be found in [And-Nem,thisVol] and [Ben,thisVol].

PROPERTIES OF CUBES: CORRESPONDENCE THEORY. It is a standard modal logic question to investigate whether important frame classes admit a modal characterization, in the form of a set of modal formulas that are valid on a frame iff that frame belongs to the class. Given the fact that the cubes are not closed under taking some of the standard frame constructions such as disjoint unions or bounded morphic images, the answer for this particular class of α-frames is negative. A natural following-up problem is to study properties of the cube frames that do have a modal characterization. The next definition gathers some of these properties.

Definition 2.1.3. Consider the following pairs of cylindric modal formulas and frame formulas:

$(CM1_i)$	$p \to \Diamond_i p$	$(N1_i)$	$\forall x T_i xx$
$(CM2_i)$	$p \to \Box_i \Diamond_i p$	$(N2_i)$	$\forall xy(T_i xy \to T_i yx)$
$(CM3_i)$	$\Diamond_i \Diamond_i p \to \Diamond_i p$	$(N3_i)$	$\forall xyz\big((T_i xy \wedge T_i yz) \to T_i xz\big)$
$(CM4_{ij})$	$\Diamond_i \Diamond_j p \to \Diamond_j \Diamond_i p$	$(N4_{ij})$	$\forall xz\big(\exists y(T_i xy \wedge T_j yz) \to \exists u(T_j xu \wedge T_i uz)\big)$
$(CM5_i)$	d_{ii}	$(N5_i)$	$\forall x E_{ii} x$
$(CM6_{ij})$	$\Diamond_i(\mathsf{d}_{ij} \wedge p) \to \Box_i(\mathsf{d}_{ij} \to p))$	$(N6_{ij})$	$\forall xyz\big((T_i xy \wedge E_{ij} y \wedge T_i xz \wedge E_{ij} z) \to y = z\big)$
$(CM7_{ijk})$	$\mathsf{d}_{ij} \leftrightarrow \Diamond_k(\mathsf{d}_{ik} \wedge \mathsf{d}_{kj})$	$(N7_{ijk})$	$\forall x\big(E_{ij} x \leftrightarrow \exists y(T_k xy \wedge E_{ik} y \wedge E_{kj} y)\big)$
$(CM8_{ij})$	$\big(\mathsf{d}_{ij} \wedge \Diamond_i(\neg p \wedge \Diamond_j p)\big) \to \Diamond_j(\neg \mathsf{d}_{ij} \wedge \Diamond_i p)$	$(N8_{ij})$	$\forall xz\big(E_{ij} x \wedge (\exists y T_i xy \wedge T_j yz \wedge y \neq z) \to \exists u(\neg E_{ij} u \wedge T_j xu \wedge T_i uz)\big)$

For finite α we set $(CM1) \equiv \bigwedge_i (CM1_i)$, etc., taking $(CM4) \equiv \bigwedge_{i,j}(CM4_{ij})$, $(CM6) \equiv \bigwedge_{i \neq j}(CM6_{ij})$, $(CM7) \equiv \bigwedge_{i,j,k}(CM7_{ijk})$ and $(CM8) \equiv \bigwedge_{i \neq j}(CM8_{ij})$. If $\alpha \geq \omega$, we let $(CM1), \ldots, (CM8)$ be the corresponding equation *schemata*.

An α-frame $\mathfrak{F}$ is called *cylindric* if $\mathfrak{F} \Vdash (CM1) \ldots (CM7)$, and *hypercylindric* if, in addition, $(CM8)$ is valid in it. The class of α-dimensional (hyper)cylindric frames is denoted as $\mathsf{(H)CF}_\alpha$.

In words, $(N1_i)$, $(N2_i)$ and $(N3_i)$ express that T_i is respectively reflexive, symmetric and transitive; together they state that T_i is an equivalence relation. $(N6_{ij})$ then means that in every T_i-equivalence class there is *at most one* element on the diagonal E_{ij} $(i \neq j)$. By $(N5_j)$ and $(N7_{jji})$ one can show that every T_i-equivalence class contains *at least one* element on the diagonal E_{ij}. Taking these observations together, we find that every world in a cylindric frame has a unique T_i-successor on the E_{ij}-diagonal. The meaning of $(N4)$ and $(N8)$ is best made clear by the following pictures:

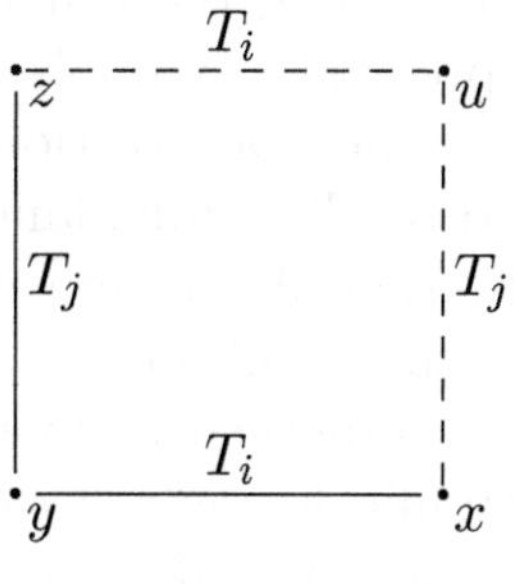

$(N4_{ij})$

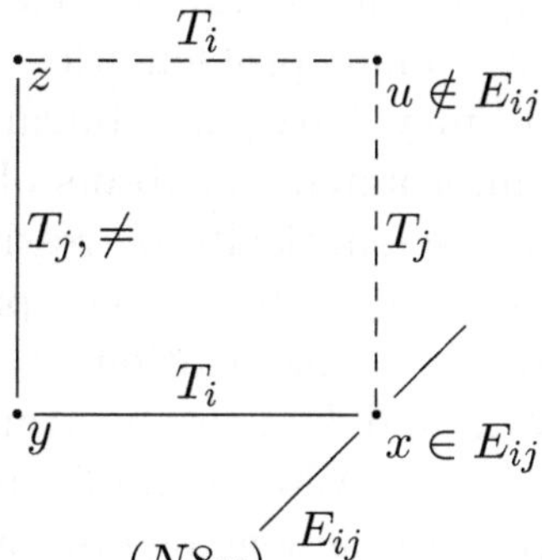

$(N8_{ij})$

The following proposition, stating that each of the above modal formulas characterizes the corresponding first-order frame condition, is a special case of a fundamental theorem in modal logic. This result, Sahlqvist's correspondence theorem, states that modal formulas of a certain syntactic shape correspond (in the sense of being equivalent as in (2.1.4) below, to a first-order formula which can be effectively computed from the modal formula.

Proposition 2.1.4. *Let $\mathfrak{F}$ be an α-frame. Then for $l = 1, \ldots, 8$ and $i, j, k < \alpha$:*

$$\mathfrak{F} \Vdash \left(CMl_{i(j(k))}\right) \iff \mathfrak{F} \models \left(Nl_{i(j(k))}\right). \tag{2.1.4}$$

Using this result, together with [Hen-Mon-Tar,85, Theorem 2.7.40] it is not hard to prove the following proposition which explains our terminology.

Proposition 2.1.5. *An α-frame $\mathfrak{F}$ is a cylindric frame iff $\mathfrak{F}^+$ is a cylindric algebra.*

Modalizing Standard First-Order Logic. At this point the reader may complain that the formalism that we have been 'modalizing' is not first-order logic at all, but at best a rather peculiar variant of it. So, to justify this section's title, let us briefly see which adaptations to make, in order to cover more standard versions of the predicate calculus.

To start with, the fact that we consider versions of first-order logic with *more than* ω many variables is not a problem at all. In fact, our only reason for allowing these was to be able to cover cylindric algebras of arbitrary dimension. We may just as well confine attention to the case $\alpha \leq \omega$.

Second, while we saw that all inductive clauses in the semantics of first-order logic are in complete accordance with the modal pattern, and that the identity formulas can be treated as modal constants, the other atomic formulas are more problematic. First of all, predicate symbols in first-order logic may have an arbitrary arity, whereas our propositional variables are interpreted as subsets of ${}^{\alpha}U$ for a fixed α. This imbalance could be corrected semantically by restricting the valuation of a proposition letter p corresponding to a k-ary predicate P, to those subsets of ${}^{\alpha}U$ that are closed under the relation $\equiv_i$ for $i \geq k$, or syntactically (in the case $\alpha = n < \omega$ and $k < n$) by translating the atomic formula $Pv_0 \cdots v_{k-1}$ as $\Diamond_k \cdots \Diamond_{n-1} p$. Another problem, however, is that, even if we restrict to the n-variable fragment of first-order logic with n-ary predicate symbols, our modal formalism can only deal with a restricted version of first-order logic. Because our valuations on n-cubes are in one-one correspondence with the interpretations of n-ary predicates, we obtain that $\mathfrak{M}, s \Vdash p$ (in the modal sense) iff $\mathfrak{M} \models Pv_0 \cdots v_{n-1}$ (in the first-order sense). But how to handle atomic formulas where the first-order variables do not occur in the fixed order $v_0 \cdots v_{n-1}$, that is, formulas of the form $Pv_{\sigma(0)} \cdots v_{\sigma(n-1)}$, for some map $\sigma : n \to n$? Atomic formulas with some *multiple* occurrence of a variable can be rewritten as non-atomic formulas involving only unproblematic atomic formulas, see [Hen-Mon-Tar,85, p. 152] (for instance: $Pv_0v_0 \iff \exists v_1(v_0 = v_1 \wedge Pv_0v_1)$.

This leaves the case what to do with atoms of the form $Pv_{\sigma(0)} \cdots v_{\sigma(n)}$ with $\sigma : n \to n$ a *permutation*. The crucial observation is that for any permutation $\sigma \in {}^{n}n$, we have

$$\mathfrak{M} \models Pv_{\sigma(0)} \cdots v_{\sigma(n-1)}[s] \iff \mathfrak{M} \models Pv_0 \cdots v_{n-1}[s \circ \sigma],$$

where $s \circ \sigma$ is the composition of $\sigma : n \to n$ and $s : n \to U$. So if we add an explicit *substitution operators* $\odot_\sigma$ to the language, with semantics

$$\mathfrak{M}, s \Vdash \odot_\sigma \varphi \iff \mathfrak{M}, s \circ \sigma \Vdash \varphi,$$

we can indeed handle any atomic formula of the form $Pv_{\sigma(0)} \cdots v_{\sigma(n-1)}$, translating it as $\odot_\sigma p$. The resulting system, which can be seen as the modal version of polyadic equality algebras, has been developed along the same lines as cylindric modal logic, see Venema [Ven,95a] for details.

TYPE-FREE AND SCHEMA VALIDITY. The ω-dimensional version of our logic may not correspond directly to the standard predicate calculus, it does encode two rather interesting notions related to first-order logic: *type-free validity*, introduced in [Hen-Mon-Tar,85, 4.3.65] (see also Simon [Sim,91]) and *scheme validity*, see Németi [Nem,87a].

Type-free logic arises if we think of CML_ω-formulas as so-called *type-free* formulas. A type is a map $\rho : Q \to \omega$ assigning a finite arity to each propositional variable/predicate letter in Q, and the ρ-instantiation φ^ρ of a formula $\varphi \in CML_\omega$ is simply the first-order formula we obtain from φ by simultaneously replacing every propositional variable p_l occurring in φ with the formula $P_l v_0 \cdots v_{\rho(p_l)-1}$ (and reading the identity constants and modal operators in the first-order way). A CML_ω-formula is *type-free valid* if each of its typed instances is valid (as a first-order formula).

Similarly, CML_ω-formulas can be seen as first-order *schemas:* formally we may define a *first-order instance* of φ as the result of uniformly substituting arbitrary first-order formulas for the propositional variables in φ. We call φ *schema valid* if each of these first-order instances is valid as a first-order formula.

It is not hard to show that these two notions are equivalent, and closely related to the cube semantics of CML_ω. More precisely, for each CML_ω-formula φ we have that

(2.1.5) $$\mathsf{C}_\omega \Vdash \varphi \iff \varphi \text{ is type-free valid } \iff \varphi \text{ is schema valid.}$$

2. Cylindric Modal Logic and Cylindric Algebras

In this section we discuss how cylindric modal logic fits in the theory of cylindric algebras. Roughly speaking, in cylindric modal logic one focuses on the *atom structures* associated with cylindric algebras. Conversely, from the perspective of modal logic, cylindric algebras provide an interesting class of *modal algebras.* Many of the notions defined here are discussed in [Hen-Mon-Tar,85, Section 2.7]; see also [Kur,thisVol] and [Bez,thisVol]. For an overview of the algebraic approach towards modal logic the reader may consult [Ven,07].

Definition 2.2.1. A cylindric-type modal algebra of dimension α, or shortly: an *α-modal algebra* is an algebra $\mathbb{A} = \langle A, +, \cdot, -, 0, 1, \Diamond_i, \mathsf{d}_{ij}\rangle_{i,j<\alpha}$ of type CML_α, where $\langle A, +, \cdot, -, 0, 1\rangle$ is a Boolean algebra, and each $\Diamond_i$ is a unary operator on $\langle A, +, \cdot, -, 0, 1\rangle$, that is, a normal ($\Diamond_i 0 = 0$) and additive ($\Diamond_i(x+y) = \Diamond_i x + \Diamond_i y$) operation.

Frames and Algebras. At the basis of the algebraic perspective on modal logic lies the following construction of an algebra from a frame.

Definition 2.2.2. The *complex algebra* of an α-frame $\mathfrak{F} = \langle W, T_i, E_{ij} \rangle_{i,j<\alpha}$ is the structure $\mathfrak{F}^+ := \langle \mathcal{P}(W), \cup, \cap, \sim_W, \varnothing, W, \langle T_i \rangle, E_{ij} \rangle$, where the map $\langle T_i \rangle : \mathcal{P}(W) \to \mathcal{P}(W)$ is given by

$$\langle T_i \rangle(X) := \{ w \in W \mid T_i wx \text{ for some } x \in X \}.$$

Given a class K of frames, we let $\mathsf{Cm}\,\mathsf{K}$ denote the associated class of complex algebras.

Intuitively, complex algebras are the algebraic encodings of frames, and are often thought of as 'concrete' α-modal algebras, in the same sense that power-set algebras are concrete Boolean algebras. An indication of their fundamental importance, and a first link between the areas of cylindric modal logic and that of cylindric algebras, is the following observation.

Theorem 2.2.3. RCA_α *is the variety generated by* $\mathsf{Cm}\,\mathsf{C}_\alpha$.

Proof. Immediate by the observation that RCA_α is the variety generated by the class of α-dimensional full cylindric set algebras, which is precisely the class $\mathsf{Cm}\,\mathsf{C}_\alpha$ of complex algebras of cubes. ∎

The original frame can be retrieved as the *atom structure* of its complex algebra, where the atom structure of an atomic α-modal algebra $\mathbb{A}$ is defined as a certain α-frame based on the collection of atoms of the (Boolean reduct of) $\mathbb{A}$. The operations of taking complex algebras and atom structures form part of a categorical duality, but we lack the space for going into detail here.

From an arbitrary (that is, not necessarily atomic) α-modal algebra we can obtain an α-frame as follows.

Definition 2.2.4. The *ultrafilter frame* of an α-modal algebra $\mathfrak{A} = \langle A, +, \cdot, -, 0, 1, \Diamond_i, \mathsf{d}_{ij} \rangle_{i,j<\alpha}$ is the α-frame $\mathfrak{A}_\bullet := \langle \mathsf{Uf}\,\mathfrak{A}, \mathsf{S}_{\Diamond_i}, \mathsf{D}_{ij} \rangle_{i,j<\alpha}$, where $\mathsf{Uf}\,\mathfrak{A}$ denotes the set of ultrafilters of the (Boolean reduct of) $\mathfrak{A}$, $\mathsf{D}_{ij} := \{ u \in \mathsf{Uf}\,\mathfrak{A} \mid \mathsf{d}_{ij} \in u \}$, and $\mathsf{S}_{\Diamond_i} \subseteq \mathsf{Uf}\,\mathfrak{A} \times \mathsf{Uf}\,\mathfrak{A}$ is given by $S_{\Diamond_i} uv$ iff $\Diamond_i a \in u$ for all $a \in v$.

The operations of taking complex algebras and ultrafilter frames can be extended to functors between the categories of α-frames (with so-called bounded morphisms as arrows), and α-modal algebras (with homomorphisms). They do not provide a full-blown categorical duality, however, unless we restrict both categories to the full subcategories of their finite members. Nevertheless, the composition of the operations $(\cdot)^+$ and $(\cdot)_\bullet$ provides one of the key notions in the area.

Definition 2.2.5. Given an α-modal algebra $\mathfrak{A}$, the algebra $(\mathfrak{A}_\bullet)^+$ is called the perfect or *canonical extension* of $\mathfrak{A}$. A class K of α-modal algebras is *canonical* if it is closed under taking canonical extensions.

As an extension of Stone's representation theorem for Boolean algebras, Jónsson and Tarski [Jon-Tar,51] proved that every α-modal algebra can be embedded in its perfect extension. This shows that every α-modal algebra $\mathfrak{A}$ can be *represented* as a subalgebra of a concrete algebra, namely, the complex algebra $(\mathfrak{A}_\bullet)^+$ of the frame $\mathfrak{A}_\bullet$. As a consequence, canonicity is a very desirable property for a variety (or class) to have because it means that every algebra in the variety can be represented as a concrete algebra *in the same variety*.

Theorem 2.2.6. *The variety* RCA_α *is canonical.*

This result, stated as Theorem 2.7.24(ii) in [Hen-Mon-Tar,85], can be seen as an instantiation of a more general result in modal logic by Fine which states that every *elementary* (first-order definable) frame class generates a canonical variety. Indeed, it is not hard to come up with a set of first-order sentences that characterize the class of α-cubes. Theorem 2.2.6 and variants are investigated from this perspective in [And-Gol-Nem,98].

FORMULAS, TERMS AND EQUATIONS. Generally, the starting point in algebraic logic is the formal identification of logical connectives with algebraic function symbols, and, consequently, of formulas with algebraic terms. Given a sufficiently expressive repertoire of connectives, the link between logic and algebra can be extended to the semantics of formulas and equations, respectively. For instance, modulo some simple translations between formulas and equations, the notion of modal validity on a frame coincides with that of equational validity on its complex algebra:

$$\mathfrak{F} \Vdash \varphi \quad \text{iff} \quad \mathfrak{F}^+ \models \varphi \approx \top, \tag{2.2.1}$$

$$\mathfrak{F} \Vdash \varphi \leftrightarrow \psi \quad \text{iff} \quad \mathfrak{F}^+ \models \varphi \approx \psi. \tag{2.2.2}$$

On the basis of (2.2.1) we will say that an algebra validates a modal formula φ if it validates its equational translation $\varphi \approx \top$.

LOGICS AND VARIETIES. Finally, we turn to axiomatization issues. In modal logic, the key concept is that of a *normal modal logic.*

Definition 2.2.7. A *normal cylindric modal logic of dimension* α, or an α-logic, is a set of CML_α-formulas containing

(CT)	all propositional tautologies
$(DB_{\Box_i})$	$\Box_i(p \to q) \to (\Box_i p \to \Box_i q)$

which is closed under the derivation rules, Modus Ponens, Universal Generalization and Substitution:

(MP)	$\vdash \varphi, \vdash \varphi \to \psi \;\Rightarrow\; \psi,$
$(UG_{\Box_i})$	$\vdash \varphi \;\Rightarrow \vdash \Box_i\varphi,$
(SUB)	$\vdash \varphi \;\Rightarrow \vdash \sigma(\varphi)$, for any substitution σ of formulas for propositional variables in φ.

Given a set Γ of formulas, we let $\mathsf{K}_\alpha.\Gamma$ denote the normal cylindric modal logic *axiomatized by* Γ, that is, the smallest α-logic containing Γ, and write K_α for $\mathsf{K}_\alpha.\varnothing$.

Algebraically, normal modal logics correspond to deductively closed sets of equations, and hence, by Birkhoff's completeness for equational logic, also to varieties of modal algebras. Given a set Γ of formulas, let V_Γ denote the variety of α-modal algebras axiomatized by the set $\{\gamma \approx \top \mid \gamma \in \Gamma\}$ of equations.

Proposition 2.2.8. *The map $L \mapsto \mathsf{V}_L$ is a dual isomorphism between the lattice of normal cylindric modal logics of dimension α, and the lattice of varieties of α-modal algebras. The variety V_L algebraizes the logic L, in the sense that*

$$\varphi \in L \quad \textit{iff} \quad \mathsf{V}_L \models \varphi \approx \top, \tag{2.2.3}$$

$$\varphi \leftrightarrow \psi \in L \quad \textit{iff} \quad \mathsf{V}_L \models \varphi \approx \psi. \tag{2.2.4}$$

Note that the algebraization given by the equivalences (2.2.3) and (2.2.4) is of a very simple nature. Generally, the notion of a class of algebras *algebraizing* a logic [Blo-Pig,89], is more sophisticated, dealing with logics as consequence relations rather than as sets of theorems, and admitting more complex translations between formulas and equations than the ones above.

AXIOMATIZING THE CUBES. Given Theorem 2.2.3 and the connections (2.2.1) and (2.2.2), axiomatizing the modal logic of the class of α-cubes, and axiomatizing the equational theory of variety of representable cylindric algebras amounts to the same thing. Unfortunately, for $\alpha > 2$, it follows from the nonfinite axiomatizability results of Monk, Andréka, and others, that if we only allow orthodox derivation systems then there is no finite set of axioms and rules that, when added to K_α, yields a complete axiomatization for the class of cubes. Nevertheless, we can see how far the modal versions of the cylindric algebra axioms bring us.

Definition 2.2.9. Let $\mathbf{CML}_\alpha$ and $\mathbf{HCML}_\alpha$ be the normal modal logics axiomatized by the formulas (CM1–7) and (CM1–8), respectively.

The following result of Venema [Ven,95b] is another immediate consequence of the Sahlqvist shape of the formulas (CM1–8).

Theorem 2.2.10. *The systems* $\mathbf{CML}_\alpha$ *and* $\mathbf{HCML}_\alpha$ *are sound and complete for the classes* CF_α *and* HCF_α*, respectively.*

DIMENSION TWO. In the case $\alpha = 2$, however, the system $\mathbf{HCML}_2$ is sound and complete for the class of squares (which is our usual term to refer to the cubes of dimension two). Consequently, the set of equations corresponding to (CM1–8) provides a finite axiomatization of the class RCA_2. Thus our axiom (CM8) provides an alternative to Henkin's equation, see [Hen-Mon-Tar,85, Theorem 3.2.65(ii)], or [Rij-Ven,95] for a more detailed discussion. We refer to [Bez,thisVol] for more information on the two-dimensional case.

3. COMPLETENESS FOR CYLINDRIC MODAL LOGIC

INTRODUCTION. As an application of modal logic in the theory of cylindric algebras, in this section we will see that if one is willing to generalize the definition of an axiomatization by admitting so-called *unorthodox* derivation rules, then the finite axiomatization problem can be overcome in a fairly simple and elegant way. Here we call a rule unorthodox if it has a side condition to the effect that the applicable instances of the rule are not closed under taking substitutions – in other words, unorthodox rules are not structural in the sense of Blok and Pigozzi [Blo-Pig,89].

Obviously, with this definition many different kinds of rules may be classified as being unorthodox – the ones we employ here are characterized by a side condition involving the occurrences of variables in the premisse of the rule. The use of such rules in modal logic goes back to the work of Gabbay [Gab,81] and Burgess [Bur,80]. More specifically, here we will focus on a method of proving completeness via a rule that involves the so-called *difference operator* and that shares with Sahlqvist's theorem the feature of automatically turning characterizations into axiomatizations; see [Ven,93] for a detailed discussion.

Before continuing, let us be a bit more precise about the distinction between logics and derivation systems. We shall call a derivation system a pair consisting of a set of formulas (called axioms), and a set of derivation rules. A *derivation* in such a system is defined as a nonempty, finite sequence $\varphi_0, \ldots, \varphi_n$ such that every φ_i is either an axiom or obtainable from $\varphi_0, \ldots, \varphi_{i-1}$ by a derivation rule. For instance, we think of the normal modal logic $\mathsf{K}_\alpha.\Gamma$ as the derivation system with as its axioms, besides Γ, all classical tautologies and the modal distribution axioms, and as its derivation rules: modus ponens, universal generalization, and substitution. A *theorem* of a derivation system is any formula that can appear as the last item of a derivation. Hence, any derivation system containing the above-mentioned triple as derivation rules, will produce, as its set of theorems, a normal modal logic. Theoremhood of a formula φ in the system $\mathbf{HCML}_\alpha$ is denoted by $\vdash_\alpha \varphi$.

CHARACTERIZING THE n-CUBES. Our unorthodox completeness theorem will be based on a rather special characterization of the n-cubes – where for the moment we fix a finite ordinal n with $2 \leq n$. The starting point for this characterization is the observation that the *inequality relation* $\neq$ on a cube can be obtained in a nice, 'modal' way, namely, as a certain composition of the cube's accessibility relations. As a corollay, we find that on the class of n-cubes, the difference operator is term definable as a compound modality. (A compound modality is a modal formula $\varphi(p)$ with one free variable p that is composed from p using diamonds, disjunctions, and conjunctions with variable-free formulas.)

The modal perspective on the inequality relation is based on two simple facts about cubes. First, two tuples are distinct iff they differ in at least one coordinate. And second, two tuples, say u and v, differ in some coordinate, say, 0, iff one can make the following 'modal walk' from $u = (u_0, \ldots, u_{n-1})$ to $v = (v_0, \ldots, v_{n-1})$ along the accessibility relations: (i) first move along T_1 to the diagonal E_{01}, arriving at $(u_0, u_0, u_2, \ldots, u_{n-1})$; (ii) continue by moving along T_0 *off* the diagonal, arriving at $(v_0, u_0, u_2, \ldots, u_{n-1})$; (iii) finally, move along $T_1, \ldots, T_{n-1}$, arriving at v. Turning to the general setting of hypercylindric n-frames, this 'modal walk' can be formalized to the following definition of a binary relation R^n.

Definition 2.3.1. For an arbitrary hypercylindric n-frame $\mathfrak{F} = (W, T_i, E_{ij})_{i,j<n}$, define $f_{ij}(u)$, H_i^n, H^n and R^n as follows: $f_{ij}(u)$ is the unique v such that $T_i uv$ and $E_{ij}v$. H^n (resp. H_i^n, $i < \alpha$) is the composition of all the T-relations, resp. all the T-relations minus T_i, i.e.

$$H^n = T_0 \circ T_1 \circ \ldots \circ T_{n-1},$$

$$H_i^n = T_0 \circ T_1 \circ \ldots \circ T_{i-1} \circ T_{i+1} \circ \ldots \circ T_{n-1}.$$

For a world u in $\mathfrak{F}$, the set $H_i^n(u) = \{v \mid \mathfrak{F} \models H_i^n uv\}$ is called the i-hyperplane through u. R^n is given by

(2.3.1)

$$R^n = \Big\{(u, v) \in W \times W \mid \mathfrak{F} \models \bigvee_i \bigvee_{j \neq i} \exists w \big(T_i f_{ji} uw \wedge \neg E_{ij} w \wedge H_i^n wv\big)\Big\}.$$

It is easily verified that in a *cube* $\mathbb{C}_n(U)$, we have that $H^n = {}^nU \times {}^nU$, that $H_i^n = \{(u, v) \in {}^nU \mid u_i = v_i\}$, and that the function f_{ij} is given by (taking $i = 1$ and $j = 0$) $f_{10}(u_0, \ldots, u_{n-1}) = (u_0, u_0, u_2, \ldots, u_{n-1})$. In a hypercylindric frame, the function f_{ij} is the projection along T_i on the E_{ij} diagonal. H^n and each H_i^n is an equivalence relation, and it does not matter in which order we compose the T_j relations to define them. The i-hyperplanes are the equivalence classes of H_i^n.

In order to understand the definition of R^n, consider the following figure depicting the case where $R^n uv$ because $T_0 f_{10} uw \wedge \neg E_{01} w \wedge H_0^n wv$ (that is, take $i = 0$ and $j = 0$ in (2.3.1)):

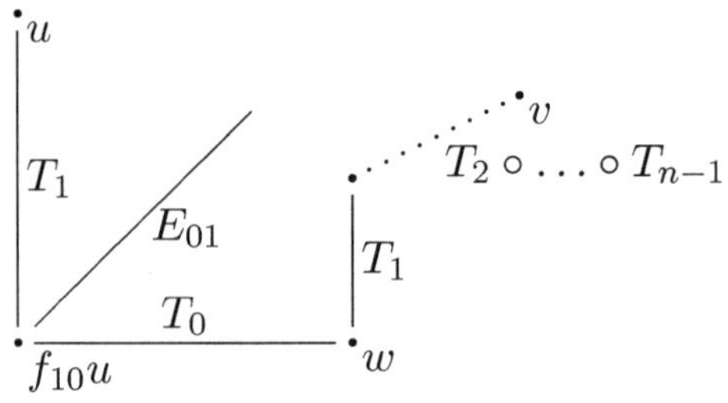

Here v lies in the hyperplane through w and 'orthogonal' to the 'line' T_0. Then, given the description, just above Definition 2.3.1, of the 'modal walk' from a tuple u to an arbitrary tuple v with $u_0 \neq v_0$, it is not hard to see that on n-cubes, R^n is the *inequality* relation. The point of our characterization is that the latter property exactly singles out the cubes among the hypercylindric frames:

Theorem 2.3.2. *Let $\mathfrak{F} = \langle W, T_i, E_{ij}\rangle_{i,j<n}$ be an n-frame. Then $\mathfrak{F}$ is isomorphic to a cube iff it is hypercylindric and R^n is the inequality relation on W.*

For a *proof* of this Theorem, which technically can be see as the frame version of Theorem 3.2.5 in [Hen-Mon-Tar,85], the reader is referred to [Ven,95b]. Concerning the result itself, one interesting observation is that it gives a *first-order* characterization of the cubes. In Theorem 2.2.6 we already saw an immediate corollary of this: the variety of representable cylindric algebras is canonical. However, as a first-order characterization, Theorem 2.3.2 is rather involved; there are certainly simpler alternatives available.

The interest in the specific form of this characterization lies in the fact that on the class of hypercylindric frames, R^n is the accessibility relation of the following compound modality.

Definition 2.3.3. Define the following abbreviated operator D_n:

$$D^n\varphi := \bigvee_i \bigvee_{j\neq i} \Diamond_j(\mathsf{d}_{ij} \wedge \Diamond_i(-\mathsf{d}_{ij} \wedge \Diamond_0 \ldots \Diamond_{i-1}\Diamond_{i+1} \ldots \Diamond_{n-1}\varphi)),$$

Note that D_n is defined to make the relation R^n act as its accessibility relation, i.e. for any hypercylindric n-model we have

$$\mathfrak{M}, u \models D_n\varphi \iff \text{there is a } v \text{ with } R^n uv \text{ and } \mathfrak{M}, v \models \varphi.$$

But then Theorem 2.3.2 states that the cubes are exactly the class of hypercylindric frames on which D_n acts as the so-called *difference operator.*

DIFFERENCE OPERATOR. The difference operator is a modality D that has the *inequality relation* $\neq$ as its intended accessibility relation:

$$\mathfrak{M}, s \models D\varphi \text{ iff } \mathfrak{M}, t \models \varphi \text{ for some } t \neq s.$$

This operator increases the expressivity of modal languages; it can be used for instance to express frame properties that are otherwise inexpressible by modal formulas, such as irreflexivity ($\Diamond p \to Dp$) or antisymmetry ($p \wedge \Diamond(\neg p \wedge \Diamond p) \to Dp$). An interesting aspect of the difference operator is that it allows one to *name* states. Consider the formula

$$\mathtt{name}_D(\varphi) := \varphi \wedge \neg D\varphi,$$

then clearly the formula $\mathtt{name}_D(\varphi)$ holds at a state s iff s is the *only* state where φ is true. For more background on the difference operator, the reader is referred to [Rij,92, Ven,93].

Now suppose that we want to *axiomatize* the behaviour of the difference modality, in a setting where D is an additional modality added to the language of, say, cylindric modal logic. Many properties of the inequality relation are easy to characterize and axiomatize: symmetry by the modal formula (D1) $p \to \neg D\neg Dp$, pseudo-transitivity ($\forall xyz\,(Rxy \wedge Ryz \to Rxz \vee x = z)$) by (D2) $DDp \to p \vee Dp$, and the inclusion property ($\forall xy\, R_ixy \to x = y \vee Rxy$) by (D3) $\Diamond_i p \to p \vee Dp$, for all diamonds $\Diamond_i$. The problem is the property of *irreflexivity* which cannot be characterized by a formula, at least not in the standard way. Consider, on the other hand, the following *derivation rule:*

$$(IR_D) \qquad \vdash \mathtt{name}_D(p) \to \varphi \Rightarrow \vdash \varphi, \text{ if } p \notin \varphi.$$

It should be obvious where this rule fails to be orthodox: any renaming which replaces a proposition letter in φ with p may transform an applicable instance of (IR_D) into a forbidden one. In order to get an understanding of what this rule does, let us first check its *soundness.* Using contraposition, assume that $\not\models \varphi$. That is, there is some model $\mathfrak{M} = (\mathfrak{F}, V)$ and a state s such that $\mathfrak{M}, s \not\models \varphi$. Now modify the valuation V to a valuation V' by changing the interpretation of p to the singleton $\{s\}$. Then $\mathtt{name}_D(p)$ is true at s by definition, whereas the formula φ remains false since its truth is not affected by changing the meaning of p. As a consequence we see that $\not\models \mathtt{name}_D(p) \rightarrow \varphi$, as required.

Often we are dealing with a situation where we do not have a *primitive* modality D, but rather some compound modality D_c which behaves like the difference operator on the intended class of frames.

Definition 2.3.4. Let D_c be some compound modality. Given a derivation system $\mathbf{\Delta}$, we define $\mathbf{\Delta}.D_c^+$ as its extension obtained by adding the D_c-versions of the axioms D(1–3) and the irreflexivity rule (IR_D).

Theorem 2.3.5. *Let S be a set of Sahlqvist axioms containing the axiom $(CM2)$ for each modality, and let D_K be a compound modality such that K is the class of n-frames that validate S and on which D_K acts as the difference operator. Then $\mathsf{K}.S.D_\mathsf{K}^+$ is sound and complete for the class K.*

Proof Sketch. To prove *soundness* is left as an exercise for the reader. The *completeness* proof falls out into three parts. First, using a multi-dimensional Lindenbaum Lemma, one may show that for every consistent formula ξ there is a set W^ξ of maximal consistent sets (MCSs) with the Properties 1–3 below (here the relation R_i between MCSs is defined as usual: $R_i\Gamma\Delta$ iff $\Diamond_i\varphi \in \Gamma$ for every $\varphi \in \Delta$):

1. There is an MCS $\Xi \in W^\xi$ containing ξ;
2. W^ξ satisfies an *Existence Lemma:* for every formula φ and for every $\Gamma \in W^\xi$ we have
$$\Diamond_i\varphi \in \Gamma \;\text{ iff }\; \varphi \in \Delta \;\text{ for some }\; \Delta \in W^\xi \;\text{ with }\; \Gamma R_i \Delta;$$
3. Every MCS has a name: for every Γ there is a proposition letter p^Γ such that
$$\mathtt{name}_{D_\mathsf{K}}\left(p^\Gamma\right) \in \Theta \;\text{ iff }\; \Theta = \Gamma.$$

Second, on the basis of these properties it makes sense to define the following variants of the canonical frame and model. $\mathfrak{F}^\xi$ is the structure $\mathfrak{F}^\xi = \left\langle W^\xi, R_i, \mathsf{D}_{ij}\right\rangle_{i,j<\alpha}$, with $\mathsf{D}_{ij} := \{\Gamma \in W^\xi \mid \mathsf{d}_{ij} \in \Gamma\}$, and $\mathfrak{M}^\xi$ is the model on $\mathfrak{F}^\xi$ determined by the valuation V given by $V(p) := \{\Gamma \in W^\xi \mid p \in \Gamma\}$. Without loss of generality we may assume that the frame $\mathfrak{F}^\xi$ is point-generated from Ξ, that is, to every $\Gamma \in W^\xi$ there is a path from Ξ following the relations R_i. The motivation behind these definitions is that they enable us to prove (by a straightforward formula induction) the following *Truth Lemma:*

$$(2.3.2) \qquad \mathfrak{M}^\xi, \Gamma \models \varphi \text{ iff } \varphi \in \Gamma,$$

for every MCS Γ and formula φ. And as an immediate corollary of the truth lemma and property (1) above it follows that the formula ξ is satisfied in the model $\mathfrak{M}^\xi$.

It remains to show that we have satisfied ξ in a model of the right *kind,* that is, we have to prove that the underlying frame $\mathfrak{F}^\xi$ belongs to the class K. By our assumption on K, this boils down to showing that (i) the axioms from S are valid on $\mathfrak{F}$, and (ii) that D_K acts as the difference operator on $\mathfrak{F}$.

For (i) we just consider a representative example: the formula $(CM4_{01})$: $\Diamond_0\Diamond_1 p \to \Diamond_1\Diamond_0 p$. By Sahlqvist correspondence (Lemma 2.1.4 above) it suffices to prove that $\mathfrak{F} \models (N4_{01})$. Assume that $\Gamma R_0 \Delta R_1 \Theta$, then by the properties established above, we find $\Diamond_0\Diamond_1 \mathtt{name}_{D_\mathsf{K}}(p^\Theta) \in \Gamma$, and so by maximal consistency, also $\Diamond_1\Diamond_0 \mathtt{name}_{D_\mathsf{K}}(p^\Theta) \in \Gamma$. This gives MCSs Π and Θ' such that $\Gamma R_1 \Pi R_0 \Theta'$ and $\mathtt{name}_{D_\mathsf{K}}(p^\Theta) \in \Theta'$. It follows that $\Theta = \Theta'$, and so Π is the required MCS such that $\Gamma R_1 \Pi$ and $\Pi R_0 \Theta$.

In order to prove (ii) one may show that the relation $R_D := \{(\Gamma, \Delta) \in W^\xi \times W^\xi \mid D\varphi \in \Gamma \text{ for all } \varphi \in \Delta\}$ is the inequality relation on $\mathfrak{F}^\xi$. For the inclusion $R_D \subseteq \neq$ one needs property 3 above, while for the opposite inclusion we use the D_c-axioms, together with the fact that $\mathfrak{F}^\xi$ is point-generated from Ξ, to show that any pair of MCSs in W^ξ is linked by the relation $R_D \cup =$. Further details are left to the reader. ∎

AXIOMATIZING THE n-CUBES. The completeness result for the n-cubes is now a fairly straightforward consequence of earlier results.

Definition 2.3.6. For finite dimensions n, $\mathbf{HCML}_n^+$ is the derivation system $\mathbf{HCML}_n$ extended with the *Irreflexivity Rule for* D_n:

$$(IR_{D_n}) \qquad \vdash \mathtt{name}_{D_n}(p) \to \varphi \Rightarrow \vdash \varphi, \text{ if } p \notin \varphi.$$

Define $\mathbf{HCML}_\omega^+$ as the system $\mathbf{HCML}_\omega$ extended with the schema of rules $\{IR_{D_n} \mid n < \omega\}$. For $\alpha > \omega$ we add besides this set, the following schema:

$$\{\vdash \varphi \Rightarrow \vdash \varphi^\tau \mid \tau : \alpha \mapsto \alpha \text{ is a bijection}\},$$

where φ^τ is the formula one obtains from φ by substituting $\Diamond_{\tau(i)}$ and $\mathsf{d}_{\tau(i)\tau(j)}$ for every occurrence of $\Diamond_i$ resp d_{ij}.

The following result is due to Venema [Ven,95b].

Theorem 2.3.7. *For any ordinal α, the derivation system* $\mathbf{HCML}_{\alpha}^{+}$ *is a sound and complete axiomatization for the class of α-dimensional cubes. That is, for all CML_{α}-formulas φ:*

$$\vdash_{\alpha}^{+} \varphi \text{ iff } \mathsf{C}_{\alpha} \models \varphi.$$

Proof. Again we leave it for the reader to prove soundness. For completeness, the case for finite α is a more or less straightforward corollary to the Theorems 2.3.2 and 2.3.5. (We omit the rather technical details of the proof that the D_n-versions of the D-axioms are derivable in the logic $\mathbf{HCML}_n$.)

When it comes to infinite dimensions we confine ourselves to the case $\alpha = \omega$. Let φ be an ω-formula such that $\mathsf{C}_{\omega} \models \varphi$. As there are only finitely many symbols occurring in φ, there is an $n < \omega$ such that φ is an n-formula. A relatively simple argument shows that for all ordinals $\beta < \gamma$, and all β-formulas ψ : $\mathsf{C}_{\beta} \models \psi$ iff $\mathsf{C}_{\gamma} \models \psi$. From this we conclude that our φ is valid in C_n, so that by finite-dimensional completeness we obtain $\vdash_n^{+} \varphi$. Now $\vdash_{\omega}^{+} \varphi$ follows as $\mathbf{HCML}_{\omega}^{+}$ is an extension of $\mathbf{HCML}_n^{+}$. ■

Given the connections between normal cylindric modal logics and equational theories, and the equivalence (2.1.5) of cube validity, type-free validity, and schema validity for CML_{α}-formulas, it is straightforward to verify that Theorem 2.3.7 also provides finite, complete axiomatizations for the equational theory of RCA_{α}, and for both type-free and first-order schema validity. The system $\mathbf{HCML}_{\omega}^{+}$ thus indicates a positive solution to Problem 4.16 of [Hen-Mon-Tar,85].

From the perspective of cylindric algebras, what goes on here can be reformulated as follows. Let HCA_n be the class of n-dimensional cylindric algebras satisfying the additional equation corresponding to the axiom ($CM8$), and consider Theorem 3.2.5 from [Hen-Mon-Tar,85] which states that an algebra in HCA_n is representable if it is *rich,* that is, it has sufficiently many elements that are satisfy a certain equation. The point is that this richness condition may be transformed into a *derivation rule* that is non-orthodox in the sense discussed at the beginning of this section. Nevertheless, this rule is sound, and when we add it to the equational axiomatization for HCA_n we obtain a finite, complete axiomatization for the variety of representable cylindric algebras.

Mutatis mutandis, this approach works in other situations as well. For instance, in [Ven,98] the author obtained a finite axiomatization for the class of representable diagonal-free algebras, using a nonorthodox derivation rule inspired by a representation result for so-called *rectangularly dense* algebras. (This and related notions of density, including the above-mentioned concept of 'richness', are discussed in detail in [And-Giv-Mik-Nem-Sim,98].) A different type of rule was used by Simon [Sim,91] to obtain a complete axiomatization for the type-free valid formulas (and hence, for the equational theory of RCA_ω).

IV. Applications of Cylindric-Like Algebras

BOLYAI SOCIETY
MATHEMATICAL STUDIES, 22

Cylindric-like Algebras and Algebraic Logic
pp. 273–301.

Crs and Guarded Logics: A Fruitful Contact

JOHAN VAN BENTHEM

Back and forth between algebra and model theory. Algebra and model theory are complementary stances in the history of logic, and their interaction continues to spawn new ideas, witness the interface of First-Order Logic and Cylindric Algebra. This chapter is about a more specialized contact: the flow of ideas between algebra and modal logic through 'guarded fragments' restricting the range of quantification over objects. Here is some general background for this topic. For a start, the connection between algebra and model theory is rather tight, since we can view universal algebra as the equational logic part of standard first-order model theory. As an illustration, van Benthem [Ben,88] has a purely model-theoretic proof of Jónsson's Theorem characterizing the equational varieties with distributive lattices of congruence relations, a major tool of algebraists. Deeper connections arise in concrete cases with categorial dualities, such as that between BAOs and the usual relational models of modal logic. An important example is the main theorem in Goldblatt and Thomason [Gol-Tho,74] characterizing the elementary modally definable frame classes through their closure under taking generated sub-frames, disjoint unions, p-morphic images, and anti-closure under ultrafilter extensions. Its original proof goes back and forth between algebras and frames, in order to apply Birkhoff's characterization of equational varieties. Later on, van Benthem [Ben,93] gave a purely model-theoretic proof, replacing trips into algebraic territory by the use of ω-saturated models. Even so, the trade between algebra and logic remains interesting, even when it is not a matter of applying concrete theorems, but exporting more general ideas. The present chapter is a case in point. So-called *relativization* started as a technique for generalizing relational and cylindric algebra, while also, in some cases, 'defusing' the undecidability of these systems. But as we shall see, mainly based on the results in Andréka, van Benthem and Németi [And-Ben-Nem,98], van Benthem [Ben,05], it has traveled well into standard logic and model theory.

1. From Crs and Relational Algebra to Modal Arrow Logic

1.1. Relativization: technique and motivation

Relativization. Relativization in relational algebra restricts the available pairs on an underlying set Y to some subset, i.e., some largest binary relation U, not necessarily the full product $Y \times Y$. Computation of values for complex algebraic terms still proceeds via the usual operations of complementation, union, composition, etc., but with their clauses relativized to work inside the set U. In particular, with compositions, we have a clause

$$x\ R; S\ y \text{ iff there exist pairs } (x, z), (z, y) \in U \text{ with } xRz \text{ and } zSy.$$

This extension of the set of models leaves several base laws of relational algebra valid. But others become invalid: Associativity $(R; S); T = R; (S; T))$ is a typical example. Crucially, in this contraction process, the set of algebraic validities becomes decidable. But if we impose additional conditions on the relation U, then undecidability may re-appear. For instance, if we require *transitivity,* relational algebra is undecidable again: the set of available pairs then looks 'too much' like the full Cartesian product $Y \times Y$. For further details, we refer to Németi [Nem,85a], [Nem,95] as well as several chapters in this Book. The system *Crs,* another product of the well-known 'Budapest School' of Hajnal Andréka, István Németi and their students develops analogous ideas for all of cylindric algebra, with similar effects – and it, too, has generated a large subsequent literature (cf. Venema [Ven,91], Marx [Mar,95], Mikulas [Mik,95], Marx and Venema [Mar-Ven,97], Venema [Ven,07], Ferenczi [Fer,thisVol,b]).

These ideas have counterparts in logic, and they have been influential in several ways. Relativization in relational algebra suggests a modal perspective where transitions are now viewed as objects in their own right ('arrows'), in addition to points or states, while algebraic terms now correspond to modal formulas defining properties of transitions. For the development of this 'Arrow Logic', cf. van Benthem [Ben,91], Venema [Ven,91], [Ven,96]. Likewise, *Crs* has influenced new logical systems, in particular 'cylindric modal logic' (Venema [Ven,95b]) and 'first-order dependence logic' (van Benthem [Ben,97b]), where gaps in the total space of variable assignments model the important phenomena of *dependence* and independence between variables, that have come to the fore recently with many authors (van den Berg [Ber,96], Hodges [Hod,01], Väänänen [Vaa,07]). Both lines from algebra to logic, arrow logic and dependence logics, will be discussed in the course of this chapter.

GENERALIZED MODELS. Extending original classes of models for logics to manipulate their properties is widespread. The famous move from 'standard models' to 'general models' in Henkin [Hen,49] turns the complex system of second-order logic into an axiomatizable two-sorted first-order logic (van Benthem and Doets [Ben-Doe,83]). Such moves are most attractive when they get an independent motivation. For relational algebra and cylindric algebra, this is provided by what van Benthem [Ben,96b] called 'content versus wrappings' in logical modeling. Intuitively, the core calculus of action embodied in relational algebra seems simple, and undecidability comes as a surprise. Thus, we want to find a semantics that gives just the bare bones of action, while additional effects of 'standard set-theoretic modeling' are separated out as negotiable decisions of formulation that engender the undecidability. This theme underlies the systems presented in this chapter.

FRAGMENTS. But there is also a quite different technical way of viewing relativization as a general logical device. Already Wadge [Wad,75] showed how relational algebra can be axiomatized smoothly by using pair notation $(x, y) : R$, making transitions explicit as objects, which suggests viewing the algebra as a fragment of first-order logic. Now it is a well-known result of Tarski's that standard relational algebra translates into the undecidable 3-*variable fragment* of first-order logic, through clauses such as

$$R;S(x,y) \longleftrightarrow \exists z\big(R(x,z) \wedge S(z,y)\big)$$

which typically use existential quantification over objects in the domain. But the clause in our earlier description replaces this by another syntactic format, namely

$$R;S(x,y) \longleftrightarrow \exists z\big(U(x,z) \wedge U(z,y) \wedge R(x,z) \wedge S(z,y)\big).$$

Thus, we end up inside a *sub-language* of the 3-variable fragment, where patterns of quantification are restricted or 'guarded' in some way by atomic formulas. Similar points hold for *Crs* and first-order dependence logics, and the result there is that we end up in a sub-language of full first-order logic known as the *Guarded Fragment.*

In this paper, we will develop this fragment view as well, and eventually, we will also address the following fundamental question about our presentation so far. What is the systematic connection between the two lines of (a) taking a logical language and extending its class of models, and (b) retaining the original model class while restricting the language?

1.2. Arrow logic in a nutshell

MOTIVATION: CORE CONTENT VERSUS WRAPPINGS. Relational algebra is a calculus of transition relations modeling actions in general. But then its undecidability raises an issue, since basic action does not seem to involve high complexity. We want a dynamic core logic avoiding spurious complexity of 'wrappings': that is, accidents of formulation. This motivates *Arrow Logic* (van Benthem [Ben,91], Venema [Ven,91], [Ven,96]), inspired by the *Crs* version of relational algebra, taking transitions seriously as objects in their own right.

MODELS AND LANGUAGE. Intuitively, binary relations denote sets of arrows. Think of 'arcs' in multi-graphs, 'transitions' for dynamic procedures in computer science, or 'preferences'. Arrows may have internal structure beyond ordered pairs $\langle$*source state, target state*$\rangle$: several arrows may share one input-output pair, but also certain pairs may not be instantiated by an arrow. This motivates the following abstract notion:

Definition 1.1.1 (Arrow Frames). *Arrow Frames* $\mathbf{F} = \langle A, C^3, R^2, I^1 \rangle$ have objects A ('arrows') with predicates C^3x, yz (x is a 'composition' of y, z), R^2x, y (y is a 'reverse' of x), I^1x (x is an 'identity' arrow).

Arrow frames do not identify transitions with ordered pairs of states. Distinct arrows may have the same pair $\langle$input, output$\rangle$, and not every such pair need have an arrow. Indeed, *Crs*-algebra suggests that arrows be ordered pairs, while giving up the idea that *all* ordered pairs are available. The resulting arrow frames need not be full Cartesian products of some state space. An even more radical version comes from category theory, with objects and morphisms. Let arrows be *functions* $f : A \to B$ inducing ordered pairs $\langle A, B \rangle$ of 'source' and 'target'. Then, the relation C expresses the partial function of composing maps, while reversal R holds between a map and its inverse, if available. Quite different interpretations and applications may be found in Kurtonina [Kur,95], which analyzes composition of linguistic expressions in the 'Lambek Calculus' of categorial grammar in arrow logic. Andréka and Mikulas [And-Mik,93] translate this system back into relational algebra.

LANGUAGE AND TRUTH DEFINITION. Arrow frames **F** support a modal language that analyzes Relational Algebra. *Arrow Models* **M** add a propositional valuation V, and one can then interpret a matching modal propositional language defining properties of arrows using two modalities reflecting the basic 'ordering operations' of relational algebra:

$\mathbf{M}, x \models p$ iff $x \in V(p)$,

$\mathbf{M}, x \models \neg\phi$ iff *not* $\mathbf{M}, x \models \phi$,

$\mathbf{M}, x \models \phi \,\&\, \psi$ iff $\mathbf{M}, x \models \phi$ *and* $\mathbf{M}, x \models \psi$,

$\mathbf{M}, x \models \phi \cdot \psi$ iff *there exist* y, z *with* Cx, yz *and* $\mathbf{M}, y \models \phi$, $\mathbf{M}, z \models \psi$,

$\mathbf{M}, x \models \phi\breve{}$ iff *there exists* y *with* Rx, y *and* $\mathbf{M}, y \models \phi$,

$\mathbf{M}, x \models Id$ iff Ix.

Eventually, one can also introduce more expressive modal operators into this basic vocabulary.

MODAL LOGIC. The *minimal logic* of arrow models is an obvious counterpart of its mono-modal version, whose key principles are the following axioms of Modal Distribution:

$$\begin{aligned}
(\phi_1 \vee \phi_2) \cdot \psi &\leftrightarrow (\phi_1 \cdot \psi) \vee (\phi_2 \cdot \psi), \\
\phi \cdot (\psi_1 \vee \psi_2) &\leftrightarrow (\phi \cdot \psi_1) \vee (\phi \cdot \psi_2), \\
(\phi_1 \vee \phi_2)\breve{} &\leftrightarrow \phi_1\breve{} \vee \phi_2\breve{}.
\end{aligned}$$

A completeness theorem is provable here along standard lines, using Henkin models, with the usual techniques as explained, for instance, in Blackburn, de Rijke and Venema [Bla-Rij-Ven,01]. The minimal logic is also decidable, again using a standard modal technique such as filtration.

1.3. Arrow logic and relational algebra via modal correspondence

LANDSCAPE OF ARROW AXIOMS AND FRAME CORRESPONDENCE. On top of this minimal system, one can analyze axioms from Relational Algebra via constraints on arrow frames via *frame correspondences.* This analysis reveals a whole landscape of options. We only state results without proof, all of them follow by standard correspondence techniques using the 'Sahlqvist form' of the relevant axioms (Blackburn, de Rijke and Venema [Bla-Rij-Ven,01]):

Example 1.1.2 (Laws for Arrow Reversal).

$$\neg(\phi)^{\breve{}} \to (\neg\phi)^{\breve{}} \text{ iff } \forall x\, \exists\, y\ Rx, y, \tag{1.1.1}$$

$$(\neg\phi)^{\breve{}} \to \neg(\phi)^{\breve{}} \text{ iff } \forall xyz : (Rx, y \,\&\, Rx, z) \to y = z. \tag{1.1.2}$$

These axioms make the binary relation R a unary *function* r of 'reversal'. Then a 'double conversion' axiom makes the function r *idempotent:*

$$(\phi)^{\breve{}\breve{}} \leftrightarrow \phi \text{ iff } \forall x : r\big(r(x)\big) = x. \tag{1.1.3}$$

With this notation, our next axioms connect reversal and composition:

Example 1.1.3 (Arrow Composition Triangles).

$$(\phi \cdot \psi)^{\breve{}} \to \psi^{\breve{}} \cdot \phi^{\breve{}} \text{ iff } \forall xyz : Cx, yz \to Cr(x), r(z)r(y), \tag{1.1.4}$$

$$\phi \cdot \neg(\phi^{\breve{}} \cdot \psi) \to \neg\psi \text{ iff } \forall xyz : Cx, yz \to Cz, r(y)x. \tag{1.1.5}$$

Together (1.1.2), (1.1.4), (1.1.5) imply the further interchange law $\forall xyz :$ $Cx, yz \to Cy, xr(z)$. Moreover, there is actually a more elegant form of axiom (1.1.5) without negation:

$$\phi \,\&\, (\psi \cdot \chi) \to \psi \cdot \big(\chi \,\&\, (\psi^{\breve{}} \cdot \phi)\big).$$

Finally, the propositional constant *Id* constrains 'identity loops'.

Example 1.1.4 (Identity Arrows).

$$Id \to Id^{\breve{}} \text{ iff } \forall x : Ix \to Ir(x), \tag{1.1.6}$$

$$Id \cdot \phi \to \phi \text{ iff } \forall xyz : (Iy \,\&\, Cx, yz) \to x = z. \tag{1.1.7}$$

CRS-STYLE CORE LOGIC. In our correspondence analysis of the basic axioms of Relational Algebra, some constraints come out as purely *universal*, making no demands on the supply of arrows. These seem the true core of action or computation. Universal frame constraints express laws for composition, converse and identity of arrows that lack existential import: by purely universal first-order sentences over arrow frames.

Fact 1.1.5. *The complete logic of arrow models satisfying all universal frame constraints valid in Relational Algebra is the set of validities for concrete Crs-style* pair arrow models *where arrows are ordered pairs – and the only change from standard models for relational set algebras is the limited supply of pairs.*

This is not difficult to see. By contrast, existential constraints force the arrow set to become more like full Cartesian spaces, i.e., the standard models leading to undecidability.

Remark 1.1.6 (Associativity is existential)**.** One perhaps counter-intuitive feature of this analysis concerns Associativity for composition. Its frame condition is existential, requiring regroupings of transitions:

$$(\phi \cdot \psi) \cdot \chi \rightarrow \phi \cdot (\psi \cdot \chi) \text{ iff } \forall\, xyzuv : (Cx, yz \,\&\, Cy, uv) \rightarrow \exists\, w : (Cx, uw \,\&\, Cw, vz))\,, \tag{1.1.8a}$$

(1.1.8b) and likewise in the opposite direction.

Now Associativity is often considered a useful core feature of dynamic logics: different orders of composition are equivalent. In contrast, *Crs*-style arrow logic distinguishes different ways of 'chunking' parts, while Associativity can lead to undecidability. Similar points are about complexity known from categorial logics (van Benthem [Ben,96a], Chapter 12).

1.4. Arrow logic over pair models

AXIOMATICS. Here is a complete axiomatization for the logic of pair arrow models with the obvious definitions for composition, reversal and identity (due to Marx [Mar,95]):

Theorem 1.1.7. *The following set of principles is complete for the logic of pair arrow models:*

(i) *the minimal arrow logic,*

(ii) *converse is an idempotent function,*

(iii) *identity laws* $Id \leftrightarrow Id \cdot Id$, $Id \leftrightarrow Id\breve{}$, $\phi \,\&\, Id \rightarrow \phi\breve{}$, $Id \cdot \phi \leftrightarrow \phi$, $\phi \cdot Id \leftrightarrow \phi$, $\phi \cdot \neg\phi\breve{} \rightarrow Id$,

(iv) *'triangle laws'* $\phi \cdot \neg(\phi^{\smile} \cdot \psi) \to \neg\psi$, $\neg(\phi \cdot \psi^{\smile}) \cdot \psi \to \neg\phi$,

(v) *the limited associativity principles*

$$\big((\phi \,\&\, Id) \cdot \psi\big) \cdot \chi \leftrightarrow (\phi \,\&\, Id) \cdot (\psi \cdot \chi),$$

$$\big((\phi \cdot (\psi \,\&\, Id)\big) \cdot \chi \leftrightarrow \phi \cdot \big((\psi \,\&\, Id) \cdot \chi\big),$$

$$(\phi \cdot \psi) \cdot (\chi \,\&\, Id) \leftrightarrow \phi \cdot \big(\psi \cdot (\chi \,\&\, Id))\big).$$

Other pleasant properties of this system include decidability and Craig interpolation. Many further results are in Marx [Mar,95], Mikulas [Mik,95], while Marx and Venema [Mar-Ven,97] is an excellent general source, also referencing the seminal work of the Budapest group.

LANGUAGE EXTENSIONS: DYNAMIC ARROW LOGIC. Pair arrow logic retains its nice properties when we extend its vocabulary with a universal modality, or other operators. These moves exemplify a program of trading deductive strength for expressive power. One important extension yields *Dynamic Arrow Logic,* with an infinitary operator for composition similar to the crucial Kleene iteration of propositional dynamic logic:

$\mathbf{M}, x \models \phi^*$ iff *x can be C-decomposed into some finite sequence* of arrows satisfying ϕ in $\mathbf{M}$.

Defined in this way, ϕ^* satisfies the following three simple laws that yield the system *DAL*:

(1.1.9) $$\phi \to \phi^*,$$

(1.1.10) $$\phi^* \cdot \phi^* \to \phi^*,$$

(1.1.11) if $\vdash \phi \to \alpha$ and $\vdash \alpha \cdot \alpha \to \alpha$, then $\vdash \phi^* \to \alpha$.

Theorem 1.1.8 (van Benthem). *DAL is complete for its intended models* ([Ben,94]).

Remark 1.1.9 (Dynamic logic as two-sorted arrow logic). One can mirror dynamic logic more closely by combining arrow logic with a modal logic of states. This involves operators of *test* **?** from propositions to programs, and *domain* $\langle\ \rangle$ from programs to propositions. These are non-homogeneous distributive modalities correlating states and arrows:

$$\mathbf{M}, x \models \phi? \text{ iff } \textit{there exists some } s \textit{ with } Tx, s \textit{ and } \mathbf{M}, s \models \phi,$$

$$\mathbf{M}, x \models \langle \pi \rangle \text{ iff } \textit{there exists some } x \textit{ with } Ds, x \textit{ and } \mathbf{M}, x \models \pi.$$

Intuitively, the relation Tx, s says that x is an identity arrow for the point s, while Ds, x says that s is a left end-point of the arrow x. Via modal frame correspondences, axioms on $?$, $\langle\ \rangle$ impose special connections between T and D. For instance, $\langle \phi ? \rangle \leftrightarrow \phi$ expresses the conjunction of $\forall s \, \exists \, x \, : Ds, x \,\&\, Tx, s$ and $\forall \, sx \, : \, Ds, x \rightarrow \forall \, s' \, : \, Tx, s' \rightarrow s = s'$.

This system is related in spirit to the abstract 'action algebras' of Pratt [Pra,95].

1.5. A fragment view?

In Section 1, we claimed that adopting a generalized semantics may sometimes be reinterpreted as a move to restricted syntax. Indeed, Arrow Logic or *Crs*-versions of relational algebra, may be translated into fragments of first-order logic involving guarded quantification. But we will explain this connection only later on, in Section 4 below.

2. From Crs to General Assignment Models for First-Order Logic

2.1. The core mechanics of first-order semantics, a modal view

The Main Recursion. The standard semantics for predicate logic has the following key clause:

$$\mathbf{M}, \alpha \models \exists \, x\phi \text{ iff } \textit{for some } d \in |\mathbf{M}| \, : \, \mathbf{M}, \alpha^x_d \models \phi.$$

The key is the use of variable assignments α that decompose quantified statements with free variables in their matrix. But looking more closely at the usual truth definition, a compositional semantics for first-order quantification only needs the following abstract core pattern:

$$\mathbf{M}, \alpha \models \exists \, x\phi \text{ iff } \textit{for some } \beta \, : \, R_x\alpha\beta \textit{ and } \mathbf{M}, \beta \models \phi.$$

Here, assignments α, β become abstract states, and the concrete relation $\alpha =_x \beta$ – which held between α and α_d^x – becomes just any binary relation R_x. This involves poly-modal models $\mathbf{M} = \langle S, \{R_x\}_{x \in VAR}, I \rangle$ with S a set of *states,* R_x a binary *update relation* for each variable x, and I an *interpretation function* giving truth values to atomic formulas $Px, Rxy, \ldots$ in each state α. Thus existential quantifiers $\exists\, x$ become existential modalities $\langle x \rangle$. This abstract modal semantics has an independent appeal: first-order evaluation is an informational process that changes computational states, as in the 'dynamic semantics' of Groenendijk and Stokhof [Gro-Sto,92], Veltman [Vel,96]. First-order logic then becomes a *dynamic logic* with a special choice of atoms and no explicit compound programs.

'DECONSTRUCTION'. From this modal point of view, standard semantics arises by insisting on three additional mathematical choices, not enforced by the core semantics:

(a) States are identified with variable assignments,

(b) Update between states must be the specific relation $=_x$, and

(c) All assignments in the function space D^{VAR} are available to evaluation.

The former are issues of implementation, but the latter is a strong existence assumption. (Actually, standard first-order logic needs only *locally finite* assignments.) Henceforth, we shall regard these further choices as negotiable. In fact, it is often felt that tricks like making predicates sets of tuples should be orthogonal to the nature of logical validity.

MINIMAL LOGIC. Our modal semantics for the first-order language validates the *minimal poly-modal logic* with

- all classical Boolean propositional laws.
- Modal Distribution: $\forall\, x(\phi \to \psi) \to (\forall\, x\phi \to \forall\, x\psi)$.
- Modal Necessitation: if $\vdash \phi$, then $\vdash \forall\, x\phi$.
- a definition of $\exists\, x\phi$ as $\neg\forall\, x\neg\phi$.

This logic is complete, and has the usual properties of first-order logic, such as Craig interpolation or the Łoš–Tarski preservation theorem. One can now pursue standard first-order model theory in tandem with its modal counterpart. For instance, consider modal *bisimulations* for these models, relating states making the same atoms true, with zigzag conditions for the relations R_x. Specializing these to standard models leads to the standard notion of *potential isomorphism* (de Rijke [Rij,93], van Benthem [Ben,96a], van Benthem and Bonnay [Ben-Bon,08]). And in all this, the modal system is *decidable.*

LANDSCAPISM. The modal perspective suggests a landscape below standard predicate logic, with a minimal modal logic at the base, ascending to standard semantics via frame constraints. In particular, this landscape contains *decidable sublogics* of predicate logic, sharing its desirable meta-properties. Thus, the 'undecidability of predicate logic' largely reflects mathematical accidents of its Tarskian modeling, encoding set-theoretic facts about function spaces D^{VAR} – beyond the core logic of quantification and variable assignment. We shall explore this view of first-order semantics in what follows, including richer languages. Abstract core models support new distinctions between various forms of quantification ('monadic' and 'polyadic') that get collapsed in standard predicate logic.

2.2. From modal state models to general assignment models

To recapitulate, we have just re-interpreted first-order logic as a modal logic on a much more general class of abstract *modal state models*

$$\mathbf{M} = \langle S, \{R_x\}_{x \in VAR}, I \rangle$$

with S a set of 'states', R_x a binary accessibility relation between states for each variable x, and I an interpretation function giving a truth value to each atomic formula in each state s. This is a huge extension of standard semantics, where no domain of 'individual objects' need now be present underpinning the states. Quantifiers become modalities:

$$\mathbf{M}, s \models \exists\, x\phi \text{ iff } \textit{for some } t : R_x st \textit{ and } \mathbf{M}, t \models \phi.$$

More concrete is the following halfway house, an intermediate semantics that retains assignments as the state space – just taking away the existential assumption of 'fullness' from standard Tarski models for first-order logic.

Definition 1.2.1 (General assignment models). A *general assignment model* is an ordered pair $(\mathbf{M}, V)$ with $\mathbf{M}$ a standard first-order model with domain D and interpretation function I, and V a non-empty set of assignments on $\mathbf{M}$, i.e., a subset of D^{VAR}. The first-order language is interpreted as usual, now at triples $\mathbf{M}, V, s$ with $s \in V$ – with the following clause for quantifiers:

$$\mathbf{M}, V, s \models \exists\, x\phi \text{ iff for some } t \in V : s =_x t \text{ and } \mathbf{M}, V, t \models \phi.$$

Here $=_x$ is the relation between assignments of identity up to x-values.

Remark 1.2.2 (Independent motivations). The 'gaps' in general assignment models are not just a trick for lowering complexity: they may be seen as modeling the natural phenomenon of *dependencies* between variables. Changes in value for one variable x may induce, or at least be correlated with, changes in value for another variable y. Hintikka and Sandu [Hin-San,97], Hodges [Hod,01], Väänänen [Vaa,07] are other recent approaches to dependence as a core notion in logic (see the website `http://www.illc.uva.nl/lint/documents.php` for later developments until 2011). Also, van Benthem and Martinez [Ben-Mar,08] point out how the above modal perspective on dependence given here may also be related to the analysis of constraint-based *information flow* in situation semantics (cf. Barwise and Seligman [Bar-Sel,95], Mares [Mar,03]).

2.3. Complete base logic

The complete set of validities for the new semantics is still well-behaved:

Theorem 1.2.3. *The logic of general assignment models is completely axiomatized by the standard axioms for the poly-modal version of the logic* $S5$ *plus all atomic 'locality principles'* $(\neg)P\mathbf{x} \to \forall\mathbf{y}(\neg)P\mathbf{x}$ *with* $\mathbf{x} \cap \mathbf{y} = \emptyset$.

This complete logic *'Crs'* has been much studied (excellent sources are Németi [Nem,95], Marx and Venema [Mar-Ven,97]). The completeness proof involves various representation arguments. Some results from van Benthem [Ben,96a], Sections 9.8 and 9.9, give the flavour. Let $\mathbf{x}$, $\mathbf{y}$ be finite sequences of variables. The notation $R_{\mathbf{x}}$ denotes the sequential composition of accessibility relations R_x as they occur in their given order in $\mathbf{x}$.

Theorem 1.2.4. *An abstract modal frame* $(S, \{R_x\}_{x \in VAR})$ *is isomorphic to the frame of a general assignment model iff the* R_x *are equivalence relations satisfying two 'Path Principles'*

(a) *if* $sR_{\mathbf{z}1}t, \ldots, sR_{\mathbf{z}k}t$, *and the only variable occurring in all* $\mathbf{z}_1, \ldots, \mathbf{z}_k$ *is* x, *then* $sR_x t$,

(b) *if no variable occurs in all of* $\mathbf{z}_1, \ldots, \mathbf{z}_k$, *then* $s = t$.

Less is needed if we accept a weaker equivalence than isomorphism:

Theorem 1.2.5. *A finite modal model is bisimilar to a general assignment model if and only if its accessibilities are all equivalence relations.*

Typically not universally valid in arbitrary modal models are the following first-order principles:

(i)	$[\mathbf{u}/\mathbf{y}]\psi \to \exists\, \mathbf{y}\psi$	with $\mathbf{u}$ free for $\mathbf{y}$ in ψ	Existential Generalization,
(ii)	$\phi(\mathbf{x}) \to \forall\, \mathbf{y}\phi(\mathbf{x})$	with no $\mathbf{y}$ free in $\phi(\mathbf{x})$	Full Locality.

Viewed positively again, these failures reflect the special handling of variables in models where not all assignments need be available. All of $x, y, z, \ldots$ then acquire a sort of 'individuality', due to interactions with other variables. As we said earlier, variables can now have or lack dependencies, which again gives them a certain life of their own.

Crs is also decidable, by modal filtration techniques. We omit details.

2.4. Language extensions

General assignment models do not just make first-order logic weaker. They also support an enrichment, in supporting *new vocabulary* reflecting distinctions that could not yet be seen in standard first-order logic. Examples are irreducibly *polyadic quantifiers* $\exists\, \mathbf{x}$ binding tuples of variables $\mathbf{x}$, with the following truth condition:

$$\mathbf{M}, V, s \models \exists\, \mathbf{x}\phi \ \text{ iff } \ \textit{for some} \ \ t \in V\ :\ s =_{\mathbf{x}} t \ \ \textit{and} \ \ \mathbf{M}, V, t \models \phi.$$

This time, $=_{\mathbf{x}}$ is identity between assignments up to values for all the variables in $\mathbf{x}$. In standard first-order logic, the notation $\exists\, xy \cdot \phi$ is just short-hand for $\exists\, x\, \exists\, y\phi$ or $\exists\, y\, \exists x\phi$ in any order. But in *GAM*-semantics, these two expressions are no longer equivalent, as not all 'intermediate assignments' for x- or y-shifts need be present – and they are both non-equivalent to $\exists\, xy$, as defined just now. Moreover, one can also interpret single or polyadic *substitution* operators directly in this modal operator style (cf. Venema [Ven,94]):

$$\mathbf{M}, V, s \models [\mathbf{y}/\mathbf{x}]\phi \text{ iff } s[\mathbf{x} := s(\mathbf{y})] \in V \;\&\; \mathbf{M}, V, s[\mathbf{x} := s(\mathbf{y})] \models \phi.$$

Further background for the systems discussed here can be found in [Ven,thisVol].

2.5. General semantics for non-first-order fixed-point languages

General assignment models also suggest new perspectives on non-first-order systems, in particular fixed-point logics of computation and action in general. Consider the *fixed-point version LFP(FO)* of first-order logic (Ebbinghaus and Flum [Ebb-Flu,95]). This language extends the usual inductive formation rules for first-order syntax with an operator

$$\mu P, \mathbf{x} \cdot \phi(P, \mathbf{Q}, \mathbf{x})$$

where P may occur only positively in $\phi(P, \mathbf{Q}, \mathbf{x})$, and $\mathbf{x}$ is a tuple of variables of the right arity for P. The relevant predicates are the smallest fixed-points of the following monotone set operation on predicates in any given model $\mathbf{M}$:

$$F_{\phi}^{M} = \lambda P \cdot \{\mathbf{d} \textit{ in } M \mid (\mathbf{M}, P),\ \mathbf{d} \models \phi(P, \mathbf{Q})\}.$$

With the fixed-point theorems underpinning this system, we see a process of successive approximation for the predicate P that involves changing assignments through ordinal stages. In this process, the full space D^{VAR} is usually taken for granted and computing the map F, may depend on the available assignments. Thus, in our present terms, the 'relativized' version of *LFP(FO)* is worth exploring. We make a few observations (cf. van Benthem [Ben,05]).

The syntax for formulas ϕ in the language *LFP(FO)* now needs a bit more care, since variables are less 'anonymous' in general assignment models, as we noted before. In particular, when defining a predicate $\mu P, \mathbf{x} \cdot \phi(P, \mathbf{Q}, \mathbf{x})$, the particular variables $\mathbf{x}$ matter. This suggests that we are only defining values for the specific atom $P\mathbf{x}$, whereas variants such as $P\mathbf{y}$ must be viewed as substitution instances $[\mathbf{y}/\mathbf{x}]P\mathbf{x}$. With this understanding, we can give a definition of semantic evaluation as before:

Definition 1.2.6 (*GAM* fixed-point evaluation). Formulas ϕ in the above language induce the following map in general assignment models $(\mathbf{M}, V)$ with some given assignment s of objects to the free variables in ϕ:

$$F_\phi^{M,s} = \lambda P \cdot \{\mathbf{d} \textit{ in } M \mid s[\mathbf{x} := \mathbf{d}] \in V \ \&\ (\mathbf{M}, P),\ s[\mathbf{x} := \mathbf{d}] \models \phi(P, \mathbf{Q})\}.$$

Smallest and greatest fixed-points are then defined as usual.

Example 1.2.7 (Transitive closure of guard predicates). Consider the fixed-point formula $\phi = \mu P, x \cdot Qx \vee \exists\, y[y/x]Px$. Its approximation sequence as defined above starts with the empty set for P, and it ends by stage ω, where iteration of the map $F_\phi^{M,s}$ produces nothing new. Here are some initial stages:

$$\begin{aligned}
P^0 &= \emptyset,\\
P^1 &= \{d \mid s[x := d] \in V \ \&\ (\mathbf{M}, P^0),\ s[x := d] \models Qx \vee \exists\, y[y/x]Px\}\\
&= \{d \mid s[x := d] \in V \ \&\ Q(d)\},\\
P^2 &= \{d \mid s[x := d] \in V \ \&\ (\mathbf{M}, P^1),\ s[x := d] \models Qx \vee \exists\, y[y/x]Px\}\\
&= \Big\{d \mid s[x := d] \in V \ \&\ \big(Q(d) \vee \textit{for some object } e :\\
&\qquad s[x := d][y := e] \in V \ \&\ s[x := e][y := e] \in V \ \&\ Q(e)\big)\Big\}.
\end{aligned}$$

Iteratively, one computes the set of all objects d for which there exists an object e satisfying Q that is reachable from d in the transitive closure of the following relation:

$$R_s ab \ \text{ iff } \ s[x := a][y := b], s[x := b][y := b] \in V.$$

We forego further details here – but note how computations in the above style bring to light fixed-point computations bring to light the hidden dependency structure of our general model.

Theorem 1.2.8. *LFP(FO) is decidable over general assignment models.*

Proof. This follows from Grädel's result on the decidability of the fixed-point version *LFP(GF)* of the Guarded Fragment defined in Section 4 below. The translation *guard* given there from arbitrary formulas to guarded ones easily extends to a language with added fixed-point operators. Next, translations are still inside the language of *LFP(GF)*, and fixed-point evaluation must stay inside the set of tuples satisfying the guard relation R. ■

What this brief exploration suggests are far more general uses for general assignment models in a generalized abstract model theory for arbitrary extensions of first-order logic (cf. van Benthem and Bonnay [Ben-Bon,08]).

3. Basics of the Guarded Fragment

Next, we look at the other way of importing relativization into logic, through the syntax of suitably chosen fragments of standard logical languages. The system that follows arose from a combination of two sources. One was cylindric relativized set algebra and its generalized models for first-order logic, the other a reflection on what makes modal logic tick as a source of decidable well-behaved fragments of first-order logic (van Benthem [Ben,95]). The two ways of thinking came together in the following large sublanguage of first-order logic, which again has an independent intuitive motivation.

3.1. Guarded syntax

The Guarded Fragment of Andréka, van Benthem and Németi [And-Ben-Nem,98] is a decidable part of first-order syntax with a semantic philosophy: quantifiers only access the total domain of individual objects 'locally' by means of predicates over objects. But there is more to the ambitions of guarding as a method, as will become clear in due course.

Here are some syntactic preliminaries. In what follows, mostly for convenience, we consider only first-order languages with predicate symbols and variables: no function symbols or identity predicates occur. But we do allow polyadic quantifiers $\exists\, \mathbf{x}\phi$, $\forall\, \mathbf{x}\phi$ over tuples of variables $\mathbf{x}$, with their obvious interpretation, which resembles the polyadic quantifiers discussed earlier in connection with general assignment models. Even though we can rewrite these in terms of successive single first-order quantifiers, we may not be able to do so *inside the fragments* we are studying. We also use polyadic simultaneous substitutions $[\mathbf{u}/\mathbf{y}]\phi$ that need not reduce to iterated single substitutions. These are taken in the standard syntactic sense that substitution is performed provided the $\mathbf{u}$ are free for the $\mathbf{y}$. If not, then some suitable alphabetic variant is taken first for ϕ.

Our key idea is that objects $\mathbf{y}$ can only be introduced relative to given objects $\mathbf{x}$, as expressed by a 'guard atom' $G(\mathbf{x}, \mathbf{y})$ where objects can occur in any order and multiplicity – and that the subsequent statement refers only to those guarded $\mathbf{x}$, $\mathbf{y}$.

Definition 1.3.1 (Guarded Formulas). *Guarded formulas* are all those constructed according to the syntax rules

$$\text{atoms } P\mathbf{x} \mid \neg \mid \vee \mid \exists\, \mathbf{y}\big(G(\mathbf{x}, \mathbf{y}) \;\&\; \phi(\mathbf{x}, \mathbf{y})\big).$$

Bold-face $\mathbf{x}$, $\mathbf{y}$ are finite tuples of variables, and G is a predicate letter.

Clearly, the well-known 'standard translation' takes basic modal logic into first-order formulas lying inside the Guarded Fragment, and the same is true for many other modal languages in the literature. For further illustrations and applications, see van Benthem [Ben,01], which also considers guarded patterns in modal neighbourhood semantics.

3.2. Decidability of GF via quasi-models

The initial motivation was that guarding quantifiers leads to decidability.

Theorem 1.3.2. *GF is decidable.*

The proof of Theorem 1 is worth stating here at least in outline, for the general ideas involved.

Proof. The first observation is that truth of first-order formulas in any model is witnessed in some finite syntactic object, called a 'quasi-model'. Let formula ϕ be true in standard model $\mathbf{M}$. Let V be the finite set of variables occurring in ϕ – free or bound. In effect, we are inside a finite-variable fragment of first-order logic here. Next, we restrict attention to the finite set Sub_ϕ consisting of ϕ and its sub-formulas, while also closing under simultaneous substitutions $[\mathbf{u}/\mathbf{y}]$ using only variables in V, that do not change syntactic forms. This is feasible because of the following simple observation, provable by some syntactic manipulation:

Lemma 1.3.3. *Finite-variable fragments are closed under performing simultaneous substitutions.*

Now each variable assignment s on $\mathbf{M}$ verifies a set Δ_s of formulas from Sub_ϕ with special properties, that we call a *type*. Note that any model realizes at most finitely many types. A 'quasi-model' is a finite set of types with some properties and mutual relations that obviously hold if the source is indeed some model $\mathbf{M}$.

Definition 1.3.4 (Quasi-models). Let F be the finite set of all formulas of length $\leq |\phi|$ that use only variables from V. Note that $\phi \in F$ while F is closed under taking sub-formulas and the above 'alphabetic variants' used with substitutions. An *F-type* is a subset Δ of F which satisfies

(i)	$\neg\psi \in \Delta$	iff	*not* $\psi \in \Delta$	whenever $\neg\psi \in F$,
(ii)	$\psi \vee \xi \in \Delta$	iff	$\psi \in \Delta$ *or* $\xi \in \Delta$	whenever $\psi \vee \xi \in F$,
(iii)	$[\mathbf{u}/\mathbf{y}]\psi \in \Delta$	only if	$\exists\, \mathbf{y}\psi \in \Delta$	whenever $\exists\, \mathbf{y}\psi \in F$.

Next, we write $\Delta =_{\mathbf{y}} \Delta'$ if Δ, Δ' share the same formulas with all their free variables disjoint from $\mathbf{y}$. A *quasi-model* is a set of F-types S such that

(iv) for each $\Delta \in S$ and each formula $\exists\, \mathbf{y}\psi \in \Delta$,
there is a type $\Delta' \in S$ with $\psi \in \Delta'$ and $\Delta =_{\mathbf{y}} \Delta'$.

ϕ *holds in a quasi-model* if $\phi \in \Delta$ for some Δ in this quasi-model.

Clearly, this definition justifies the following assertion:

Lemma 1.3.5. *If a first-order formula has a model, then it is also true in some quasi-model.*

The converse is not true for all first-order formulas, but it does hold for *GF*.

Lemma 1.3.6. *If a guarded formula has a quasi-model, then it has a standard model.*

The key fact is that quasi-models can be 'unraveled' to tree-like standard models without affecting truth values of guarded formulas in their set F: details of the proof are in Andréka, van Benthem and Németi [And-Ben-Nem,98]. Decidability of *GF* now follows because we can test satisfiability for arbitrary guarded formulas ϕ by testing for the existence of a quasi-model for ϕ whose size is effectively bounded by the length of ϕ. ■

This decision procedure can be adapted easily to give an optimal computational complexity result (Grädel [Gra,99b]). Satisfiability is *2EXPTIME*-complete for guarded formulas, and it is *EXPTIME*-complete for *GF* with a fixed bound on the arities of predicates.

3.3. Other meta-properties

The Guarded Fragment was meant to serve several purposes at once. On the one hand its complexity is low enough to be decidable, while it is expressive enough to generalize most common modal languages. This demonstrates the balance sought in all good modal-like languages. Another desirable feature concerns its meta-theory.

Basic modal logic resembles first-order logic in all its meta-properties, partly thanks to the role of modal bisimulation as a generalization of the key invariance of potential isomorphism. For the case of *GF*, let a *partial isomorphism* be a finite one-to-one partial map between models which preserves relations both ways. In any model $\mathbf{M}$, call a set X of objects *guarded* if there is a relation symbol R, say k-ary, and objects $a_1, \ldots, a_k \in M$ (possibly with repetitions) with $R^{\mathbf{M}}(a_1, \ldots, a_k)$ and $X = \{a_1, \ldots, a_k\}$.

Definition 1.3.7 (Guarded Bisimulations). A *guarded bisimulation* is a non-empty set $\mathbf{F}$ of finite partial isomorphisms between two models $\mathbf{M}$ and $\mathbf{N}$ that satisfies the following two back-and-forth conditions for any $f : X \to Y$ in $\mathbf{F}$:

(i) for any guarded $Z \subseteq M$, there is a $g \in \mathbf{F}$ with domain Z such that g and f agree on the intersection $X \cap Z$,

(ii) for any guarded $W \subseteq N$, there is a $g \in \mathbf{F}$ with range W such that the inverses g^{-1} and f^{-1} agree on $Y \cap W$.

The point of the definition shows in this semantic invariance:

Fact 1.3.8. *Let* $\mathbf{F}$ *be a guarded bisimulation between models* $\mathbf{M}$ *and* $\mathbf{N}$ *with* $f \in \mathbf{F}$. *For all guarded formulas* ϕ *and variable assignments* α *into the domain of* f, *we have that* $\mathbf{M}, \alpha \models \phi$ *iff* $\mathbf{N}, f \circ \alpha \models \phi$.

The next result follows an analogue for modal logic and bisimulation:

Theorem 1.3.9. *Let* ϕ *be any first-order formula. The following two assertions are equivalent:*

(i) *ϕ is invariant for guarded bisimulations,*

(ii) *ϕ is equivalent to a GF formula.*

Techniques based on this invariance establish even meta-properties of *GF* that do not follow from just being a sublanguage of first-order logic, such as the standard model-theoretic preservation theorems. *GF* shares this good behaviour to a large extent, witness the Łoš–Tarski theorem for *GF*-formulas that are preserved under taking sub-models given in Andréka, van Benthem and Németi [And-Ben-Nem,98]. Cf. also van Benthem [Ben,01] on *GF* as an instrument for finding out 'what makes modal logic tick'. But subsequent work has shown that the picture is somewhat mixed:

Theorem 1.3.10. *Beth Definability holds, but Craig Interpolation fails for GF. Interpolation remains valid when we view guard predicates as part of the logical vocabulary.*

Proofs are in Hoogland, Marx and Otto [Hoo-Mar-Ott,99], Hoogland and Marx [Hoo-Mar,02b].

3.4. Excursion: quasi-models per se

The methods around *GF* may have a broader spin-off. In particular, *quasi-models* are a mix of modal filtration, semantic tableaus for first-order logic, and the 'mosaics' of algebraic logic. Right now, mosaics – introduced in Németi's 1986 dissertation [Nem,86a], with Németi [Nem,95] as a more up-to-date reference – seem the method of choice for proving decidability in modal and algebraic logics. But quasi-models may also be appreciated on their own. First, a quasi-model for some initial formula ϕ is a modal model $\mathbf{M}_\phi$ for a first-order language as it stands. The types are the worlds, there are accessibility relations $=_\mathbf{x}$ of agreeing on all formulas having no free variables in $\mathbf{x}$, and for atoms, $V(\Delta, P\mathbf{x}) = 1$ iff $P\mathbf{x} \in \Delta$. The following Truth Lemma is then easily proved by induction:

Lemma 1.3.11. *For all formulas $\alpha \in \mathrm{SUB}_\phi$, and all types Δ in the model $\mathbf{M}_\phi$, $\mathbf{M}, \Delta \models \alpha$ iff $\alpha \in \Delta$.*

Thus, quasi-models are models by themselves, and this may lead to new links between modal and first-order logic. Modal models may be related to quasi-models modulo forms of bisimulation (van Benthem [Ben,96a]).

Still, having a quasi-model – finite by definition – does not imply having a finite standard model. It is easy to find a quasi-model for the formula $\forall\, xyz\big((Rxy\ \&\ Ryz) \to Rxz\big)\ \&\ \forall\, x\ \exists\, y\ Rxy\ \&\ \forall\, x\ \neg Rxx$ which only has infinite models. In fact, having a quasi-model need not imply standard satisfiability at all. The predicate-logically inconsistent formula $\exists\, x\ \exists\, y\ Rxy\ \&\ \neg\exists\, y\ \exists\, x\ Ryx$ is clearly satisfiable in the general assignment model $\mathbf{M}$ with domain $\{1,2\}$, relation $\{\langle 1,2\rangle\}$, and just one admissible assignment s, viz. $\{(x,1),(y,2)\}$. This model also satisfies the earlier Existential Generalization and Full Locality. The single type of $\mathbf{M}$ induced by s is therefore a quasi-model for $\exists\, x\ \exists\, y\ Rxy\ \&\ \neg\exists\, y\ \exists\, x\ Rxy$. This 'inconsistency' in a set of types may seem strange – but it also shows that quasi-models are intriguing structures.

3.5. Extensions

GF is not yet the end of the road. Analyzing the earlier proof of decidability, van Benthem [Ben,97a] noticed that it goes through for the following extension.

Definition 1.3.12. *Loosely guarded formulas* extend the syntax of *GF* by allowing a conjunction of atoms $\gamma(\mathbf{x},\mathbf{y})$ instead of $G(\mathbf{x},\mathbf{y})$ in the quantifier clause, provided each variable from $\mathbf{y}$ co-occurs with each variable from $\mathbf{x}$, $\mathbf{y}$ in at least one atom of $\gamma(\mathbf{x},\mathbf{y})$. This yields the *fragment LGF.*

The point here is that the guarding of objects does not take place all at once, but two-by-two. This suffices for several earlier results:

Theorem 1.3.13. *Both LGF and its fixed-point extension are decidable.*

As an application, modal logics like that of temporal *Since* and *Until* are decidable, since their truth conditions are typically loosely guarded (van Benthem [Ben,97a], [Ben,01]). For instance, *UNTIL pq* is defined by the *LGF*-formula

$$\exists\, y\big(x < y\ \&\ \forall\, z\big((x < z \wedge z < y) \to Qz\big)\big).$$

Another striking application is the decidability of the earlier complete logic of pair arrow models. The translation in Section 2 that relativized to the 'top relation' U does not take formulas '$(x,y)\ :\ R$' into *GF* itself, but it does take them into *LGF*!

Next, by way of contrast, consider the first-order property of *transitivity*, which can lead to undecidable fragments of *FOL*. It has 'one guard too few':

$$\forall xyz\big((x < z \wedge z < y) \rightarrow x < y\big).$$

Next, consider non-first-order extensions, like we had before in Sections 2, 3. A striking positive result concerns the extension *LFP*(*GF*) of *GF* with *fixed-point operators* μ, ν:

Theorem 1.3.14 (Grädel). *LFP*(*GF*) *is decidable* ([Gra,99c]).

By contrast, validity for the fixed-point extension *LFP*(*FO*) of full first-order logic is of high non-arithmetical complexity, as it can define the natural numbers categorically. Grädel [Gra,99a] also determines the computational complexity for *LFP*(*GF*).

As a small application, the preceding result explains the validity of many modal logics over transitive models, even though transitivity by itself is dangerously non-guarded. Instead of working over transitive models, take models with arbitrary binary relations. Then a modality for a special transitive relation becomes an iteration modality for the transitive closure of the given arbitrary relation, and this can be defined inside *LFP*(*GF*).

New guarded-style fragments keep appearing, cf. [Bar-Got-Ott,10], [Bar-Cat-Seg,11] on query languages for data bases – see also [Dun,thisVol]. Further uses of guarding have been proposed in other settings such as game logics, cf. [Man,thisVol] on Crs and IF logic.

3.6. Border line: confluence

A useful alternative way of understanding the guarding technique looks 'from above'. What expressive resources will typically lead to un-decidability? Here is a natural comparison with a related, though subtly different fragment of first-order logic.

BOUNDING VERSUS GUARDING. *Bounded formulas* have all their quantifiers relativized to an atomic predicate, as in the ubiquitous pattern $\exists\, x\ :\ x \in y$ in set theory. Feferman and Kreisel [Fef-Kre,69] show that the characteristic semantic feature of bounded first-order formulas is their *invariance for generated sub-models.* The *Bounded Fragment BF* differs from the Guarded Fragment in allowing the more general format of quantification

$$\exists\, \mathbf{y}\big(G(\mathbf{x}, \mathbf{y})\ \&\ \phi(\mathbf{x}, \mathbf{y}, \mathbf{z})\big)$$

where the formula at the end may contain new free variables. *BF* is undecidable, but it has applications in arithmetic and set theory, as a way of defining 'absolute' properties not affected by the difference between standard models and generalized models. Ten Cate [Cat,05] has a modern treatment with new results, including the one that *BF* equals the first-order definable part of basic modal logic with added propositional quantifiers. Finally, van Benthem [Ben,05] shows how, following Montague [Nem,86a], Gallin [Gal,75], bounding serves as a general technique for lowering complexity in second-order logic.

TILING PROBLEMS, GRID STRUCTURE, AND CONFLUENCE. One way of seeing that *BF* is undecidable is by noting that it can define the following geometrical *Tiling Problem.* Let a finite set of 'tiles' be given, with colours on each of their four sides. 'Matching tiles' have the same colours on adjacent sides. Now the Tiling Problem asks:

Can we cover the plane $N \times N$ with matching tiles from the given set?

Fact 1.3.15. *The Tiling Problem is* undecidable.

Tiling has the same complexity as the Non-Halting Problem: deciding if a given Turing machine will keep computing forever on given input. (To see the Π^0_1-complexity, note that one can tile a plane iff one can tile all finite sub-planes, using Koenig's Lemma.) Indeed, the two problems are equivalent, as one can code successive tape contents as horizontal rows, with the vertical sequence as the 'computation'. It is essential here that positions on successive rows can be compared in the right way, and this is what the grid structure of $N \times N$ does. Since tiling problems are easier to encode than Turing machine computations, they have gained popularity as a way of showing high complexity.

Now it is easy to show the undecidability of first-order logic. Consider a set T of square tiles $\{t_1, \dots, t_m\}$, *TP* is the task of putting one tile on each point in the grid $N \times N$ giving adjacent edges the same colour. We can reduce this task effectively to a *satisfiability problem* in first-order logic, by constructing a formula ϕ_T with the following property:

Fact 1.3.16. *ϕ_T is satisfiable iff the given set T can tile the $\mathbb{N} \times \mathbb{N}$-plane.*

The formula ϕ_T is constructed as follows. We choose unary predicates P_t for each tile t, and binary relations *NORTH* and *EAST* for moving around in the grid, whose one-step immediate successor versions are $NORTH^+$,

$EAST^+$. Now we write up what tiling amounts to. Note that the 'adjacent colours' condition just amount to giving an ordering from tiles to a finite set of 'fitting tiles' in each direction. More concretely, each point then has to satisfy the following properties, involving only finite formulas in total:

(a) the exhaustive finite disjunction T of all tiling predicates P_t,

(b) $(P_t x \wedge NORTH^+ xy) \rightarrow$ 'disjunction of all atomic formulas $P_{t'}$ with t' fitting to the north' (y), $(P_t x \wedge EAST^+ xy) \rightarrow$ 'disjunction of all $P_{t'}$ with t' fitting to the east' (y),

(c) *NORTH, EAST* are transitive irreflexive relations with successors,

(d) $\forall xyz : (NORTH^+ xy \wedge EAST^+ yz) \rightarrow \exists\, u : (EAST^+ xu \wedge NORTH^+ uz)$,

(e) there is a unique initial point for the whole structure.

We prefix a universal quantifier to make sure this holds everywhere.

It is clear how to satisfy ϕ_T given a tiling. But also conversely, if ϕ_T is satisfiable at some point s in a grid-like model $\mathbf{M}$, we can use the unary predicates P_t in $\mathbf{M}$ to tile the plane $\mathbb{N} \times \mathbb{N}$. Working from the origin, first read off the tiling for the initial point, and then proceeding inductively, tile in triangles, using the grid property of the model to place the next edge in such a way that no conflicts arise in the placement pattern.

The preceding reduction shows that satisfiability for first-order logic is undecidable, since the Tiling Problem is. Now, the crucial feature behind this reduction is the grid structure, defined by the *confluence* property (d). While this formula employs bounded quantifiers, putting it in the fragment *BF*, it is typically not in *GF* or even *LGF*: not all pairs of objects come with an atomic bound, witness the case of y and z. This is significant. Grid structure tends to involve high complexity, a fact also known from modal logics with two modalities for two relations which satisfy a commutation axiom expressing grid structure (cf. Halpern and Vardi [Hal-Var,89], van Benthem and Pacuit [Ben-Pac,06] for technical details).

Remark 1.3.17 (Trees versus Grids)**.** By contrast with the realm of grids, many ordinary modal logics are decidable since their semantics is based on *trees* rather than grids. In that case, we are in the realm of *Rabin's Theorem* saying that the complete monadic second-order logic over a countable tree with finitely many successor relations plus the relation of 'precedence' between nodes is decidable.

4. Two Perspectives: Fragments or Generalized Semantics

Now we need to compare generalized assignment-based semantics for *FOL* with standard semantics for *GF*.

4.1. Restricted syntax versus generalized semantics

Giving each quantifier a guard is a syntactic restriction banning unbounded quantifiers. In this sense, *GF* is a fragment of the full language *FOL*. But there is also another perspective, where this move rather represents a semantic generalization. We now assume that quantification will normally take place in 'structured domains', where access from one group of objects to another must go via some connecting relation R of some appropriate arity. Binary modal accessibility is a typical example. Standard models are the special case with R the universal relation. Informally, then, there seems to be an analogy between

(a) using guarded formulas over standard models, and

(b) using arbitrary first-order formulas over suitably generalized models.

We will now show how these two approaches are equivalent in our setting.

4.2. Reducing *GAM* logic to *GF*

The following translation result is proved in Andréka, van Benthem and Németi [And-Ben-Nem,98], Section 5.

Definition 1.4.1 (Guarded translation). Consider a k-variable first-order language $L\{x_1, \ldots, x_k\}$, with R a new k-ary predicate. The translation *guard* takes k-variable first-order formulas ϕ to guarded first-order formulas guard (ϕ) by relativizing all quantifiers to one and the same atom $Rx_1 \ldots x_k$. This translation works for polyadic first-order quantifiers just as well as single ones – and it even extends to the above substitution operators, if desired. There is also a matching semantic operation of model expansion. Let $(\mathbf{M}, V)$ be any general assignment model for $L\{x_1, \ldots, x_k\}$ – without the new predicate R. The standard model GUARD $(\mathbf{M}, V)$ is $\mathbf{M}$ viewed as a standard model, and expanded with the following interpretation:

$$R(d_1, \ldots, d_k) \text{ iff the assignment } x_i := d_i \ (1 \leq i \leq k) \text{ is in } V.$$

The following is easy to prove by induction on first-order formulas:

Lemma 1.4.2. *For all available assignments s in the family V, and for all k-variable formulas ϕ,*

$$\mathbf{M}, V, s \models \phi \ \textit{iff} \ \text{GUARD}(\mathbf{M}, V),\ s \models \text{guard}(\phi).$$

Here is a reduction of *GAM*-semantics to the Guarded Fragment.

Theorem 1.4.3. *For all first-order k-variable formulas ϕ, the following assertions are equivalent:*

(i) *ϕ is satisfiable in general assignment models,*

(ii) $Rx_1 \ldots x_k \wedge \text{guard}(\phi)$ *is satisfiable in standard models.*

Proof. From (i) to (ii), the Lemma supplies the reason. For the converse, suppose that $Rx_1 \ldots x_k \wedge \text{guard}(\phi)$ has a standard model $\mathbf{M}$ under some variable assignment s. Now define a general assignment model $(\mathbf{N}, V)$ by retaining only those variable assignments on $\mathbf{M}$ whose values for $x_1, \ldots, x_k$ stand in the relation R_M. These include the assignment s itself. Then it is easy to see that $\mathbf{N}, V, s \models \phi$ – as with the Lemma. ■

The translation also works directly for the full first-order language without the k-restriction, by a slightly modified translation. The converse direction was left open in Andréka, van Benthem and Németi [And-Ben-Nem,98]. Marx [Mar,01], van Benthem [Ben,05] have solutions.

This translation is at the same time a faithful embedding of the earlier complete logic of general assignment models (Section 3) into the logic of the Guarded Fragment, which provides another explanation of its decidability.

4.3. Reducing *GF* to *GAM* logic

We need a translation again. But this time, it is not compositional in the earlier sense. The reason is the earlier failure of Existential Generalization (i) and Full Locality (ii) in general assignment models. We need these principles for some finite set of relevant formulas in the proof to follow, and hence we put them into the translation.

Definition 1.4.4 (*GAM* translation). Let ϕ be any guarded first-order formula with a total set of variables $\mathbf{x} = x_1, \ldots, x_k$. Let *set-up* (ϕ) be the finite conjunction of all formulas of the following form:

(i)′ $\forall \mathbf{x}([\mathbf{u}/\mathbf{y}]\psi \to \exists\, \mathbf{y}\psi)$ where $\mathbf{u}, \mathbf{y} \subseteq \mathbf{x}$ and $\psi(\mathbf{z})$ is a subformula of ϕ,

(ii)′ $\forall \mathbf{x}(\psi(\mathbf{z}) \to \forall \mathbf{y}\ \psi(\mathbf{z}))$ where $\mathbf{z}, \mathbf{y} \subseteq \mathbf{x}$ with $\mathbf{z}$ disjoint from $\mathbf{y}$, and $\psi(\mathbf{z})$ is a subformula of ϕ.

The, not necessarily guarded, formula *gam* (ϕ) is the conjunction $\phi \wedge$ *set-up* (ϕ).

In particular, the prefixed polyadic universal quantifier $\forall \mathbf{x}$ running over all relevant variables makes sure that the implications (i)′, (ii)′ hold throughout any general assignment model which has the formula *set-up* (ϕ) true at any assignment at all.

Theorem 1.4.5. *For all guarded formulas ϕ, the following are equivalent:*

(i) *ϕ is satisfiable in standard models,*

(ii) *gam (ϕ) is satisfiable in general assignment models.*

Proof. From (i) to (ii), it suffices to note that any standard model for ϕ also satisfies *gam* (ϕ), since the formulas in the second conjunct are universally valid. Also recall that standard models are general assignment models with a full set of assignments.

Next, from (ii) to (i), let $\mathbf{M}, V, s \models$ *gam* (ϕ). As in Section 3, this situation induces a quasi-model for ϕ. Recall that the relevant formulas are all sub-formulas of ϕ plus their alphabetic variants with variables from $\mathbf{x}$. The types of the quasi-model are now all sets of relevant formulas true at the assignments in V. We must check the four clauses of the Definition. The first two follow directly by the truth definition for Boolean operations. The existential generalization clause holds for all types by the truth of conjunct (i)′ in *set-up* (ϕ). And finally, the 'witness clause' (d) for existential quantifiers in suitably related types holds because of the truth condition for the existential quantifier in general assignment models plus the true transfer condition (ii)′ in *set-up* (ϕ). Thus, the given guarded formula ϕ has a quasi-model – and hence it also has a standard model. ∎

The same reasoning extends to the loosely guarded fragment *LGF*. Andréka and Németi have some interesting variations and extensions. Marx [Mar,01] proves several further results, including a characterization of the 'packed fragment' (a slight extension of *LGF*). This is the largest fragment of first-order logic that is insensitive between evaluation in standard models and models relativized to some 'tolerance relation'.

5. Discussion

The two main perspectives in this chapter are generalized models and static guards. Given the connections in Section 5, we merely discuss a few final points about the latter.

The Scope of Guarding. Guarding still has not become a fully general method. For instance, in modal logic, a lot of generality may be missed, since many results might work even when basic modal logic is replaced by *GF*, the μ-calculus by *LFP*(*GF*), etc. Can the method of guarded fragments also be stated as a more general operation on logics? Van Benthem [Ben,05] has an analysis of related methods for lowering complexity in second-order logic, going back to the fragment of 'persistent formulas' dating back to Orey in the 1950s (cf. Gallin [Gal,75]). These are invariant between Henkin general models and their associated standard models. Another challenge is guarding *generalized quantifiers.* We do not even know if the basic modal logic with a quantifier "for most successors of the current world" is decidable, despite partial results in Van der Hoek and de Rijke [Hoe-Rij,93] and Pacuit and Salame [Pac-Sal,04]. Perhaps most intriguingly, guards may help lower complexity in *fixed-point logics,* as we have seen in a number of cases.

Guarding Lower Down? Guards make sense, not just higher up from first-order logic, but also lower down in restricted formalisms. Kerdiles [Ker,01] considers a language *CG* of *conceptual graphs* which has only atoms, conjunction, and existential quantifiers. The complexity of the general consequence problem between such formulas is *NP*, but consequence between guarded *CG*-formulas is in *P*. This suggests that guarding can take the '*N*' out of '*NP*' sometimes, but the precise extent of this is unknown.

Coda: algebra and logic once more. We have seen in this chapter how ideas from algebra and algebraic logic can influence research in modal logic and first-order logic. Is there also a converse stream? It would be good to see which of the various model-theoretic topics in this chapter make sense at an algebraic level.[1] With many concrete systems of modal logic being developed today (cf. the information dynamics of van Benthem [Ben,11]), one feels that insights found there really live at some higher 'generic' abstraction level that can often be brought out better in an algebraic approach[2]. Thus, in their own special compass, the topics of this paper illustrate that the interface of logic and algebra is as important as ever.

[1]For instance, it would be useful to have a better algebraic take on the *fixed-point logics* that crop up everywhere in contemporary pure and applied modal logic.

[2]One recent attempt in this line is in Sadrzadeh, Palmigiano and Ma [Sad-Pal-Ma,11].

BOLYAI SOCIETY
MATHEMATICAL STUDIES, 22

Cylindric-like Algebras and Algebraic Logic
pp. 303–320.

Cylindric Probability Algebras

RADOSAV. S. ĐORĐEVIĆ and MIODRAG. D. RAŠKOVIĆ

Sometimes it is not enough to prove that a formula $\varphi(x)$ is satisfied or not by an element $a \in A$ in the model $\mathfrak{A}$ we are interested in. It may happen that we are looking for the quantity of those $a \in A$ for which $\varphi[a]$ is true. Of course, the mathematical discipline for these considerations is probability theory. The logic suitable for this kind of reasoning was introduced by H. J. Keisler in 1976. This logic has formulas similar to those of $L_{\mathcal{A}} \subseteq L_{\omega_1\omega}$ ($\mathcal{A}$ is a countable admissible set), except that the quantifiers $(Px \geq r)$ ($r \in \mathcal{A} \cap [0,1]$ is a real number) are used instead of the usual quantifiers $(\forall x)$ and $(\exists x)$. A model for this logic is a pair $\langle \mathfrak{A}, \mu \rangle$, where $\mathfrak{A}$ is a classical structure and μ is a probability measure defined in such a way that definable subsets of the universe A are measurable. The quantifiers are interpreted in the natural way, i.e.

$$\langle \mathfrak{A}, \mu \rangle \models (Px \geq r)\varphi(x) \qquad \text{iff} \qquad \mu\{a \in A : \langle \mathfrak{A}, \mu \rangle \models \varphi[a]\} \geq r.$$

In [Kei,85] Keisler introduced several probability logics and developed model theory for them together with Hoover (see [Hoo,78]). These logics are essentially infinitary. That means that if we allow only finite formulas we still must have an infinite rule of inference in order to prove even the weak form of the completeness theorem (see [Kei,85]). For more on probabilistic logics we refer the reader to [Dor,92, Dor,93, Dor-Ras-Ogn,04, Iko-Ras-Mar-Ogn,07, Mar-Ogn-Ras,04, Ogn-Ras,00, Ogn-Per-Ras,08, Per-Ogn-Ras,08a, Per-Ogn-Ras,08b, Rad-Per-Ogn-Ras,08, Ras,86, Ras-Dor,96, Ras-Mar-Ogn,08, Ras-Ogn-Mar,04].

The notion of a cylindric probability algebra was introduced in [Ras-Dor-Bra,97], [Mar-Ras-Dor,01] and [Ras-Dor,00] as a common algebraic abstraction from the theory of deductive systems of probability logics on the one hand, and the geometry associated with basic set-theoretic notions on the other hand. These two sources are connected because models of deductive systems of probability logics give rise in a natural way to probability structures within set-theoretical algebras. A natural correspondence that exists between Boolean algebra and propositional logic, cylindric algebra and first-order predicate logic (see [Hen-Mon-Tar,85], Theorem 1.1.10.), Keisler's logic with infinitary predicates (see [Kei,63]) and polyadic algebras has motivated us to introduce cylindric probability algebras as an algebraic counterpart of some probability logics.

There are many concepts which emphasize the algebraic and measure theoretical aspects rather than the probability logical. For example, M. Ferenczi (see [Fer,05]) deals with some general constructions of probabilities and measurable functions defined on cylindric algebras. He treats probabilities defined specially on cylindric set algebras.

The application of cylindric algebras in the study of probability logics is based on the concept of probability cylindrification. In Section 1 we introduce the notion of a weak cylindric probability algebra as an algebraic abstraction from the theory of deductive systems of the weak probability logic $L_{\mathcal{A}P\forall}$ (see [Ras,87]). The notion of a weak polyadic probability algebra is introduced in Section 2. An apparatus for an algebraic study of the graded probability logic will be presented in Section 3. As it is known Boolean algebras represent the propositional logic in a sense while locally finite dimensional cylindric algebras represent first order logic (see [Hen-Mon-Tar,85], 3.2. Representation theory). So, we introduce locally finite dimensional cylindric probability algebras and completeness (Boolean representation) theorems are proven for these algebras.

1. Weak Cylindric Probability Algebras

Let $\mathcal{A}$ be a countable admissible set such that $\omega \in \mathcal{A}$ (see [Bar,75]). Let L be a countable $\mathcal{A}$-recursive set of finitary relation, function and constant symbols. The logic $L_{\mathcal{A}P\forall}$ is the minimal extension of the infinitary logic $L_{\mathcal{A}}$ and the probability logic $L_{\mathcal{A}P}$ (see [Kei,85] and [Hoo,78]). The set Fm_L of all formulas of $L_{\mathcal{A}P\forall}$ is closed under countable disjunctions and conjunctions, negation, usual quantifiers $(\forall, \exists)$ and probability quantifiers $(P\overrightarrow{v} \geq r)$, where $\overrightarrow{v}$ is a finite tuple of distinct variables and $r \in \mathcal{A} \cap [0,1]$. The structure

$$\mathfrak{Fm}_L = \langle Fm_L, \vee, \wedge, \neg, F, T, \exists v_i, P\overrightarrow{v} \geq r, v_p = v_q \rangle$$

is the free algebra of formulas of $L_{AP\forall}$, which contains as distinguished elements the expressions: false (F), true (T) and $v_p = v_q$ for any $p, q < \omega$.

Axioms and rules of inference for $L_{AP\forall}$ are those for $L_{\mathcal{A}}$ and the weak L_{AP}, as listed in [Kei,71] and [Kei,85], together with the axioms listed in [Ras,87].

Let Σ be any set of sentences of $L_{AP\forall}$ and let $\equiv_\Sigma$ be the relation on Fm_L defined by

$$\varphi \equiv_\Sigma \psi \quad \text{iff} \quad \Sigma \vdash \varphi \leftrightarrow \psi.$$

Then the relation $\equiv_\Sigma$ is a congruence relation on Fm_L and the quotient algebra

$$\mathfrak{Fm}_L^\Sigma = \big\langle Fm_L/\!\equiv_\Sigma, \vee^\Sigma, \wedge^\Sigma, \neg^\Sigma, F^\Sigma, T^\Sigma, (\exists v_i)^\Sigma, (P\overrightarrow{v} \geq r)^\Sigma, (v_p = v_q)^\Sigma \big\rangle$$

will be called a weak cylindric probability algebra of formulas.

Let $\mathfrak{A} = \langle A, R_i^{\mathfrak{A}}, f_j^{\mathfrak{A}}, c_k^{\mathfrak{A}}, \mu_n \rangle_{n<\omega}$ be a weak probability structure for $L_{AP\forall}$ such that μ_n's are finitely additive probability measures defined on the sets of all definable subsets of A^n and let φ be any formula of $L_{AP\forall}$. Then, for $\varphi^{\mathfrak{A}} = \big\{ a \in A^\omega : \mathfrak{A} \models \varphi[a] \big\}$, we have

$$\big((\exists v_i)\varphi\big)^{\mathfrak{A}} = \big\{ a \in A^\omega : a \restriction \omega \setminus \{i\} = b \restriction \omega \setminus \{i\} \text{ for some } b \in \varphi^{\mathfrak{A}} \big\},$$

$$\big((P\overrightarrow{v} \geq r)\varphi\big)^{\mathfrak{A}} = \Big\{ a \in A^\omega : \mu_n\big\{ (b_{k_1}, \ldots, b_{k_n}) : b \in \varphi^{\mathfrak{A}}, \\ b_j = a_j,\ j \notin K \big\} \geq r \Big\},$$

where $\overrightarrow{v} = v_{k_1}, \ldots, v_{k_n}$ and $K = \{k_1, \ldots, k_n\}$. As usual, a unary cylindric set operation C_i is defined on the subsets of A^ω by setting, for any $X \subseteq A^\omega$,

$$\mathsf{C}_i(X) = \big\{ y \in A^\omega : y \restriction \omega \setminus \{i\} = x \restriction \omega \setminus \{i\} \text{ for some } x \in X \big\}.$$

A unary cylindric probability set operation $\mathsf{C}^r_{\langle K \rangle}$ is defined on the subsets of A^ω by setting, for any $X \subseteq A^\omega$,

$$\mathsf{C}^r_{\langle K \rangle}(X) = \Big\{ y \in A^\omega : \mu_n\big\{ (x_{k_1}, \ldots, x_{k_n}) : x \in X \\ \& \ (j \notin K \rightarrow x_j = y_j) \big\} \geq r \Big\},$$

where $\langle K \rangle$ is a tuple of distinct integers corresponding to a finite subset $\{k_1, \ldots, k_n\}$ of ω and $r \in [0, 1]$. By means of $\mathsf{C}^r_{\langle K \rangle}$ we obtain a cylinder generated by translating only the section of X whose measure is not less than r parallelly to the $(k_1, \ldots, k_n)$-axis of A^ω. It follows from $\mathsf{C}_i(\varphi^{\mathfrak{A}}) = \big((\exists v_i)\varphi\big)^{\mathfrak{A}}$ and $\mathsf{C}^r_{\langle K\rangle}(\varphi^{\mathfrak{A}}) = \big((P\overrightarrow{v} \geq r)\varphi\big)^{\mathfrak{A}}$ that the function $f(\varphi^{\Sigma}) = \varphi^{\mathfrak{A}}$ is a natural homomorphic transformation from the weak cylindric probability algebra of formulas $\mathfrak{Fm}^{\Sigma}_L$ onto the weak cylindric probability set algebra

$$\big\langle \mathbb{A}, \cup, \cap, \sim, \emptyset, A^\omega, \mathsf{C}_i, \mathsf{C}^r_{\langle K\rangle}, \mathsf{D}_{pq}\big\rangle,$$

where $\mathbb{A}$ is the collection of all sets of the form $\varphi^{\mathfrak{A}}$ and $\mathsf{D}_{pq} = (v_p = v_q)^{\mathfrak{A}}$.

The abstract notion of a weak cylindric probability algebra is defined by equations which hold in both algebras mentioned above. We shall assume throughout that a fixed indexation by hereditarily countable sets (from $\mathcal{A} \subseteq \mathrm{HC}$) is given. So let $A = \{x_i : i \in I\}$ and $I \subseteq \mathcal{A}$. If for any $\{x_j : j \in J\} \subseteq A$, where $J \subseteq I$ and $J \in \mathcal{A}$, we have $\sum_{j\in J} x_j \in A$, then we say that a Boolean algebra $\langle A, +, \cdot, -, 0, 1\rangle$ is $\mathcal{A}$-complete.

Definition 2.1.1. A *weak cylindric probability algebra* is a structure

$$\mathfrak{A} = \big\langle A, +, \cdot, -, 0, 1, \mathsf{c}_i, \mathsf{c}^r_{\langle K\rangle}, \mathsf{d}_{pq}\big\rangle,$$

such that $\langle A, +, \cdot, -, 0, 1\rangle$ is an $\mathcal{A}$-complete Boolean algebra, c_i and $\mathsf{c}^r_{\langle K\rangle}$ are unary operations on A for each $i < \omega$ and each finite tuple $K \subseteq \omega$ of distinct integers, $\mathsf{d}_{pq} \in A$ for all $p, q < \omega$, and the following postulates hold:

(WCP_0) $\langle A, +, \cdot, -, 0, 1, \mathsf{c}_i, \mathsf{d}_{pq}\rangle$ is a cylindric algebra of dimension ω.

(WCP_1) (i) $\mathsf{c}^r_{\langle\emptyset\rangle} x = x$, (ii) $\mathsf{c}^r_{\langle K\rangle} 0 = 0$, where $r > 0$.

(WCP_2) $\mathsf{c}^0_{\langle K\rangle} x = 1$.

(WCP_3) If $r \geq \mathsf{s}$, then $\mathsf{c}^r_{\langle K\rangle} x \leq \mathsf{c}^{\mathsf{s}}_{\langle K\rangle} x$.

(WCP_4) $\mathsf{c}^r_{\langle K\rangle}\big(x + \mathsf{c}^{\mathsf{s}}_{\langle L\rangle} y\big) = \mathsf{c}^r_{\langle K\rangle} x + \mathsf{c}^{\mathsf{s}}_{\langle L\rangle} y$, where $K \subseteq L$.

(WCP_5) (i) $\mathsf{c}^r_{\langle K\rangle} x \cdot \mathsf{c}^{\mathsf{s}}_{\langle K\rangle} y \leq \mathsf{c}^{r+\mathsf{s}-1}_{\langle K\rangle}(x \cdot y)$,

(ii) $\mathsf{c}^r_{\langle K\rangle} x \cdot \mathsf{c}^{\mathsf{s}}_{\langle K\rangle} y \cdot \mathsf{c}^1_{\langle K\rangle} - (x \cdot y) \leq \mathsf{c}^{r+\mathsf{s}}_{\langle K\rangle}(x + y)$.

(WCP_6) $\mathsf{c}^r_{\langle K\rangle} - x = -\sum_{m>0} \mathsf{c}^{1-r+1/m}_{\langle K\rangle} x$.

(WCP_7) $-\mathsf{c}^1_{\langle\{k\}\rangle} - x \leq \mathsf{c}_k x$.

(WCP_8) $\mathsf{c}^r_{\langle K\rangle} x \leq \mathsf{c}^r_{\langle \pi(K)\rangle} x$, where π is a permutation of ω and $\big\langle \pi(K)\big\rangle$ is $k_{\pi 1}, \ldots, k_{\pi n}$.

(WCP_9) If $i \in K$, then: (i) $\mathsf{c}_i c^r_{\langle K\rangle} x = \mathsf{c}^r_{\langle K\rangle} x$,
(ii) $\mathsf{c}^r_{\langle K\rangle} \mathsf{c}_i x = \mathsf{c}^r_{\langle K\setminus\{i\}\rangle} \mathsf{c}_i x$.

(WCP_{10}) If $i, j \notin K$, then: (i) $\mathsf{c}^r_{\langle K\rangle} \mathsf{c}_i(\mathsf{d}_{ij} \cdot x) = \mathsf{c}_i \mathsf{c}^r_{\langle K\rangle}(\mathsf{d}_{ij} \cdot x)$,
(ii) $\mathsf{c}^r_{\langle K\cup\{i\}\rangle} \mathsf{c}_j(\mathsf{d}_{ij} \cdot x) = \mathsf{c}^r_{\langle K\cup\{j\}\rangle} \mathsf{c}_i(\mathsf{d}_{ij} \cdot x)$.

We point out that the axioms (WCP_2)–(WCP_6) express well-known properties of finitely additive measures and the axioms (WCP_7) and (WCP_8) express the conditions $(\forall x)\varphi \to (Px \geq 1)\varphi$ and $(Px_1 \ldots x_n \geq r)\varphi \to (Px_{\pi 1} \ldots x_{\pi n} \geq r)\varphi$ of $L_{\mathcal{A}P\forall}$, respectively. The properties of c_i and the substitution operation s^i_j defined by $\mathsf{s}^i_j x = \begin{cases} x, & \text{if } i = j \\ \mathsf{c}_i(\mathsf{d}_{ij} \cdot x), & \text{if } i \neq j \end{cases}$, are well-known (see [Hen-Mon-Tar,85], 1.5. Substitutions). Now we point out several properties of probability cylindrification operations (see [Ras-Dor-Bra,97]).

Theorem 2.1.2. *If* $\left\langle A, +, \cdot, -, 0, 1, \mathsf{c}_i, \mathsf{c}^r_{\langle K\rangle}, \mathsf{d}_{pq} \right\rangle$ *is a weak cylindric probability algebra, then:*

(i) $\mathsf{c}^r_{\langle K\rangle} 1 = 1$.

(ii) *If* $r > 0$ *and* $\mathsf{s} > 0$, *then* $\mathsf{c}^r_{\langle K\rangle} x = x$ *iff* $\mathsf{c}^{\mathsf{s}}_{\langle K\rangle} - x = -x$.

(iii) *If* $r > 0$ *or* $r = \mathsf{s} = 0$ *and* $K \subseteq L$, *then* $\mathsf{c}^r_{\langle K\rangle}\left(x \cdot \mathsf{c}^{\mathsf{s}}_{\langle L\rangle} y\right) = \mathsf{c}^r_{\langle K\rangle} x \cdot \mathsf{c}^{\mathsf{s}}_{\langle L\rangle} y$.

(iv) $\mathsf{c}^r_{\langle K\rangle} x \cdot -\mathsf{c}^r_{\langle K\rangle} y \leq \sum_{m>0} \mathsf{c}^{1/m}_{\langle K\rangle}(x \cdot -y)$.

(v) *If* $x \leq y$, *then* $\mathsf{c}^r_{\langle K\rangle} x \leq \mathsf{c}^r_{\langle K\rangle} y$.

(vi) $\mathsf{c}^r_{\langle K\rangle} x + \mathsf{c}^r_{\langle K\rangle} y \leq \mathsf{c}^r_{\langle K\rangle}(x + y)$.

(vii) $\mathsf{c}^r_{\langle K\rangle}(x \cdot y) \leq \mathsf{c}^r_{\langle K\rangle} x \cdot \mathsf{c}^r_{\langle K\rangle} y$.

(viii) $\mathsf{c}^1_{\langle K\rangle} x \cdot \mathsf{c}^1_{\langle K\rangle} y = \mathsf{c}^1_{\langle K\rangle}(x \cdot y)$.

(ix) *If* $K = \{k_1, \ldots, k_n\}$, *then* $\mathsf{c}^1_{\langle K\rangle} x = x$ *iff* $\mathsf{c}_{k_1} \ldots \mathsf{c}_{k_n} x = x$.

(x) *If* $K = \{k_1, \ldots, k_n\}$ *and* $r > 0$, *then* $\mathsf{c}^r_{\langle K\rangle} x \leq \mathsf{c}_{k_1} \ldots \mathsf{c}_{k_n} x$.

(xi) *If* $p, q \notin K$, *then* $\mathsf{c}^1_{\langle K\rangle} \mathsf{d}_{pq} = \mathsf{d}_{pq}$.

(xii) *If* $i \in K$, *then* $\mathsf{s}^i_j c^r_{\langle K\rangle} x = \mathsf{c}^r_{\langle K\rangle} x$ *and* $\mathsf{s}^i_j s^m_i \mathsf{c}^r_{\langle K\rangle} x = \mathsf{s}^m_j c^r_{\langle K\rangle} x$.

(xiii) *If* $i, j \notin K$, *then* $\mathsf{s}^i_j c^r_{\langle K\rangle} x = \mathsf{c}^r_{\langle K\rangle} \mathsf{s}^i_j x$ *and* $\mathsf{c}^r_{\langle K\cup\{i\}\rangle} \mathsf{s}^j_i x = \mathsf{c}^r_{\langle K\cup\{j\}\rangle} \mathsf{s}^i_j x$.

The algebraic notion of an ideal in a weak cylindric probability algebra can be modified using specific properties of these algebras.

Definition 2.1.3. An *ideal* in a weak cylindric probability algebra $\mathfrak{A}$ is a non-empty set $I \subseteq A$ such that the following conditions hold:

(i) I is a Boolean ideal of $\mathfrak{A}$; i.e.,

  (a) $0 \in I$,

  (b) If $\{a_j : j \in J\} \subseteq I$ and $J \in \mathcal{A}$, then $\sum_{j\in J} a_j \in I$,

  (c) If $x \in I$ and $y \leq x$, then $y \in I$.

(ii) For all $i < \omega$, if $x \in I$, then $\mathsf{c}_i x \in I$.

It follows from Definition 2.1.3 and the part (x) of Theorem 2.1.2 that, for any finite $K \subseteq \omega$ and $r \in (0, 1]$, if $x \in I$, then $\mathsf{c}^r_{\langle K\rangle} x \in I$. An ideal I of a weak cylindric probability algebra $\mathfrak{A}$ determines the congruence relation $\sim = \big\{ (x, y) : x \cdot -y + y \cdot -x \in I \big\}$. Indeed, as usual, if $x \sim y$, then $\mathsf{c}_i x \sim \mathsf{c}_i y$ and, for $r > 0$,

$$\mathsf{c}^r_{\langle K\rangle} x \cdot -\mathsf{c}^r_{\langle K\rangle} y + \mathsf{c}^r_{\langle K\rangle} y \cdot -\mathsf{c}^r_{\langle K\rangle} x \leq \sum_{m>0} \mathsf{c}^{1/m}_{\langle K\rangle} (x \cdot -y) + \sum_{m>0} \mathsf{c}^{1/m}_{\langle K\rangle} (y \cdot -x)$$

by the part (iv) of Theorem 2.1.2, so $\mathsf{c}^r_{\langle K\rangle} x \sim \mathsf{c}^r_{\langle K\rangle} y$. It is not difficult to see that the quotient algebra $\mathfrak{A}/I$ is a weak cylindric probability algebra and that there is a natural homomorphism from $\mathfrak{A}$ onto $\mathfrak{A}/I$.

As in the case of arbitrary algebras, there is a natural correspondence between homomorphisms and ideals in weak cylindric probability algebras. The following theorem gives a connection between weak cylindric probability algebras of formulas and ideals.

Theorem 2.1.4. *Let* I *be an ideal in* $\mathfrak{Fm}^{\Sigma}_L$ *and let* Δ *be the set of all sentences* φ *of* $L_{AP\forall}$ *such that* $(\neg\varphi)^{\Sigma} \in I$. *Then* $\Sigma \subseteq \Delta$ *and* $\mathfrak{Fm}^{\Sigma}_L/I$ *is isomorphic to* $\mathfrak{Fm}^{\Delta}_L$.

Weak cylindric probability algebras of formulas have some special properties that other weak cylindric probability algebras might not have. We shall mention one of them which is important for our purpose.

Definition 2.1.5. The *dimension set* Δx of an element $x \in A$ is the set of all indices $k < \omega$ such that $\mathsf{c}_k x \neq x$. A weak cylindric probability algebra $\mathfrak{A}$ is locally finite dimensional if Δx is finite for all $x \in A$.

It follows from the clause (ix) of Theorem 2.1.2 that $\Delta x = \{k : \mathsf{c}_k^1 x \neq x\}$, i.e., the coordinates in which x is not a cylinder can be obtained also by applying probability cylindrifications of the form c_k^1 (we write c_k^1 instead of $\mathsf{c}_{\langle\{k\}\rangle}^1$).

Every formula φ of $L_{\mathcal{A}P\forall}$ has only finitely many free variables. If v_k is a variable not occurring in φ, then $\models (\exists v_k)\varphi \leftrightarrow \varphi$ and $\models (Pv_k > 0)\varphi \leftrightarrow \varphi$. As a consequence, for any given set Σ of sentences of $L_{\mathcal{A}P\forall}$, there are at most finitely many indices $k < \omega$ such that φ is equivalent under Σ neither to $(\exists v_k)\varphi$ nor to $(Pv_k > 0)\varphi$. Thus, any weak cylindric probability algebra of formulas $\mathfrak{Fm}_L^\Sigma$ is locally finite dimensional.

The following theorem gives some elementary properties of Δ.

Theorem 2.1.6. *If $\mathfrak{A}$ is a weak cylindric probability algebra, then:*

(i) $\Delta 0 = \Delta 1 = \emptyset$;

(ii) $\Delta\big(\sum_{j\in J} x_j\big) \subseteq \bigcup_{j\in J} \Delta x_j$, $J \in \mathcal{A}$;

(iii) $\Delta\big(\prod_{j\in J} x_j\big) \subseteq \bigcup_{j\in J} \Delta x_j$, $J \in \mathcal{A}$;

(iv) $\Delta - x = \Delta x$;

(v) $\Delta \mathsf{d}_{pq} = \{p, q\}$;

(vi) $\Delta \mathsf{c}_i x \subseteq \Delta x \setminus \{i\}$;

(vii) $\Delta \mathsf{s}_j^i x \subseteq \big(\Delta x \setminus \{i\}\big) \cup \{j\}$;

(viii) $\Delta \mathsf{c}_{\langle K\rangle}^r x \subseteq \Delta x \setminus K$.

Proof. The clauses (i)–(vii) are well-known properties of Δ from the classical theory of cylindric algebras (see [Hen-Mon-Tar,85], 1.6. Dimension sets). Let i be any integer such that $i \notin \Delta x \setminus K$. If $i \in K$, then $\mathsf{c}_i \mathsf{c}_{\langle K\rangle}^r x = \mathsf{c}_{\langle K\rangle}^r x$ by (WCP_9)(i). If $i \notin \Delta x \cup K$, then

$$\mathsf{c}_i \mathsf{c}_{\langle K\rangle}^r x = \mathsf{c}_i \mathsf{c}_{\langle K\rangle}^r \mathsf{c}_i x = \mathsf{c}_i \mathsf{c}_{\langle K\cup\{i\}\rangle}^r \mathsf{c}_i x = \mathsf{c}_{\langle K\cup\{i\rangle}^r \mathsf{c}_i x = \mathsf{c}_{\langle K\rangle}^r x$$

by (WCP_9). So, $i \notin \Delta \mathsf{c}_{\langle K\rangle}^r x$. ■

We prove the following analog of the Boolean representation theorem for locally finite dimensional weak cylindric probability algebras. Also, this result shows that algebras of the form $\mathfrak{Fm}_L^\emptyset$ (the set Σ is empty) have a certain freeness property.

Theorem 2.1.7. *If $\mathfrak{A}$ is a locally finite dimensional weak cylindric probability algebra and $|A| > 1$, then there is a homomorphism from $\mathfrak{A}$ onto a weak cylindric probability set algebra.*

Proof. First, we prove that $\mathfrak{A}$ is isomorphic to a weak cylindric probability algebra of formulas $\mathfrak{Fm}_L^\Sigma$ for some L and Σ.

For each $a \in A$, the set Δa is finite, i.e. $\Delta a \subseteq \{1, \ldots, n\}$, $n < \omega$. Let R_a be an n-ary relation symbol corresponding to $a \in A$. We fix the language $L = \{R_a : a \in A\}$. By induction on the complexity of formulas of the logic $L_{\mathcal{A}P\forall}$ we define a function $f : Fm_L \mapsto A$ satisfying $\vdash \varphi$ implies $f(\varphi) = 1$ as follows:

(i) If φ is an atomic formula $R_a(v_{k_1}, \ldots, v_{k_n})$ and $j_1, \ldots, j_n$ are the first n integers in $\omega \setminus \{1, \ldots, n, k_1, \ldots, k_n\}$, then $f(\varphi) = \mathsf{s}_{k_1}^{j_1} \cdots \mathsf{s}_{k_n}^{j_n} \mathsf{s}_{j_1}^{1} \cdots \mathsf{s}_{j_n}^{n} a$;

(ii) $f(v_p = v_q) = \mathsf{d}_{pq}$;

(iii) $f(\neg\varphi) = -f(\varphi)$;

(iv) $f(\bigvee \Phi) = \Sigma_{\varphi\in\Phi} f(\varphi)$, $\Phi \in \mathcal{A}$;

(v) $f(\bigwedge \Phi) = \prod_{\varphi\in\Phi} f(\varphi)$, $\Phi \in \mathcal{A}$;

(vi) $f((\exists v_i)\varphi) = \mathsf{c}_i f(\varphi)$;

(vii) $f((P\overrightarrow{v} \geq r)\varphi) = \mathsf{c}_{\langle K\rangle}^{r} f(\varphi)$,

where $\overrightarrow{v} = v_{k_1}, \ldots, v_{k_m}$ and $K = \{k_1, \ldots, k_m\}$.

By induction on the complexity of formulas of $L_{\mathcal{A}P\forall}$, we prove the following *substitution property*:

$$\text{(S)} \qquad f(\varphi) = \mathsf{s}_{k_1}^{j_1} \cdots \mathsf{s}_{k_n}^{j_n} \mathsf{s}_{j_1}^{m_1} \cdots \mathsf{s}_{j_n}^{m_n} f(\varphi^*),$$

where φ^* is a formula obtained by the substitution of some free variables $v_{k_1}, \ldots, v_{k_n}$ of φ with $v_{m_1}, \ldots, v_{m_n}$ respectively, and $j_1, \ldots, j_n$ are some distinct integers in $\omega \setminus \{1, \ldots, n, k_1, \ldots, k_n, m_1, \ldots, m_n\}$. We show only the case of the probability quantification, because other cases are easy using appropriate properties of s_j^i (see [Hen-Mon-Tar,85], 1.5. Substitutions).

Suppose φ is $(Pv_{l_1}, \ldots, v_{l_m} \geq r)\psi(v_{k_1}, \ldots, v_{k_n}, v_{l_1}, \ldots, v_{l_m})$, where $L = \{l_1, \ldots, l_m\}$ and $L \cap \{m_1, \ldots, m_n, k_1, \ldots, k_n\} = \emptyset$. For some distinct integers $j_1, \ldots, j_n$ in $\omega \setminus \{1, \ldots, n, k_1, \ldots, k_n, m_1, \ldots, m_n, l_1, \ldots, l_m\}$ we have:

$$\begin{aligned}
f(\varphi) &= \mathsf{c}_{\langle L\rangle}^{r} \mathsf{s}_{k_1}^{j_1} \cdots \mathsf{s}_{k_n}^{j_n} \mathsf{s}_{j_1}^{m_1} \cdots \mathsf{s}_{j_n}^{m_n} f(\psi^*) && \text{by induction assumption}\\
&= \mathsf{s}_{k_1}^{j_1} \cdots \mathsf{s}_{k_n}^{j_n} \mathsf{s}_{j_1}^{m_1} \cdots \mathsf{s}_{j_n}^{m_n} \mathsf{c}_{\langle L\rangle}^{r} f(\psi^*) && \text{by (xiii) of Theorem 2.1.2}\\
&= \mathsf{s}_{k_1}^{j_1} \cdots \mathsf{s}_{k_n}^{j_n} \mathsf{s}_{j_1}^{m_1} \cdots \mathsf{s}_{j_n}^{m_n} f(\varphi^*).
\end{aligned}$$

Next, by induction on the complexity of formulas of the logic $L_{\mathcal{A}P\forall}$, we shall prove the following *dimension property:*

(D) if v_i does not occur free in φ, then $i \notin \Delta f(\varphi)$.

As usual, we show only the case of the probability quantification. So, let φ be the formula $(Pv_{l_1}, \ldots, v_{l_m} \geq r)\psi(v_{k_1}, \ldots, v_{k_n}, v_{l_1}, \ldots, v_{l_m})$ such that v_i does not occur free in φ, i.e., $i \notin \{k_1, \ldots, k_n\}$. Then

$$\begin{aligned} \Delta f(\varphi) &\subseteq \Delta f(\psi) \setminus \{l_1, \ldots, l_m\} && \text{by (viii) of Theorem 2.1.6} \\ &\subseteq \{k_1, \ldots, k_n\} && \text{by induction hypothesis,} \end{aligned}$$

i.e., $i \notin \Delta f(\varphi)$.

Now we shall prove that each logical axiom of $L_{\mathcal{A}P\forall}$ is in the set

$$\Gamma = \{\varphi \in Fm_L : f(\varphi) = 1\}.$$

It follows from the classical theory of cylindric algebras that each logical axiom of $\mathcal{A} \cap L_{\omega\omega}$ is in Γ. Also, each axiom of weak logic $L_{\mathcal{A}P}$ is in Γ. We show only the following cases.

Let φ be $(P\vec{v} > r)\psi \leftrightarrow \bigvee_{m>0}(P\vec{v} \geq r + 1/m)\psi$ (the Archimedean property). Then for $\vec{v} = v_{k_1}, \ldots, v_{k_n}$ and $K = \{k_1, \ldots, k_n\}$ we have

$$\begin{aligned} f\big((P\vec{v} > r)\psi\big) &= -\mathsf{c}_{\langle K\rangle}^{1-r} - f(\psi) = \sum\nolimits_{m>0} \mathsf{c}_{\langle K\rangle}^{r+1/m} f(\psi) \quad \text{by } (\mathsf{WCP}_6) \\ &= f\left(\bigvee\nolimits_{m>0}(P\vec{v} \geq r + 1/m)\psi\right); \end{aligned}$$

so, $f(\varphi) = 1$.

Let φ be $(\forall v_i)\psi \to (Pv_i \geq 1)\psi$. Then

$$\begin{aligned} f(\varphi) &= - - \mathsf{c}_i - f(\psi) + \mathsf{c}_i^1 f(\psi) \\ &\geq \mathsf{c}_i - f(\psi) + -\mathsf{c}_i - f(\psi) \quad \text{by } (\mathsf{WCP}_8) \\ &= 1. \end{aligned}$$

Finally, we shall prove that each logical theorem of $L_{\mathcal{A}P\forall}$ is in Γ. Obviously Γ is closed under Modus Ponens and under Conjunction rule. We have two Generalization rules.

If $\varphi \to \psi(v_i) \in \Gamma$ and v_i is not free in φ, then

$$\begin{aligned} f(\varphi \to (\forall v_i)\psi) &= -f(\varphi) + -\mathsf{c}_i - f(\psi) \\ &= -\big(\mathsf{c}_i f(\varphi) \cdot \mathsf{c}_i - f(\psi)\big) && \text{by } (\mathsf{D}) \\ &= -\mathsf{c}_i\big(\mathsf{c}_i f(\varphi) \cdot -f(\psi)\big) && \text{by } (\mathsf{WCP}_0) \\ &= 1 && \text{by assumption.} \end{aligned}$$

So, $\varphi \to (\forall v_i)\psi \in \Gamma$.

If $\varphi \to \psi(v_{k_1}, \dots, v_{k_n}) \in \Gamma$ and $\overrightarrow{v} = v_{k_1}, \dots, v_{k_n}$ are not free in φ, then

$$\begin{aligned} f(\varphi \to (P\overrightarrow{v} \geq 1)\psi) &= -f(\varphi) + \mathsf{c}^1_{\langle K \rangle} f(\psi) \\ &= -\mathsf{c}^1_{\langle K \rangle} f(\varphi) + \mathsf{c}^1_{\langle K \rangle} f(\psi) && \text{by } (\mathsf{D}) \text{ and Theorem 2.1.2} \\ &= \mathsf{c}^1_{\langle K \rangle}\big(-f(\varphi) + f(\psi)\big) && \text{by } (\mathsf{WCP}_4) \\ &= 1 && \text{by assumption.} \end{aligned}$$

So, $\varphi \to (P\overrightarrow{v} \geq 1)\psi \in \Gamma$.

It follows that $\vdash \varphi \leftrightarrow \psi$ implies $f(\varphi) = f(\psi)$. So, a well-defined function $g : Fm_L / \equiv_\emptyset \mapsto A$ introduced by $g(\varphi^\emptyset) = f(\varphi)$ is a homomorphism from $Fm^\emptyset_L$ onto $\mathfrak{A}$ such that $g\big(R_a(v_1, \dots, v_n)^\emptyset\big) = a$. Let I be a subset of $Fm_L / \equiv_\emptyset$ such that $\varphi^\emptyset \in I$ iff $g(\varphi^\emptyset) = 0$, and let Σ be a set of all sentences φ of $L_{AP\forall}$ such that $(\neg\varphi)^\emptyset \in I$. Then I is an ideal in $\mathfrak{Fm}^\emptyset_L$ and $\mathfrak{A} \cong (\mathfrak{Fm}^\emptyset_L)/I \cong \mathfrak{Fm}^\Sigma_L$ from Theorem 2.1.4. Moreover, Σ is consistent, since $|A| > 1$. If $\mathfrak{A}$ is a weak probability model of Σ (see [Ras,87]), then we have a "natural" homomorphism from $\mathfrak{Fm}^\Sigma_L$ onto the weak cylindric probability set algebra

$$\big\langle \{\varphi^{\mathfrak{A}} : \varphi \in Fm_L\}, \cup, \cap, \sim, \emptyset, A^\omega, \mathsf{C}_i, \mathsf{C}^r_{\langle K \rangle}, \mathsf{D}_{pq} \big\rangle.$$

This completes the proof. ■

2. Weak Polyadic Probability Algebras

Let V and Pr be disjoint sets of variables and predicate symbols, respectively, ν a function from the set Pr to the class of all ordinals and let $\mathfrak{m}$ be a cardinal. In [Kei,63] Keisler introduces a formal system $L(V, \nu, \mathfrak{m})$ which has predicates with infinitely many argument places and quantifiers over infinite sets of variables, but which has only finite propositional connectives and no identity symbol, and whose proofs are finite. We assume that V, ν, $\mathfrak{m}$ satisfy the conditions I, II and III from [Kei,63].

Let $\mathcal{R}$ be a Lukasiewicz chain $\left\{0, \frac{1}{n}, \dots, \frac{n-1}{n}, 1\right\}$ together with the operations

$$x \oplus y = \min\{x+y, 1\} \qquad \text{and} \qquad \neg x = 1 - x.$$

In [Mar-Ras-Dor,99] we construct similar weak probability logic with infinitary predicates $L(V, \nu, \mathfrak{m}, \mathcal{R})$ by adding probability quantifiers $(Px \geq r)$, where $r \in \mathcal{R}$ and $x \in V^\alpha$ is a sequence of distinct variables of length α, $\overline{\overline{\alpha}} < \mathfrak{m}$. The set $\mathcal{R}$ is taken to be finite in order to preserve the finiteness of proofs.

Motivated by the connection between Keisler's $L(V, \nu, \mathfrak{m})$ logic and polyadic algebra (see [Kei,63]), we introduce an algebra called *weak polyadic probability algebra* which corresponds to the weak probability logic with infinitary predicates $L(V, \nu, \mathfrak{m}, \mathcal{R})$ (see [Mar-Ras-Dor,01]). As before, we need the concept of probability cylindrification operations. Also, for each $\tau \in V^V$, the substitution $\mathsf{S}_f(\tau)\Phi$ of free variables in a formula Φ by τ is defined by

$$\mathsf{S}_f(\tau)(\forall x)\Phi = (\forall x)\mathsf{S}_f(\sigma)\Phi),$$

$$\mathsf{S}_f(\tau)(Px \geq r)\Phi = \big((Px \geq r)\mathsf{S}_f(\sigma)\Phi\big),$$

where $\sigma \in V^V$ and $\sigma(v) = \begin{cases} \tau(v), & v \in V \setminus \operatorname{range} x \\ v, & v \in \operatorname{range} x, \end{cases}$ i.e., we can write $\sigma = \big(\tau \restriction (V \setminus \operatorname{range} x)\big) \restriction V$ ($\restriction$ is defined as in [Kei,63]). Thus, we need the concept of substitution operations.

Definition 2.2.1. A *weak polyadic probability algebra* of dimension β, briefly WPP_β, is a structure

$$\left\langle A, +, \cdot, -, 0, 1, \mathsf{c}_{(K)}, \mathsf{c}^r_{(K)}, \mathsf{s}_\tau \right\rangle,$$

such that $\langle A, +, \cdot, -, 0, 1\rangle$ is a Boolean algebra, $\mathsf{c}_{(K)}$, $\mathsf{c}^r_{(K)}$ and s_τ are unary operations on A for each sequence (K) of ordinals from β of length α, $\overline{\overline{\alpha}} \leq \overline{\overline{\beta}}$, each $r \in \mathcal{R}$ and $\tau \in \beta^\beta$, if the following postulates hold:

(WPP$_0$) $\langle A, +, \cdot, -, 0, 1, \mathsf{c}_{(K)}, \mathsf{s}_\tau \rangle$ is a polyadic algebra of dimension β.

(WPP$_1$) (i) $\mathsf{c}^r_{(\emptyset)} x = x$, (ii) $\mathsf{c}^r_{(K)} 0 = 0$, where $r > 0$.

(WPP$_2$) $\mathsf{c}^0_{(K)} x = 1$.

(WPP$_3$) If $r \geq \mathsf{s}$, then $\mathsf{c}^r_{(K)} x \leq \mathsf{c}^{\mathsf{s}}_{(K)} x$.

(WPP$_4$) If $r > 0$, then $\mathsf{c}^r_{(K)} \mathsf{c}^{\mathsf{s}}_{(K)} x = \mathsf{c}^{\mathsf{s}}_{(K)} x$.

(WPP$_5$) (i) $-\mathsf{c}^r_{(K)} x \cdot \mathsf{c}^{1-\mathsf{s}}_{(K)} - y \leq -\mathsf{c}^{\min\{1, r+\mathsf{s}\}}_{(K)} (x \cdot y)$,

(ii) $\mathsf{c}^r_{(K)} x \cdot \mathsf{c}^{\mathsf{s}}_{(K)} y \cdot \mathsf{c}^1_{(K)} - (x \cdot y) \leq \mathsf{c}^{\min\{1, r+\mathsf{s}\}}_{(K)} (x + y)$.

(WPP$_6$) (i) $\mathsf{c}^r_{(K)} - x \geq -\mathsf{c}^{1-r}_{(K)} x$,

(ii) $\mathsf{c}^{\mathsf{s}}_{(K)} - x \leq -\mathsf{c}^{1-r}_{(K)} x$, $\mathsf{s} > r$,

(iii) $-\mathsf{c}^{1-\mathsf{s}}_{(K)} - x \leq \mathsf{c}^{\mathsf{s}^+}_{(K)} x$, where $\mathsf{s}^+ = \min\{r \in \mathcal{R} : r > \mathsf{s}\}$.

(WPP$_7$) If $\sigma \restriction (\beta \setminus \operatorname{rang} K) = \tau \restriction (\beta \setminus \operatorname{rang} K)$, then $\mathsf{s}_\sigma \mathsf{c}^r_{(K)} x = \mathsf{s}_\tau \mathsf{c}^r_{(K)} x$.

(WPP$_8$) If $\sigma \restriction \sigma^{-1}(K)$ is $1-1$ function, then $\mathsf{s}_\sigma \mathsf{c}^r_{\sigma^{-1}(K)} x = \mathsf{c}^r_{(K)} \mathsf{s}_\sigma x$.

(WPP$_9$) (i) $\mathsf{c}^r_{(K)} x \leq \mathsf{c}_{(K)} x$, $r > 0$,

(ii) $\mathsf{c}_{(K)} \mathsf{c}^r_{(K_1)} x = \mathsf{c}^r_{(K_1)} x$, $\operatorname{rang} K \subseteq \operatorname{rang} K_1$,

(iii) $\mathsf{c}^r_{(K)} \mathsf{c}_{(K_1)} x = \mathsf{c}_{(K_1)} x$, $r > 0$ and $\operatorname{rang} K \subseteq \operatorname{rang} K_1$.

Let V_1 be a set of new variables such that $V_1 \cap V = \emptyset$, $V_1 \cap Pr = \emptyset$ and $\overline{\overline{V_1}} = \overline{\overline{V}}$. Let $V^* = V \cup V_1$, $L^* = L(V^*, \nu, \mathfrak{m}, \mathcal{R})$ and let Σ be any set of sentences in L^*. Let $H \subseteq Fm_{L^*}$ be a set of formulas whose free variables are in the set V and bound in V_1. Then, for each $\Phi \in H$ we define $\Phi/\Sigma = \{\Psi : \Sigma \vdash_{L^*} \Psi \leftrightarrow \Phi, \Psi \in H\}$, $H_\Sigma = \{\Phi/\Sigma : \Phi \in H\}$, $0_\Sigma = \bot/\Sigma$, $1_\Sigma = \top/\Sigma$ and operations $+_\Sigma, \cdot_\Sigma, -_\Sigma$ on H_Σ as usual. Also, for each sequence x of distinct variables from V of length α, $\overline{\overline{\alpha}} < \mathfrak{m}$, $r \in \mathcal{R}$, $\tau \in V^V$ and $\Phi \in H$ we define

$$(\exists^\Sigma x)(\Phi/\Sigma) = \big((\exists x)\Phi\big)/\Sigma,$$

$$(P^\Sigma x \geq r)(\Phi/\Sigma) = \big((Px \geq r)\Phi\big)/\Sigma,$$

$$\mathsf{S}^\Sigma(\tau)(\Phi/\Sigma) = \big(\mathsf{S}_f(\tau)\mathsf{S}_f(\tau_0^{-1})\mathsf{S}(\tau_0)\Phi\big)/\Sigma,$$

where $\tau_0 : V^* \overset{1-1}{\mapsto} V_1$ (see [Mar-Ras-Dor,99, Mar-Ras-Dor,01]).

If $\mathfrak{m} = \overline{\overline{V}}^+$ and Σ is consistent in L^*, by routine checking, we obtain that the following algebra of formulas

$$\mathbb{H}_\Sigma = \big\langle H_\Sigma, +_\Sigma, \cdot_\Sigma, -_\Sigma, 0_\Sigma, 1_\Sigma, \exists^\Sigma x, P^\Sigma x \geq r, \mathsf{S}^\Sigma(\tau) \big\rangle$$

is a weak polyadic probability algebra of dimension V.

A weak probability structure of type ν is

$$\mathfrak{A} = \langle A, R_p, \mu_\alpha \rangle_{p \in Pr,\ \overline{\overline{\alpha}} < \overline{\overline{V}}},$$

where A is a nonempty set, $R_p \subset A^{\nu(p)}$, μ_α is a finitely additive probability measure on A^α with range $\mathcal{R}$, such that the set

$$\{ b \circ x : \models_{\mathfrak{A}} \Phi[b],\ b \restriction (V \setminus \operatorname{range} x) = a \restriction (V \setminus \operatorname{range} x),\ b \in A^V \}$$

is μ_α measurable, for each sequence of variables $x \in V^\alpha$, any $a \in A^V$ and any formula Φ (for details see [Kei,63] and [Mar-Ras-Dor,99]).

Let $\mathfrak{A}$ be a weak probability structure for L and $V = \beta$, $\Phi^{\mathfrak{A}} = \{ a \in A^\beta : \models_{\mathfrak{A}} \Phi[a \restriction V^*] \}$ and $\mathbb{A} = \{ \Phi^{\mathfrak{A}} : \Phi \in H \}$. For each sequence (K) of ordinals from β of length α, $\overline{\overline{\alpha}} \leq \overline{\overline{\beta}}$, each $r \in \mathcal{R}$ and each $\sigma \in \beta^\beta$, we define

$$\begin{aligned} \mathsf{C}_{(K)}(\Phi^{\mathfrak{A}}) = \{ & a \in A^\beta : b \restriction (\beta \setminus \operatorname{rang}(K)) = a \restriction (\beta \setminus \operatorname{rang}(K)), \\ & \text{for some } b \in \Phi^{\mathfrak{A}} \}, \end{aligned}$$

$$\begin{aligned} \mathsf{C}^r_{(K)}(\Phi^{\mathfrak{A}}) = \{ & a \in A^\beta : \mu_\alpha \{ b \circ (K) : b \restriction (\beta \setminus \operatorname{rang}(K)) \\ & = a \restriction (\beta \setminus \operatorname{rang}(K)),\ \text{for some } b \in \Phi^{\mathfrak{A}} \} \geq r \}, \end{aligned}$$

$$\mathsf{S}_\sigma(\Phi^{\mathfrak{A}}) = \{ a \in A^\beta : a_\sigma \in \left(\mathsf{S}_f(\tau_0^{-1}) \mathsf{S}(\tau_0) \Phi \right)^{\mathfrak{A}} \},$$

where $a_\sigma = \left(a_{\sigma(\alpha)} \right)_{\alpha < \beta}$ for $a = (a_\alpha)_{\alpha < \beta}$. It is easy to verify that the above operations are well defined and that the sets from the definition of $\mathsf{C}^r_{(K)}$ are measurable. We may show that $\mathbb{A}$ is closed for these operations and that the set algebra

$$\langle \mathbb{A}, \cup, \cap, \sim, \emptyset, A^\beta, \mathsf{C}_{(K)}, \mathsf{C}^r_{(K)}, \mathsf{S}_\sigma \rangle$$

is a weak polyadic probability algebra of dimension β.

Theorem 2.2.2. *If $\mathfrak{A} = \langle A, +, \cdot, -, 0, 1, \mathsf{c}_{(K)}, \mathsf{c}^r_{(K)}, \mathsf{s}_\sigma \rangle$ is a weak polyadic probability algebra of infinite dimension β, then there exists a set of sentences Σ of $L = L(V, \nu, \mathfrak{m}, \mathcal{R})$ such that $\mathbb{H}_\Sigma \cong \mathfrak{A}$.*

Proof. We shall put $V = \beta$, $Pr = A$, $\nu(p) = \beta$ for each $p \in Pr$ and $x \in \beta^\beta$ "1-1" and "onto". Let Σ be the next set of sentences:

$$(\forall x)\left((-p)(x) \leftrightarrow \neg(p(x)) \right), \qquad (\forall x)\left((\mathsf{s}_\sigma p)(x) \leftrightarrow \mathsf{S}(\sigma) p(x) \right),$$

$$(\forall x)\left((p+q)(x) \leftrightarrow p(x) \vee q(x) \right), \quad (\forall x)\left((\mathsf{c}_{(K)} p)(x) \leftrightarrow (\exists y) p(x) \right),$$

$$(\forall x)\left((p \cdot q)(x) \leftrightarrow p(x) \wedge q(x) \right), \quad (\forall x)\left((\mathsf{c}^r_{(K)} p)(x) \leftrightarrow (Py \geq r) p(x) \right),$$

where $p, q \in Pr$, $\sigma \in V^V$, $r \in \mathcal{R}$ and $(K) = y$ is a sequence of distinct variables of length α, $\overline{\overline{\alpha}} < \mathfrak{m}$. Then $H_\Sigma = \left\{ \left(p(x)\right)^\Sigma : p \in Pr \right\}$ and a function $g : A \mapsto H_\Sigma$ defined by

$$g(p) = \left(p(x)\right)^\Sigma$$

is the desired isomorphism (see [Mar-Ras-Dor,01]). ∎

Finally, we prove the following representation theorem.

Theorem 2.2.3. *If $\mathfrak{A}$ is a weak polyadic probability algebra of infinite dimension β, then there is a homomorphism of $\mathfrak{A}$ onto a weak polyadic probability set algebra.*

Proof. By the previous theorem there exists the set of sentences Σ in $L = L(\beta, \nu, \overline{\overline{\beta}}^+, \mathcal{R})$ such that $\mathbb{H}_\Sigma \cong \mathfrak{A}$. Since $\overline{\overline{B}} > 1$, it follows that Σ is consistent in L. By the completeness theorem, Σ has a weak probability model $\mathfrak{B}$ which generate the desired weak polyadic probability set algebra

$$\left\langle \mathbb{B}, \cup, \cap, \sim, \emptyset, A^\beta, \mathsf{C}_{(K)}, \mathsf{C}^r_{(K)}, \mathsf{S}_\sigma \right\rangle$$

as usual. ∎

3. Cylindric Probability Algebras

To show that the theory of cylindric probability algebras is rooted in probability logic, we shall now describe the relationship between the graded probability logic L_{AP} (see [Kei,85], [Hoo,78]) and cylindric probability algebras introduced in [Ras-Dor,00]. First we shall describe cylindric probability set algebras.

Let $\langle A, \mu_n \rangle_{n<\omega}$ be a graded probability space and let μ_ω be the completion of the measure on A^ω determined by the μ_n's. As before, we define a unary *probability cylindrification* operation $\mathsf{C}^r_{\langle K \rangle}$ on the subsets of A^ω (see Section 1). It follows from the Fubini property that for any μ_ω-measurable set X, the section $\left\{ (x_{k_1}, \ldots, x_{k_n}) : x \in X \;\&\; (j \notin K \to x_j = y_j) \right\}$ is μ_n-measurable for each $y \in A^\omega$, and also that $\mathsf{C}^r_{\langle K \rangle}(X)$ is μ_ω-measurable. Further, for each permutation σ of ω such that $\sigma = Id$ almost surely, i.e.,

σ fixes all but finitely many elements of ω, a unary *permutation* operation S_σ on the subsets of A^ω is defined by setting, for any $X \subseteq A^\omega$,

$$\mathsf{S}_\sigma(X) = \big\{\big(x_{\sigma(1)}, x_{\sigma(2)}, \ldots\big) : x \in X\big\}.$$

For any μ_ω-measurable set X, the set $\mathsf{S}_\sigma(X)$ is μ_ω-measurable. We do not need the full class of permutations, because the number of free variables is finite even in infinite formulas of $L_{\mathcal{A}P}$ logic.

Let $\mathcal{A}$ be a countable admissible set such that $\omega \in \mathcal{A}$. We suppose that a Boolean set algebra $\langle A, \cup, \cap, \sim, \emptyset, A^\omega\rangle$, where $A \subseteq \mathcal{P}(A^\omega)$, is $\mathcal{A}$-*complete*, i.e., for any $\{A_j : j \in J\} \subseteq A$, where $J \subseteq I$ and $J \in \mathcal{A}$ we have $\bigcup_{j\in J} A_j \in A$. So, a structure

$$\big\langle A, \cup, \cap, \sim, \emptyset, A^\omega, \mathsf{C}^r_{\langle K\rangle}, \mathsf{S}_\sigma, \mathsf{D}_{pq}\big\rangle_{K,\sigma,r\in[0,1],\ p,q<\omega}$$

in which A is a collection of subsets of A^ω closed under all operations of an $\mathcal{A}$-complete Boolean algebra, under all probability cylindrifications and under all permutations, having all diagonal hyperplanes $\mathsf{D}_{pq} = \{x \in A^\omega : x_p = x_q\}$ as distinguished members, is called *cylindric probability set algebra.*

As before, for any set Σ of sentences of $L_{\mathcal{A}P}$, the quotient algebra

$$\mathfrak{Fm}^\Sigma_L = \big\langle Fm_L /{\equiv_\Sigma}, \vee^\Sigma, \wedge^\Sigma, \neg^\Sigma, F^\Sigma, T^\Sigma, (P\overrightarrow{v} \geq r)^\Sigma, \mathsf{S}^\Sigma_\sigma, (v_p = v_q)^\Sigma\big\rangle$$

is called a *cylindric probability algebra of formulas of $L_{\mathcal{A}P}$ associated to* Σ. In this way, basic metalogical problems are interpreted as algebraic problems concerning the associated algebra of formulas.

Let $\mathfrak{A} = \langle A, R^\mathfrak{A}, \mathsf{c}^\mathfrak{A}, \mu_n\rangle$ be a graded probability model for Σ. Then, for any formula φ of $L_{\mathcal{A}P}$, we have

$$\mathsf{C}^r_{\langle K\rangle}(\varphi^\mathfrak{A}) = \big((P\overrightarrow{r} \geq r)\varphi\big)^\mathfrak{A} \qquad \text{and} \qquad \mathsf{S}_\sigma(\varphi^\mathfrak{A}) = (\mathsf{S}_\sigma\varphi)^\mathfrak{A},$$

where $\overrightarrow{v} = v_{k_1}, \ldots, v_{k_n}$ and $K = \{k_1, \ldots, k_n\}$. Therefore, we get a cylindric probability set algebra by a homomorphic transformation of a cylindric probability algebra of formulas.

As before, the abstract notion of a *cylindric probability algebra* is defined by equations which hold in both algebras mentioned above, where the emphasis is on the axioms of graded probability logic [Kei,85].

Definition 2.3.1. A *cylindric probability algebra* is a structure

$$\mathfrak{A} = \big\langle A, +, \cdot, -, 0, 1, \mathsf{c}^r_{(K)}, \mathsf{s}_\sigma, \mathsf{d}_{pq}\big\rangle,$$

such that $\langle A, +, \cdot, -, 0, 1\rangle$ is an $\mathcal{A}$-complete Boolean algebra, $\mathsf{c}^r_{(K)}$ are unary operations on A called probability cylindrifications for each finite $K \subseteq \omega$, s_σ is an unary operation on A called permutation for each permutation σ of ω such that $\sigma = Id$ almost surely, $\mathsf{d}_{pq} \in A$ for all $p, q < \omega$, and the following postulates hold:

(CP$_1$) (i) $\mathsf{c}^r_{\langle\emptyset\rangle}x = x$, (ii) $\mathsf{c}^r_{\langle K\rangle}0 = 0$, where $r > 0$.

(CP$_2$) $\mathsf{c}^0_{\langle K\rangle}x = 1$.

(CP$_3$) If $r \geq \mathsf{s}$, then $\mathsf{c}^r_{\langle K\rangle}x \leq \mathsf{c}^\mathsf{s}_{\langle K\rangle}x$.

(CP$_4$) $\mathsf{c}^r_{\langle K\rangle}\big(x + \sum_{j\in J}\mathsf{c}^\mathsf{s}_{\langle L\rangle}\mathsf{c}^t_{\langle M\rangle}y_j\big) = \mathsf{c}^r_{\langle K\rangle}x + \sum_{j\in J}\mathsf{c}^\mathsf{s}_{\langle L\rangle}\mathsf{c}^t_{\langle M\rangle}y_j$, where $J \in \mathcal{A}$ and $K \subseteq L \cup M$.

(CP$_5$) (i) $\mathsf{c}^r_{\langle K\rangle}x \cdot \mathsf{c}^\mathsf{s}_{\langle K\rangle}y \leq \mathsf{c}^{\max\{0, r+\mathsf{s}-1\}}_{\langle K\rangle}(x \cdot y)$,

(ii) $\mathsf{c}^r_{\langle K\rangle}x \cdot \mathsf{c}^\mathsf{s}_{\langle K\rangle}y \cdot \mathsf{c}^1_{\langle K\rangle} - (x \cdot y) \leq \mathsf{c}^{\min\{r+\mathsf{s},1\}}_{\langle K\rangle}(x + y)$.

(CP$_6$) $\mathsf{c}^r_{\langle K\rangle} - x = -\sum_{m>0}\mathsf{c}^{1-r+1/m}_{\langle K\rangle}x$.

(CP$_7$) $\prod_{J_2\subseteq J_1}\mathsf{c}^r_{\langle K\rangle}\prod_{j\in J_2}x_j \leq \mathsf{c}^r_{\langle K\rangle}\prod_{j\in J_1}x_j$, where J_2 ranges over the finite subsets of J_1 and $J_1 \in \mathcal{A}$.

(CP$_8$) $\mathsf{c}^r_{\langle K\rangle}x = \mathsf{c}^r_{\langle(\pi(K))\rangle}x$, where π is a permutation of $\{1, \ldots, n\}$ and $\big(\pi(K)\big)$ is $k_{\pi 1}, \ldots, k_{\pi n}$.

(CP$_9$) If $K \cap L = \emptyset$, then $\mathsf{c}^r_{\langle K\rangle}\mathsf{c}^\mathsf{s}_{\langle L\rangle}x \leq \mathsf{c}^{r\cdot\mathsf{s}}_{\langle K\rangle,\langle L\rangle}x$.

(CP$_{10}$) (i) $\mathsf{s}_{Id}x = x$,

(ii) $\mathsf{s}_\sigma\mathsf{s}_\tau x = \mathsf{s}_{\sigma\circ\tau}x$,

(iii) $\mathsf{s}_\sigma\sum_{j\in J}x_j = \sum_{j\in J}\mathsf{s}_\sigma x_j$, where $J \in \mathcal{A}$,

(iv) $\mathsf{s}_\sigma - x = -\mathsf{s}_\sigma x$.

(CP$_{11}$) (i) $\mathsf{s}_\sigma\mathsf{c}^r_{\langle K\rangle}\mathsf{c}^\mathsf{s}_{\langle L\rangle}x = \mathsf{s}_\tau\mathsf{c}^r_{\langle K\rangle}\mathsf{c}^\mathsf{s}_{\langle L\rangle}x$, where $\sigma \restriction (K \cup L)^\mathsf{c} = \tau \restriction (K \cup L)^\mathsf{c}$,

(ii) $\mathsf{c}^r_{\langle K\rangle}\mathsf{s}_\sigma x = \mathsf{s}_\sigma\mathsf{c}_{\langle\sigma^{-1}(K)\rangle}x$, where $\big\langle\sigma^{-1}(K)\big\rangle$ is $\sigma^{-1}(k_1), \ldots, \sigma^{-1}(k_n)$.

(CP$_{12}$) (i) $\mathsf{d}_{pp} = 1$,

(ii) $x \cdot \mathsf{d}_{pq} \leq \mathsf{s}_\sigma x$, where $\sigma(p) = q$,

(iii) $\mathsf{s}_\sigma\mathsf{d}_{pq} = \mathsf{d}_{\sigma(p)\sigma(q)}$,

(iv) $\mathsf{c}^1_{\langle K\rangle}\mathsf{d}_{pq} = \mathsf{d}_{pq}$, where $p, q \notin K$.

We point out that the axioms (CP_2), (CP_3), (CP_5)–(CP_9) express non-negativity, monotonicity, finite additivity, the Archimedean property, countable additivity, symmetry and product independence of probability measures $\{\mu_n : n < \omega\}$ (see [Kei,85]).

As before, some necessary properties of probability cylindrification operations (see Theorem 2.1.2) can be obtained by routine checking.

Let $I \subseteq A$ be a Boolean ideal of $\mathfrak{A}$ such that for any finite $K \subseteq \omega$ and $r \in (0,1]$ and for any permutation σ of ω such that $\sigma = Id$ a.s., if $x \in I$, then $\mathsf{c}^r_{\langle K\rangle} x \in I$ and $\mathsf{s}_\sigma x \in I$. It is not difficult to see that the quotient algebra $\mathfrak{A}/I$ is a cylindric probability algebra and that there is a natural homomorphism from $\mathfrak{A}$ onto $\mathfrak{A}/I$.

The notions of the dimension set Δx of an element $x \in A$ and the locally finite dimensional cylindric probability algebra are defined as usual (see Section 1., where we write c^1_k instead of c_k). It is easy to see that $k \in \Delta x$ if and only if $\mathsf{c}^r_k x \neq x$ for any $r > 0$. Also, any cylindric probability algebra of formulas is locally finite dimensional. We point out that $\Delta \mathsf{c}^r_{\langle K\rangle} x \subseteq \Delta x \backslash K$ and $\Delta \mathsf{s}_\sigma x \subseteq \sigma(\Delta x)$ express a relationship between the dimension set and probability cylindrification operations and permutations respectively (see Theorem 2.1.6).

The algebra $\mathfrak{Fm}^{\emptyset}_L$ (the set Σ is empty) has a certain freeness property.

Theorem 2.3.2. *Let $L = \{R_i : i \in I_0\}$ be a set of finitary relation symbols and let $\mathfrak{A}$ be a cylindric probability algebra. Let f be a function from L into A such that $\Delta f(R_i) \subseteq n_i$ for each n_i-ary relation R_i. Then there is a homomorphism $g : Fm_L / \equiv_\emptyset \mapsto A$ such that $g\big(R_i(v_1, \ldots, v_{n_i})^\emptyset\big) = f(R_i)$ for each $i \in I_0$.*

Proof. Using the proof of Theorem 2.1.7, a function $h : Fm_L \mapsto A$ satisfying $\vdash \varphi$ implies $h(\varphi) = 1$ can be defined first for atomic formula as follows:

If φ is an atomic formula $R_i(v_{k_1}, \ldots, v_{k_{n_i}})$, $j_1, \ldots, j_{n_i}$ are the first n_i integers in $\omega \setminus \{1, \ldots, n_i, k_1, \ldots, k_{n_i}\}$ and σ, τ are the permutations of ω such that $\sigma = \binom{1,\ldots,n_i}{j_1,\ldots,j_{n_i}}$, $\tau = \binom{j_1,\ldots,j_{n_i}}{k_1,\ldots,k_{n_i}}$, $\sigma \restriction \{1, \ldots, n_i, j_1, \ldots, j_{n_i}\}^{\mathsf{c}} = Id$ and $\tau \restriction \{k_1, \ldots, k_{n_i}, j_1, \ldots, j_{n_i}\}^{\mathsf{c}} = Id$, then we define

$$h(\varphi) = \mathsf{s}_\tau \mathsf{s}_\sigma f(R_i).$$

Then by induction on formulas of the graded probability logic we define h as usual (see the proof of Theorem 2.1.7).

By induction on the complexity of formulas, we prove the *substitution property:*

$$\text{(S)} \qquad h(\mathsf{S}_\sigma \varphi) = \mathsf{s}_\sigma h(\varphi),$$

and the *dimension property:*

$$\text{(D)} \qquad \text{if } v_k \text{ does not occur free in } \varphi, \text{ then } k \notin \Delta h(\varphi).$$

Now we shall prove that each logical axiom of the graded probability logic L_{AP} is in the set

$$\Gamma = \{\varphi \in Fm_L : h(\varphi) = 1\}.$$

For example, we show only the following *product independence* case: If φ is $(P\overrightarrow{v} \geq r)(P\overrightarrow{w} \geq \mathsf{s})\psi \to (P\overrightarrow{v}, \overrightarrow{w} \geq r \cdot \mathsf{s})\psi$, where $K \cap L = \emptyset$ for $K = \{k_1, \ldots, k_n\}$ and $L = \{l_1, \ldots, l_m\}$, $\overrightarrow{v} = v_{k_1}, \ldots, v_{k_n}$ and $\overrightarrow{w} = v_{l_1}, \ldots, v_{l_m}$, then

$$\begin{aligned} h(\varphi) &= -\mathsf{c}^{r}_{\langle K\rangle}\mathsf{c}^{\mathsf{s}}_{\langle L\rangle}h(\psi) + \mathsf{c}^{r\cdot\mathsf{s}}_{\langle K\rangle,\langle L\rangle}h(\psi) \\ &\geq -\mathsf{c}^{r\cdot\mathsf{s}}_{\langle K\rangle,\langle L\rangle}h(\psi) + \mathsf{c}^{r\cdot\mathsf{s}}_{\langle K\rangle,\langle L\rangle}h(\psi) \;\; \text{by } (\mathsf{CP}_9) \\ &= 1. \end{aligned}$$

Obviously Γ is closed under Modus Ponens, Conjunction rule and under probability Generalization (similarly as in Theorem 2.1.7). Thus, each logical theorem of the graded probability logic is in Γ. It follows that $\vdash \varphi \leftrightarrow \psi$ implies $h(\varphi) = h(\psi)$. So, a well-defined function $g : Fm_L/\!\equiv_\emptyset \mapsto A$ introduced by $g(\varphi^\emptyset) = h(\varphi)$ for any $\varphi \in Fm_L$, is the desired homomorphism. ■

As a consequence (see Section 1), we obtain the following representation theorem for locally finite dimensional cylindric probability algebras.

Theorem 2.3.3 (Boolean representations). *If $\mathfrak{A}$ is a locally finite dimensional cylindric probability algebra and $|A| > 1$, then there is a homomorphism of $\mathfrak{A}$ onto a cylindric probability set algebra.*

BOLYAI SOCIETY
MATHEMATICAL STUDIES, 22

Cylindric-like Algebras and Algebraic Logic
pp. 321–331.

Cylindric Algebras and Relational Databases

IVO DÜNTSCH

The success of the relational data model introduced by Codd [Cod,70] can – at least partly – be attributed to its clarity and succinctness. The semantics of the model are constraints among and within the base relations which, as it turned out, could be formulated in various fragments of first order logic. Therefore, a natural approach seems to be the general framework of cylindric structures. The two main examples of these are concrete algebras of n-ary relations on the one hand, and algebras of first order sentences on the other. It is interesting to note that throughout most of the research on the theory of relational databases, this dichotomy can be observed, albeit in a different terminology. Codd's relational algebra and domain- or tuple calculus are well known examples for this phenomenon, and, keeping in mind the connection between cylindric algebras and first order logic, it comes as no surprise that the resulting query languages are of the same expressive power. The extended relations of [Yan-Pap,82] are, loosely speaking, nothing else than an embedding of some finite cylindric algebra into an infinite algebra of formulae, and a transformation from finite dimensional relations to formulae (and infinite dimensional cylindric algebras) is implicitly present in [Cos,87].

The plan for the current chapter is as follows: After a brief introduction to Codd's data model, I will present an embedding of relational data tables into cylindric structures similar to the first such construction by Imilienski and Lipski [Imi-Lip,84a]. The rest of the chapter will be taken up by the discussion of various forms of data dependencies in algebraic setting. Most results are given for cylindric algebras of dimension ≥ 3. I would like to thank the referee for pointing out that it would be interesting to generalize the results to dimensions < 3. I will use the definitions and notation of [Hen-Mon-Tar,85].

1. The Relational Data Model

The two components of Codd's original concept of a relational database are objects and operators [Cod,70]:

1. Suppose that α is a countable cardinal, and $\{D_n : n < \alpha\}$ is a collection of nonempty sets which are called *domains*. Intuitively, α is a set of attribute names, and a domain D_n consists of those values which the n-th attribute of a data table may take. Even though an instance of a database is finite, the domains are usually allowed to be infinite so as to have enough choices for a field entry. If $\emptyset \neq X \subseteq \alpha$, a subset r of $\prod_{i \in X} D_i$ is called a *data table over the schema* X, written as $\langle X, r\rangle$; the largest data table with schema X is denoted by $\mathbf{1}_X$, i.e. $\mathbf{1}_X = \prod_{n \in X} D_n$. Note that data tables are (heterogeneous) relations. The elements of a data table are called *tuples* or *records*; observe that tuples are choice functions with domain X. If r is a data table over X, we set $a(r) = X$, and call it the *schema of* r. A *database* is a structure $\mathcal{D} = \langle D_n, r_j\rangle_{n<\alpha,\ j \in J}$ such that, for every $j \in J$, r_j is a data table with schema $a(r_j)$.

2. Besides the Boolean set operators $\cup$ and $\setminus$ which are defined on data tables with the same schema X, there are four operations specific to databases:

   (a) *Join:* The *(natural) join* $\bowtie$ of two data tables r and s is defined as

   $$r \bowtie s = \left\{f \in \mathbf{1}_{a(r)\cup a(s)} : f \restriction a(r) \in r,\ f \restriction a(s) \in s\right\}.$$

   Observe that $a(r \bowtie s) = a(r) \cup a(s)$, and that the join operator glues r and s along their common attributes.

   (b) *Project:* If $Y \subseteq X$, then the *projection operator* is defined by

   $$\pi_Y(r) = \begin{cases} \{f \restriction Y : f \in r\}, & \text{if } Y \neq \emptyset, \\ \mathbf{1}_{a(r)}, & \text{otherwise.} \end{cases}$$

   So, π_Y picks all columns from r whose attribute name is in Y.

(c) *Rename:* Suppose that $i \in a(r)$, $j \in \alpha \setminus a(r)$, and $D_i = D_j$. We need an operation which changes the name of column i to that of column j: Set $Y = \big(a(r) \setminus \{i\}\big) \cup \{j\}$, and define

$$\delta_{i \leftarrow j}(r) = \Big\{ f \in \mathbf{1}_Y : f \restriction \big(a(r) \setminus \{i\}\big) = g \restriction \big(a(r) \setminus \{i\}\big), \\ f(j) = g(i) \text{ for some } g \in r \Big\}.$$

(d) *Select:* Select is a unary operator on relations, or, more precisely, a class of operators. For each attribute $n \in a(r)$ and each $d \in D_n$ let

$$\sigma_{n=d}(r) = \big\{ f \in r : f(n) = d \big\}.$$

In other words, $\sigma_{n=d}(r)$ selects from r all those tuples which have entry d in column m.

2. An Embedding

Suppose that $\mathcal{D} = \langle D_n, r_j \rangle_{n<\alpha,\, j \in J}$ is a database, and $\mathcal{R}$ the set of all its data tables. It is our aim to embed $\mathcal{D}$ into a Cs_α over an appropriate base set, and to translate the operations among data tables into Cs_α operations. To avoid the messiness of typed structures – and since we are not concerned with computational efficiency –, we shall assume throughout this section that the domains are all equal, say, $D_i = D$. This may entail problems with respect to the selection operators when comparison of domain elements is involved. However, these can, in theory, be resolved, by restricting the comparisons on D to appropriate elements, so that there is no restriction of generality.

One difficulty we encounter when considering a data table r is that $a(r)$ can be a proper subset of α. However, in the translation we have in mind, we want to consider all occurring relations as α-ary; in other words, data tables need to be interpreted as α-ary relations without losing the significant columns $a(R)$. The most natural way to do this is to follow the philosophy that "all information is no information", and introduce dummy columns into r, see e.g. [Nem,91]. For example, if $a(r) = \alpha$, $X = \alpha \setminus \{0\}$, and $r = D \times \pi_X(r)$ then $\pi_X(r)$ carries the same information as r.

This method, however, has the drawback that it works only if $D \times r \notin \mathcal{R}$ for $r \in \mathcal{R}$. For example, let $D = \{a, b\}$, $\alpha = \mathbf{3}$, $r_0 = D \times \{a\} \times \{b\}$, and $r_1 = \langle a, b \rangle$ with. $\alpha(r_1) = \{1, 2\}$. The dummy embedding of r_1 into ${}^{\mathbf{3}}D$ is now equal to r_0 which is not what we want. If $\alpha = \omega$ and $a(r) < \omega$ for all $r \in \mathcal{R}$, then this problem does not occur. However, this does not seem a satisfactory solution, since it unnecessarily restricts our choice of domains. It is proposed in [Imi-Lip,84a] to flag the irrelevant columns by an extra symbol; however, there it is not clear how the cylindric operations act on these extended relations. In the embedding below we shall also make use of a flag to identify irrelevant columns, though, in a different way.

Let $U = D \cup \{u\}$, where $u \notin D$, and let V be the set of all α-termed sequences over U. Define a mapping $e : \mathcal{R} \to 2^V$ by

$$e(r) = \{ f \in V : f \restriction a(r) \in r \}.$$

Theorem 3.2.1. *Suppose that* $\mathcal{D} = \langle D_n, r_j \rangle_{n<\alpha, j\in J}$ *is a database such that* $D_i = D$ *for all* $i < \alpha$, *and let* $\mathcal{R}$ *be a collection of data tables over* $\mathcal{D}$. *Furthermore,* U *and* e *are defined as above. Then,* e *is injective, and for all* $r, s \in \mathcal{R}$, $Y \subseteq \alpha$,

$$e(r \cup y) = e(r) \cup e(s) \qquad \text{if } a(r) = a(s). \tag{3.2.1}$$

$$e(r \setminus s) = e(r) \setminus e(s) \qquad \text{if } a(r) = a(s). \tag{3.2.2}$$

$$e\big(\pi_Y(r)\big) = \mathsf{c}_{(\alpha \setminus Y)} e(r). \tag{3.2.3}$$

Here, $\mathsf{c}_{(\alpha\setminus Y)}$ *is a generalized cylindrification, see* [Hen-Mon-Tar,85], *Ch.* 1.7.

$$e(r \bowtie s) = e(r) \cap e(s). \tag{3.2.4}$$

$$e\big(\delta_{i\leftarrow j}(r)\big) = \mathsf{s}^i_j\big(e(r)\big), \tag{3.2.5}$$

where s^i_j *denotes the "j for i substitution" of* [Hen-Mon-Tar,85], *Ch.* 1.5.

Proof. Let $r, s \in \mathcal{R}$ and $r \neq s$. If $a(r) \neq a(s)$, say, $n \in a(r) \setminus a(s)$, then there is some $f \in e(s)$ such that $f(n) = u$. On the other hand, $n \in a(r)$ implies that $u \neq f(n)$ for all $f \in e(r)$, and thus, $e(r) \neq e(s)$. If $a(r) = a(s)$ say, $f \in r \setminus s$. If $a(r) = a(s)$, then $r \neq s$ implies that w.l.o.g. there is some $f \in r \setminus s \neq \emptyset$. Thus, for all $g \in s$ there is some $n \in a(r) = a(s)$ with $f(n) \neq g(n)$. Hence, $e(r) \neq e(s)$.

The proofs of (3.2.1)–(3.2.5) are analogous to the corresponding properties of [Imi-Lip,84a] and can safely be omitted. ∎

Recall that for any $r \in V$, *the dimension set of* r is $\Delta(r) = \{n < \alpha : \mathsf{c}_n(r) \neq r\}$ ([Hen-Mon-Tar,85], Ch. 1.6). The definition of e shows that

$$n \in a(r) \Longleftrightarrow (\forall f \in V)\big[f \in e(r) \Rightarrow f(n) \neq u\big], \tag{3.2.6}$$

and

$$\begin{aligned} & n \in \Delta\big(e(r)\big) \\ & \Longleftrightarrow \mathsf{c}_n\big(e(r)\big) \neq e(r), \\ & \Longleftrightarrow (\exists f \in V)\big[f \notin e(r) \text{ and } \big(\exists g \in e(r)\big(f(m) = g(m),\ m \neq n\big)\big], \\ & \Longleftrightarrow (\exists f \in V)[f(n) = u \text{ and } \big(\exists g \in e(r)\big(f(m) = g(m),\ m \neq n\big)\big], \\ & \Longleftrightarrow n \in a(r). \end{aligned}$$

By flagging irrelevant columns by a new element, we have, in a way, introduced a new constant, the algebraic version of which are regular thin elements ([Hen-Mon-Tar,85], p. 60): For each $n < \alpha$ let $v_n^u = \{f \in V : f(n) = u\}$. Then, each $e(r)$ satisfies

$$\mathsf{c}_n\big(e(r)\big) \neq r \Rightarrow e(r) \cap v_n^u = \emptyset. \tag{3.2.7}$$

Observe that $\Leftarrow$ is always true for any nonzero $e(r)$. Generally, given a set U and some $u \in U$, a set r of α-termed sequences over U is called a *db-relation (with respect to u)*, if $\mathsf{c}_n(r) \neq r$ implies $r \cap v_n^u = \emptyset$ for all $n < \alpha$. We now have

Theorem 3.2.2.

(i) *Suppose that* $\mathcal{D} = \langle D_n, r_j \rangle_{n<\alpha, j \in J}$ is a database *such that* $D_i = D$ *for all* $i < n$, *and let* $\mathcal{R}$ *be a collection of data tables over* $\mathcal{D}$. *Furthermore,* U *and* e *are defined as above. Then, each* $e(r)$ *is a db-relation.*

(ii) *Conversely, if* $r \subseteq {}^{\alpha}U$ *is a db-relation with respect to some* $u \in U$, *then* $r' = r \restriction \Delta(r)$ *is a subset of* ${}^{\Delta(r)}\big(U \setminus \{u\}\big)$, *and* $e(r') = r$.

Theorem 3.2.3. *Suppose that* $\alpha > 2$ *and* $\mathcal{R}$ *is the collection of db-relations (of dimension* α*) with respect to some* $u \in U$. *Then,*

(i) *If* $r, s \in \mathcal{R}$, *then* $r \cap s \in \mathcal{R}$, *and, if* $\Delta(r) = \Delta(s)$, *then* $r \cup s \in \mathcal{R}$ *and* $r \setminus s \in \mathcal{R}$.

(ii) *If* $r \in \mathcal{R}$ *and* $i < \alpha$, *then* $\mathsf{c}_i(r) \in \mathcal{R}$.

(iii) *If* $r \in \mathcal{R}$, *then* $\mathsf{s}^i_j(r) \in \mathcal{R}$ *for all* $i, j < \alpha$.

Proof. 1. and 2. are straightforward to prove, so we only show 3: Let $k < \alpha$, and suppose that $\mathsf{s}^i_j(r) \cap v^u_k \neq \emptyset$; we need to show that $\mathsf{c}_k\big(\mathsf{s}^i_j(r)\big) = \mathsf{s}^i_j(r)$. If $k = i$, then we are done, since $\mathsf{c}_i\big(\mathsf{s}^i_j(r)\big) = \mathsf{s}^i_j(r)$. If $\mathsf{c}_k(r) \neq r$, then $r \cap v^u_k = \emptyset$, i.e. $r \subseteq -v^u_k$, since r is a db-relation. But then, $\mathsf{s}^i_j(r) = \mathsf{c}_i(r \cap \mathsf{d}_{ij}) \leq \mathsf{c}_i(r) \leq \mathsf{c}_i(-v^u_k) = -v^u_k$, which contradicts our hypothesis; thus, $\mathsf{c}_k(r) = r$. If $k \neq j$, then, by 1.5.8. of [Hen-Mon-Tar,85], we have $\mathsf{c}_k\big(\mathsf{s}^i_j(r)\big) = \mathsf{s}^i_j\big(\mathsf{c}_k(r)\big) = \mathsf{s}^i_j(r)$. If $k = j$, then $\mathsf{c}_j(r) = r$; furthermore, $r \cap \mathsf{d}_{i,j} \cap v^u_k \neq \emptyset$, since $\mathsf{c}_i(-v^u_j) = -v^u_j$. Hence, there is some $f \in r$ such that $f(j) = u = f(i)$, and therefore, $\mathsf{c}_i(r) = r$. Again from 1.5.8. of [Hen-Mon-Tar,85] it follows that $\mathsf{s}^i_j(r) = \mathsf{s}^i_j\big(\mathsf{c}_i(r)\big) = \mathsf{c}_i(r) = r$, and $\mathsf{c}_j\big(\mathsf{s}^i_j(r)\big) = \mathsf{c}_j(r) = r$. ■

The db-relations have some more simple but remarkable properties:

Theorem 3.2.4. *Let* r, s *be non zero db-relations. Then,*

(i) *If* $r \subseteq s$, *then* $\Delta(s) \subseteq \Delta(r)$.

(ii) *If* $X, Y \subseteq \Delta(r)$ *and* $\mathsf{c}_{(X)}(r) \subseteq \mathsf{c}_{(Y)}(r)$, *then* $X \subseteq Y$.

(iii) *If* $X \subseteq \Delta(r)$, *then* $\Delta\mathsf{c}_{(X)}(r) = \Delta r \backslash X$.

A special case of Theorem 3.2.4.2 is that $\mathsf{c}_i(r) = \mathsf{c}_j(r)$ implies that $i = j$, whenever $r > 0$ and $i, j \in \Delta(r)$. This does not usually happen in arbitrary Cs; however, it seems necessary to model data base relations more realistically: Suppose that r corresponds to a concrete relation of a database with scheme X, and that $i, j \in X$. Now, c_i corresponds to the projection of r onto the scheme $X \backslash \{i\}$, similarly for j. If i and j name different columns of r, one would certainly want different projections of r when one blanks out the columns i and j respectively. In other words, removing one column should not remove another column as well.

3. Data Dependencies

A central concern of database design is to avoid redundancy and ensure the integrity of the database, and various *dependencies* have been considered. These may be viewed as specifications of the semantics of a database, and describe constraints among and within the data tables. For example, the earliest dependencies were the *functional dependencies* [Cod,79]: If $A_0, \ldots, A_k, B$ are attribute names and r is a relation of the database, the functional dependency $A_0, \ldots, A_k \to_r B$ means that whenever two tuples from a relation r agree on the attributes named by $A_0, \ldots, A_k$, then they agree also on attribute B. Since many such constraints are meaningful in various situations, many types of dependencies have been investigated; Thalheim [Tha,96] mentions in 1996 that over 100 have been considered in the literature.

Most of the research on dependency theory has centred around two main questions. The first of these is the *implication problem* for a given class of dependencies: For a given finite set Σ of dependencies we say that Σ implies a dependency d, and write $\Sigma \models d$, if every relation which satisfies all constraints in Σ also satisfies d. The implication problem (a semantical concept) is to find an algorithm which decides $\Sigma \models d$. If there is such an algorithm, we say that the *implication problem is solvable* for the class of dependencies considered.

The second, closely related, problem is to find a complete axiom system for a class of dependencies, or else show that such a system does not exist. There is a large amount of literature on the subject – for an overview see e.g. [Tha,96] – which to explore in detail space does not permit.

In formulating constraints, the project and join operator play a significant role; indeed, the very general algebraic dependencies are stated in terms of these operators (see Table 1, where we assume that all expressions are well defined).

1. $\pi_X\big(\pi_X(r)\big) = \pi_X(r),\ \pi_{a(r)}(r) = r,$	2. $r \bowtie \pi_X(r) = r,\ \pi_{a(r)}(r \bowtie s) \subseteq r,$
3. $r \subseteq s \Rightarrow \pi_X(r) \subseteq \pi_X(s),$	4. $r \subseteq s \Rightarrow r \bowtie t \subseteq s \bowtie t,$
5. $r \bowtie s = s \bowtie r,$	6. $(r \bowtie s) \bowtie t = r \bowtie (s \bowtie t),$
7. $X \subseteq a(r)$ and $Y \subseteq a(s) \Rightarrow \pi_{X\cap Y}(r \bowtie s) \subseteq \pi_{X\cap Y}\big(r \bowtie \pi_Y(s)\big).$	

Table 1: Axioms for algebraic dependencies [Yan-Pap,82]

By way of example, we shall concentrate in the sequel on general project-join dependencies and cylindric dependencies. We suppose that all data tables of a database have the same schema; this is no significant restriction by Theorem 3.2.1.

3.1. Project-join dependencies

Formally, a language for project-join expressions over a set of generators contains symbols for both operators and a set of predicates $\{P_j : j \in J\}$ all having the same finite arity, say, α. The set of *project-join expressions over the generators* $\{P_j : j \in J\}$ (pjes) is defined as follows:

(i) Each P_j is a pje.

(ii) If τ is a pje and $Y \subseteq \alpha$, then $\pi_Y(\tau)$ is a pje.

(iii) If τ and σ are pjes, then $\tau \bowtie \sigma$ is a pje.

(iv) No other expression is a pje.

Evaluation of pjes in databases is done in the obvious way: Suppose that $\mathcal{D} = \langle D_n, r_j \rangle_{n<\alpha,\, j\in J}$ is a database. If τ is a pje we define its value $e'(\tau)$ in $\mathcal{D}$ inductively:

- $e'(P_j) = r_j$
- $e'(\pi_Y \sigma) = \pi_Y\big(e'(\sigma)\big)$
- $e'(\sigma \bowtie \rho) = e'(\sigma) \bowtie e'(\rho)$.

We can relate equations among pjes to equations in a reduct of Cs_α: A *(diagonal free) cylindric (lower) semilattice of dimension* α, in short, a csl_α, is an algebra $< \mathfrak{A} = \langle A, \cdot, \mathsf{c}_i, \rangle_{i<\alpha}$ such that $\langle A, \cdot \rangle$ is a semilattice, and for all $x, y \in A$ and $i, j < \alpha$,

(C1) $$x \cdot \mathsf{c}_i x = x$$

(C2) $$\mathsf{c}_i \mathsf{c}_j x = \mathsf{c}_j \mathsf{c}_i x$$

(C3) $$\mathsf{c}_i(x \cdot \mathsf{c}_i y) = \mathsf{c}_i x \cdot \mathsf{c}_i y.$$

If $\mathfrak{A}$ is a subalgebra of a Cs_α, then $\mathfrak{A}$ is called a *set cylindric (lower) semilattice of dimension* α, in short, a scsl_α. With some abuse of language we denote the class of all set cylindric lower semilattices of dimension α also by scsl_α.

Theorem 3.3.1. *The equational theory of pjes over α-dimensional databases is equivalent to the equational theory of the class* scsl_α ([Dun-Mik,01]).

Proof. Consider the following translation tr of pjes onto cylindric expressions:

1. $\mathrm{tr}\,(P_j) = P_j$;
2. $\mathrm{tr}\,(\pi_Y \sigma) = \mathsf{c}_{(\alpha \smallsetminus Y)}\,\mathrm{tr}\,(\sigma)$;
3. $\mathrm{tr}\,(\sigma \bowtie \rho) = \mathrm{tr}\,(\sigma) \cdot \mathrm{tr}\,(\rho)$.

It is shown in [Dun-Mik,01] that for all pjes τ and σ, $\tau = \sigma$ holds in every α-dimensional database if and only if $\mathrm{tr}\,(\tau) = \mathrm{tr}\,(\sigma)$ is valid in every scsl_α.
■

A *project-join dependency* (pjd) is a formula $\tau = \sigma$ with τ and σ pjes. A database $\mathcal{D}$ *satisfies* the pjd $\tau = \sigma$, written as $\mathcal{D} \models \tau = \sigma$, if and only if $e'(\tau) = e'(\sigma)$.

A set Σ of pjds *implies* a pjd δ, written as $\Sigma \models \delta$ if and only if for every database $\mathcal{D}$, $\mathcal{D} \models \Sigma$ implies $\mathcal{D} \models \delta$. Note that we can express that a pje σ follows from a pje τ $(\tau \leq \sigma)$, since $e'(\tau) \subseteq e(\sigma)$ if and only if $e'(\tau \bowtie \sigma) = e'(\sigma)$.

Suppose that $\Sigma \cup \{\tau\}$ is a finite set of pjds. Since valid pjds are equivalent to valid equations of scsl_n by Theorem 3.3.1, it follows that the implication $\Sigma \models \tau$ is equivalent to a quasi-equation of cylindric expressions. The following result implies that there is no finite set of formulae axiomatizing α-dimensional pjds if $\alpha > 2$:

Theorem 3.3.2. *If $\alpha > 2$, then the quasi-equational theory of* scsl_α *is not finitely axiomatizable* ([Hod-Mik,00]).

3.2. Cylindric dependencies

Cylindric dependencies were introduced in [Cos,87], and they are an extension of project-join dependencies to the positive reduct of first order logic with equality. The corresponding query language $\mathcal{L}_\omega^+$ has the connectives $\wedge$, $\vee$, $\exists$, **T**, **F**, $=$, an infinite set $V = \{v_i : i < \omega\}$ of variables, and a set of predicates $\mathcal{P} = \{P_i : i \in I\}$; to simplify matters we may suppose that all P_i have the same finite arity, say, k, using dummy embeddings. Variables are not typed, i.e. every variable is allowed to appear in any column of a predicate P_i.

In this setting, a database is a pair $\langle D, \mathcal{R}\rangle$, where D is a nonempty set, and $\mathcal{R} = \{R_i : i \in I\}$ is a set of relations over D with finite arity $\gamma(i)$.

Furthermore, the objects of the theory are valuations, i.e. mappings from the set V of variables of the language to the domain D. A relation r of arity k of a database then corresponds to the set $\rho_r = \{f \in {}^V D : f[v_0, \ldots, v_{k-1}] \in r\}$.

A *cylindric dependency* (CD) now is just a positive equation of Cs_ω, i.e. it does not contain complementation. The following axiom system is given in [Cos,87]:

C1. A set of equations which say that $\langle CE, \cdot, +, 0, 1\rangle$ is a distributive lattice with smallest element 0 and largest element 1.

C2. $\mathsf{c}_i 0 = 0$,

C3. $\tau \cdot \mathsf{c}_i \tau = \tau$,

C4. $\mathsf{c}_i(\tau \cdot \mathsf{c}_i \sigma) = \mathsf{c}_i \tau \cdot \mathsf{c}_i \sigma$,

C5. $\mathsf{c}_i(\tau + \sigma) = \mathsf{c}_i \tau + \mathsf{c}_i \sigma$,

C6. $\mathsf{c}_i \mathsf{c}_j \tau = \mathsf{c}_j \mathsf{c}_i \tau$,

C7. $\mathsf{d}_{ii} = 1$,

C8. If $j \neq i, k$, then $\mathsf{d}_{ik} = \mathsf{c}_j(\mathsf{d}_{ij} \cdot \mathsf{d}_{jk})$,

C9. $\mathsf{c}_j P_m = P_m$ for all $j \geq \gamma(m)$,

C10. $\mathsf{d}_{ij} \cdot \mathsf{c}_i(\mathsf{d}_{ij} \cdot \tau) \leq \tau$.

for all $i, j, k \in \omega$, $m \in I$, and $\tau, \sigma \in CE$[1]. Implication of a CD φ from a set of CDs is defined in the usual way. An abstract algebra $\mathfrak{C}$ which satisfies these axioms is called a *cylindric lattice with generators* P_k. If each element of $\mathfrak{C}$ is finite dimensional, we call $\mathfrak{C}$ *locally finite.* If the set of generators is understood or unimportant, we just speak of $\mathfrak{C}$ as a cylindric lattice. A cylindric lattice $\mathfrak{C}$ is called *representable* if it can be embedded into the product of algebras of the form

$$\left\langle 2^{^\omega D}, \cap, \cup, \emptyset, {}^\omega D, \mathsf{C}_i, \mathsf{D}_{ij} \right\rangle_{i,j\in\omega}.$$

Contrary to a claim in [Cos,87] the system C1.–C10. is not complete for the intended models. Indeed, the situation is more complicated:

Theorem 3.3.3. *Let Σ be a set of universal formulae axiomatizing (the quasi-equational theory of) representable cylindric lattices of dimension ω. Then Σ contains infinitely many variables* ([Dun-Mik,01]).

This result has the following consequence for cylindric dependencies:

Corollary 3.3.4. *There is no finite set of (universal) axiom schemas such that $\Sigma \models \tau$ if and only if $\Sigma \vdash \tau$ for every set $\Sigma \cup \{\tau\}$ of cylindric dependencies.*

There is no way to equationally repair the claim, since it is shown in [And-Nem,91] that, unlike representable cylindric algebras, the representable cylindric lattices do not form an equational class. Furthermore, it does not help to restrict ourselves to finite dimensions because of the following result:

Theorem 3.3.5. *The quasi-equational and the equational theories of representable cylindric lattices of dimension n is not finitely axiomatizable if $n > 2$* ([Hod-Mik,00]).

Note that in this case we have a stronger negative result than in the case of pjds, since we cannot finitely axiomatize even the valid equations between cylindric expressions, let alone valid inferences between equations.

[1] Axiom C10. was actually omitted from [Cos,87].

BOLYAI SOCIETY
MATHEMATICAL STUDIES, 22

Cylindric-like Algebras and Algebraic Logic
pp. 333–347.

Probability Measures and Measurable Functions on Cylindric Algebras

MIKLÓS FERENCZI*

The concept of probability (measure) defined on a Boolean algebra and that of a measurable function with respect to Boolean set algebras are basic concepts of probability theory (measure theory). In probability theory, the concept of algebras of *events* (as Boolean algebra) plays the role of propositional logic.

Here we generalize the concepts of finitely additive probability and measurable function from Boolean algebras to cylindric algebras. The relation between locally finite cylindric algebras and first order logic is analogous to that of Boolean algebras and propositional logic.

We use ordinary Boolean algebras instead of Boolean σ-algebras. The probabilities to be introduced here are supposed to be partially continuous (partially σ-additive). These probabilities are between the finitely additive and σ-additive probabilities,they have some remarkable, unusual properties.

The topic investigated here can be treated from many points of view. It can be approached from the viewpoint of algebra, probability theory, measure theory, mathematical logic, probability logic, etc. We are going to emphasize the algebraic and measure theoretical aspects rather than the mathematical logical (probability logical) ones. The main purpose of the researches sketched here is to develop the foundations for the probability and measure theory, with character of first order logic instead of propositional logic.

One of the basic concepts in probability and measure theory is the concept of a *measurable function*. As is known, if $\mathfrak{A}$ and $\mathfrak{B}$ are Boolean set algebras with units V and X respectively, then a function $f : X \to V$ is measurable if the inverse $\left(f^{-1}\right)^*$ maps A into B. This function will be

*Dedicated to the memory of Paul Halmos.

denoted by $\widehat{f}$ so $\widehat{f}a = \{x : fx \in a, x \in X\}$, $a \in A$. So the function $f : X \to V$ is measurable if $\widehat{f}$ maps A to B, i.e. $\widehat{f}a \in B$ for every $a \in A$. As is known if this property is satisfied, then $\widehat{f}$ is a homomorphism between $\mathfrak{A}$ and $\mathfrak{B}$. So we can say that the measurable functions are those point functions $f : X \to V$ which are suitable to induce homomorphisms from $\mathfrak{A}$ to $\mathfrak{B}$, where the homomorhism is $\widehat{f}$. From now on, *by measurable function* we mean a point function f which is, in a sense, suitable to induce a homomorphism between the respective algebras.

How to generalize the concept of measurable function to cylindric set algebras? The situation is more complicated of course than the Boolean one, for example, the condition $\widehat{f}a \in B$ is not sufficient for $\widehat{f}$ to be a cylindric homomorphism.

The definition of homomorphism induced by point function will depend on the kind of the cylindric set algebras (Cs_α, Gs_α, Gws_α, etc.) for which the point function is defined. For example, if the cylindric set algebras are *generalized cylindric set algebras* (algebras in Gs_α), then the generator function is defined on the bases of the algebras and if it induces a cylindric homomorphism we talk about *base-homomorphism* (see [Hen-Mon-Tar,85] Def. 3.1.50). [Hen-Mon-Tar,85] Theorem 3.1.52 characterizes those functions f which induce base-homomorphisms. In [Hen-Mon-Tar,85] Theorem 3.1.54 (Ferenczi) the base-homomorphisms between power set algebras in Gs_α are characterized.

Some words on the *history* of the topic. The pioneers of the modern treatment of probabilities (measures) defined on first-order formulas are Gaifman and Łoš (see [Gai,64], [Los,73]). They focused on the model theoretical aspects. Scott and Krauss generalized essentially these results, they set out from σ-additive probabilities presenting a detailed description on the connection with probability logic and probability theory (see [Sco-Kra,66]). The pioneers of application of algebraic logic to this area are Fenstad, Georgescu, Ferenczi. Fenstad and Georgescu used polyadic algebras and they applied a weakened concept of continuity (see [Fen,67], [Geo,78]). Fenstad emphasized the probability theoretical aspects (see [Fen,67]), further the definition of the probability was extended to formulas with free variables. The concept of measure defined on cylindric algebras was introduced in [Fer,83], and cylindric homomorphisms induced by point functions were investigated in [Fer,86a]. Rašković and Đorđević applied cylindric algebras to algebraize a kind of probability logic. They generalized and developed Keisler's graded probability logic ([Kei,85]). They algebraized this logic and they introduced the concept of cylindric probability algebra, probability cylindrification and proved representation theorems (see [Ras-Dor-Bra,97], [Ras-Dor,00] and [Dor-Ras,thisVol]). Nowadays, non-standard analysis has an essential influence to the topic (see [Fer,10a]).

Most of the concepts and results included in this paper are due to the present author, the expections will be indicated.

In Section 1, we investigate probabilities on *abstract cylindric algebras* in Lf_α (or in Dc_α). A strong and a weak kind of the partial continuity of the foregoing probabilities are defined: the concept of cylindric probability and that of the weak cylindric probability. Theorem 4.1.6 includes a sufficient condition for a weak cylindric probability to be a cylindric probability. The main theorem of the section, Theorem 4.1.7, is connected with the construction of cylindric probabilities, with extensions of probabilities. At the end of the section, there are some results on the analogies between cylindric algebraic and measure properties. In Section 2 probabilities defined on cylindric *set algebras* are investigated. In Theorem 4.2.1 a construction is given for cylindric probability defined on cylindric set algebras associated with countable first-order models. It is shown that using these kinds of probabilities, cylindric probabilities can be composed on Lindenbaum algebras by transformation or by integral. In Section 3, inducing homomorphisms by point functions, i.e. measurable functions are investigated. In addition to base-homomorphisms, new kinds of homomorphisms induced by point functions are introduced: the homomorphism induced in a unit set sense and the homomorphism induced in a weak set sense. These kinds of homomorphisms are characterized in Theorems 4.3.3 and 4.3.4.

1. Probability Measures on Cylindric Algebras

Throughout the paper, we assume that the cylindric algebras are *locally finite* ones, i.e. they are in Lf_α (however, the results are valid for the larger class Dc_α). The concepts and results in this section are included in [Fer,83], [Fer,05] and [Fer,10a].

First, we list some basic definitions which are connected with probabilities defined on abstract cylindric algebras.

Definition 4.1.1. The real function p is a *probability* (or *probability measure*) on the cylindric algebra $\mathfrak{A}$ if the following properties hold for every $x, y \in A$

(i) $0 \leq px \leq 1$;

(ii) $p0 = 0$;

(iii) $p(x + y) = px + py$ if $x \cdot y = 0$.

Recall that the following property is true for cylindric algebras in Lf_α:

$$\mathsf{c}_\kappa x = \sum_{\lambda \in \Gamma} \mathsf{s}^\kappa_\lambda x$$

for every $\kappa < \alpha$ and countable set Γ of ordinals. These suprema are called cylindric suprema.

Definition 4.1.2. A probability p defined on a cylindric algebra $\mathfrak{A}$ is a *cylindric probability* (or Q-probability) if the following additional supremum property (iv) holds for every $x \in A$, $\kappa < \alpha$

$$\text{(4.1.1)} \qquad \text{(iv)} \qquad p(\mathsf{c}_\kappa x) = \sup_{n \in w} p\Bigg(\sum_{k=1}^{n} \mathsf{s}^\kappa_{\lambda_k} x\Bigg)$$

for every ω-sequence of ordinals $\lambda_1, \lambda_2, \ldots \lambda_k, \ldots$ with infinite range ($\lambda_k < \alpha$).

Remark 4.1.3. Property (4.1.1) reflects the above Boolean property. (4.1.1) means a kind of *partial continuity* of the probability.

In the case of Lindenbaum–Tarski algebras (for formula algebras), property (4.1.1) takes the following form:

$$\text{(4.1.2)} \qquad p[\exists\, x\alpha] = \sup_{n \in \omega} p \bigvee_{i=1}^{n} [\alpha(x/y_i)]$$

for *any* infinite sequence $y_1, y_2, \ldots y_i, \ldots$ of individual variables, with infinite range, where $[\beta]$ denotes the equivalence class containing the formula β.

Property (4.1.2) means a kind of *extensionality* of probability: probabilities of the quantifier free formulas *determine* the probability of any formula (while probability logic, in general, is not extensional). This property is useful because the probabilities of quantifier free formulas *can be measured by statistical methods.*

Definition 4.1.4. A probability p defined on a cylindric algebra $\mathfrak{A}$ is *symmetrical* if $px = p(\mathsf{s}_\tau x)$ for every finite (simultaneous) substitution operator s_τ. A probability p defined on a cylindric algebra $\mathfrak{A}$ has the *product property* (or it is a product probability) if $p(x \cdot y) = px \cdot py$ for every $x, y \in A$ such that $\Delta x \cap \Delta y = \emptyset$, where Δx, Δy are the dimension sets of x and y.

Definition 4.1.5. A probability p defined on a cylindric algebra $\mathfrak{A}$ is a *weak cylindric probability* (or q-probability) if property (4.1.1) holds for *some* ω-sequence of distinct ordinals $\lambda_1, \lambda_2, \ldots \lambda_k, \ldots$ $(\lambda_k < \alpha)$ for every $x \in A$, $\kappa < \alpha$.

This latter version of probability is due to Gaifman (see [Gai,64]). A weak cylindric probability is not necessary is a cylindric probability. By the following theorem, symmetry of the probability is a sufficient condition for a weak cylindric probability to be a cylindric probability.

Theorem 4.1.6. *If p is a symmetrical q-probability, then p is a cylindrical probability (Q-probability)* [Fer,83].

First, we consider probabilities on abstract cylindric algebras. We investigate the problem of constructing probabilities.

The next theorem is about extending probabilities from a generator system G to the whole algebra $\mathfrak{A}$ (see [Fer,05]). Assume that G is a *Boolean* generator system, i.e. it is a generator system and it is a Boolean algebra. The real function p is said to be a probability on G if its restriction to G, satisfies the axioms (i), (ii) and (iii) restricting it to G. The probability p is said to be σ-additive if it is σ-additive, restricting it to G.

Theorem 4.1.7. *Assume that G is a Boolean generator system of the cylindric algebra $\mathfrak{A} \in \mathsf{Lf}_\alpha$ and p is a σ-additive probability on G. Then p can be extended to $\mathfrak{A}$ as cylindric probability and the extension is unique (see* [Fer,83]*).*

Proof. We are going to apply the following theorem of Vladimirov ([Vla,02], Ch. 7.): Let $\mathcal{H}$ be a Boolean σ-algebra, let $\mathcal{H}_0$ be an arbitrary *Boolean* subalgebra of $\mathcal{H}$ and let p be a σ-additive probability on $\mathcal{H}_0$. In this case, there is a Boolean σ-algebra $\widetilde{\mathcal{H}}$ which includes $\mathcal{H}_0$ so that, as σ-algebra, $\widetilde{\mathcal{H}}$ is a subalgebra of $\mathcal{H}$ and there is a σ-additive probability $\overline{p}$ on $\widetilde{\mathcal{H}}$ which is an extension of p.

To apply this theorem, first, let us extend $\mathfrak{A}$ to a Boolean σ-algebra, this can be done obviously, let us denote the σ-algebra obtained by $\mathcal{H}$. Let us choose G as the Boolean algebra $\mathcal{H}_0$ occurring in the cited theorem. By the assumption of the theorem, p is a σ-additive probability on G. Thus the conditions of Vladimirov's theorem are satisfied.

So there is a σ-subalgebra $\widetilde{\mathcal{H}}$ of $\mathcal{H}$ which includes G and there is a σ-additive probability $\overline{p}$ on $\widetilde{\mathcal{H}}$ which is an extension of p. But $\widetilde{\mathcal{H}}$ includes the suprema in (4.1.1), since $\widetilde{\mathcal{H}}$ is a σ-subalgebra of the σ-algebra $\mathcal{H}$. $\overline{p}$ is σ-additive in $\mathcal{H}$, i.e. it is also continuous, thus the supremum property in (4.1.1) is valid.

The uniqueness of the extension of p can be checked by induction. ■

Remark 4.1.8. A special case of the above extension is when $\mathfrak{A}$ is the basic formula algebra $\mathfrak{Fm}^{\Lambda}$ corresponding to a countable first-order language Λ (theory is not specified) and G is the Boolean subalgebra corresponding to the quantifier free formulas. In this case, every probability p defined on G is σ-additive. This latter is a consequence of the compactness theorem of first-order logic (see [Gai,64]). Consequently, every probability p defined on G can be extended to $\mathfrak{A}$ as cylindric probability.

There is a version of the theorem for non-standard stochastics, too, see [Fer,10a].

An important method to obtain probabilities on cylindric algebras is the transformation from other algebras. The next theorem concerns transformations of cylindric probabilities.

Assume that $\mathfrak{A}, \mathfrak{B} \in \mathsf{Lf}_\alpha$.

Theorem 4.1.9. *Assume that $h \in Hom\,(\mathfrak{A}, \mathfrak{B})$. The following propositions (i) and (ii) are true:*

(i) *if p is a cylindric probability on $\mathfrak{B}$, then the following function p' is a cylindric probability on $\mathfrak{A}$*

$$p'x := p(hx) \quad (x \in A).$$

(ii) *if p is a cylindric probability on $\mathfrak{A}$ and p vanishes on the kernel of h, then the following function p' is a cylindric probability on $\mathfrak{B}$*

$$p'y := p(x_y),$$

where x_y is an element of A such that $h(x_y) = x$ $(x \in A,\ y \in B)$.

We do not detail here how can we obtain product probabilities on free (or on amalgamated) products of cylindric algebras (see [Fer,91]).

Now we come to some *analogies* between measure and algebraic properties. There are well-known *algebra-measures analogies* in the theory of Boolean algebras. Such a connection is that there is a one-to-one connection between the Boolean ultrafilters and the finitely additive 0–1 measures. First we investigate the generalization of this connection to cylindric algebras.

Definition 4.1.10. A Boolean ultrafilter F of a cylindric algebra is said to be *cylindric* one if for every x and $\kappa < \alpha$, $\mathsf{c}_\kappa x \in F$ *implies* $\mathsf{s}^\kappa_\lambda x \in F$ for some $\lambda < \alpha$.

Assume that $\mathfrak{A} \in \mathsf{Lf}_\alpha$.

Theorem 4.1.11. *There is a one-to-one connection between the cylindric ultrafilters of* $\mathfrak{A}$ *and the* 0–1 *q-probabilities defined on* $\mathfrak{A}$.

Proof. Assume that F is a cylindric ultrafilter. We can in an obvious way associate with F a finitely additive 0–1 probability p. We show that this probability has the q-property. F is cylindric one so if $\mathsf{c}_\kappa x \in F$, then $\mathsf{s}^\kappa_\lambda x \in F$ for some $\lambda < \alpha$. $\mathsf{c}_\kappa x \in F$ implies that $p(\mathsf{c}_\kappa x) = 1$. Let us choose the sequence $\lambda_1, \lambda_2, \dots \lambda_k, \dots$ ($\lambda_k < \alpha$) in the definition of q-probability so that it should include the ordinal λ. $\mathsf{s}^\kappa_\lambda x \in F$ implies that $p(\mathsf{s}^\kappa_\lambda x) = 1$. Obviously, if the sequence $\lambda_1, \lambda_2, \dots \lambda_n$ contains λ, then $p\big(\sum_{k=1}^n \mathsf{s}^\kappa_{\lambda_k} x\big) \geq p(\mathsf{s}^\kappa_\lambda x)$. If $\mathsf{c}_\kappa x \notin F$, i.e. $p(\mathsf{c}_\kappa x) = 0$, then $\mathsf{s}^\kappa_\lambda x \notin F$ because $\mathsf{s}^\kappa_\lambda x \leq \mathsf{c}_\kappa x$, thus $p(\mathsf{s}^\kappa_\lambda x) = 0$ for every λ. p is indeed a q-probability.

Assume that p is a q-probability on $\mathfrak{A}$. We know that the elements with probability 1 form a Boolean ultrafilter. We have to prove that it is a cylindric ultrafilter. Assume that $\mathsf{c}_\kappa x \in F$ for some x and κ. Let us apply the q-probability property. By (4.1.1), $p\big(\sum_{k=1}^n \mathsf{s}^\kappa_{\lambda_k} x\big) = 1$ for some $\lambda_1, \lambda_2, \dots \lambda_n$ and n. The 0–1 property implies that $p(\mathsf{s}^\kappa_{\lambda_k} x) = 1$ for some $k \leq n$. By definition of F, it follows that $\mathsf{s}^\kappa_{\lambda_k} x \in F$. ■

Let us consider the basic formula algebra $\mathfrak{Fm}^\Lambda$ occurring in Remark 3.1.8 and let $\mathfrak{B}$ be any cylindric algebra. The property described by the proposition below is analogous to that in Theorem 4.1.7 in the case of the algebra $\mathfrak{Fm}^\Lambda$. It follows from the fact that $\mathfrak{Fm}^\Lambda$ is a dimension restricted free cylindric algebra.

Let G be the Boolean subalgebra of $\mathfrak{Fm}^\Lambda$ corresponding to the quantifier-free formulas and let $g : G \to B$ be a Boolean homomorphism. In this case, g can be extended to $\mathfrak{Fm}^\Lambda$ as a cylindric homomorphism, and the extension is unique.

2. Probability Measures on Cylindric Set Algebras

In cylindric set algebras the operation c_κ is a *projection* parallel to the κ'th axis. Property (4.1.1) means that the measures of certain transformations of a set x *determine the measure of the projection of the set.* This is an interesting and unusual property in the measure and probability theory. Furthermore, property (4.1.1) is unusual, since the suprema are *different* from the ordinary infinite σ-unions for sets, hence the features of the usual σ-additivity of probabilities are not applicable, special technics are needed.

The results in this section are included in [Fer,05]. There are special methods for constructing cylindric probabilities on cylindric *set* algebras. We consider set algebras in $\mathsf{Cs}_\alpha^{\mathrm{reg}} \cap \mathsf{Lf}_\alpha$. It is known that such an algebra can be considered as a cylindric set algebra associated with a first-order model (see Serény's paper, [Ser,thisVol]). For technical reasons we consider set algebras of these kinds only.

Assume that a first-order language Λ is given and let $\mathfrak{M}$ be a model of type of Λ. Let $\mathfrak{A}^{\mathfrak{M}}$ denote the cylindric set algebra corresponding to the model $\mathfrak{M}$. Let $|\beta|$ denote the element (the set) in $\mathfrak{A}^{\mathfrak{M}}$ corresponding to the formula β and $\mathfrak{M}$.

In the case of a cylindric set algebra $\mathfrak{A}^{\mathfrak{M}}$, the property (4.1.1) takes the following form:

$$(4.2.1) \qquad p|\exists x\alpha| = \sup_{n\in\omega} p \bigcup_{i=1}^{n} |\alpha(y_i)|$$

for *any* x and infinite sequence $y_1, y_2, \ldots y_i, \ldots$ of individual variables, with infinite range, where $\cup$ denotes the usual union of sets and $\alpha(y_i)$ means $\alpha(x/y_i)$.

Let us assume that the universe of the model $\mathfrak{M}$ is countable, i.e. the base U of the cylindric set algebra $\mathfrak{A}^{\mathfrak{M}}$ is countable. Let us denote $\mathfrak{A}^{\mathfrak{M}}$ by $\mathfrak{A}$ for short. Let $\mathcal{H}$ denote the Boolean part of $\mathfrak{A}$.

Let us consider a *strictly positive* σ-additive probability distribution on U (a distribution is strictly positive if there is no element with zero probability). It can be considered as a distribution on the power set $\mathfrak{Sb}(U)$ considering this latter as a Boolean σ set algebra. Let p be the Tth power of this distribution on the Tth power ${}^T\mathfrak{Sb}(U)$ of $\mathfrak{Sb}(U)$, where T is the set of the individual variables in Λ.

Theorem 4.2.1. *The restriction of p to $\mathfrak{A}$ is a cylindric probability (Q-probability), it is symmetrical and has the product property* (see [Fer,05]).

Proof. The elements of A are finite dimensional cylinder sets of ${}^TU$, because these elements form truth sets (interpretations) of formulas with finitely many free variables, so they are subsets of some finite dimensional space. U is countable, consequently any finite dimensional cylinder set P belongs to ${}^T\mathfrak{Sb}(U)$, because P is a countable union of points in a given finite dimensional space. Thus $\mathcal{H} \subset {}^T\mathfrak{Sb}(U)$. The restriction of p can be considered as a probability on $\mathcal{H}$, thus also on $\mathfrak{A}$. The symmetry and product properties follow because p is a power of a distribution.

We have to prove the supremum property (4.2.1). We prove the following infimum variant of (4.1.1):

$$p|\forall x\alpha| = \inf_{n\in\omega} p\bigcap_{i=1}^{n}\big|\alpha(y_i)\big|. \tag{4.2.2}$$

We should use the elements of U, thus let us extend the language Λ by a set of constants corresponding to the elements of U and let us interpret the constants by the original elements. Denote Λ' the expanded language, from now, we work in Λ'. Let $\mathfrak{A}^*$ be the cylindric set algebra corresponding to the extended language Λ' and model $\mathfrak{M}'$. Obviously, $\mathcal{H} \subset \mathcal{H}^* \subset {}^T\mathfrak{Sb}(U)$, where $\mathcal{H}^*$ is the Boolean part of $\mathfrak{A}^*$. p can be considered also on $\mathfrak{A}^*$ and it has the symmetry and product properties, too. It is enough to prove the infimum property (4.2.2) for $\mathfrak{A}^*$.

So consider an element of A of the form $|\forall x\alpha|$ and let $y_1, y_2, \dots y_i, \dots$ be an arbitrary sequence occuring in (4.2.2). Let $z_1, z_2, \dots z_k$ be the free variables of α in addition to x.

Let $u_1, u_2, \dots u_k$ be arbitrary but fixed constants corresponding to the elements in U which are among those the language has been extended. Our first claim is as follows:

$$p\big|\forall x\alpha(x, u_1, u_2, \dots u_k)\big| = \inf_{n\in\omega} p\Big(\bigcap_{i\in n}\big|\alpha(y_i, u_1, u_2, \dots u_k)\big|\Big). \tag{4.2.3}$$

$\forall x\alpha(x, u_1, u_2, \ldots u_k)$ is a closed formula, thus it is true or false in $\mathfrak{M}$, and by the definition of probability, the left-hand side is equal to either 0 or 1. In the second case, there is no u_0 such that $\alpha(u_0, u_1, u_2, \ldots u_k)$ is false, since $\forall x\alpha(x, u_1, u_2, \ldots u_k)$ is true on $\mathfrak{M}'$. Thus the set $\left|\alpha(y_i, u_1, u_2, \ldots u_k)\right|$ is the whole space ${}^TU$ for every i, hence the right-hand side of (4.2.2) is 1. If the left-hand side of (4.2.2) equals 0, then the formula $\forall x\alpha(x, u_1, u_2, \ldots u_k)$ is false in the model, hence there exists a u_0 such that $\alpha(u_0, u_1, u_2, \ldots u_k)$ is false. Thus the set $\left|\alpha(y_i, u_1, u_2, \ldots u_k)\right|$ is different from the whole space by the strict positivity of the original distribution on U. Therefore the probability m of $\left|\alpha(y_i, u_1, u_2, \ldots u_k)\right|$ is strictly less than 1. It follows from symmetry that p does not depend on i because of symmetry. The product property of p implies

$$p\Bigl(\bigcap_{i\in n}\left|\alpha(y_i, u_1, u_2, \ldots u_k)\right|\Bigr) = m^n$$

where $m = p\left|\alpha(y_i, u_1, u_2, \ldots u_k)\right|$, thus the infimum in (4.2.2) is zero, that is (4.2.2) is true.

Now consider the formulas $\vartheta(x, z)$ with two free variables. Applying the summation property concerning conditional distributions in the space $\bigl\{(x, z) : (x, z) \in U \times U\bigr\}$ for the sets $\left|\vartheta(x, z)\right|$, $\left|\vartheta(x, u_j)\right|$ and $|u_j|$, we have

$$p\left|\vartheta(x, z)\right| = \sum_{j=1}^{\infty} p|u_j|\, p^{u_j}\left|\vartheta(x, u_j)\right|, \tag{4.2.4}$$

where p^{u_j} denotes the conditional distribution of p with respect to the condition $z = u_j$ $(j = 1, 2, \ldots)$, and the u_j's run over the constant symbols corresponding to the elements in U.

Relation (4.2.3) can be generalized to formulas of the form $\vartheta(y_1, \ldots y_n, z_1, \ldots z_k)$ having $n + k$ free variables:

$$p\left|\vartheta(y_1, \ldots y_n, z_1, \ldots z_k)\right| = \sum_{j=1}^{\infty} p|v_j|\, p^{v_j}\left|\vartheta(y_1, \ldots y_n, v_j)\right|, \tag{4.2.5}$$

where v_j denotes a k-tuple of constants in U, and the sequence of v_j's $(j = 1, 2, \ldots)$ are an enumeration of the finitely dimensional points in the space ${}^kU$.

In particular, if ϑ in (4.2.4) is of the form $\forall x\alpha(x, z_1, \ldots z_k)$, then

$$p\left|\forall x\alpha(x, z_1, \ldots z_k)\right| = \sum_{j=1}^{\infty} p|v_j|\, p^{v_j}\left|\forall x\alpha(x, v_j)\right|. \tag{4.2.6}$$

Returning to the original infimum condition in (4.2.2), assume that $x, z_1, \dots z_k$ are the free variables of α. By (4.2.5) and (4.2.3), we have that the *left-hand side* is the following:

$$p\big|\forall x\alpha(x, z_1, \dots z_k)\big| = \sum_{j=1}^{\infty} p|\,v_j|\; p^{v_j}\big|\forall x\alpha(x, v_j)\big|$$

$$= \sum_{j=1}^{\infty} p|v_j| \inf_{n\in\omega} p^{v_j}\Big(\bigcap_{i\in n} |\alpha(y_i, v_j)|\Big)$$

(note that (4.2.3) is also true for p^{v_j}, since, for fixed v_j, p and p^{v_j} are different only in a constant).

Let us consider the *right-hand side* of the infimum part of (4.2.2). Let us denote the formula $\bigwedge_{i\in n} \alpha(y_i)$ by $\vartheta(y_1, \dots y_n, z_1, \dots z_k)$. By the definition of Q-probability, we can assume without the loss of generality that $\{z_1, \dots z_k\} \cap \{(y_1, \dots y_n\} = \emptyset$.

Using that $\big|\bigwedge_{i\in n} \alpha(y_i)\big| = \bigcap_{i\in n} |\alpha(y_i)|$, apply (4.2.4) to the right-hand side of the infimum condition (4.2.2):

$$\inf_{n\in\omega} p \bigcap_{i\in n} |\alpha(y_i)| = \inf_{n\in\omega} p\Big|\bigwedge_{i\in n} \alpha(y_i)\Big|$$

$$= \inf_{n\in\omega} \sum_{j=1}^{\infty} p|v_j|\; p^{v_j}\big|\vartheta(y_1, \dots y_n, v_j)\big|$$

$$= \sum_{j=1}^{\infty} p|v_j| \inf_{n\in\omega} p^{v_j}\big|\vartheta(y_1, \dots y_n, v_j)\big|$$

$$= \sum_{j=1}^{\infty} p|v_j| \inf_{n\in\omega} p^{v_j}\Big(\bigcap_{i\in n} |\alpha(y_i, v_j)|\Big).$$

We got the right-hand side of (4.2.6) and the proof is finished. ■

Remark 4.2.2. If the language Λ is countable, then the following consequences of Theorem 4.2.1 are true, applying the Löwenheim–Skolem theorem:

a) By the Löwenheim–Skolem theorem, in the case of a *countable* language Λ, arbitrary infinite model $\mathfrak{N}$ has an elementary submodel $\mathfrak{M}$ with countable universe. It is known that cylindric set algebras corresponding to these models are isomorphic. Thus, constructing a cylindric probability on $\mathfrak{A}^{\mathfrak{N}}$ can be reduced to constructing on $\mathfrak{A}^{\mathfrak{M}}$.

b) By Theorem 4.1.9, the canonical homomorphism transforms the cylindric probability from $\mathfrak{A}^{\mathfrak{M}}$ to the Lindenbaum–Tarski algebra of the language. It follows that, in the case of a countable language and a consistent theory, we can construct a cylindric probability on any Lindenbaum–Tarski algebra in this way.

There are concrete constructions yielding cylindric probabilities for cylindric set algebras with also *uncountable* universes (see [Fer,05]). The construction for non-standard stochastics in [Fer,10a] is closely related to the construction included in Theorem 4.2.1.

If an infinite collection of models corresponding to a given language Λ, theory Σ is assumed and cylindric probabilities are given on them, then, under certain conditions, it is possible to *compose* a cylindric probability on the respective formula algebra $\mathfrak{Fm}_{\Sigma}^{\Lambda}$. The following easy theorem is due to Łoš, while its generalization is the result of Fenstad (see [Los,73], [Fen,67]).

Assume that $\{\mathfrak{M}_i : i \in \Gamma\}$ is an infinite set of first-order models of a theory Σ and let $\left\{\mathfrak{A}_i^{\mathfrak{M}_i} : i \in \Gamma\right\}$ be the respective cylindric set algebras corresponding to these models. Assume that p_i are cylindric probabilities on $\mathfrak{A}_i^{\mathfrak{M}_i}$ and λ is a σ-additive probability on some σ-algebra $\mathcal{C}$ with universe Γ such that the functions $p_i|\varphi|$ are $\mathcal{C}$-integrable for every formula φ.

Theorem 4.2.3. *Let the function* s *be defined as follows:*

$$\mathsf{s}|\varphi| = \int_{\Gamma} p_i|\varphi|\, d\lambda. \tag{4.2.7}$$

The function s is a cylindric probability on $\mathfrak{Fm}_{\Sigma}^{\Lambda}$.

Remark 4.2.4. Theorem 4.2.3 can be generalized from $\mathfrak{Fm}_{\Sigma}^{\Lambda}$ for suitable products of cylindric set algebras. There are also some conditions under which the converse of the theorem, a decomposition theorem is also true (see [Los,73], [Fen,67]).

3. Homomorphisms and Measurable Functions

In this section we introduce some new kinds of homomorphisms which are induced by point functions (like base-homomorphisms (see [Hen-Mon-Tar,85] Def. 3.1.50). The concepts and results in this section are from [Fer,86a] and [Fer,99].

Let $V = \dot{\bigcup}_{r\in R} Z^{(r)}$ and $X = \dot{\bigcup}_{k\in K} U^{(k)}$ be the units of some cylindric algebras $\mathfrak{A}, \mathfrak{B} \in \mathsf{Crs}_\alpha \cap \mathsf{CA}_\alpha$ respectively, where $X \subseteq {}^\alpha T$ and $V \subseteq {}^\alpha P$ for some sets T and P, $Z^{(r)}$ and $U^{(k)}$ are the subunits.

Assume that f maps X to V. Let $\widehat{f}a = \{x : fx \in a, x \in X\}$, so the function $\widehat{f}$ maps A to $\mathfrak{Sb}\, X$, where $\mathfrak{Sb}\, X$ denotes the power set of X.

Definition 4.3.1. The function $f : X \to V$ induces a *homomorphism in the unit set sense* if $\widehat{f} \in Hom\,(\mathfrak{A}, \mathfrak{B})$. The homomorphism induced in this way is said to be a unit set homomorphism.

Assume that $\mathfrak{A}$ and $\mathfrak{B}$ are set algebras in Gws_α. So the unit V of $\mathfrak{A}$ is of the form $\dot{\bigcup}_{r\in R} {}^\alpha Z_r^{(pr)}$, where $pr \in {}^\alpha Z_r$ and ${}^\alpha Z_{r_1}^{(pr_1)} \cap {}^\alpha Z_{r_2}^{(pr_2)} = \emptyset$ if $r_1 \neq r_2$, and the unit X of $\mathfrak{B}$ is of the form $\dot{\bigcup}_{k\in K} {}^\alpha U_k^{(pk)}$, where $pk \in {}^\alpha U_k$ and ${}^\alpha U_{k_1}^{(pk_1)} \cap {}^\alpha U_{k_2}^{(pk_2)} = \emptyset$ if $k_1 \neq k_2$.

(4.3.1)

Assume that for every $k \in K$, $r \in R$, a function $t^k : U_k \to Z_r$ is given. Let the function $f : X \to {}^\alpha P$ be defined as follows: $fx = \langle t^k x_\kappa : \kappa < \alpha \rangle$ if $x \in {}^\alpha U_k^{(pk)}$.

Definition 4.3.2. The functions $\{t^k : k \in K\}$ induce a *homomorphism in the weak set sense* if $\widehat{f} \in Hom\,(\mathfrak{A}, \mathfrak{B})$. The homomorphism induced in this way is said to be a weak set homomorphism.

The next theorem is a characterization of the functions t^k which are suitable to induce a weak set homomorphism between the power set algebras $\mathfrak{A}$ and $\mathfrak{B}$ with units V and X respectively, considering them as algebras in Gws_α.

Theorem 4.3.3. *The functions $\{t^k : k \in K\}$ induce a homomorphism from $\mathfrak{A}$ to $\mathfrak{B}$ in the weak set sense if and only if the functions $t^k : U_k \to Z_r$ have the one-to-one property* ([Fer,86a]).

The proof is similar to that of [Hen-Mon-Tar,85] Theorem 3.1.52.

There are characterizations of the general weak set homomorphisms (homomorphisms between arbitrary set algebras $\mathfrak{A}$ and $\mathfrak{B}$), too (see [Fer,86a]).

The next theorem is a characterization of base and weak set homomorphisms among unit set homomorphisms.

Let $V = \dot{\bigcup}_{r\in R} Z^{(r)}$ and $X = \dot{\bigcup}_{k\in K} U^{(k)}$ be the units of some cylindric algebras $\mathfrak{A}$, $\mathfrak{B} \in \mathsf{Crs}_\alpha \cap \mathsf{CA}_\alpha$ respectively, where $Z^{(r)}$'s and $U^{(k)}$'s are the subunits.

Let $\mathfrak{A}$, $\mathfrak{B}$, V and X be the same as in (4.3.1).

Theorem 4.3.4. *Assume that f induces a unit set homomorphism, i.e. $\widehat{f} \in Hom(\mathfrak{A}, \mathfrak{B})$ is true for the function $f : X \to V$. Then the following propositions* (i), (ii) *hold:*

(i) *if V and X are units of some generalized weak set algebras, i.e. units of some $\mathfrak{A}$, $\mathfrak{B} \in \mathsf{Gws}_\alpha$, then f induces a homomorphism in the weak set sense if and only if, for every fixed $x, y \in X$ and fixed $\lambda < \alpha$ the following property is true:* a) *if $x_\kappa = y_\kappa$ for every $\kappa \neq \lambda$, then $(fx)_\kappa = (fy)_\kappa$ for every $\kappa \neq \lambda$.*

(ii) *if V and X are units of some generalized set algebras, i.e. units of some $\mathfrak{A}, \mathfrak{B} \in \mathsf{Gs}_\alpha$, then f induces a base-homomorphism if and only if, for every fixed $x, y \in X$, $\lambda < \alpha$, the following property b) is true: b) for any $\mu < \alpha$ if $x_\mu = y_\mu$, then $(fx)_\mu = (fy)_\mu$.*

(see [Fer,86a]).

Proof. (i) If f induces a weak set homomorphism, then a) is true by definition because the condition implies that x and y belong to the same weak set. The proof of the converse:

First, we claim that if $x, y \in U^{(k)}$ are arbitrary, then $x_\mu = y_\mu$ implies $(fx)_\mu = (fy)_\mu$ for any $\mu < \alpha$, i.e. the value of $(fx)_\mu$ depends only on the value x_μ in $U^{(k)}$. $x, y \in U^{(k)}$ implies that $y \in \mathsf{C}_{\lambda_1,\lambda_2,\ldots,\lambda_f}\{x\}$ for some ordinals $\lambda_1, \lambda_2, \ldots \lambda_f$ because x and y are included in the weak space $U^{(k)}$ (so x and y can be connected in $U^{(k)}$), i.e. we can obtain y from x in finitely many steps by replacing the elements of y one-by-one in such a way that (i) we change only a single element of the sequence at every step, (ii) every element is changed at most once during the process, (iii) all the intermediary sequences are in $U^{(k)}$, i.e. we remain within $U^{(k)}$. So if $x_\mu = y_\mu$, then the value x_μ cannot change. The application of property a) in a step-by-step manner shows that $(fx)_\mu$ can not change, thus $(fx)_\mu = (fy)_\mu$ as we claimed. Let the function be $t^{k,\mu}x_\mu$ defined as $(fx)_\mu$.

We claim that the value of $t^{k,\mu}x_\mu$ does not depend on μ, i.e. if $x, o \in U^{(k)}$, $u = x_\mu = o_\lambda$ for some $u \in U_k$, then $t^{k,\mu}x_\mu = t^{k,\lambda}o_\lambda$, where the base of $U^{(k)}$ is U_k and $\mu, \lambda < \alpha$. Let us consider the point x_u^λ and let us denote it by x', $x' \in U^{(k)}$. $x'_\mu = x_\mu$ implies that $t^{k,\mu}x'_\mu = t^{k,\mu}x_\mu$ by the definition of the function $t^{k,\mu}$. $u = x_\mu$, $x'_\mu = x_\mu$, therefore $x'_\mu = x'_\lambda$. f induces a homomorphism, so $\widehat{f}\mathsf{D}^V_{\lambda\mu} = \mathsf{D}^X_{\lambda\mu}$, i.e. $x'_\mu = x'_\lambda$ implies $(fx')_\mu = (fx')_\lambda$. Hence $t^{k,\mu}x'_\mu = t^{k,\mu}x'_\lambda$ by the definitions of $t^{k,\mu}$ and $t^{k,\lambda}$.

But as it is proven above, $x'_\lambda = o_\lambda$ implies $(fx')_\lambda = (fo)_\lambda$. But, $(fx')_\lambda = t^{k,\mu}x_\mu$ and $(fo)_\lambda = t^{k,\lambda}o_\lambda$. Thus $t^{k,\mu}x_\mu = t^{k,\lambda}o_\lambda$, as stated.

$t^{k,\mu}$ does not depend on μ, thus we can denote it by t^k. So if $x \in U^{(k)}$, then $t^k x_\mu = (fx)_\mu$. If $fx \in Z^{(r)}$ for some $r \in R$ and $y \in U^{(k)}$, then $fy \in Z^{(r)}$ by condition a), so t^k maps U_k into Z_r.

(ii) If f induces a base-homomorphism, then b) is true by definition. The proof of the converse is similar to that of (i). Taking into consideration that the subbases are disjoint in Gs_α, b) implies that if $x, y \in X$ are arbitrary, then $x_\mu = y_\mu$ implies $(fx)_\mu = (fy)_\mu$ for any fixed $\mu < \alpha$, i.e. the value of $(fx)_\mu$ depends only on the value x_μ. Denote the corresponding function by t^μ, i.e. $t^\mu x_\mu = (fx)_\mu$ should be true for the function f.

t^μ does not depend on μ, i.e. if $x, o \in X$ and $u = x_\mu = o_\lambda$ for some $u \in b(X)$, and $\mu, \lambda < \alpha$, then $t^\mu x_\mu = t^\lambda o_\lambda$. Namely, $\mathsf{C}^X_\lambda \mathsf{D}^X_{\lambda\mu} = X$, therefore $x' = x_u^\lambda \in X$ and $x'_\mu = x_\mu$ imply that $t^\mu x'_\mu = t^\mu x_\mu$. Similarly, as in (i), it follows that $x'_\mu = x'_\lambda$ imply that $(fx')_\mu = (fx')_\lambda$. Therefore, $t^\mu x'_\mu = t^\lambda x'_\lambda = t^\lambda o_\lambda = t^\lambda u$, so $t^\mu x_\mu = t^\mu x'_\mu = t^\lambda o_\lambda$ as we stated. ■

Condition a) says that f maps any λ-straight line, into λ-straight line while condition b) says that f maps any μ hyperplane in X into a μ hyperplane in V.

The relational algebraic analogies of the above concepts and results can be found in [Fer,99].

V. Other Algebraic Versions of Logic

BOLYAI SOCIETY
MATHEMATICAL STUDIES, 22

Cylindric-like Algebras and Algebraic Logic
pp. 351–366.

Cylindric Set Algebras and IF Logic*

ALLEN L. MANN

Independence-friendly logic (IF logic) [Hin,96, Hin-San,89] is a conservative extension of first-order logic that can be viewed as a generalization of Henkin's branching quantifiers [Hen,61]. For example, in the branching quantifier sentence

$$\begin{pmatrix} \forall v_0 \exists v_1 \\ \forall v_2 \exists v_3 \end{pmatrix} \varphi(v_0, v_1, v_2, v_3), \tag{1.0.2}$$

the rows of the branching quantifier indicate that v_1 depends (only) on v_0, while v_3 depends (only) on v_2. It is a result due to Ehrenfeucht that in general such sentences cannot be expressed in ordinary first-order logic [Hen,61]. However, sentence (1.0.2) can be expressed in IF logic as

$$\forall v_0 \exists v_1 \forall v_{2/\{v_0,v_1\}} \exists v_{3/\{v_0,v_1\}} \varphi(v_0, v_1, v_2, v_3), \tag{1.0.3}$$

where the slashes indicate that v_2 and v_3 do not depend on v_0 or v_1. For our purposes, it will be more convenient to work with dependence-friendly logic (DF logic), in which we indicate on which variables each quantifier is dependent rather than independent. For example, in DF logic we would write sentence (1.0.2) as

$$\forall v_{0\backslash\emptyset} \exists v_{1\backslash\{v_0\}} \forall v_{2\backslash\emptyset} \exists v_{3\backslash\{v_2\}} \varphi(v_0, v_1, v_2, v_3), \tag{1.0.4}$$

where the backslashes indicate that v_0 and v_2 do not depend on any other variables, v_1 depends on v_0, and v_3 depends on v_2.

Although Henkin mentions semantic games, he chooses not to use them to interpret branching quantifiers. Instead, the meaning of a branching

*Preparation of this article was supported by the Institut d'histoire et de philosophie des sciences et des techniques (UMR 8590—Paris 1/CNRS/ENS) and the Academy of Finland (project 129208) in the context of the European Science Foundation EUROCORES project Logic for Interaction (LINT).

quantifier sentence is determined by its Skolemization. Sentence (1.0.2) is true if and only if there exist functions f and g satisfying

$$\forall v_0 \forall v_2 \varphi(v_0, f(v_0), v_2, g(v_2)),$$

that is, if and only if the Σ^1_1 sentence

$$\exists f \exists g \forall v_0 \forall v_2 \varphi(v_0, f(v_0), v_2, g(v_2)) \tag{1.0.5}$$

is true. In contrast, the meaning of a DF sentence is defined via a game between two players, Eloïse ($\exists$) and Abélard ($\forall$). Given a structure $\mathfrak{A}$ and a DF sentence φ, Eloïse attempts to verify φ by picking elements of the universe to be the values of existentially quantified variables. For disjunctions, she chooses which disjunct to verify. Dually, Abélard attempts to falsify φ by picking the values of universally quantified variables and the conjuncts he wishes to falsify. The game ends when the players reach an atomic formula. Eloïse wins if the final assignment satisfies the atomic formula, and Abélard wins if it does not. The sentence is true if Eloïse has a winning strategy, and it is false if Abélard has a winning strategy.

Unlike the game for ordinary first-order logic, the semantic game for DF logic is a game of imperfect information. Just as the different rows of a branching quantifier indicate on which variables the Skolem functions should depend, the backslashes on the quantifiers (and connectives) of a DF sentence indicate the values of which variables the players are aware as they play. For example, in the semantic game for the sentence

$$\forall v_{0\backslash\emptyset} \exists v_{1\backslash\{v_0\}} R v_0 v_1 \tag{1.0.6}$$

Eloïse is allowed to see the value of v_0 when choosing v_1. Hence she has a winning strategy in any structure $\mathfrak{A}$ whose interpretation of R contains a function; otherwise Abélard has a winning strategy. In contrast, in the semantic game for

$$\forall v_{0\backslash\emptyset} \exists v_{1\backslash\emptyset} R v_0 v_1 \tag{1.0.7}$$

Eloïse must choose v_1 without knowing the value of v_0. Hence (1.0.7) is true if $R^{\mathfrak{A}}$ is the total relation, false if $R^{\mathfrak{A}}$ does not contain a function, and undetermined otherwise.

The major difference between first-order logic with branching quantifiers and DF logic is their treatment of negation. Enderton [End,70] and Walkoe [Wal,70] proved independently that every branching quantifier sentence is equivalent to a Σ_1^1 sentence, and vice versa. Since the contradictory negation of a Σ_1^1 sentence is in general a Π_1^1 sentence, the contradictory negation of a branching quantifier sentence is not always expressible as a branching quantifier sentence. The same is true for DF sentences. However, there is a natural way to interpret negation as the role-reversal of the players. Eloïse attempts to verify $\sim \varphi$ by falsifying φ, while Abélard attempts to falsify $\sim \varphi$ by verifying φ. Thus $\sim \varphi$ is true if and only if φ is false, $\sim \varphi$ is false if and only if φ is true, and $\sim \varphi$ is undetermined if and only if φ is. This suggests that the propositional logic underlying DF logic is Kleene's strong three-valued logic, which we will prove below.

The fact that Eloïse has a winning strategy for $G(\mathfrak{A}, \varphi)$ is expressible by a Σ_1^1 sentence. For example, in (1.0.5) the functions f and g encode Eloïse's winning strategy for (1.0.4). Likewise, the fact that Abélard has a winning strategy. Thus we can treat one Σ_1^1 sentence as the truth-condition for φ, and another as its falsity-condition. Hence a DF sentence has the same expressive power as a pair of contrary Σ_1^1 sentences. Moreover, working with branching quantifiers, Burgess [Bur,03] proved that every pair of contrary Σ_1^1 sentences can be viewed as the truth- and falsity-conditions for some DF sentence.

1. Dependence-Friendly Logic

To simplify our notation, we assume that all variables come from a fixed set $\{v_n : n < \alpha\}$ for some ordinal α.

Definition 1.1.1. Given a first-order signature σ, $\mathscr{L}^{\mathrm{DF}_\alpha}(\sigma)$ is the language generated by the following grammar:

$$\varphi := \Phi \mid \sim \varphi \mid \{\varphi \vee_J \psi\}_{J \subseteq \alpha} \mid \{\exists v_{n \setminus J} \varphi\}_{J \subseteq \alpha}$$

where Φ consists of all atomic σ-formulas. An element of $\mathscr{L}^{\mathrm{DF}_\alpha}(\sigma)$ is called a DF_α *formula*. When the signature is implicit, we will simply write $\mathscr{L}^{\mathrm{DF}_\alpha}$.

We adopt the standard abbreviations of $\varphi \wedge_J \psi$ for $\sim (\sim \varphi \vee_J \sim \psi)$ and $\forall v_{n \setminus J} \varphi$ for $\sim \exists v_{n \setminus J} \sim \varphi$. We suppress the backslash symbol on connectives for aesthetic reasons, and we are content to keep track of which variables are in each dependence set by their indices.

We must extend the game-theoretical semantics for DF sentences presented in the introduction to all DF formulas. To do so, we need a way to keep track of the information available to the players at each position of the semantic game. In ordinary first-order logic, it is sufficient to use a single assignment to encode the values of the variables. In DF logic, we must use sets of assignments, called *teams*, to encode the information available to the players. Given a structure $\mathfrak{A}$, a DF_α formula φ, and a team $V \subseteq {}^\alpha A$, the semantic game $G(\mathfrak{A}, \varphi, V)$ is played as usual except that at the beginning of the game a third player, whom we will call Nature, randomly chooses an assignment $\vec{a} \in V$ which is used to assign values to the free variables in φ, the bound variables in φ, and even the variables that do not appear in φ.

Definition 1.1.2. Let $\mathfrak{A}$ be a structure, φ a DF_α formula, and $V, W \subseteq {}^\alpha A$.

(+) $\mathfrak{A} \models^+_V \varphi$ if and only if Eloïse has a winning strategy for $G(\mathfrak{A}, \varphi, V)$.

(−) $\mathfrak{A} \models^-_W \varphi$ if and only if Abélard has a winning strategy for $G(\mathfrak{A}, \varphi, W)$.

In the first case we say that V is a *winning team*[1] for φ in $\mathfrak{A}$, and in the second case we say that W is a *losing team* for φ in $\mathfrak{A}$. We write $\mathfrak{A} \models^+ \varphi$ when $\mathfrak{A} \models^+_{{}^\alpha A} \varphi$, and write $\mathfrak{A} \models^- \varphi$ when $\mathfrak{A} \models^-_{{}^\alpha A} \varphi$. The previous statement can be summarized by saying $\mathfrak{A} \models^\pm \varphi$ if and only if $\mathfrak{A} \models^\pm_{{}^\alpha A} \varphi$.

For a fuller treatment of game-theoretical semantics in the context of IF logic, see [Man,09, Section 1.2].

Game-theoretical semantics gives us an intuitive definition of truth for DF formulas. However, it is not compositional, which makes it unsuitable for proving theorems by induction. Luckily, Wilfrid Hodges discovered a compositional semantics for IF logic called *trump semantics* that we can adapt to DF logic [Hod,97a, Hod,97b].

Definition 1.1.3. Two assignments $\vec{a}, \vec{b} \in {}^\alpha A$ are *indistinguishable on* J, denoted $\vec{a} \approx_J \vec{b}$, if $\vec{a} \upharpoonright J = \vec{b} \upharpoonright J$.

Definition 1.1.4. Define a partial operation $\cup_J$ on pairs of teams by declaring $V_1 \cup_J V_2 = V_1 \cup V_2$ whenever

- $V_1 \cap V_2 = \emptyset$,
- if $\vec{a} \in V_1$ and $\vec{b} \in V_2$, then $\vec{a} \not\approx_J \vec{b}$,

[1] Hodges [Hod,97a, Hod,97b] calls a winning team a *trump*, and a losing team a *cotrump*. The term *team* was introduced by Väänänen [Vaa,07].

and letting $V_1 \cup_J V_2$ be undefined otherwise. Thus the formula $V = V_1 \cup_J V_2$ asserts that $\{V_1, V_2\}$ is a pair of disjoint teams that cover V in such a way that assignments that are indistinguishable on J are always in the same cell.

Definition 1.1.5. A function $f : V \to A$ is *determined by* J, denoted $f : V \underset{J}{\to} A$, if $f(\vec{a}) = f(\vec{b})$ whenever $\vec{a} \approx_J \vec{b}$.

Definition 1.1.6. The assignment $\vec{a}(n : b)$ is defined by

$$\vec{a}(n:b)_i = \begin{cases} b & \text{if } i = n, \\ a_i & \text{otherwise.} \end{cases}$$

Furthermore, if $V \subseteq {}^\alpha A$ and $f : V \to A$,

$$V(n:f) = \{\vec{a}(n : f(\vec{a})) : \vec{a} \in V\},$$

$$V(n:A) = \{\vec{a}(n:b) : \vec{a} \in V,\ b \in A\}.$$

Theorem 1.1.7. *Let $\mathfrak{A}$ be a structure, let φ be a DF_α formula, and let $V, W \subseteq {}^\alpha A$.*

- *If φ is atomic, then*

  (+) $\mathfrak{A} \models^+_V \varphi$ *if and only if* $\mathfrak{A} \models_{\vec{a}} \varphi$ *for all* $\vec{a} \in V$,

  (−) $\mathfrak{A} \models^-_W \varphi$ *if and only if* $\mathfrak{A} \not\models_{\vec{b}} \varphi$ *for all* $\vec{b} \in W$.

- *If φ is $\sim \psi$, then $\mathfrak{A} \models^\pm_V \sim \psi$ if and only if $\mathfrak{A} \models^\mp_V \psi$.*

- *If φ is $\psi_1 \vee_J \psi_2$, then*

  (+) $\mathfrak{A} \models^+_V \psi_1 \vee_J \psi_2$ *if and only if* $\mathfrak{A} \models^+_{V_1} \psi_1$ *and* $\mathfrak{A} \models^+_{V_2} \psi_2$ *for some* $V = V_1 \cup_J V_2$,

  (−) $\mathfrak{A} \models^-_W \psi_1 \vee_J \psi_2$ *if and only if* $\mathfrak{A} \models^-_W \psi_1$ *and* $\mathfrak{A} \models^-_W \psi_2$.

- *If φ is $\exists v_{n \setminus J} \psi$, then*

  (+) $\mathfrak{A} \models^+_V \exists v_{n \setminus J} \psi$ *if and only if* $\mathfrak{A} \models^+_{V(n:f)} \psi$ *for some* $f : V \underset{J}{\to} A$,

  (−) $\mathfrak{A} \models^-_W \exists v_{n \setminus J} \psi$ *if and only if* $\mathfrak{A} \models^-_{W(n:A)} \psi$

([Hod,97a, Theorem 7.5], see also [Dec,05, Theorem 5.3.5]).

Dependence-friendly formulas have two important properties that reflect the fact that it is impossible for both players to have winning strategies for the same game. Also, if one player has a winning strategy given a certain amount of information, then the same player has a winning strategy given more information.

Proposition 1.1.8. *Let* $V \subseteq {}^{\alpha}A$.

(i) $\mathfrak{A} \models^{+}_{V} \varphi$ *and* $\mathfrak{A} \models^{-}_{V} \varphi$ *if and only if* $V = \emptyset$.

(ii) *If* $V' \subseteq V$, *then* $\mathfrak{A} \models^{\pm}_{V} \varphi$ *implies* $\mathfrak{A} \models^{\pm}_{V'} \varphi$

([Hod,97b, page 57], *see also* [Dec,05, Lemma 6.2.1]*).*

2. Dependence-Friendly Cylindric Set Algebras

The *meaning* of a DF_{α} formula φ in a structure $\mathfrak{A}$ is a pair whose first coordinate is the set of winning teams for the formula, and whose second coordinate is the set of losing teams. That is, $\|\varphi\|_{\mathfrak{A}} = \left\langle \|\varphi\|^{+}_{\mathfrak{A}}, \|\varphi\|^{-}_{\mathfrak{A}} \right\rangle$, where

$$\|\varphi\|^{+}_{\mathfrak{A}} = \{V \subseteq {}^{\alpha}A \,:\, \mathfrak{A} \models^{+}_{V} \varphi\},$$

$$\|\varphi\|^{-}_{\mathfrak{A}} = \{W \subseteq {}^{\alpha}A \,:\, \mathfrak{A} \models^{-}_{W} \varphi\}.$$

Analogous to Definition 4.3.4 in [Hen-Mon-Tar,85], the universe of the α-dimensional dependence-friendly cylindric set algebra over $\mathfrak{A}$ consists of the meanings of all the DF_{α} formulas expressible in the language of $\mathfrak{A}$. In symbols,

$$\mathsf{Cs}_{\mathrm{DF}_{\alpha}}(\mathfrak{A}) = \left\{ \|\varphi\|_{\mathfrak{A}} \,:\, \varphi \in \mathscr{L}^{\mathrm{DF}_{\alpha}} \right\}.$$

We then define constants 1, 0, and diagonal elements D_{ij}, as well as operations ${}^{\cup}$, $+_J$, $\cdot_J$, and $C_{n,J}$ corresponding to $\top$, $\bot$, $v_i = v_j$, $\sim$, $\vee_J$, $\wedge_J$, and $\exists v_{n\setminus J}$, respectively (see Definition 1.2.1). We denote the *α-dimensional dependence-friendly cylindric set algebra over* $\mathfrak{A}$ by $\mathfrak{Cs}_{\mathrm{DF}_{\alpha}}(\mathfrak{A})$. As in the case of ordinary cylindric set algebras, we can define dependence-friendly cylindric set algebras without reference to a base structure $\mathfrak{A}$.

Definition 1.2.1. A *dependence-friendly cylindric power set algebra* is an algebra whose universe is $\mathscr{P}\big(\mathscr{P}({}^{\alpha}U)\big) \times \mathscr{P}\big(\mathscr{P}({}^{\alpha}U)\big)$, where U is a set called the *base,* and α is an ordinal called the *dimension* of the algebra. Every element X of a dependence-friendly cylindric power set algebra is an ordered pair of sets of teams. We will use the notation X^+ to refer to the first coordinate of the pair, and X^- to refer to the second coordinate. There are a finite number of operations:

- the constant $1 = \big\langle \mathscr{P}({}^{\alpha}U), \{\emptyset\} \big\rangle$;
- the constant $0 = \big\langle \{\emptyset\}, \mathscr{P}({}^{\alpha}U) \big\rangle$;
- for all $i, j < \alpha$, the constant D_{ij} is defined by

  (+) $\mathsf{D}^+_{ij} = \mathscr{P}\big(\{\vec{a} \in {}^{\alpha}U \,:\, a_i = a_j\}\big)$,

  (−) $\mathsf{D}^-_{ij} = \mathscr{P}\big(\{\vec{a} \in {}^{\alpha}U \,:\, a_i \neq a_j\}\big)$;
- if $X = \langle X^+, X^- \rangle$, then $X^{\cup} = \langle X^-, X^+ \rangle$;
- for every $J \subseteq \alpha$, the binary operation $+_J$ is defined by

  (+) $V \in (X +_J Y)^+$ if and only if $V = V_1 \cup_J V_2$ for some $V_1 \in X^+$ and $V_2 \in Y^+$,

  (−) $(X +_J Y)^- = X^- \cap Y^-$;
- for every $J \subseteq \alpha$, the binary operation $\cdot_J$ is defined by

  (+) $(X \cdot_J Y)^+ = X^+ \cap Y^+$,

  (−) $W \in (X \cdot_J Y)^-$ if and only if $W = W_1 \cup_J W_2$ for some $W_1 \in X^-$ and $W_2 \in Y^-$;
- for every $n < \alpha$ and $J \subseteq \alpha$, the unary operation $C_{n,J}$ is defined by

  (+) $V \in C_{n,J}(X)^+$ if and only if $V(n : f) \in X^+$ for some $f : V \underset{J}{\to} U$,

  (−) $W \in C_{n,J}(X)^-$ if and only if $W(n : U) \in X^-$.

Definition 1.2.2. A *dependence-friendly cylindric set algebra* (or DF *algebra*) is any subalgebra of a dependence-friendly cylindric power set algebra. A DF$_\alpha$ *cylindric set algebra* (or DF$_\alpha$ *algebra*) is a DF cylindric set algebra of dimension α.

Independence-friendly cylindric set algebras (IF algebras) were defined in [Man,09, Definition 2.1].

In view of Proposition 1.1.8, not every element of a dependence-friendly cylindric power set algebra can be the meaning of a DF formula. Only those elements X with the property that $X^+ \cap X^- = \{\emptyset\}$, and $V' \subseteq V \in X^\pm$ implies $V' \in X^\pm$ can be meanings. Such elements are called *double suits* [Cam-Hod,01, Section 3]. A *double-suited* DF *algebra* is one in which every element is a double suit.

For the rest of the paper we will focus on double-suited algebras. The reader should verify that the DF algebra generated by a set of double suits is a double-suited algebra. A similar phenomenon occurs with ordinary cylindric set algebras. The meaning $\varphi^{\mathfrak{A}}$ of an ordinary first-order formula has the property that for all $\vec{a} \in \varphi^{\mathfrak{A}}$ and $\vec{b} \in {}^{\alpha}A$, if $\vec{b}$ agrees with $\vec{a}$ on the free variables of φ, then $\vec{b} \in \varphi^{\mathfrak{A}}$. An element of a cylindric set algebra with the corresponding property is called *regular*, where the set of free variables is replaced by the dimension set of the element. A cylindric set algebra is *regular* if all its elements are regular [Hen-Mon-Tar,85, Definition 3.1.1(viii)], and the cylindric set algebra generated by a set of regular elements with finite dimension sets is regular [Hen-Mon-Tar,85, Corollary 3.1.64]. See also [Hen-Mon-Tar-And-Nem,81, pages 145–149].

3. The Perfect Subreduct

Naturally, one wonders about the relationship between ordinary cylindric set algebras and their dependence-friendly brethren. Every element of an ordinary cylindric set algebra has the form $V \subseteq {}^{\alpha}U$, while every double suit has the form

$$\Big\langle \bigcup_{i<\beta} \mathscr{P}(V_i), \bigcup_{j<\gamma} \mathscr{P}(W_j) \Big\rangle,$$

for some $V_i, W_j \subseteq {}^{\alpha}U$ such that $V_i \cap W_j = \emptyset$ for all i, j. Furthermore, we may assume that $\{V_i : i < \beta\}$ and $\{W_j : j < \gamma\}$ are antichains when partially ordered by $\subseteq$. Double suits of the form $\langle \mathscr{P}(V), \mathscr{P}({}^{\alpha}U \setminus V)\rangle$ are called *perfect* because they are the meanings of those DF formulas whose semantic game is one of perfect information. There is a natural mapping

$$V \mapsto \langle \mathscr{P}(V), \mathscr{P}({}^{\alpha}U \setminus V)\rangle$$

that sends meanings of ordinary first-order formulas to the meanings of their dependence-friendly analogues. In fact, this mapping embeds any α-dimensional cylindric set algebra $\mathfrak{C}$ into the reduct of the DF_α algebra $\mathfrak{D}$ generated by

$$\{\langle \mathscr{P}(V), \mathscr{P}({}^\alpha U \setminus V)\rangle : V \in \mathfrak{C}\}$$

to the signature $\langle 1, 0, \mathsf{D}_{ij}, +_\alpha, \cdot_\alpha, C_{n,\alpha}\rangle$. The image of $\mathfrak{C}$ under this mapping is called the *perfect subreduct* of $\mathfrak{D}$. In particular, $\mathfrak{Cs}_\alpha(\mathfrak{A})$ is isomorphic to the perfect subreduct of $\mathfrak{Cs}_{\mathrm{DF}_\alpha}(\mathfrak{A})$, which captures algebraically the fact that DF logic is a conservative extension of first-order logic [Man,08].

4. The De Morgan Reduct

When restricted to perfect double suits, the perfect-information operations $+_\alpha$ and $\cdot_\alpha$ correspond to the Boolean operations $\cup$ and $\cap$ of an ordinary cylindric set algebra. Unfortunately, when applied to all double suits, $+_\alpha$ and $\cdot_\alpha$ may not even be lattice operations. Consider the one-dimensional double suit $X = \big\langle \mathscr{P}\big(\{0\}\big) \cup \mathscr{P}\big(\{1\}\big), \{\emptyset\}\big\rangle$, and observe that

$$X +_1 X = \big\langle \mathscr{P}\big(\{0,1\}\big), \{\emptyset\}\big\rangle \neq X.$$

At the other extreme, we consider the zero-information operations $+_\emptyset$ and $\cdot_\emptyset$, which correspond to positions in the semantic game where the players must move in complete ignorance of the current assignment. Unlike $+_\alpha$ and $\cdot_\alpha$, the operations $+_\emptyset$ and $\cdot_\emptyset$ are lattice operations on double suits. The present section and the next are adapted from [Man,09].

Definition 1.4.1. A *De Morgan algebra* $\mathfrak{A} = \langle A; 1, 0, \sim, \vee, \wedge\rangle$ is a bounded distributive lattice with an additional unary operation $\sim$ that satisfies

$$\sim\sim x = x \quad \text{and} \quad \sim(x \vee y) = \sim x \wedge \sim y.$$

Unlike a Boolean algebra, it is possible for a De Morgan algebra to have an element such that $\sim a = a$. Such an element is called a *center* or *fixed point*. A *centered De Morgan algebra* is a De Morgan algebra with a center.

Definition 1.4.2. A *Kleene algebra* is a De Morgan algebra that satisfies the additional condition

$$x \wedge \sim x \leq y \vee \sim y.$$

Proposition 1.4.3. *A Kleene algebra has at most one center.*

Proof. Suppose a and b are both centers. Then $a = a \wedge \sim a \leq b \vee \sim b = b$ and $b = b \wedge \sim b \leq a \vee \sim a = a$. ■

For the elementary theory of De Morgan and Kleene algebras, we refer the reader to the excellent book [Bal-Dwi,74].

Definition 1.4.4. The *De Morgan reduct* of a DF_α algebra is the reduct of the algebra to the signature $\langle 1, 0, ^{\cup}, +_\emptyset, \cdot_\emptyset \rangle$.

Proposition 1.4.5. *Let X and Y be double suits.*

(i) $X +_\emptyset Y = \langle X^+ \cup Y^+, X^- \cap Y^- \rangle$.

(ii) $X \cdot_\emptyset Y = \langle X^+ \cap Y^+, X^- \cup Y^- \rangle$.

Proof. (i) Suppose $V \in (X +_\emptyset Y)^+$. Then $V = V_1 \cup_\emptyset V_2$ for some $V_1 \in X^+$ and $V_2 \in Y^+$. However, since $\approx_\emptyset$ is the total relation, either $V_1 = V$ and $V_2 = \emptyset$ or vice versa. Thus $V \in X^+ \cup Y^+$. Conversely, suppose $V \in X^+ \cup Y^+$. Without loss of generality, assume $V \in X^+$. Then $V = V \cup_\emptyset \emptyset$, where $V \in X^+$ and $\emptyset \in Y^+$, so $V \in (X +_\emptyset Y)^+$. Thus $(X +_\emptyset Y)^+ = X^+ \cup Y^+$. Observe that $(X +_\emptyset Y)^- = X^- \cap Y^-$ by definition.

The proof of (ii) is similar. ■

Definition 1.4.6. Let X and Y be double suits. Define $X \leq Y$ if $X^+ \subseteq Y^+$ and $Y^- \subseteq X^-$.

Theorem 1.4.7. *The class of De Morgan reducts of double-suited* DF_α *algebras generates the variety of all Kleene algebras.*

Proof. Let $\mathfrak{D}$ be the De Morgan reduct of a double-suited DF_α algebra. First, we will prove that $\mathfrak{D}$ is a Kleene algebra. Using Proposition 1.4.5, it is easy to show that $\mathfrak{D}$ is a bounded distributive lattice and that the lattice order agrees with $\leq$ as defined above. We verify the other three axioms. For all $X, Y \in \mathfrak{D}$,

$$(X^{\cup})^{\cup} = \langle X^-, X^+ \rangle^{\cup} = \langle X^+, X^- \rangle = X,$$

$$(X +_\emptyset Y)^{\cup} = \langle X^- \cap Y^-, X^+ \cup Y^+ \rangle = X^{\cup} \cdot_\emptyset Y^{\cup},$$

$$X \cdot_\emptyset X^{\cup} = \big\langle \{\emptyset\}, X^- \cup X^+ \big\rangle \leq \big\langle Y^+ \cup Y^-, \{\emptyset\} \big\rangle = Y +_\emptyset Y^{\cup}.$$

Therefore $\mathfrak{D}$ is a Kleene algebra.

To prove that the class of De Morgan reducts of double-suited DF_α algebras generates the variety of all Kleene algebras, it suffices to show that every nontrivial, subdirectly irreducible Kleene algebra is isomorphic to the De Morgan reduct of some double-suited DF_α algebra. Kalman [Kal,58] proved that, up to isomorphism, the only nontrivial, subdirectly irreducible De Morgan algebras are:

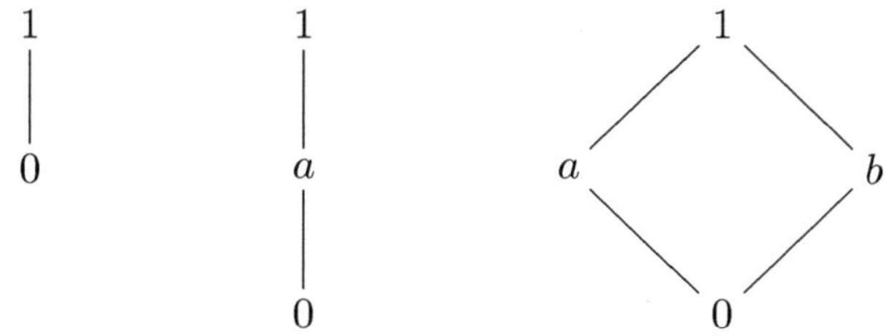

$\underline{\mathbf{B}}$ $\underline{\mathbf{K}}$ $\underline{\mathbf{M}}$

where $\sim a = a$ and $\sim b = b$. Of these, only $\underline{\mathbf{B}}$ and $\underline{\mathbf{K}}$ are Kleene algebras. For any base U, let

$$1 = \left\langle \mathscr{P}({}^{\alpha}U), \{\emptyset\} \right\rangle,$$

$$\Omega = \left\langle \{\emptyset\}, \{\emptyset\} \right\rangle,$$

$$0 = \left\langle \{\emptyset\}, \mathscr{P}({}^{\alpha}U) \right\rangle.$$

Then $\{0, 1\}$ and $\{0, \Omega, 1\}$ are double-suited DF_α algebras whose De Morgan reducts are isomorphic to $\underline{\mathbf{B}}$ and $\underline{\mathbf{K}}$, respectively. ■

It can be shown that 0, Ω, and 1 are the only possible meanings of DF sentences. Furthermore, if $X, Y \in \{0, \Omega, 1\}$, then for any $J \subseteq \alpha$ we have $X +_J Y = X +_\emptyset Y$ and $X \cdot_J Y = X \cdot_\emptyset Y$. Therefore, the propositional logic underlying DF logic is Kleene's strong three-valued logic [Kle,52, Section 64] (see also [Pri,08]). The connection between IF logic and Kleene's strong three-valued logic was first observed by Hintikka and Sandu [Hin-San,97, Proposition 5.1].

5. The Monadic De Morgan Reduct

Generalizing Halmos' [Hal,62] work on monadic Boolean algebras, Cignoli [Cig,91] and Petrovich [Pet,96] equip bounded distributive lattices with an additional unary operation to obtain what they call Q-distributive lattices. Soon after, Petrovich [Pet,99] extended their investigations by reintroducing negation, defining monadic De Morgan algebras.

Definition 1.5.1. A *quantifier* on a De Morgan algebra is a unary operation ∇ such that:

(Q1) $\nabla 0 = 0$,

(Q2) $x \leq \nabla x$,

(Q3) $\nabla(x \vee y) = \nabla x \vee \nabla y$,

(Q4) $\nabla(x \wedge \nabla y) = \nabla x \wedge \nabla y$.

(Q5) $\nabla(\sim \nabla x) = \sim \nabla x$.

A De Morgan algebra equipped with a quantifier is called a *monadic De Morgan algebra.*

Definition 1.5.2. The *monadic De Morgan reduct* of a DF_1 algebra is the reduct of the algebra to the signature $\langle 1, 0, {}^{\cup}, +_{\emptyset}, \cdot_{\emptyset}, C_{0,\emptyset} \rangle$.

Lemma 1.5.3. *In any* DF_1 *algebra we have* $C_{0,J}(1) = 1$ *and* $C_{0,J}(0) = 0$. *In any double-suited* DF_1 *algebra with* Ω,

$$C_{0,J}(X) = \begin{cases} 1 & \textit{if } X \nleq \Omega, \\ \Omega & \textit{if } 0 < X \leq \Omega, \\ 0 & \textit{if } X = 0. \end{cases}$$

Proof. $C_{0,J}(1) = \langle \mathscr{P}(U), \{\emptyset\} \rangle$ because $V \in C_{0,J}(1)^+$ if and only if there exists an $f : V \underset{J}{\rightarrow} U$ such that $V(0 : f) \in \mathscr{P}(U)$, which is always true. Also, $W \in C_{0,J}(1)^-$ if and only if $W(0 : U) = \emptyset$ if and only if $W = \emptyset$.

$C_{0,J}(0) = \langle \{\emptyset\}, \mathscr{P}(U) \rangle$ because $V \in C_{0,J}(0)^+$ if and only if $V(0 : f) = \emptyset$ for some $f : V \underset{J}{\rightarrow} U$ if and only if $V = \emptyset$, while $C_{0,\emptyset}(0)^- = \mathscr{P}(U)$ because $W \in C_{0,J}(0)^-$ if and only if $W(0 : U) \in \mathscr{P}(U)$, which is always true.

For the rest of the proof, assume X and $C_{0,J}(X)$ are double suits. Suppose $X \nleq \Omega$. Then there is a $\emptyset \neq V \in X^+$. Let $b \in V$, and let $f : U \underset{J}{\to} U$ be the constant function that sends every element to b. Then $U(0 : f) = \{b\} \in X^+$, so $U \in C_{0,J}(X)^+$. Hence $C_{0,J}(X) = \langle \mathscr{P}(U), \{\emptyset\} \rangle$.

Suppose $0 < X \leq \Omega$. Then $X^+ = \emptyset$ and $U \notin X^-$. It follows that $C_{0,J}(X) = \langle \{\emptyset\}, \{\emptyset\} \rangle$ because $V \in C_{0,J}(X)^+$ if and only if there exists a function $f : V \underset{J}{\to} U$ such that $V(0 : f) = \emptyset$ if and only if $V = \emptyset$. Also, $W \in C_{0,J}(X)^-$ if and only if $W(0 : U) \in X^-$ if and only if $W = \emptyset$. ■

Theorem 1.5.4. *The class of monadic De Morgan reducts of double-suited* DF_1 *algebras generates a proper subvariety of the variety of all monadic Kleene algebras.*

Proof. Let $\mathfrak{D}$ be the monadic De Morgan reduct of a double-suited DF_1 algebra with base U. First we verify that $\mathfrak{D}$ satisfies the axioms. We have already checked (Q1).

(Q2) Suppose $V \in X^+$. If $V = \emptyset$, then $V \in C_{0,\emptyset}(X)^+$ because $C_{0,\emptyset}(X)$ is a double suit. If $b \in V$, let $f : V \underset{\emptyset}{\to} U$ be the constant function that sends every element to b. Then $V(0 : f) = \{b\} \in X^+$. Thus $V \in C_{0,\emptyset}(X)^+$.

Now suppose $W \in C_{0,\emptyset}(X)^-$. If $W = \emptyset$, then $W = \emptyset = W(0 : U) \in X^-$. If $W \neq \emptyset$, then $U = W(0 : U) \in X^-$. Hence $W \in \mathscr{P}(U) = X^-$.

(Q3) For the rest of the proof we will assume that V and W are non-empty. Suppose $V \in C_{0,\emptyset}(X +_\emptyset Y)^+$. Then there is a function $f : V \underset{\emptyset}{\to} U$ such that

$$V(0 : f) \in (X +_\emptyset Y)^+ = X^+ \cup Y^+.$$

Without loss of generality, assume $V(0 : f) \in X^+$. It follows that

$$V \in C_{0,\emptyset}(X)^+ \cup C_{0,\emptyset}(Y)^+ = \big(C_{0,\emptyset}(X) +_\emptyset C_{0,\emptyset}(Y)\big)^+.$$

Conversely, suppose $V \in \big(C_{0,\emptyset}(X) +_\emptyset C_{0,\emptyset}(Y)\big)^+$. Then without loss of generality $V \in C_{0,\emptyset}(X)^+$, so there is an $f : V \underset{\emptyset}{\to} U$ such that

$$V(0 : f) \in X^+ \subseteq (X +_\emptyset Y)^+.$$

Hence $V \in C_{0,\emptyset}(X +_\emptyset Y)^+$.

Also, $W \in C_{0,\emptyset}(X +_\emptyset Y)^-$ if and only if

$$W(0:U) \in X^- \cap Y^- = (X +_\emptyset Y)^-$$

if and only if

$$W \in C_{0,\emptyset}(X)^- \cap C_{0,\emptyset}(Y)^- = \left(C_{0,\emptyset}(X) +_\emptyset C_{0,\emptyset}(Y)\right)^-.$$

(Q4) Suppose we have $V \in C_{0,\emptyset}\left(X \cdot_\emptyset C_{0,\emptyset}(Y)\right)^+$. Then there is a constant function $f : V \underset{\emptyset}{\to} U$ and an element $b \in U$ such that

$$\{b\} = V(0:f) \in \left(X \cdot_\emptyset C_{0,\emptyset}(Y)\right)^+ = X^+ \cap C_{0,\emptyset}(Y)^+,$$

and another constant function $g : \{b\} \underset{\emptyset}{\to} U$ and element $b' \in U$ such that $\{b'\} = V(0:f)(0:g) \in Y^+$. It follows that $g \circ f : V \underset{\emptyset}{\to} U$ is a constant function such that $V(0:g \circ f) = \{b'\} \in Y^+$. Hence

$$V \in C_{0,\emptyset}(X)^+ \cap C_{0,\emptyset}(Y)^+ = \left(C_{0,\emptyset}(X) +_\emptyset C_{0,\emptyset}(Y)\right)^+.$$

Conversely, suppose $V \in \left(C_{0,\emptyset}(X) +_\emptyset C_{0,\emptyset}(Y)\right)^+ = C_{0,\emptyset}(X)^+ \cap C_{0,\emptyset}(Y)^+$. Then there exist constant functions $f : V \underset{\emptyset}{\to} U$ and $h : V \underset{\emptyset}{\to} U$ as well as elements $b, b' \in U$ such that $\{b\} = V(0:f) \in X^+$ and $\{b'\} = V(0:h) = Y^+$. Let $g : \{b\} \underset{\emptyset}{\to} U$ be the function that maps b to b'. Then $h = g \circ f$, $V(0:f)(0:g) = V(0:h) \in Y^+$, and $V(0:f) \in C_{0,\emptyset}(Y)^+$. Thus

$$V(0:f) \in X^+ \cap C_{0,\emptyset}(Y)^+ = \left(X \cdot_\emptyset C_{0,\emptyset}(Y)\right)^+,$$

so $V \in C_{0,\emptyset}\left(X \cdot_\emptyset C_{0,\emptyset}(Y)\right)^+$.

Now suppose $W \in C_{0,\emptyset}\left(X \cdot_\emptyset C_{0,\emptyset}(Y)\right)^-$. Then

$$U = W(0:U) \in \left(X \cdot_\emptyset C_{0,\emptyset}(Y)\right)^- = X^- \cup C_{0,\emptyset}(Y)^-.$$

If $U(0:U) = U \in X^-$, then $U \in C_{0,\emptyset}(X)^-$. Hence

$$U \in C_{0,\emptyset}(X)^- \cup C_{0,\emptyset}(Y)^- = \left(C_{0,\emptyset}(X) \cdot_\emptyset C_{0,\emptyset}(Y)\right)^-.$$

If $U \in C_{0,\emptyset}(Y)^-$, then

$$U \in C_{0,\emptyset}(X)^- \cup C_{0,\emptyset}(Y)^- = \left(C_{0,\emptyset}(X) \cdot_\emptyset C_{0,\emptyset}(Y)\right)^-.$$

In either case, $W \in \big(C_{0,\emptyset}(X) \cdot_\emptyset C_{0,\emptyset}(Y)\big)^-$.

Conversely, suppose

$$W \in \big(C_{0,\emptyset}(X) \cdot_\emptyset C_{0,\emptyset}(Y)\big)^- = C_{0,\emptyset}(X)^- \cup C_{0,\emptyset}(Y)^-.$$

If $W \in C_{0,\emptyset}(X)^-$, then

$$W(0:U) \in X^- \cup C_{0,\emptyset}(Y)^- = \big(X \cdot_\emptyset C_{0,\emptyset}(Y)\big)^-,$$

so $W \in C_{0,\emptyset}\big(X \cdot_\emptyset C_{0,\emptyset}(Y)\big)^-$. If $W \in C_{0,\emptyset}(Y)^-$, then

$$W(0:U)(0:U) = W(0:U) \in Y^-.$$

Hence $W(0:U) \in C_{0,\emptyset}(Y)^-$. Thus

$$W(0:U) \in X^- \cup C_{0,\emptyset}(Y)^- = \big(X \cdot_\emptyset C_{0,\emptyset}(Y)\big)^-.$$

Therefore $W \in C_{0,\emptyset}\big(X \cdot_\emptyset C_{0,\emptyset}(Y)\big)^-$.

(Q5) Suppose $V \in C_{0,\emptyset}\big(C_{0,\emptyset}(X)^\cup\big)^+$. Then $V(0:f) \in C_{0,\emptyset}(X)^-$ for some $f : V \underset{\emptyset}{\to} U$. Hence

$$U(0:U) = V(0:f)(0:U) \in X^-.$$

Thus $U \in C_{0,\emptyset}(X)^- = \big(C_{0,\emptyset}(X)^\cup\big)^+$, which implies $V \in \big(C_{0,\emptyset}(X)^\cup\big)^+$.

Conversely, suppose $V \in \big(C_{0,\emptyset}(X)^\cup\big)^+ = C_{0,\emptyset}(X)^-$. Then

$$U(0:U)(0:U) = V(0:U) \in X^-,$$

so

$$U(0:U) \in C_{0,\emptyset}(X)^- = \big(C_{0,\emptyset}(X)^\cup\big)^+.$$

Thus $U \in C_{0,\emptyset}\big(C_{0,\emptyset}(X)^\cup\big)^+$, which implies $V \in C_{0,\emptyset}\big(C_{0,\emptyset}(X)^\cup\big)^+$.

Now suppose $W \in C_{0,\emptyset}\big(C_{0,\emptyset}(X)^\cup\big)^-$. Then

$$U = W(0:U) \in \big(C_{0,\emptyset}(X)^\cup\big)^-,$$

which implies $W \in \big(C_{0,\emptyset}(X)^\cup\big)^-$. Conversely, suppose $W \in \big(C_{0,\emptyset}(X)^\cup\big)^- = C_{0,\emptyset}(X)^+$. Then there is a constant function $f : W \underset{\emptyset}{\to} U$ and an element $b \in U$ such that $\{b\} = W(0:f) \in X^+$. Let $g : U \underset{\emptyset}{\to} U$ be the constant function that sends every element to b. Then $U(0 : g) = \{b\} \in X^+$, so $U(0 : U) = U \in C_{0,\emptyset}(X)^+ = \big(C_{0,\emptyset}(X)^{\cup}\big)^-$. Thus $U \in C_{0,\emptyset}\big(C_{0,\emptyset}(X)^{\cup}\big)^-$. Therefore $\mathfrak{D}$ is a monadic Kleene algebra.

Next we will prove that $\mathfrak{D}$ satisfies the inequality

$$\text{(Q6)} \qquad \nabla(x \wedge \sim x) \le \sim \nabla(x \wedge \sim x),$$

which is equivalent to an equation. Observe that for any double suit X,

$$X \cdot_{\emptyset} X^{\cup} = \big\langle \{\emptyset\}, X^+ \cup X^- \big\rangle \le \Omega,$$

so by Lemma 1.5.3 we have

$$C_{0,\emptyset}(X \cdot_{\emptyset} X^{\cup}) \le \Omega \le \big(C_{0,\emptyset}(X \cdot_{\emptyset} X^{\cup})\big)^{\cup}.$$

Finally, we exhibit a monadic Kleene algebra that does not satisfy (Q6). Let $\langle \underline{\mathbf{K}}, \nabla \rangle$ be the three-element Kleene algebra $\{0, a, 1\}$ equipped with the quantifier ∇ defined by

$$\nabla x = \begin{cases} 1 & \text{if } x > 0, \\ 0 & \text{if } x = 0. \end{cases}$$

It is easy to check that $\langle \underline{\mathbf{K}}, \nabla \rangle$ is a monadic Kleene algebra. However,

$$\nabla(a \wedge \sim a) = 1 \not\le 0 = \sim \nabla(a \wedge \sim a). \qquad \blacksquare$$

Conjecture 1.5.5. The class of monadic De Morgan reducts of double-suited DF_1 algebras generates the variety of all monadic Kleene algebras that satisfy

$$\text{(Q6)} \qquad \nabla(x \wedge \sim x) \le \sim \nabla(x \wedge \sim x).$$

BOLYAI SOCIETY
MATHEMATICAL STUDIES, 22

Cylindric-like Algebras and Algebraic Logic
pp. 367–389.

Polyadic Algebras

GÁBOR SÁGI

Polyadic algebras were introduced and intensively studied by Halmos, after having studied cylindric algebras in Tarski's seminar in Berkeley; we refer to Section 5.4 of [Hen-Mon-Tar,85], see also [Hal,62]. This class of algebras can be regarded as an alternative approach to algebraize first order logic. After a thorough reformulation of Henkin, Monk, and Tarski, polyadic algebras also can be regarded as certain generalizations of cylindric algebras. On one hand, polyadic algebras have nice representation properties, on the other, their languages are rather large (in the ω-dimensional case the cardinality of their set of operations is continuum), which makes their equational theory recursively undecidable for trivial reasons. This is undesirable from metalogical point of view, hence, during the last decades, certain countable (even finite) reducts of polyadic algebras have also been intensively studied. The goal of this research direction is to find a countable reduct of polyadic algebras which has nice representation properties, and, at the same time, their equational theory is recursively enumerable.

This section is closely related to, and is based on Section 5.4 of [Hen-Mon-Tar,85]. In more detail, in Section 1 we recall the definition of certain classes of polyadic algebras. In Section 2 we are dealing with the representation theory of polyadic algebras; in Section 3 we are establishing some connections between polyadic and cylindric algebras. Finally, Section 4 is devoted to study the recursion theoretic complexity of the equational theories of certain classes of (reducts of) polyadic algebras.

1. Basic Properties of Polyadic Algebras

We start by recalling the definitions of polyadic and polyadic equality algebras and their representable subclasses. These definitions are the same as Definitions 5.4.1 and 5.4.22 of [Hen-Mon-Tar,85].

Definition. Let α be an ordinal. An algebra

$$\mathfrak{A} = \left\langle A; \cdot, -, 0, 1, \mathsf{c}_{(\Gamma)}, \mathsf{s}_\tau \right\rangle_{\Gamma \subseteq \alpha,\ \tau \in {}^\alpha\alpha}$$

is defined to be a *polyadic algebra* of dimension α iff the following equational stipulations hold for any $x, y \in A$, for any $\Gamma, \Delta \subseteq \alpha$ and for any $\sigma, \tau \in {}^\alpha\alpha$.

(P_0) $\langle A; \cdot, -, 0, 1\rangle$ is a Boolean Algebra;

(P_1) $\mathsf{c}_{(\Gamma)}0 = 0$;

(P_2) $x \leq \mathsf{c}_{(\Gamma)}x$;

(P_3) $\mathsf{c}_{(\Gamma)}\big(x \cdot \mathsf{c}_{(\Gamma)}(y)\big) = \mathsf{c}_{(\Gamma)}(x) \cdot \mathsf{c}_{(\Gamma)}(y)$;

(P_4) $\mathsf{c}_{(\emptyset)}x = x$;

(P_5) $\mathsf{c}_{(\Gamma)}\mathsf{c}_{(\Delta)}x = \mathsf{c}_{(\Gamma\cup\Delta)}x$;

(P_6) $\mathsf{s}_{id}x = x$;

(P_7) $\mathsf{s}_\sigma \mathsf{s}_\tau x = \mathsf{s}_{\sigma\circ\tau}x$;

(P_8) $\mathsf{s}_\sigma(x \cdot y) = \mathsf{s}_\sigma(x) \cdot \mathsf{s}_\sigma(y)$;

(P_9) $\mathsf{s}_\sigma(-x) =\sim \mathsf{s}_\sigma(x)$;

(P_{10}) if $\sigma|_{\alpha\sim\Gamma} = \tau|_{\alpha\sim\Gamma}$ then $\mathsf{s}_\sigma\mathsf{c}_{(\Gamma)}x = \mathsf{s}_\tau\mathsf{c}_{(\Gamma)}x$;

(P_{11}) if $\Delta = \tau^{-1}[\Gamma]$ and $\tau|_\Delta$ is one-one then $\mathsf{c}_{(\Gamma)}\mathsf{s}_\tau x = \mathsf{s}_\tau\mathsf{c}_{(\Delta)}x$.

The class of all α dimensional polyadic algebras will be denoted by PA_α.

Definition. Let α be an ordinal. An algebra

$$\mathfrak{A} = \left\langle A; \cdot, -, 0, 1, \mathsf{c}_{(\Gamma)}, \mathsf{s}_\tau, \mathsf{d}_{ij} \right\rangle_{\Gamma \subseteq \alpha,\ \tau \in {}^\alpha\alpha,\ i,j\in\alpha}$$

is defined to be a *polyadic equality algebra* of dimension α iff its PA_α-type reduct is a polyadic algebra of dimension α, for all $i, j \in \alpha$ we have $\mathsf{d}_{ij} \in A$ and, in addition, the following equational stipulations hold for any $x \in A$, for any $i, j \in \alpha$ and for any $\tau \in {}^\alpha\alpha$.

(E_1) $\mathsf{d}_{ii} = 1$;

(E_2) $x \cdot \mathsf{d}_{ij} \leq \mathsf{s}_{[i/j]}x$;

(E_3) $\mathsf{s}_\tau \mathsf{d}_{ij} = \mathsf{d}_{\tau(i)\tau(j)}$.

The class of all α dimensional polyadic equality algebras will be denoted by PEA_α.

Next we recall the definition of representable polyadic algebras; this definition is the same as Definition 5.4.22 of [Hen-Mon-Tar,85].

Definition. Let α be an ordinal, U a set and let $W \subseteq {}^\alpha U$. For $\tau \in {}^\alpha\alpha$, $\Gamma \subseteq \alpha$, $i, j \in \alpha$ and $x \subseteq W$ the operations of cylindrification $\mathsf{C}^W_{(\Gamma)}$, substitution S^W_τ and diagonal elements D^W_{ij} are defined as follows.

$$\mathsf{C}^W_{(\Gamma)}(x) = \big\{ z \in W \, : \, (\exists r \in x)\big(z|_{\alpha - \Gamma} = r|_{\alpha-\Gamma}\big) \big\};$$

$$\mathsf{S}_\tau(x) = \{ z \in W \, : \, z \circ \tau \in x \} \quad \text{and}$$

$$\mathsf{D}_{ij} = \{ z \in W \, : \, z_i = z_j \}.$$

The structure

$$\mathfrak{A} = \big\langle \mathcal{P}(W); \cap, \sim, \emptyset, W, \mathsf{C}^W_{(\Gamma)}, \mathsf{S}^W_\tau, \mathsf{D}^W_{ij} \big\rangle_{\Gamma \subseteq \alpha,\ \tau \in {}^\alpha\alpha,\ i,j \in \alpha}$$

is called the *full α-dimensional relativized polyadic set algebra* of W. If $W = {}^\alpha U$ then $\mathfrak{A}$ is called the α-dimensional *full polyadic set algebra* of U. In addition, according to Definition 5.4.22 of [Hen-Mon-Tar,85],

(i) the class Pse_α of *polyadic equality set algebras* of dimension α consists of all subalgebras of full polyadic set algebras (of appropriate dimension);

(ii) the class Rppe_α of *representable polyadic equality algebras* of dimension α consists of all subdirect products of full polyadic set algebras (of appropriate dimension);

(iii) the class Gp_α of α dimensional *generalized polyadic set algebras* consists of all subalgebras of relativized polyadic set algebras of $W = \bigcup_{i \in I} {}^\alpha U_i$ (of appropriate dimension).

Note, that according to the last item of the previous definition, generalized polyadic set algebras are relativied polyadic set algebras of some W, where W is the union of the αth direct power of some sets U_i. We emphasize, that we do not require the different U_i to be disjoint from each other. The class of relativized subalgebras of the disjoint unions of ${}^{\alpha}U_i$ is called the class Gwp_α of generalized weak polyadic set algebras. It is easy to see, that $\mathsf{Rppe}_\alpha = \mathbf{I}\,\mathsf{Gwp}_\alpha$.

Remark 2.1.1. By a representation of a polyadic (equality) algebra $\mathfrak{A}$ we mean an isomorphism between $\mathfrak{A}$ and an Rppe (or RPA, respectively). Representability with relativized algebras has also deserved considerable attention - this research direction can be well motivated by the Resek–Thompson theorem in cylindric algebra theory. In this connection we refer to Remark 3.2.88 of [Hen-Mon-Tar,85].

Definition. The classes of SPA_α and RPA_α consist of the diagonal-free reducts of elements of Pse_α and Rppe_α respectively.

It is routine to check that $\mathsf{RPA}_\alpha \subseteq \mathsf{PA}_\alpha$ and $\mathsf{Rppe}_\alpha \subseteq \mathsf{PEA}_\alpha$. We will see in Section 2 below, that the aim of representation theory of polyadic algebras is establishing results related to the converse inclusions.

As we mentioned, if $\alpha \geq \omega$, then the cardinality of operations of PA_α is uncountable, making the equational theory undecidable for trivial reasons. Hence, finite and countably infinite reducts of polyadic algebras has also been intensively studied. Next, we recall the definitions of some countable reducts of polyadic algebras.

Definition 2.1.2. Let α be given and let $\mathsf{G} \subseteq {}^{\alpha}\alpha$ be a semigroup (under composition of functions). Then the class $\mathsf{G}-\mathsf{PA}_\alpha$ consists of all subreducts of elements of PA_α having the following set of operations:

- the Boolean operations;
- $\big\{\mathsf{c}_{(\Gamma)} : \Gamma \subseteq \alpha \text{ is finite}\big\}$ and
- $\{\mathsf{s}_\tau : \tau \in \mathsf{G}\}$.

If $\mathsf{G} = \big\{\tau \in {}^{\alpha}\alpha : \big\{i \in \alpha : \tau(i) \neq i\big\} \text{ is finite}\big\}$ then the class $\mathsf{G}-\mathsf{PA}_\alpha$ is denoted by QPA_α and called the class of α dimensional quasi-polyadic algebras.

$\mathsf{G}-\mathsf{PEA}_\alpha$ and QPEA_α are defined similarly.

Remark 2.1.3. Let α be an ordinal, and as usual, for $i, j \in \alpha$ let $[i/j] : \alpha \to \alpha$ be the function mapping i onto j and leaving every other element fixed. Similarly, $[i, j]$ denotes the function that maps i onto j, maps j onto i and leaves every other element of α fixed. It is easy to check, that the semigroup $\mathsf{G} = \{\tau \in {}^\alpha\alpha : \{i \in \alpha : \tau(i) \neq i\} \text{ is finite}\}$ can be generated by $\mathsf{G}_0 = \{\, [i/j], [i, j] : i, j \in \alpha\}$. Hence, in QPA_α, for $\tau \in \mathsf{G}$ the operation s_τ is term definable by the operations $\{\mathsf{s}_\tau : \tau \in \mathsf{G}_0\}$.

Ferenczi proved in [Fer] that in QPEA_α, axiom (P_{10}) is *not independent* of the other axioms.

Suppose α is given. Then QPA_α can be regarded as the "minimalistic polyadic extension" of CA_α. Particularly, the language of QPA_ω contains countably many operation symbols only. However, there is an essential difference between the definition of QPA_α and CA_α.

For different α, the classes CA_α can be defined as a system of varieties. In more detail, this means the following. If α is given, and $\xi \in {}^\alpha\alpha$ is a function, then ξ acts on the equations of the language of CA_α in the natural way: if e is a CA_α-equation, then $\xi(e)$ can be obtained from e by replacing each occurrence of c_i by $\mathsf{c}_{\xi(i)}$ and d_{ij} by $\mathsf{d}_{\xi(i)\xi(j)}$, respectively. Then there is a finite set E of equations such that for any α, CA_α is the class of all models of $\{\, \xi[E] : \xi \in {}^\alpha\alpha \text{ is a permutation}\}$. This uniform definability may be useful, because it makes accessible some techniques of universal algebra.

The set of instances of the equations in the definition of QPA_α is not closed under all permutations of α; in addition, it is not obvious, if there exists an alternative definition of QPA_α containing schemas of equations whose set of instances is closed under permutations. To study the situation, Sain and Thompson in [Sai-Tho,91] introduced the classes FPA_α and FPEA_α of *Finitary Polyadic (Equality) Algebras* of dimension α.

Definition 2.1.4. A *finitary polyadic equality algebra* of dimension α is an algebra

$$\mathfrak{A} = \left\langle A; \cdot, -, 0, 1, \mathsf{c}_i, \mathsf{s}^i_j, \mathsf{p}_{ij}, \mathsf{d}_{ij} \right\rangle_{i,j\in\alpha}$$

where d_{ij} are constants, and the following equational stipulations hold for any $i, j, k \in \alpha$:

(F_0) $\langle A; \cdot, -, 0, 1\rangle$ is a Boolean algebra, $\mathsf{s}^i_i = \mathsf{p}_{ii} = \mathsf{d}_{ii} = Id|_A$ and $\mathsf{p}_{ij} = \mathsf{p}_{ji}$;

(F_1) $x \leq \mathsf{c}_i x$;

(F_2) $\mathsf{c}_i(x \vee y) = \mathsf{c}_i(x) \vee \mathsf{c}_i(y)$;

(F$_3$) $\mathsf{s}^i_j\mathsf{c}_i(x) = \mathsf{c}_i(x)$;

(F$_4$) $\mathsf{c}_i\mathsf{s}^i_j(x) = \mathsf{s}^i_j\mathsf{c}_i(x)$ if $i \neq j$;

(F$_5$) $\mathsf{s}^i_j\mathsf{c}_k(x) = \mathsf{c}_k\mathsf{s}^i_j(x)$ if $k \notin \{i, j\}$;

(F$_6$) s^i_j and p_{ij} are Boolean endomorphisms;

(F$_7$) $\mathsf{p}_{ij}\mathsf{p}_{ij}x = x$;

(F$_8$) $\mathsf{p}_{ij}\mathsf{p}_{ik}(x) = \mathsf{p}_{jk}\mathsf{p}_{ij}(x)$ if i, j, k are distinct;

(F$_9$) $\mathsf{p}_{ij}\mathsf{s}^i_j(x) = \mathsf{s}^j_i(x)$;

(F$_{10}$) $\mathsf{s}^i_j\mathsf{d}_{ij} = 1$;

(F$_{11}$) $x \cdot \mathsf{d}_{ij} \leq \mathsf{s}^i_j x$.

The class of finitary polyadic equality algebras of dimension α is denoted by FPEA_α; its diagonal free subreduct is denoted by FPA_α.

It is easy to check, that the set of instances of defining equations of FPEA_α is closed under permutations of α.

Theorem 2.1.5 (Sain, Thompson). *Let $\alpha > 2$.*

(i) *The varieties FPEA_α and QPEA_α are term definitionally equivalent.*

(ii) *The varieties FPA_α and QPA_α are term definitionally equivalent.*

The proof can be found in [Sai-Tho,91].

2. Representation Theory of Polyadic Algebras

We will cut this section into two parts: first we survey results from the representation theory of polyadic algebras (without diagonal elements) and next, we will deal with the case of polyadic equality algebras (containing diagonal elements). Both parts can be further divided to the finite dimensional and to the infinite dimensional case or to "positive" and "negative" results.

2.1. Representation theory, the diagonal-free case

The first theorem we should mention is the following celebrated result of Daigneault and Monk. The original proof can be found in [Dai-Mon,63]; see also Remark 5.4.41 of [Hen-Mon-Tar,85].

Theorem (Daigneault, Monk).*For infinite α we have $\mathsf{PA}_\alpha = \mathsf{RPA}_\alpha$.*

We also note, that independently, Keisler in [Kei,63] proved a completeness theorem for a version of first order logic with infinitary predicates; this completeness theorem may be considered as the logical version of the Daignault-Monk representation theorem.

On one hand, the Daigneault–Monk Theorem is elegant: it describes a finite schema of equations axiomatizing RPA_α (for infinite α). On the other hand, these equational schemas have continuum many instances for the smallest, $\alpha = \omega$ case. This is necessary, because the cardinality of the set of operations of RPA_ω is the continuum.

PA's of infinite dimension also have the superamalgamation property [Say,10a], and their atomic algebras are completely representable [Say,08c], but, this is not the case with cylindric algebras. In fact it is commonly accepted that polyadic algebras and cylindric algebras belong to different paradigms. The algebras introduced by Sain in [Sai,00], see Section 2.3, enjoy the positive aspects of both paradigms.

If α is finite, then the polyadic axiom schemas (P_0)–(P_{11}) have finitely many instances only. For finite α, the class RPA_α is a variety; however, as the next theorem indicates, its equational theory is rather complicated.

Theorem 2.2.1. *For finite $\alpha \geq 2$ the variety RPA_α is not finitely axiomatizable. In addition, Rppe_α "cannot be axiomatized by finitely many variables": if Σ is a set of polyadic equations such that $\mathsf{Mod}\,(\Sigma) = \mathsf{Rppe}_\alpha$ then, for every $n \in \omega$ there is an equation $e_n \in \Sigma$ containing at least n distinct variables.*

For a proof and more details, see [And,97a], [And,97b] and [Joh,69].

As we mentioned, QPA_α may be considered as the "minimalistic polyadic extension" of cylindric algebras. Even, this minimalistic class cannot be finitely axiomatized, as the following theorem says.

Theorem 2.2.2 (Sain, Thomson). *For $\alpha > 2$, the class RQPA_α of representable quasi-polyadic algebras of dimension α cannot be axiomatized by finitely many equations.*

The proof, and stronger related results can be found in [Sai-Tho,91]. In fact, in [Sai-Tho,91] it was shown, that for $\alpha > 2$, the class of representable FPA_α (or equivalently, QPA_α) cannot be defined by finitely many equational schemas closed under permutations of the dimension set (i.e. these classes cannot be defined by finitely many Monk-type equational schemas).

By Theorem 2.2.2, for finite α, RPA_α is a proper subclass of PA_α. If we take smaller reducts, some positive results may be obtained.

Definition 2.2.3. For a set U, the structure

$$\mathfrak{A} = \left\langle \mathcal{P}({}^\alpha U); \cap, -, 0, 1, \mathsf{S}_{[i/j]} \right\rangle_{i,j \in \alpha}$$

is called the α-dimensional *full substitution set algebra* of U.

(i) The class SetSA_α of substitution set algebras of dimension α consists of subalgebras of full substitution set algebras (of appropriate dimension);

(ii) the class RSA_α of representable substitution algebras of dimension α is defined to be $\mathsf{RSA}_\alpha = \mathbf{ISP}\,\mathsf{SetSA}_\alpha$.

These classes first was studied by Pinter (see [Hen-Mon-Tar,85], page 267). In [Sag,02] and in [Sag,99a] the following were proved for RSA_n.

Theorem 2.2.4 (Sági).

(i) *For finite $n \geq 2$ the class RSA_n is a finitely axiomatizable quasi-variety, but not a variety;*

(ii) *the generated variety is also finitely axiomatized and it consists of (isomorphic copies of) the appropriate reducts of Gp_α;*

(iii) *the (quasi-)equational theory of RSA_α is decidable for $\alpha \leq \omega$.*

We emphasize, that in (iii) above, $\alpha = \omega$ is allowed, as well.

Proof. We give a sketch only. Let $n \geq 2$ be finite and fixed. Let V be the variety generated by SetSA_n. Suppose, that $\mathfrak{A} \in \mathsf{SetSA}_n$ with base set U and $W \subseteq U$. Let $\mathfrak{B} \in \mathsf{SetSA}_n$ be the full set algebra over W. Then it is easy to see, that the function $\varphi_W : \mathfrak{A} \to \mathfrak{B}$ satisfying $\varphi_W(x) = x \cap {}^nW$ is a SetSA_n-homomorphism. Now let $a \in A - \{0\}$ be arbitrary. Then there exists $s \in A$. Let $W = \mathrm{ran}\,(s)$. Then φ_W is a homomorphism from $\mathfrak{A}$ which maps a to a non-zero element and the base set of its image is of cardinality at most n. Consequently, every $\mathfrak{A} \in \mathsf{SetSA}_n$ can be embedded into a direct product $\Pi_{i \in I}\mathfrak{B}_i$ such that the base set of each $\mathfrak{B}_i$ is of cardinality at most n. It follows, that V can be generated by finitely many finite algebras. In addition, V has a Boolean reduct, hence it is congruence distributive. Thus, by Baker's theorem, V is finitely axiomatizable. This proves the first part of (ii) and (iii) for finite α and for the equational theory of RSA_n.

Next, we show that RSA_n is not a variety. Let σ be the quasi-equation

$$\mathsf{s}_{[0/1]}(x) \cdot \mathsf{s}_{[1/0]}(y) = 0 \Rightarrow \mathsf{s}_{[1/0]}(x) \cdot \mathsf{s}_{[0/1]}(y) = 0.$$

It is easy to see, that σ holds in every SetSA_n, hence in RSA_n. Let $\mathfrak{A} \in \mathsf{SetSA}_n$ be the full set algebra on base set $\{0, \ldots, n-1\}$. Then $\mathfrak{A}$ has a homomorphic image, in which σ is not true; for the details, see Theorem 3 of [Sag,02]. This shows, that RSA_n is not a variety.

Next, we show, that SetSA_n is closed under ultraproducts. Let $\mathfrak{A}_i \in \mathsf{SetSA}_n$; suppose, that the base set of $\mathfrak{A}_i$ is U_i and let F be an ultrafilter over I. Let $U = \Pi_{i \in I}U_i/F$ and let $\varphi : \Pi_{i \in I}\mathfrak{A}_i/F \to \mathsf{P}({}^nU)$ defined to be

$$\varphi\big(\langle x(i) \,:\, i \in I\rangle/F\big)$$

$$= \Big\{\langle \mathsf{s}_0, \ldots, \mathsf{s}_{n-1}\rangle/F \,:\, \big\{i \in I \,:\, \big\langle \mathsf{s}_0(i), \ldots, \mathsf{s}_{n-1}(i)\big\rangle \in x(i)\big\} \in F\Big\}.$$

It is easy to check that φ is an embedding of $\Pi_{i \in I}\mathfrak{A}_i/F$. It follows, that SetSA_n, is closed under ultraproducts. Consequently RSA_n is also closed under ultraproducts, hence it is a quasi-variety. This proves the last two parts of (i).

The proofs of the remaining parts of (i), (ii) and (iii) are much more longer, hence we omit them. They can be found in [Sag,02] and in [Sag,99a]. ■

Parts of Theorem 2.2.4 not proved above, are based on a semigroup-theoretic observation. For a given set U, $\mathsf{NP}(U)$ denotes the semigroup of finite, non-bijective selfmaps (that is,

$$\mathsf{NP}(U) = \Big\{t \in {}^UU \,:\, \big\{x \in U \,:\, x \neq t(x)\big\} \text{ is finite and } t \text{ is not bijective}\Big\}$$

and the operation of $\mathsf{NP}(U)$ is composition of functions). It was shown in [Tho,93] and [Sag,99b] that there is a finite equational axiom system Σ such that, $\mathsf{NP}(U)$ is presented by (the set of all instances of) Σ. Thus, $\mathsf{NP}(U)$ can be presented by a finite set of "presentation schemas". These schemas can be used to describe the action of the substitution operations – in the finite dimensional case they have finitely many instances, and this fact implies that RSA_α is finitely axiomatizable for finite α. For completeness, we note, that Jónsson in [Jon,62] provided a finite schema presentation for the semigroup of finite transformations $F(U) = \big\{t \in {}^U U : \{x \in U : x \neq t(x)\}$ is finite$\big\}$.

It seems, that cylindrifications are the responsible for the negative results.

2.2. Representation theory, the diagonal case, negative results

It is natural to ask, whether the analogue of the Daigneault–Monk Theorem for Rppe_ω remains true. It will turn out, that the situation is essentially different from the diagonal free case.

We start by showing, that Rppe_ω is not closed under ultraproducts. As described in Remark 5.4.41 of [Hen-Mon-Tar,85], this was first proved by Monk. Later Johnson showed that certain ultraproducts of polyadic equality set algebras are not in Rppe_ω, hence Rppe_ω cannot be axiomatized by any set of first order formulas (particularly, it is not a variety).

Theorem. Rppe_ω *is not closed under ultraproducts.*

Proof. Let pred : $\omega \to \omega$ defined to be pred$(0) = 0$ and pred$(n+1) = n$ for each $n \in \omega$. Observe, that in each $\mathfrak{A} \in \mathsf{Rppe}_\omega$ the following "infinitary quasi-equation" is true for all $x \in A$:

$$(*) \qquad \Big(\bigwedge_{i,j\in\omega} x \leq \mathsf{d}_{ij}\Big) \quad \Rightarrow \quad x = \mathsf{s}_{\text{pred}}(x).$$

Now let $\mathfrak{C} \in \mathsf{Cs}_\omega$ be countable, whose base set U is infinite and let $\mathfrak{A}$ be the full ω dimensional polyadic equality set algebra of U. Let F be a nonprincipal ultrafilter over ω, let $\mathfrak{B} = {}^\omega\mathfrak{A}/F$ and finally let $b = \langle b_n : n \in \omega\rangle/F \in B$ where

$$b_n = \Big(\prod_{i,j<n} \mathsf{d}_{ij}\Big) \cdot \Big(\prod_{i<n} -\mathsf{d}_{in}\Big).$$

Then clearly, for every $i, j \in \omega$ we have $\mathfrak{B} \models b \leq \mathsf{d}_{ij}$ but $s^{\mathfrak{B}}_{\text{pred}}(b) \neq b$. So $(*)$ does not hold in $\mathfrak{B}$ and consequently, $\mathfrak{B} \notin \mathsf{Rppe}_\omega$. ■

Remark 2.2.5. Now, by the previous theorem, there exists a nonrepresentable $\mathfrak{B} \in \mathsf{PEA}_\omega$. Using the notation of the previous proof, we can conclude, that the CA_ω-type reduct of $\mathfrak{B}$ is in RCA_ω (i.e., it is representable). Indeed, $\mathfrak{B}$ has been constructed as an ultrapower of a polyadic set-algebra $\mathfrak{A}$ such that $\mathfrak{B}$ is not representable. Since taking ultraproducts and forming reducts are commuting operations, it follows, that the CA_ω-type reduct of $\mathfrak{B}$ coincides with the appropriate ultrapower of the CA_ω-type reduct of $\mathfrak{A}$. Since this latter is a representable CA_ω, it follows from Theorem 3.1.109 of [Hen-Mon-Tar,85], that the CA_ω-type reduct of $\mathfrak{B}$ is also in RCA_α, but according to the previous theorem, $\mathfrak{B}$ itself is not in Rppe_ω.

After the previous theorem, the next problem is to axiomatize the variety generated by (or, equivalently, the equational theory of) Rppe_ω. A finite axiomatization is impossible, because of the set of operations of Rppe_ω is of infinite cardinality. So a finite schema of equations would be desirable. A natural candidate for such a finite schema axiomatization is (P_0)–(E_3).

It turned out, that there is an equation valid in **HSP** Rppe_ω but does not valid in PEA_ω. One could hope, that adding new, similar schemas to (P_0)–(E_3) might lead to an axiomatization of Rppe_ω. We will see below, that this also cannot be done. We start by fixing the precise definition for "schemas similar to (P_0)–(E_3)".

Having a look for (P_0)–(E_3), one can realize, that these schemas contain variables Γ, Δ ranging over subsets of ω and σ, τ ranging over ${}^\omega\omega$. Sometimes there is a condition between the sets and functions occurring in the names of operations in the schema, but such conditions always expressible in a certain first order language.

Consider (P_7) as a typical example. It can be rephrased as follows: if $\varrho, \tau, \sigma \in {}^\omega\omega$ are such that $\varrho = \sigma \circ \tau$ then the equation $\mathsf{s}_\varrho(x) = \mathsf{s}_\sigma \mathsf{s}_\tau(x)$ is an instance of (P_7). Here ϱ, σ and τ can be treated as "variables" ranging over the "names" of certain polyadic operations; and, at the same time, they denote functions. So the names of polyadic operations have a structure, and one can use this structure to describe a general equational schema which applies for many polyadic operations. This motivates the next three definitions (originally introduced in Németi–Sági [Nem-Sag,00]; see also [Sag,99a]).

Definition 2.2.6.

(i) Let L be a first order language containing countable many unary function symbols $f_0, f_1, \ldots$, countably many unary relation symbols

$r_0, r_1, \dots$ and countably many constant symbols $n_0, n_1, \dots$ (and nothing more).

(ii) Let L_{PT} be the similarity type (in the algebraic sense) of Boolean algebras endowed with unary operation symbols $\mathsf{s}_{f_0}, \mathsf{s}_{f_1}, \dots, \mathsf{c}_{r_0}, \mathsf{c}_{r_1}, \dots$ and constant symbols $\mathsf{d}_{n_0 n_0}, \mathsf{d}_{n_0 n_1}, \dots$. Here the indices of the symbols c, s and d are the same as the corresponding symbols in L.

(iii) By a *Halmos schema* we mean a pair $\langle s, e \rangle$ where s is a first order sentence of L and e is an equation of L_{PT}.

In order to keep notation closer to intuition, we will write $s \Rightarrow e$ in place of $\langle s, e \rangle$.

Definition 2.2.7. Let $s \Rightarrow e$ be a Halmos schema and let g be an equation in the language of PEA_α. Then g is defined to be an (α dimensional) instance of $s \Rightarrow e$ iff there are

$$f_0^M, f_1^M, \dots \in {}^\alpha\alpha, \quad r_0^M, r_1^M, \dots \subseteq \alpha \quad \text{and} \quad n_0^M, n_1^M, \dots \in \alpha \quad \text{such that}$$

$$\left\langle \alpha; r_0^M, r_1^M, \dots, f_0^M, f_1^M, \dots, n_0^M, n_1^M, \dots \right\rangle \models s$$

and g can be obtained from e by replacing r_i, f_i and n_i by r_i^M, f_i^M and n_i^M, respectively.

For example, the set of all instances of (P_7) coincides with the set of all instances of the Halmos schema

$$(\forall v)\big(f_0(v)\big) = f_1\big(f_2(v)\big) \quad \Rightarrow \quad \mathsf{s}_{f_1}\mathsf{s}_{f_2}(x) = \mathsf{s}_{f_0}(x).$$

It is easy to see, that all elements of (P_0)–(E_3) can be expressed by a suitable Halmos schema in this sense.

The use of set theoretic structure of the names of polyadic operations makes the axiom system (P_0)–(E_3) so elegant (and, as we will see, also the structure of the names of operations makes the equational theory of Rppe_ω so complicated).

Definition 2.2.8. Let $\mathfrak{A} \in \mathsf{PEA}_\alpha$. A Halmos schema is defined to be *valid* in $\mathfrak{A}$ iff every α dimensional instance of it is valid in $\mathfrak{A}$. A Halmos scehma is *valid* in a class of algebras iff it is valid in all elements of the class.

Let PEA_α^+ be the class of all models of all α-dimensional instances of Halmos schemas valid in Rppe_α. Note, that PEA_α^+ is the smallest variety containing Rppe_α and defined by Halmos schemas.

Now we are ready to state our non-axiomatizability result.

Theorem 2.2.9 (Németi, Sági). $\mathsf{PEA}_\omega^+ \neq \mathbf{HSP}\,\mathsf{Rppe}_\omega$. *That is, the equational theory of* Rppe_ω *is not axiomatizable by Halmos schemas.*

The proof can be found in [Nem-Sag,00], see also [Sag,99a]. ∎

In these papers it is shown, that there is an equation e_{CM} valid in Rppe_ω but not in PEA_ω^+. Although it is rather complicated, e_{CM} is explicitly given.

Next, one could try to axiomatize Rppe_ω by some kind of equation-schemas different from Halmos schemas. If Rppe_ω would be finitely axiomatizable by some kind of schemas Σ, then, as we will see in Section 4, the equational consequences of Σ would form a Π^1_1-hard set (in the recursion theoretic sense). Since finite (schema) axiomatizability of a theory usually implies recursive enumerability, we can conclude, that Rppe_ω cannot be finitely axiomatized by any kind of "reasonable" schemas.

After these negative results it is natural to ask, what happens, if one takes small subreducts of PEA_ω. The analogue of Theorem 2.2.2 remains true for RQPEA_α, $\alpha > 2$, as well.

Theorem 2.2.10 (Sain, Thomson). *For* $\alpha > 2$, *the class* RQPEA_α *of representable quasi-polyadic equality algebras of dimension* α *cannot be axiomatized by finitely many Monk type schemas.*

The proof can also be found in [Sai-Tho,91]. Of course, by Theorem 2.2.10, the class of representable quasi-polyadic equality algebras cannot be axiomatized by a finite set of equations.

2.3. Representation theory, the diagonal case, positive results

We can obtain positive results when we do not insist on square representations, and commutativity of cylindrifications, see [Fer,thisVol,b], but the problem is really hard when we seek classical representations. We start this subsection by a positive result due to I. Sain which is in a sharp contrast of the non-finite axiomatizability results presented so far and it has a definite knowledge theoretical significance.

Theorem 2.2.11 (Sain). *There is a finite reduct L of the language of* PEA_ω *such that*

(i) *all the* CA_ω*-operations are term-definable in L and*

(ii) *the class of L-subreduct of* Rppe_ω *is a finitely axiomatizable variety.*

The proof can be found in [Sai,00] and it has also been based on semigroup theoretic investigations. At the technical level, the cornerstone was to find a finitely generated, finitely presented subsemigroup of ${}^\omega\omega$ with further nice properties. We emphasize, that in Theorem 2.2.11 the set Σ of equations axiomatizing the representable algebras is not only described by a finite set of equational schemas – Σ itself is a finite set of equations.

Next we give a sufficient condition which implies representability of a QPEA_ω. Let $\mathfrak{A} \in \mathsf{PA}_\alpha$. As usual, the *dimension set* $\Delta(a)$ of $a \in A$ is defined to be $\Delta(a) = \{ i \in \alpha : \mathsf{c}_i(a) \neq a \}$. In addition, $\mathfrak{A}$ is defined to be *locally finite-dimensional* iff every $a \in A$ has a finite-dimension set. A PEA_α is defined to be locally finite-dimensional iff its PA_α-reduct is locally finite-dimensional.

Our goal is to show, that every locally finite-dimensional QPEA_ω is representable. Although this is a classical result, for completeness we include here a proof, because other classical representation theorems for locally finite dimensional cylindric algebras can be quickly derived from this one. To present the proof, we need further preparations.

Remark 2.2.12. We recall a method of constructing homomorphisms from certain reducts of PEA_α into (relativized) set algebras. The idea comes from Andréka–Németi [And-Nem,75] (see also Remark 3.2.9 of [Hen-Mon-Tar,85]) where it was developed for locally finite-dimensional cylindric algebras. Below we adapt the method to certain elements of PEA_α.

Let α be any set, $\Gamma \subseteq \mathcal{P}(\alpha)$ and $\Lambda \subseteq {}^\alpha\alpha$. Let

$$\mathfrak{A} = \left\langle A; \cdot, -, 0, 1, \mathsf{c}_{(\gamma)}, \mathsf{d}_{ij}, \mathsf{s}_\tau \right\rangle_{\gamma\in\Gamma,\ i,j\in\alpha,\ \tau\in\Lambda}$$

be a reduct of a PEA_α. Let F be any ultrafilter on $\mathfrak{A}$. Then the kernel $\ker(F)$ of F is defined to be

$$\ker(F) = \{ \langle i, j\rangle \in \omega \times \omega : \mathsf{d}_{ij} \in F \}.$$

It is easy to see, that $\ker(F)$ is an equivalence relation. For any $\tau \in {}^\alpha\alpha$ we will denote by τ/E the function satisfying $\tau/E(i) = \tau(i)/E$ for every $i \in \alpha$. Finally, for each $a \in A$ let

$$\mathrm{rep}_F(a) = \{ \tau/E : \tau \in \Lambda, \mathsf{s}_\tau(a) \in F \}.$$

Our aim is to show, that if $\mathfrak{A}$ is locally finite dimensional, then rep_F is a QPEA_α-homomorphism for some carefully chosen F. To do so, we still need some further preparations.

Lemma 2.2.13. *Let $\mathfrak{A}$ be a locally finite-dimenisonal QPEA_α and let F be an ultrafilter over I. Then*

(i) *The set $A_{lf} := \{a \in {}^I A/F : \Delta(a) \text{ is finite}\}$ is closed under the QPEA_α-operations.*

(ii) *If $\mathfrak{A}$ is an α-dimensional quasi-polyadic equality set algebra, then the QPEA_α generated by A_{lf} is isomorphic to an α-dimensional quasi-polyadic equality set algebra.*

Proof. To see (i), let $a, b \in A_{lf}$ and let $\mathsf{c}_{(\Gamma)}, \mathsf{s}_\tau$ be QPEA_α-operations. Then it is easy to see, that

$$\Delta(a \cdot b) \subseteq \Delta(a) \cup \Delta(b);$$

$$\Delta(\mathsf{d}_{ij}) \subseteq \{i, j\} \quad \text{for every} \quad i, j \in \alpha;$$

$$\Delta\big(\mathsf{c}_{(\Gamma)}(a)\big) \subseteq \Delta(a) \quad \text{and}$$

$$\Delta\big(\mathsf{s}_\tau(a)\big) \subseteq \tau^{-1}\big[\Delta(a)\big].$$

The right hand side is finite in all cases (for the last case we note, that $\tau^{-1}\big[\Delta(a)\big]$ is finite because $\{k \in \alpha : \tau(k) \neq k\}$ is finite).

The idea of the proof of (ii) is similar to that of Theorem 2.2.4(i). Assume, that the base set of $\mathfrak{A}$ is U. Define $\varphi : A_{lf} \to \mathcal{P}\big({}^\alpha({}^I U/F)\big)$ to be

$$\varphi(a) = \Big\{\langle \mathsf{s}_k/F : k \in \alpha\rangle \in {}^\alpha\big({}^I U/F\big) : \\ \big\{j \in I : \big\langle \mathsf{s}_k(j) : k \in \alpha\big\rangle \in a_j\big\} \in F\Big\}$$

where $a = \langle a_j : j \in I\rangle/F$. It is easy to check, that φ is an embedding from the QPEA_α generated by A_{lf} into the full α-dimensional quasi-polyadic equality set algebra on the base set ${}^I U/F$. ■

Definition 2.2.14. Let $\mathfrak{A}$ be a Boolean algebra (possibly with extra operations). The set of ultrafilters of $\mathfrak{A}$ will be denoted by $\mathcal{U}(\mathfrak{A})$. For any $a \in A$ we define N_a to be

$$N_a = \{F \in \mathcal{U}(\mathfrak{A}) : a \in F\}.$$

Remark 2.2.15. The following facts are well known: $\{N_a : a \in A\}$ is a basis of a topology on $\mathcal{U}(\mathfrak{A})$; we will denote the generated topology by τ. $\mathcal{U}(\mathfrak{A})$ endowed with τ is called the *Stone dual* space of $\mathfrak{A}$ and is denoted by $\mathfrak{A}^*$. It is a compact Hausdorff space.

Definition 2.2.16. Suppose $X \subseteq A$, $F \in \mathcal{U}(\mathfrak{A})$ and $a \in A$ such that $a = \sup(X)$. Then we say, that F *preserves* X iff

$$a \in F \quad \Rightarrow \quad (\exists b \in X)(b \in F).$$

Note, that the converse implication always holds.

Lemma 2.2.17. *Suppose $X \subseteq A$ and $a \in A$ such that $a = \sup(X)$. Then*

$$\mathcal{U}_X := \{F \in \mathcal{U}(\mathfrak{A}) : F \text{ does not preserve } X\}$$

is nowhere dense in $\mathfrak{A}^*$.

Proof. Let G be a nonempty open set of $\mathfrak{A}^*$. By shrinking it, if necessary, we may assume that G is basic open, that is, $\mathsf{G} = N_b$ for some $0 \neq b \in A$. It is enough to show that there exists $0 < c < b$ such that $N_c \cap \mathcal{U}_X = \emptyset$. To do so, we will distinguish two cases.

Case 1: $b \cdot (-a) \neq 0$. In this case $c = b \cdot (-a)$ is suitable.

Case 2: $b \cdot (-a) = 0$. In this case $b \leq a$. Assume, seeking a contradiction, that for every $x \in X$ we have $b \cdot x = 0$. It follows, that $-b$ is an upper bound for X and hence $a \leq -b$. Consequently, $b \leq a \leq -b$, so $b = 0$, a contradiction.

By the previous paragraph, there exists $x \in X$ with $b \cdot x \neq 0$. Then, for $c = b \cdot x$ we have $N_c \cap \mathcal{U}_X = \emptyset$, as desired. ∎

Lemma 2.2.18. *Let $\mathfrak{A} \in \mathsf{QPEA}_\omega$ be countable and locally finite-dimensional. Let $a \in A - \{0\}$. For each $i \in \omega$ let $X_i \subseteq A$ and $b_i \in A$ be such that $b_i = \sup(X_i)$. Then there exists $F \in \mathcal{U}(\mathfrak{A})$ such that F preserves all X_i, $i \in \omega$, moreover, $a \in F$ and every equivalence class of $\ker(F)$ is infinite.*

Proof. Let $\delta : \omega \to \omega$ be a function such that for every $k \in \omega$ the set $\{n \in \omega : \delta(n) = k\}$ is infinite. In addition, let $B_i := \{F \in \mathcal{U}(\mathfrak{A}) : F$ does not preserve $X_i\}$. We will modify the standard proof of the Baire Category theorem. More concretely, by recursion we will define a sequence of elements $\langle a_n : n \in \omega \rangle$ of $\mathfrak{A}$ such that the following hold for all $n, m \in \omega$:

(a) $a_0 = a$ and $a_n \neq 0$;

(b) if $n < m$ then $a_m \leq a_n$;

(c) $N_{a_{n+1}} \cap B_n = \emptyset$;

(d) $(\exists k \in \omega) n \leq k$, $a_{n+1} \leq \mathsf{d}_{\delta(n)k}$.

Let $a_0 = a$; then (a)–(d) are clearly true. Next, suppose, that $n \in \omega$ and a_m has already been defined for all $m < n$. Then $\Delta(a_{n-1})$ is finite, hence exists $k \in \omega$ with $k \geq n - 1$ and $k \notin \Delta(a_{n-1})$. Then

$$\begin{aligned} & \mathsf{c}_k\big(a_{n-1} \cdot \mathsf{d}_{\delta(n-1)k}\big) \\ &= \mathsf{c}_k\big(\mathsf{c}_k(a_{n-1}) \cdot \mathsf{d}_{\delta(n-1)k}\big) \\ &= \mathsf{c}_k(a_{n-1}) \cdot \mathsf{c}_k\big(\mathsf{d}_{\delta(n-1)k}\big) \\ &= a_{n-1} \cdot 1 \\ &= a_{n-1}, \end{aligned}$$

hence by (a), $a_{n-1} \cdot \mathsf{d}_{\delta(n-1)k} \neq 0$. By Lemma 2.2.17, the set B_{n-1} is nowhere dense, hence there exists a nonzero $a_n \in A$ with $a_n \leq a_{n-1} \cdot \mathsf{d}_{\delta(n-1)k}$ such that $N_{a_n} \cap B_{n-1} = \emptyset$. Clearly, (a)-(d) remains true. In this way, the sequence $\langle a_n, n \in \omega \rangle$ can be completely defined.

Combining (a) and (b), one obtains, that $\{N_{a_n} : n \in \omega\}$ has the finite intersection property. Since $\mathfrak{A}^*$ is a compact space, it follows, that there exists $F \in \bigcap_{n \in \omega} N_{a_n}$. Then by (c), F preserves X_i, for every $i \in \omega$. In addition, by (a), we have $a \in F$. Finally, let $i, m \in \omega$ be arbitrary. We will show, that there exists $k \geq m$ such that $\mathsf{d}_{ik} \in F$. Let $n \in \omega$ be such that $n \geq m$ and $i = \delta(n)$. Then by (d), there exists $k \in \omega$ such that $m \leq n \leq k$ and $a_{n+1} \leq \mathsf{d}_{\delta(n)k}$ Therefore $\mathsf{d}_{\delta(n)k} = \mathsf{d}_{ik} \in F$, as desired. It follows, that $i/\ker(F)$ is unbounded in ω. ■

Theorem 2.2.19. *Let $\mathfrak{A} \in \mathsf{QPEA}_\omega$ be locally finite-dimensional. Then*

(i) *For each $0 \neq a \in A$ there exist an α-dimensional quasi-polyadic equality set algebra $\mathfrak{B}_a$ and a homomorphism $\varphi_a : \mathfrak{A} \to \mathfrak{B}_a$ such that $\varphi_a(a) \neq 0$.*

(ii) *$\mathfrak{A}$ is representable.*

Proof. Let $0 \neq a \in A$ be fixed. Let $\mathfrak{A}_0$ be a countable elementary substructure of $\mathfrak{A}$ containing a. For any $\tau \in {}^\omega\omega$ define $\mathsf{s}_\tau : A \to A$ to be $\mathsf{s}_\tau(x) = \mathsf{s}_{\tau'}(x)$, where $\tau|_{\Delta(x)} = \tau'|_{\Delta(x)}$ and $\tau'|_{\omega-\Delta(x)}$ is the identity function. This is meaningful, since $\mathsf{s}_{\tau'}$ is a quasi-polyadic operation. Throughout this proof, we assume that s_τ is a basic operation of $\mathfrak{A}_0$ for every $\tau \in {}^\omega\omega$.

Now we turn to the proof of (i). For each $i \in \omega$ and $b \in A_0$ let $X_{i,b} = \{\mathsf{s}_{[i/j]}(b) : j \in \omega - \Delta(b)\}$. Then, by item 1.11.6(i) of [Hen-Mon-Tar,85] we have $b = \sup(X_{i,b})$. By Lemma 2.2.18 there exists an ultrafilter $F \in N_a$ preserving every $X_{i,b}$ such that every equivalence class of $\ker(F)$ is infinite. Clearly, $\mathrm{rep}_F(a) \neq \emptyset$.

We claim, that, in fact, rep_F is a homomorphism. Here is a sketch for a proof of this claim. One can verify by a straightforward computation, that rep_F preserves $\cdot$ and d_{ij} for any $i, j \in \omega$. Next, one can show, that if $\tau, \tau' \in {}^\omega\omega$ are such that, for any $i \in \omega$ we have $\langle \tau(i), \tau'(i) \rangle \in \ker(F)$ then for any $x \in A_0$

$$(**) \qquad \mathsf{s}_\tau(x) \in F \quad \text{iff} \quad \mathsf{s}_{\tau'}(x) \in F,$$

this may be established with an induction on $n := \left|\Delta(x) \cap \{i \in \omega : \tau(i) \neq \tau'(i)\}\right|$. Then $(**)$ implies, that rep_F preserves complementation and all the s_σ. Finally, combining $(**)$ with the fact, that F preserves each $X_{i,b}$ and using, that each equivalence class of $\ker(F)$ is infinite, one obtains, that rep_F preserves c_i, for all $i \in \omega$.

Thus, rep_F is a homomorphism from $\mathfrak{A}_0$ into some $\mathsf{C}_a \in \mathsf{Pse}_\omega$. Now let $\mathcal{U}$ be an $|A|$-regular ultrafilter[1] and let $\mathsf{D}_a = \left({}^I\mathfrak{A}_0/\mathcal{U}\right)_{lf}$. Then $\mathfrak{A}$ can be embedded into D_a. Let $\mathfrak{B}_a = \left({}^I\mathsf{C}_a/\mathcal{U}\right)_{lf}$. By Lemma 2.2.13, $\mathfrak{B}_a \in \mathsf{Pse}_\omega$; it is also a homomorphic image of D_a. Hence, there exists a homomorphism φ_a from $\mathfrak{A}$ into $\mathfrak{B}_a$ mapping a to a nonzero element, as desired.

Now we turn to prove (ii). By (i), for each $0 \neq a \in A$ there exist an α-dimensional quasi-polyadic equality set algebra $\mathfrak{B}_a$ and a homomorphism $\varphi_a : \mathfrak{A} \to \mathfrak{B}_a$ with $\varphi_a(a) \neq 0$. Define $\varphi : \mathfrak{A} \to \Pi_{0\neq a\in A}\mathfrak{B}_a$ to be $\varphi(x) = \langle \varphi_a(x), a \in A - \{0\} \rangle$. Then φ is the desired embedding. ■

[1]For the definition and basic properties of regular ultrafilters we refer to [Cha-Kei,90].

3. Connections with Cylindric and Quasi-polyadic Algebeas

In this section we are comparing PEA_α with CA_α and QPEA_α. Clearly, every PEA_α has a CA_α-type and a QPEA_α-type reduct. In addition, the following facts are true:

(i) The Df_α-type reduct of a PA_α is a Df_α;

(ii) The CA_α-type reduct of a PEA_α is a CA_α. In addition, $\mathsf{s}_{[i/j]}(x) = \mathsf{c}_i(\mathsf{d}_{ij} \cdot x)$;

(iii) If $\beta \geq \omega$, $\mathfrak{A} \in \mathsf{PEA}_\beta$ then the CA_β-type reduct of $\mathfrak{A}$ is a representable CA_β.

For the proofs, see Theorem 5.4.3 and Corollary 5.4.18 of [Hen-Mon-Tar,85].

Remark 2.3.1. We note, that the converse of the above (ii) is not true: there exists an $\mathfrak{A} \in \mathsf{CA}_\alpha$ which cannot be obtained as the cylindric reduct of a suitable PEA_α. Indeed, as was shown in Section 5.4 of [Hen-Mon-Tar,85], every $\mathfrak{A} \in \mathsf{PEA}_\omega$ satisfies the marry-go-round properties, but there exists a cylindric algebra $\mathfrak{B}$, which does not satisfies these properties. The same argument shows, that there exists a CA_ω which cannot be embedded into the cylindric reduct of a QPEA_ω (because every QPEA_ω also satisfies the marry-go-round properties). This supports the view, that quasi-polyadic equality algebras are "between" cylindric and polyadic algebras.

After Remark 2.3.1, the next natural question is: if a cylindric algebra $\mathfrak{A}$ satisfies the marry-go-round properties (i.e. $\mathfrak{A} \in \mathsf{CA}_\alpha^+$) then does it follow, that $\mathfrak{A}$ is isomorphic to the CA_α-type reduct of a suitable $\mathfrak{B} \in \mathsf{QPEA}_\alpha$? The answer is negative:

Theorem 2.3.2 (Sayed Ahmed). *There exists $\mathfrak{A} \in \mathsf{RCA}_\omega$ such that $\mathfrak{A}$ is not isomorphic to the CA_ω-type reduct of any $\mathfrak{B} \in \mathsf{QPEA}_\omega$. (Since $\mathfrak{A}$ is representable, it obviously satisfies the marry-go-round properties).*

For the proof and further details, see [Say,09a]. On the other hand, in [Fer,07a], Ferenczi proved the following.

Theorem 2.3.3. *There is a weakening QPEA_α^- of the axioms of QPEA_α such that if $\mathfrak{A} \in \mathsf{CA}_\alpha^+$ (i.e. $\mathfrak{A}$ is a CA_α satisfying the merry-go-round properties) then $\mathfrak{A}$ is isomorphic to the CA_α-type reduct of a suitable $\mathfrak{B} \in \mathsf{QPEA}_\alpha^-$.*

It is also natural to search subclasses of CA_α (or, subclasses of CA_α^+) whose elements can be obtained as CA_α-type reducts of certain QPEA_α. Of course, Lf_ω, Dc_ω and the class included in [Hen-Mon-Tar,85], Theorem 3.2.52 are such classes. The following result is a generalization of this latter theorem from CA_α to CA_α^+ due to Ferenczi (see, [Fer,07b], Theorem 3.5).

An algebra $\mathfrak{A}$ in FPEA_α is said to be $\overline{\mathrm{R}}$-representable if $\mathfrak{Rd}_{ca}\,\mathfrak{A}$ is isomorphic to some $\mathfrak{A}$'$\in \mathsf{Crs}_\alpha \cap \mathsf{CA}_\alpha$ and $\mathfrak{A}$' endowed with the the natural transposition $[i,j]$ as the operation p_{ij} is in FPEA_α. The class of the $\overline{\mathsf{R}}$-representable algebras is denoted by $\overline{\mathsf{R}}\mathsf{FPEA}_\alpha$.

An algebra $\mathfrak{A}$ in CA_α *can be supplemented* to an algebra $\widetilde{\mathfrak{A}}$ in FPEA_α if supplementing $\mathfrak{A}$ by the usual substitution operators s_j^i and by certain operators p_{ij} $(i,j<\alpha)$ the algebra obtained is in FPEA_α.

The following theorem is Theorem 3.5 in [Fer,07b].

Theorem 2.3.4. *Suppose that $\mathfrak{A} \in \mathsf{CA}_\alpha^+$ and $\mathfrak{A} = \mathfrak{Nr}_\alpha\,\mathfrak{B}$ for some $\mathfrak{B} \in \mathsf{CA}_{\alpha+1}^+$ $(\alpha \geq 4)$. Then $\mathfrak{A}$ can be supplemented to an algebra $\widetilde{\mathfrak{A}}$ so that $\widetilde{\mathfrak{A}} \in \overline{\mathsf{R}}\mathsf{FPEA}_\alpha$ and $\mathsf{p}_{ij}a = {}_ks(i,j)a$ if $a \in A$, for any fixed $k \notin \Delta a$ and for $k=\alpha$ if $\Delta a = \alpha$.*

Proof. As a consequence of the merry-go-round properties, the condition $\mathfrak{B} \in \mathsf{CA}_{\alpha+1}^+$ implies that the operator $\mathsf{p}_{ij} = {}_ks(i,j)$ for any $a \in B$, $k \notin \Delta a$, $k \leq \alpha$ satisfies axioms (F_6), (F_7), (F_8), (F_9). By [Hen-Mon-Tar,85], Theorem 1.5.15 $\mathsf{s}_k(i,j)$ does depend on k if $k \notin \Delta a$. Further, if $a \in A$, then ${}_ks(i,j)a \in A$ for $k \leq \alpha$. This latter statement is true by definition if $k \notin \Delta a$ and by $\mathfrak{A} = \mathfrak{Nr}_\alpha\,\mathfrak{B}$ if $\Delta a = \alpha$, $k = \alpha$. So $\mathfrak{A}$ can be supplemented to an algebra $\widetilde{\mathfrak{A}}$ in FPEA_α with $\mathsf{p}_{ij}a = {}_ks(i,j)a$, $a \in A$ *for any* $k \notin \Delta a$ and for $k=\alpha$ if $\Delta a = \alpha$.

We claim that $\widetilde{\mathfrak{A}}$ is $\overline{\mathrm{R}}$-representable. By the Resek–Thompson theorem $\mathfrak{B}$ is representable by an algebra $\mathfrak{B}^* \in \mathsf{Crs}_{\alpha+1} \cap \mathsf{CA}_{\alpha+1}$. By [Hen-Mon-Tar,85] 3.1.125 $\mathfrak{Nr}_\alpha\,\mathfrak{B}^*$ is in $\mathsf{Crs}_\alpha \cap \mathsf{CA}_\alpha$. Let us denote $\mathfrak{Nr}_\alpha\,\mathfrak{B}^*$ by $\mathfrak{A}$'. Obviously, $\mathfrak{A}$ and $\mathfrak{A}$' are isomorphic, let us denote by a' the element in A' corresponding to $a, a \in A$.

Considering the isomorphism between $\mathfrak{B}$ and $\mathfrak{B}^*$ and the fact that ${}_ks(i,j)$ can be expressed by cylindrifications and diagonals included in $\mathfrak{B}$, $\mathfrak{A}$' can also be supplemented to an algebra $\widetilde{\mathfrak{A}'}$ in FPEA_α with

$$\mathsf{P}_{ij}a' = {}_k\mathsf{S}^V(i,j)a' \tag{2.3.1}$$

for any $a \in A$, $k \notin \Delta a$ and $k=\alpha$ if $\Delta a = \alpha$, where V is the unit of $\mathfrak{A}'$ (so $\mathfrak{A}'$ is closed under ${}_k\mathsf{S}^V(i,j)$, too) and $\widetilde{\mathfrak{A}} \simeq \widetilde{\mathfrak{A}'}$.

The operator ${}_k\mathsf{S}^V(i,j)$ is defined in terms of cylindrifications an diagonals, therefore by the Lemma 3.2 in [Fer,07b]

$$(2.3.2) \qquad {}_k\mathsf{S}^V(i,j)a' = [i,j]a'$$

if $a \in A$. Comparing (2.3.1) and (2.3.2) we get that $\mathfrak{A}' \in \overline{\mathsf{R}}\mathsf{FPEA}_\alpha$. The proof is completed. ■

Moreover, Andréka and Németi proved in [And-Nem,84] that for $4 \leq \alpha < \omega$ there exists a nonrepresentable $\mathfrak{A} \in \mathsf{PEA}_\alpha$ such that its CA_α-type reduct is representable (as a cylindric algebra).

We close this section by recalling the infinite dimensional analogs of this result. As we have seen in Remark 2.2.5 above, there exists a nonrepresentable PEA_ω whose CA_ω-type reduct is representable. Moreover, by a recent result of T. Sayed Ahmed [And-Nem-Say,12], there exists a nonrepresentable QPEA_ω with a representable CA_ω-type reduct. Related investigations can also be found in Sági [Sag,11].

4. Complexity of the Equational Theories of Certain Classes of Polyadic Algebras

In this section we study the recursion theoretic complexity of the equational theories of polyadic algebras of dimension α. We start by the finite dimensional case.

Theorem 2.4.1. *For $3 \leq \alpha < \omega$ the equational theory of Rppe_α is undecidable.*

This theorem may be derived from the analogous result for cylindric algebras, see e.g. [Hen-Mon-Tar,85].

As we mentioned, if α is infinite, then the langauge of PA_α contains continuum many operation symbols, hence, the equational theory of Rppe_α is not recursively enumerable for trivial reasons. We will see below, that the situation remains the same, if we study "rich enough" finite reducts. Again, the rest of this section is divided into two parts: first we will deal with polyadic algebras without diagonal elements and then with polyadic equality algebras.

4.1. Complexity of equational theories, the diagonal-free case

After the Daigneault–Monk Theorem one could think, that if L is any finite reduct of the language of PA_ω, then the set of equations written in L and valid in RPA_ω forms a recursively enumerable set. Indeed, usually, representation theorems imply completeness theorems, and completeness theorems usually imply recursive enumerability. It turned out, that this commonsense reasoning breaks down in the case of RPA_ω.

Theorem 2.4.2 (Sági). *There is a finite reduct L of the language of PA_ω such that the set of equational consequences of* (P_0)–(P_{11}) *written in L is not recursively enumerable.*

The proof can be found in [Sag,01], see also [Sag,99a].

There are some positive results, as well.

Theorem 2.4.3 (Sain–Gyuris). *There is a finite reduct L of the language of PA_ω such that all the CA_ω operations are term definable in L and the variety generated by the L-reducts of RPA_ω can be axiomatized by a recursive set Σ of equations. In addition, although Σ is infinite, it may be described by finitely many schemas.*

We note, that the schemas occurring in Theorem 2.4.3 are essentially simpler than Halmos schemas in general. The proof of Theorem 2.4.3 can be found in [Sai-Gyu,97].

4.2. Complexity of equational theories, the diagonal case

As we have seen in Section 2, the equational theory of Rppe_ω is rather complex in the "axiomatic sense", that is, there is no way to axiomatize it by Halmos schemas. In this section we survey some results on the recursion theoretic complexity of the equational theory of Rppe_ω.

We start by recalling some notions from recursion theory. Throughout $\mathcal{N}$ denotes the standard model of number theory. By a Π^1_1 formula we mean a second order formula in prenex form in which every second order variable is quantified universally. A set A of natural numbers is defined to be

- arithmetical iff A is definable in $\mathcal{N}$ by a first order formula of arithmetic;

- A is called a Π^1_1-set iff it is definable by a Π^1_1-formula of arithmetic.

As it is well known, recursively enumerable sets and their complements are arithmetical sets, and the family of arithmetical sets contains sets much more complicated than any recursively enumerable set. Clearly, arithmetical sets are Π^1_1, as well.

Theorem 2.4.4 (Németi–Sági). *There is a strictly finite reduct L of the language of Rppe_ω and a recursive function* tr *mapping Π^1_1 formulas of arithmetic to equations of L such that for any Π^1_1 sentence σ*

$$\mathcal{N} \models \sigma \quad \textit{iff} \quad \mathsf{Rppe}_\omega \models \mathrm{tr}\,(\sigma).$$

The proof can be found in [Nem-Sag,00], see also [Sag,99a].

Theorem 2.4.4 may be interpreted as follows: there is a finite reduct L of the language of Rppe_ω such that the set P of (Gödel numbers of) equations written in L and valid in Rppe_ω is at least as complicated as the set S of (Gödel numbers of) Π^1_1 formulas of arithmetic true in $\mathcal{N}$. By a (version of) Tarski's theorem of undefinability of truth, S is not Π^1_1. Hence P is not Π^1_1, as well. Consequently, $Eq(\mathsf{Rppe}_\omega)$ cannot be axiomatized by any kind of finite equational schemas Σ whose set of consequences is recursively enumerable (or at least Π^1_1).

Although Theorems 2.4.2 and 2.4.4 have very strong consequences, the reducts L in them are rather artificial: the indices of the substitution operations are carefully chosen, tricky recursive functions on ω. Hence, it is natural to ask what can be said about "more natural" reducts of Rppe_ω. The following theorem is due to R. McKenzie and it shows, that even, some "natural reducts" of Rppe_ω may have a complicated equational theory.

Recall, that pred, suc : $\omega \to \omega$ are the functions defined by

$$\mathrm{pred}\,(0) = 0, \quad \mathrm{pred}\,(n+1) = n \quad \text{and} \quad \mathrm{suc}\,(n) = n+1$$

for every $n \in \omega$.

Theorem 2.4.5 (McKenzie). *Let L be any countable reduct of the language of Rppe_ω containing the set of operations $\big\{\cdot, \sim, \mathsf{c}_{(\omega)}, \mathsf{s}_{\mathrm{suc}}, \mathsf{s}_{\mathrm{pred}}, \mathsf{s}_{[0,1]}, \mathsf{s}_{[i/j]}, \mathsf{d}_{ij} \,:\, i,j \in \omega\big\}$. Then the set of equations valid in Rppe_ω and written in L is not recursively enumerable.*

For the details, see Chapter 11 of [Cra,74].

VI. Connections with Abstract Algebraic Logic and Universal Logic

BOLYAI SOCIETY
MATHEMATICAL STUDIES, 22

Cylindric-like Algebras and Algebraic Logic
pp. 393–419.

Definability Issues in Universal Logic

ILDIKÓ SAIN*

Besides Universal Logic (J-Y. Béziau [Bez,10], [Bez,05]) we focus on its predecessor Universal Algebraic Logic going back to the 70's ([And-Ger-Nem,77], [And-Ger-Nem,73]) and originating with Tarski's school, cf. [Hen-Mon-Tar,85, Part II, Sec. 5.6, p. 255].

In the course of the development of logic, there have been developed a great number of various logical systems, e.g. propositional logic, classical first order logic and its variants (like finite-variable fragments of it or its rank-free version), many versions of modal- and multimodal logic, to mention just some of the most traditional systems.

As described e.g. in Goguen–Burstall [Gog-Bur,92], starting from the 60-s of the 20th century, the development of theoretical computer science also brought to the light a huge number of further logical systems (e.g. logics of programs, lambda calculus). When investigating this group of logical systems, it became clear, because of the applications in mind, that the *semantical aspects* of the logical systems must not be ignored.

After a while it became apparent that, when checking some logical properties of these logical systems (from now on "logics", for short), certain patterns of ideas, concepts, proofs kept being repeated with only slight differences. It was time to develop appropriate abstract levels of the subject. Several schools have been formed (like Abstract Model Theory, Institutions Theory, Universal Logic and others). Cf. also the recent success story of Universal Logic [Bez,10], Béziau's school, [And-Nem-Sai,01], or [Nem-And,94] for growing importance of this direction.

Some of these schools benefited from using universal algebraic methods. The most outstanding of these schools was led by Alfred Tarski. First they

*Thanks are due to an anonymous referee, Hajnal Andréka, and István Németi for their careful reading and useful suggestions. Also thanks are due to Miklós Ferenczi and Dezső Miklós for strong support in their rôles as editors.

concentrated on the algebraic counterpart of first order logic, developing this way the theories of cylindric-, polyadic- and relation algebras. These studies naturally led to finding the algebraic counterparts of some other logics (e.g. that of first order logic with infinitary conjunction, modal- and multi modal logics). The theories of these classes of algebras have been developed in the style of developing just any class in abstract algebra (like group theory or ring theory). Indeed, in Henkin–Monk–Tarski [Hen-Mon-Tar,85, Part I] the theory of cylindric algebras has been built up in such a fashion. At the same time the logical motivation can also be felt strongly, throughout the monograph.

Some researchers wished to make this feeling more explicit via concretely describing and investigating the process of "turning logics into algebras"; and concentrating on a *two way connection* between the "country" of LOGIC and that of ALGEBRA (*"bridge"*). Already in [Hen-Mon-Tar,85, Part II], Chapter 5.6 has been devoted to this subject.

The ambition here is to find, via a *general* method or algorithm:

(1) the specific class(es) of algebras corresponding to a given logic (e.g., to propositional logic, this class is the class of Boolean algebras);

(2) the algebraic counterparts / characterizations of distinguished properties of logics.

As it is described e.g. in [And-Nem-Sai,01], [Sai-Gyu,97], [Nem-And,94], the point in looking at the algebraic form of a logic is that properties of a logic can be checked via investigating algebraic properties of the class of algebras we had associated to our logic.

A basic example for such a correspondence is the following *equivalence theorem.*

Let **L** denote a logic and let $\mathsf{Alg}\,(\mathbf{L})$ denote the class of universal algebras obtained from **L** according to item (1). Beth definability property can be generalised, from classical first order logic, to our general concept of a logic. Now, under some natural conditions on **L**, we have that

L *has the Beth definability property*

iff

all the epimorphisms are surjective in the category $\mathsf{Alg}\,(\mathbf{L})$ *with the class of objects:* $\mathsf{Alg}\,(\mathbf{L})$, *and with morphisms: all the homomorphisms between them.*[1]

[1]A precise formulation of this theorem will be given in Section 4.

Two groups (among others[2]) devoted much attention to this logic-algebra connection. Willem Blok, Don Pigozzi, and their followers started out from syntactical systems called *deductive systems*, to turn them into algebras, while Hajnal Andréka, István Németi and their followers put a strong emphasis on algebraizing the *semantical* aspects of logics as well.[3] The work of these two schools is compared in a paper by Font and Jansana [Fon-Jan,94], and a short comparison is given also in Andréka–Németi–Sain [And-Nem-Sai,01, Remark 41, p. 199] and [And-Nem-Sai,09, relevant remark in subsection "Defining the framework"]. In [And-Blo-Nem-Pig-Sai,97], the two groups above as co-authors, started incorporating the two points of views into a joint perspective. In the present writing we are concerned with the semantical approach. Recently the subject matter of the present paper is gaining popularity and importance under the name "Universal Logic". It even has a prestigious journal: Logica Universalis, devoted entirely to this subject, see [Bez,10].

We start out from a rather general concept of a logic. The examples for logics mentioned above, and many other logics, are special cases of this general concept. In Section 1 we formulate the definition of this general logic. In Section 2 we show that some distinguished logics are special cases of the general concept introduced in Section 1. In Section 3 we give an algorithm of turning such a general logic into two closely related classes of algebras. In Section 4 we introduce definability concepts in our general framework and recall some distinguished results about them. In Section 5 we give a category theoretic characterisation of weak Beth definability property. Section 6 is concerned with some additional motivation for our overview as well as with new trends.

1. A General Concept of a Logic with Semantics

We start out from a fairly general mathematical framework for logics, the generality of which is in between the lovely "flatland" of concrete logical systems and the highest level (of abstraction) of institutions (see e.g.

[2]We have in mind Institutions Theory, cf. [Mar-Mes,01], the recent success of Universal Logic, to mention a few.

[3]When dealing with strongly complete and sound logics, then, of course, the Blok–Pigozzi approach is giving information about the semantical aspects in an implicit way. But there are questions (like definability) when one has to start right away from semantical systems.

[Mar-Mes,01], [Gog-Bur,92]). In fact, our framework can be considered to be a concrete example of an institution (cf. the end of the present section and [And-Nem-Sai,01, p. 201]). In the present section we sketch our general framework, similarly as in e.g. [And-Nem-Sai,01], [And-Nem-Sai,09], [Blo-Pig,89], [Hoo,01].

Definition 1.1.1 (Logic). By a *logic* $\mathcal{L}$ we mean an ordered quadruple

$$\mathcal{L} \stackrel{\text{def}}{=} \langle F, \vdash, M, \models \rangle, \text{ where}$$

- F (called the set of *formulas* of $\mathcal{L}$) is a set of finite sequences (called *words*) over some set X (called the *alphabet* of $\mathcal{L}$), that is, $F \subseteq X^*$;
- $\vdash$ (called the *provability relation* of $\mathcal{L}$) is a relation between sets of formulas and formulas, that is, $\vdash \subseteq \mathcal{P}(F) \times F$;
- M is a class (called the class of *models* of $\mathcal{L}$);
- $\models$ (called the *validity relation*) is a relation between M and F, that is, $\models \subseteq M \times F$.

Instead of "$\langle \mathfrak{M}, \varphi \rangle \in \models$" and "$\langle \Sigma, \varphi \rangle \in \vdash$" we write, respectively, "$\mathfrak{M} \models \varphi$" and "$\Sigma \vdash \varphi$".

Intuitively, F is the collection of "texts" or "sentences" or "formulas" that can be "said" in the language $\mathcal{L}$. For $\Gamma \subseteq F$ and $\varphi \in F$, the intuitive meaning of $\Gamma \vdash \varphi$ is that φ is provable (or derivable) from Γ with the syntactical inference system (or deductive mechanism) $\vdash$ of $\mathcal{L}$. In all important cases, $\vdash$ is subject to certain (well-known) conditions like $\Gamma \vdash \varphi$ and $\Gamma \cup \{\varphi\} \vdash \psi$ imply $\Gamma \vdash \psi$ for any $\Gamma \subseteq F$ and $\varphi, \psi \in F$. The validity relation $\models$ tells us which texts are "true" in which models under what conditions.

Usually F and $\vdash$ are defined by what is called a grammar in mathematical linguistics. $\langle F, \vdash \rangle$ together with the grammar defining them is called the *syntactical part* of $\mathcal{L}$, while $\langle M, \models \rangle$ is the *semantical part* of $\mathcal{L}$ (cf. e.g. [Gab,94, Def. 14.2.4, p. 359]). It is usual and natural to assume that F is decidable, and that the provability relation $\vdash_0 \stackrel{\text{def}}{=} \big\{ \langle \Gamma, \varphi \rangle \ : \ \Gamma \vdash \varphi, \ \Gamma \subseteq F, \ |\Gamma| < \omega \big\}$ is r.e., but we will not need these here.

As a binary relation between M and F, $\models$ induces a Galois-connection between M and F, and in particular, it defines two closure operators, one on M and one on F. Next we collect some of the relevant definitions.

We extend $\models$ to $\mathcal{P}(M) \times \mathcal{P}(F)$: Let $K \subseteq M$ and $\Sigma \subseteq F$. Then

$$K \models \Sigma \text{ iff } (\forall \mathfrak{M} \in K)(\forall \varphi \in \Sigma)\mathfrak{M} \models \varphi.$$

We will write $K \models \varphi$ in place of $K \models \{\varphi\}$, and $\mathfrak{M} \models \Sigma$ when $K = \{\mathfrak{M}\}$. The *theory* of K is $Th(K) \stackrel{\text{def}}{=} \{\varphi \in F : K \models \varphi\}$, and the *class of models* of Σ is $\mathsf{Mod}\,(\Sigma) \stackrel{\text{def}}{=} \{\mathfrak{M} \in M : \mathfrak{M} \models \Sigma\}$.

$$Th(\mathfrak{M}) \stackrel{\text{def}}{=} Th\left(\{\mathfrak{M}\}\right) \quad \text{and} \quad \mathsf{Mod}\,(\varphi) \stackrel{\text{def}}{=} \mathsf{Mod}\left(\{\varphi\}\right).$$

Let $\Sigma \cup \{\varphi\} \subseteq F$. We say that φ is a *semantical consequence* of Σ, in symbols $\Sigma \models \varphi$, iff $\mathsf{Mod}(\Sigma) \models \varphi$. We say that φ is a *valid formula* of $\mathcal{L}$, in symbols $\models \varphi$, iff $M \models \varphi$. $\mathsf{Mod}\ Th(K)$ is called the *axiomatizable hull* of K.

Provability, or derivability: $\vdash_{\mathcal{L}} \varphi$ iff $\emptyset \vdash_{\mathcal{L}} \varphi$, in this case we say that φ is *derivable*. If $\Sigma \vdash_{\mathcal{L}} \varphi$, then we say that φ is *derivable from* Σ (in $\mathcal{L}$).

$Th\,\mathsf{Mod}$ and $\mathsf{Mod}\ Th$ are the two closure operators induced by $\models$. The semantical consequence relation is a binary relation between $\mathcal{P}(F)$ and F, just like $\vdash$ is. To treat derivability and semantical consequence uniformly, in some papers a logical system is defined to be $\langle F, \models\!\!\!\equiv \rangle$ where $\models\!\!\!\equiv\, \subseteq \mathcal{P}(F) \times F$. E.g. Blok–Pigozzi [Blo-Pig,89] uses this definition. In this notion, $\models\!\!\!\equiv$ can mean either the derivability relation $\vdash$ or the semantical consequence relation $\models$. Tarski–Givant [Tar-Giv,87, p. 16 (§1.6)] uses practically the same definition of a logic as we do, but instead of calling $\mathcal{L}$ a logic, they call $\mathcal{L}$ a "formalism".

Our definition of a logic is too broad for proving interesting theorems about logics. Therefore next we will define a subclass of logics which we will call here *algebraizable (semantical) logics*[4]. Our notion of an algebraizable logic is broad enough to cover the majority of the logics investigated in the literature. On the other hand, it is narrow enough for proving interesting theorems about them, that is, we will be able to establish typical logical facts that hold for most logics studied in the literature.[5]

In items **(A)**–**(D)** below we collect some properties of logics (cf. e.g. [And-Nem-Sai,01]). Here, and throughout the paper, ω denotes the set of all natural numbers.

[4][And-Nem-Sai,94a], [Fon-Jan,94] and several related papers use the phrase *"strongly nice logic"* for what we call an algebraizable semantical logic in this paper.

[5]For completeness, we note that the literature does study logics which are more general than algebraizable, cf. e.g. [Pig,91].

(A) The set F of formulas is usually defined by fixing a set Cn of logical connectives and a set P of atomic formulas, as follows:

(i) Assume that two sets, P and Cn are given, such that every element of Cn has a finite rank. Then $F(P, Cn)$ denotes the smallest set H satisfying

(1) $P \subseteq H$, and

(2) for every $\varphi_1, \ldots, \varphi_k \in H$ and $c \in Cn$ of rank k, $c(\varphi_1, \ldots, \varphi_k) \in H$.

Note that $F(P, Cn)$ is the universe of the word-algebra of type Cn generated by P.

(ii) Let $\mathcal{L}$ be a logic, that is, $\mathcal{L} = \langle F, \vdash, M, \models \rangle$. We say that F is given by $\langle P, Cn \rangle$ if $F = F(P, Cn)$. In this case we say that P is the set of *atomic formulas* of $\mathcal{L}$, and Cn is the set of *logical connectives* of $\mathcal{L}$. The word-algebra generated by P and using the logical connectives of Cn as algebraic operations is denoted by $\mathfrak{F}$, and is called the *formula algebra* of $\mathcal{L}$. Thus $\mathfrak{F} = \langle F, c^{\mathfrak{F}} \rangle_{c \in Cn}$ where $c^{\mathfrak{F}}(\varphi_1, \ldots, \varphi_k) \stackrel{\text{def}}{=} c(\varphi_1, \ldots, \varphi_k) \in F_{\mathcal{L}}$ for all $\varphi_1, \ldots, \varphi_k \in F_{\mathcal{L}}$ and k-ary connective $c \in Cn$.

(iii) We say that $\underline{\mathcal{L}\ \textit{has connectives}}$ if F is given by $\langle P, Cn \rangle$ for some sets P, Cn as above. In this case we assume that $\langle P, Cn \rangle$ is given together with $\mathcal{L}$.

(B) Usually, validity of formulas in models is defined indirectly by first defining the meanings or denotations of formulas in models. The idea is that the meaning of some syntactical entity (like a noun-phrase, or a sentence) need not always be a truth value. Therefore first we define a so-called meaning function which, to each syntactical entity φ and each model $\mathfrak{M}$, associates some semantical entity $mng(\varphi, \mathfrak{M})$ called the meaning of φ in $\mathfrak{M}$. After knowing what the basic syntactical entities mean in the models, one may be able to derive information about which sentences are true or valid in which models.

(i) Let mng be an arbitrary function with domain $F \times M$; and let us call $mng(\varphi, \mathfrak{M})$ *the meaning of* φ *in* $\mathfrak{M}$. For a fixed $\mathfrak{M}$, the function $mng_{\mathfrak{M}}$ mapping F to the set of meanings in $\mathfrak{M}$ is defined by letting for all $\varphi \in F$, $mng_{\mathfrak{M}}(\varphi) \stackrel{\text{def}}{=} mng(\varphi, \mathfrak{M})$. We say that mng is a *meaning function* for $\mathcal{L}$, if the validity of a formula depends only on its meaning, i.e., if

$$mng_{\mathfrak{M}}(\varphi) = mng_{\mathfrak{M}}(\psi) \Longrightarrow (\mathfrak{M} \models \varphi \text{ iff } \mathfrak{M} \models \psi).$$

(ii) Assume that $\mathcal{L}$ has connectives. We say that the meaning function mng is *compositional* if the meanings of formulas are built up from the meanings of their sub-formulas, i.e., if the condition below is satisfied for all $\varphi_i, \psi_i \in F$, $1 \leq i \leq k$ and k-ary connective $c \in Cn$:

$$\bigwedge_{i=1}^{k} mng_{\mathfrak{M}}(\varphi_i) = mng_{\mathfrak{M}}(\psi_i) \Rightarrow$$
$$\Rightarrow mng_{\mathfrak{M}}\big(c(\varphi_1, \ldots, \varphi_k)\big) = mng_{\mathfrak{M}}\big(c(\psi_1, \ldots, \psi_k)\big).$$

This condition says exactly that the kernel $ker\,(mng_{\mathfrak{M}})$ of $mng_{\mathfrak{M}}$

$$\big(\text{where } ker\,(mng_{\mathfrak{M}}) \overset{\text{def}}{=} \{\langle\varphi, \psi\rangle : mng_{\mathfrak{M}}(\varphi) = mng_{\mathfrak{M}}(\psi)\} \subseteq F \times F\big)$$

is a congruence relation, thus $mng_{\mathfrak{M}}$ is a homomorphism on the formula algebra. We say that $\mathcal{L}$ is *compositional* if it has connectives and a compositional meaning function (w.r.t. these connectives). This property is traditionally called *Frege's principle of compositionality.*

(C) In many logics we have a derived connective $\leftrightarrow$ and a formula denoted by *True* which establish a strong connection between mng and $\models$, namely

$$\begin{array}{lll}(1) & mng_{\mathfrak{M}}(\varphi) = mng_{\mathfrak{M}}(\psi) & \text{iff} \quad \mathfrak{M} \models \varphi \leftrightarrow \psi \quad \text{and} \\ (2) & mng_{\mathfrak{M}}(\varphi) = mng_{\mathfrak{M}}(True) & \text{iff} \quad \mathfrak{M} \models \varphi.\end{array}$$

In these logics there is a strong connection between $Th(\mathfrak{M}) \subseteq F$ and $ker\,(mng_{\mathfrak{M}}) \subseteq F \times F$: $mng_{\mathfrak{M}}$ and $Th(\mathfrak{M})$ recapture one another. We will say that $\mathcal{L}$ has the filter-property iff there are derived connectives that generalize the above situation as follows.

(i) Assume that F is given by $\langle P, Cn\rangle$. Let $FV = \{\phi_i : i < \omega\}$ be a fixed set called the set of *formula-variables.* The elements of Fs $\overset{\text{def}}{=} F(FV, Cn)$ are called the *formula-schemes* of $\mathcal{L}$. An *instance* of a formula-scheme is given by substituting formulas for the formula variables in it. A formula-scheme is called *valid* if all of its instances are valid.

(ii) By a k-ary *derived connective* we mean a formula-scheme $\Delta \in$ Fs using the formula variables $\phi_0, \ldots, \phi_{k-1}$ only. If $\varphi_0, \ldots, \varphi_{k-1}$ are formulas, then $\Delta(\varphi_0, \ldots, \varphi_{k-1})$ denotes the instance of Δ when we replace $\phi_0, \ldots, \phi_{k-1}$ by $\varphi_0, \ldots, \varphi_{k-1}$ respectively.

(iii) We say that $\mathcal{L}$ has the *filter-property* iff there are derived connectives $\varepsilon_0, \ldots, \varepsilon_{m-1}$ and $\delta_0, \ldots, \delta_{m-1}$ (unary) and $\Delta_0, \ldots, \Delta_{n-1}$ (binary)

$(m, n \in \omega)$ of $\mathcal{L}$ with the following properties: For all $\varphi, \psi \in F$ and for all $\mathfrak{M} \in M$.

$$(1) \qquad mng_{\mathfrak{M}}(\varphi) = mng_{\mathfrak{M}}(\psi) \iff (\forall i < n)(\mathfrak{M} \models \varphi \Delta_i \psi),$$
$$(2) \ (\forall j < m)\big(mng_{\mathfrak{M}}\left(\varepsilon_j(\varphi)\right) = mng_{\mathfrak{M}}\left(\delta_j(\varphi)\right)\big) \iff \mathfrak{M} \models \varphi.$$

In the example at the beginning of item **(C)**, $m = n = 1$, $\varepsilon = Id$ (the identity), $\delta = True$, $\Delta = \leftrightarrow$.

We will often abbreviate $(\forall i < n)$ $(\mathfrak{M} \models \varphi \Delta_i \psi)$ by $\mathfrak{M} \models \varphi \Delta \psi$ and $(\forall j < m)$ $\big(mng_{\mathfrak{M}}\left(\varepsilon_j(\varphi)\right) = mng_{\mathfrak{M}}\left(\delta_j(\varphi)\right)\big)$ by $mng_{\mathfrak{M}}(\varepsilon\varphi) = mng_{\mathfrak{M}}(\delta\varphi)$. This way, the above items look like:

$$(1) \quad mng_{\mathfrak{M}}(\varphi) = mng_{\mathfrak{M}}(\psi) \iff \mathfrak{M} \models \varphi \Delta \psi,$$
$$(2) \quad mng_{\mathfrak{M}}(\varepsilon\varphi) = mng_{\mathfrak{M}}(\delta\varphi) \iff \mathfrak{M} \models \varphi.$$

(D) Logics have some substitution properties. Usually we expect a logic to be substitutional. If it is not, then we rather treat it as a "theory" of a substitutional logic.

(i) By a *substitution* s we mean a function $s : P \to F$. If $\varphi \in F$ then $\varphi\big(\overline{p}/s(\overline{p})\big)$ denotes the formula we obtain from φ by simultaneously substituting $s(p)$ for every occurrence of p, for all $p \in P$ in φ. In other words, $\varphi\big(\overline{p}/s(\overline{p})\big) \stackrel{\text{def}}{=} \hat{s}(\varphi)$, where $\hat{s}$ is the (unique) extension of $s : P \to F$ to a homomorphism $\hat{s} : \mathfrak{F} \to \mathfrak{F}$.

(ii) $\mathcal{L}$ has the *(syntactical) substitution property* (or $\mathcal{L}$ is *substitutional*) iff for any formula $\varphi \in F$ and substitution $s : P \to F$,

$$\models \varphi \quad \text{implies} \quad \models \varphi\big(\overline{p}/s(\overline{p})\big).$$

This means that a formula of $\mathcal{L}$ is valid iff the corresponding formula scheme of $\mathcal{L}$ is valid (where we get the corresponding formula-scheme by substituting atomic formulas $p_i \in P$ with formula variables $\phi_i \in FV$).

(iii) $\mathcal{L}$ has the *semantical substitution property* iff for any model $\mathfrak{M} \in M$ and substitution $s : P \to F$ there is a(nother) model $\mathfrak{N} \in M$ such that $mng_{\mathfrak{N}}(p) = mng_{\mathfrak{M}}\big(\hat{s}(p)\big)$ for all $p \in P$. Intuitively, the model $\mathfrak{N}$ is the substituted version of $\mathfrak{M}$ along s.

The semantical substitution property says that the atomic formulas can have the meanings of any other formulas. (This statement will be made precise in Proposition 1.3.2.)

Finally, it would be in order here to discuss properties of the provability relation $\vdash$, e.g., that $\vdash$ is most often given via axioms and rules (Hilbert style inference systems). We do not go into this because of space limitations, but see e.g. [And-Nem-Sai,01, items 33 (p. 192), 48–52 (pp. 209–210)], [Gab,94, chapters 14, 15], or [Mik,95].

Thus, a "full-fledged" logic $\mathcal{L} = \langle F, \vdash, M, \models \rangle$ sometimes is given in the form

$$\mathcal{L} = \big\langle \langle P, Cn \rangle, \vdash, M, mng, \models \big\rangle$$

as in [Gab,94, p. 359 or p. 398]. Often not all parts of a logic are given. Sometimes we have only $\langle F, \vdash \rangle$ and we are searching for (an "adequate") semantics $\langle M, \models \rangle$ for it such that, e.g., $\langle F, \vdash, M, \models \rangle$ is complete. (This happened, e.g., in studying Lambek-calculus, or **S5**, or Floyd's method for proving correctness of programs. See e.g. [And-Mik,93], [And-Nem-Sai,79], [And-Nem-Sai,82], [Mak-Sai,89].) Or, even more often, we have $\langle F, M, \models \rangle$ and we are searching for a provability relation $\vdash$ such that $\langle F, \vdash, M, \models \rangle$ would be complete. Sometimes $\langle F, \vdash \rangle$ is called a "syntactical logic", and $\langle F, M, \models \rangle$ is called a "semantical logic"[6]. (One should keep in mind, though, that $\langle F, \models \rangle$ is a "syntactical logic" in this sense.) From now on we will often omit some parts of a logic. Most often we will deal with $\langle F, M, mng, \models \rangle$ and we will call $\mathcal{L} = \langle F, M, mng, \models \rangle$ a logic. All the logic properties defined above are meaningful for it, e.g. that $\mathcal{L}$ is compositional etc.

Definition 1.1.2 (Algebraizable semantical logic, structural logic). Let $\mathcal{L} = \langle F, M, mng, \models \rangle$ be a logic in the above sense. We say that $\mathcal{L}$ is an *algebraizable semantical logic* iff $\mathcal{L}$ is compositional, $\mathcal{L}$ has the filter-property, and $\mathcal{L}$ has both substitution properties. We say that $\mathcal{L}$ is *structural* iff $\mathcal{L}$ is compositional and has the semantical substitution property.

In most cases, the set P of atomic formulas is a parameter in the definition of a logic $\mathcal{L}$. Namely, it is a fixed but *arbitrary* set. This way $\mathcal{L}$ is a function of P, and we should write $\mathcal{L}^P$ (instead of $\mathcal{L}$) to make this explicit. Most often the choice of P has only limited influence on the behaviour of $\mathcal{L}$.

Definition 1.1.3 (General logic, algebraizable (general) logic). A *general logic* is defined to be a function $\mathbf{L} \stackrel{\text{def}}{=} \langle \mathcal{L}^P : P \text{ is a set} \rangle$, where for each

[6]The literature of syntactical logics is very extensive, cf. e.g. [Bez,05, pp. 3–17] for an overview demonstrating that it goes back to Tarski's work of the 1920's. Semantical logics are often discussed under the name "institutions", cf. the end of our Section 1.

set P, $\mathcal{L}^P = \langle F^P, \vdash^P, M^P, mng^P, \models^P \rangle$ is a logic in the sense of Def. 1.1.1. We will sometimes refer to elements of M^P as P-*models*.

All the logic properties introduced so far extend, in a natural way, to general logic, cf. [And-Nem-Sai,01, Def. 39, pp. 198–199]. Basically, **L** has a logic property iff all its parametrized parts $\mathcal{L}^P$ have that property in the old sense (of items **(A–D)** above).

In the cases of the properties "**L** has connectives" and "**L** has the filter property", all the logics $\mathcal{L}^P$ in **L** share the same set Cn of connectives, and derived connectives ε_i, δ_i, Δ_j, respectively. The substitution properties are formalized in such a way that, in the conditions defining them, possibly different parameters P and Q appear, as follows.

L *has the substitution property* iff $\models^P \varphi$ implies $\models^Q \varphi(\overline{p}/s(\overline{p}))$ for all P, Q, $s : P \to F^Q$, and $\varphi \in F^P$.

L *has the semantical substitution property* iff for all sets P, Q, $s : P \to F^Q$ and $\mathfrak{M} \in M^Q$ there exists $\mathfrak{N} \in M^P$ such that $mng^Q_{\mathfrak{M}} \circ \hat{s} = mng^P_{\mathfrak{N}}$.

L is called an *algebraizable (general) logic*, or *algebraizable logic* for short, iff **L** is compositional, has the filter-property, and has both substitution properties. **L** is called *structural* iff it is compositional and has the semantical substitution property (cf. [And-Nem-Sai,01, Def. 39(v), pp. 198–199]).

We note that **L** is an algebraizable (general) logic iff $\mathcal{L}^P$ is an algebraizable semantical logic for each P, the connectives and the derived connectives for the filter-property are the same for each P, and the condition below holds for each P, Q:

$$(1.1.1) \quad P \subseteq Q \Longrightarrow \{mng^P_{\mathfrak{M}} : \mathfrak{M} \in M^P\} = \big\{(mng^Q_{\mathfrak{M}}) \restriction F^P : \mathfrak{M} \in M^Q\big\}.$$

Intuitively, this condition says that $\mathcal{L}^P$ is the natural restriction of $\mathcal{L}^Q$. ($P \subseteq Q \Rightarrow mng^P_{\mathfrak{M}} \subseteq mng^Q_{\mathfrak{M}}$, cf. condition **(D)**(iii) way above).

In the rest of this paper we will omit provability $\vdash^P$, because we concentrate on semantical properties.

The interested reader may notice that a general logic is an example for an *institution* in the sense of e.g. [Gog-Bur,92]. See also [And-Nem-Sai,01, p. 201], [Nem-And,94, Sec. 14, p. 358].

For a second we let

$$\text{Vocabularies} \stackrel{\text{def}}{=} \text{“the class of possible choices of the parameter } P \text{ of } \mathbf{L}\text{”}.$$

In Universal Logic and Institutions Theory our notion of a general logic $\mathbf{L} = \langle \mathcal{L}^P : P \in \text{Vocabularies}\rangle$ corresponds to what is called there an institution or a logic, see Kutz et al [Kut-Mos-Luc,10, Def. 2.1, p. 268], Gabbay [Gab,94, Sec. 14.2.4, Def. 2.4 (pp. 358–359), pp. 396–400], Goguen–Burstall [Gog-Bur,92], Mossakowski–Goguen–Diaconescu–Tarlecki [Mos-Gog-Dia-Tar,05]. Our notation corresponds to that of Institutions Theory as follows:

P	Σ	signature
F^P	$Sent(\Sigma)$	sentences of signature Σ
Vocabularies	Sign (= Sign^I)	the collection of all signatures
$\mathbf{L}$	I	institution

Thus $\mathbf{L} = I = \langle \mathcal{L}^\Sigma : \Sigma \in \text{Sign}\rangle$, and

$$\mathcal{L}^P = \mathcal{L}^\Sigma = \big\langle \text{Sent}\,(\Sigma), \vdash(\Sigma), \mathsf{Mod}\,(\Sigma), \models(\Sigma)\big\rangle .$$

In particular, our results herein apply to institutions in general. Cf. [And-Nem-Sai,01, pp. 201–202] for more on this.

2. Examples for (Semantical) Logics

In this section we show that some of the basic/well-known logical systems fit into our general concept of a logic, outlined in the previous section.

Example 1.2.1 (Propositional logic $\mathbf{L}_S$). We call a quadruple $\mathcal{L}_S \stackrel{\text{def}}{=} \langle F_S, M_S, mng_S, \models_S\rangle$ *propositional* (or *sentential*) *logic* iff conditions (i)–(iii) below hold.

(i) In accordance with item **(A)**(i) in the previous section, the set F_S of formulas of $\mathcal{L}_S$ is defined to be $F\big(P, \{\land, \neg\}\big)$, where P is an arbitrary set (the set of atomic formulas), and the logical connectives $\land$, $\neg$ have ranks 2 and 1, respectively.

(ii) The class M_S of *models* of $\mathcal{L}_S$ is defined by

$$M_S \stackrel{\text{def}}{=} \big\{\langle W, v\rangle \mid W \text{ is a set, and } v : P \to \mathcal{P}(W)\big\}.$$

If $\mathfrak{M} = \langle W, v\rangle \in M_S$ then W is called the set of *possible states* (or *worlds* or *situations*) of $\mathfrak{M}$. The function v can be viewed as a valuation of the atomic formulas to subsets of the universe W.

(iii) Let $\langle W, v\rangle \in M_S$, $w \in W$, and $\varphi \in F_S$. We define the binary relation $\Vdash_v \subseteq W \times F_S$ by recursion on the complexity of the formulas:

- if $p \in P$ then $\left(w \Vdash_v p \overset{\text{def}}{\Longleftrightarrow} w \in v(p)\right)$,
- if $\psi_1, \psi_2 \in F_S$, then

$$w \Vdash_v \neg\psi_1 \overset{\text{def}}{\Longleftrightarrow} w \nVdash_v \psi_1$$

$$w \Vdash_v (\psi_1 \wedge \psi_2) \overset{\text{def}}{\Longleftrightarrow} w \Vdash_v \psi_1 \text{ and } w \Vdash_v \psi_2.$$

If $w \Vdash_v \varphi$ then we say that φ is *true in* w, or w *forces* φ.

Now $mng_S\left(\varphi, \langle W, v\rangle\right) \overset{\text{def}}{=} \{w \in W : w \Vdash_v \varphi\}$. We may notice that, for a fixed $\mathfrak{M} = \langle W, v\rangle$, $mng_{\mathfrak{M}} \overset{\text{def}}{=} mng_{S,\mathfrak{M}}$ $\left(= mng_{\langle W,v\rangle}\right)$ is nothing but the unique extension of $v : P \to \mathcal{P}(W)$ to a (Boolean) homomorphism $mng_{\mathfrak{M}} : \mathfrak{F} \to \mathfrak{P}(W)$ (the powerset algebra over W).

$$\langle W, v\rangle \models_S \varphi \quad (\varphi \text{ is } \textit{valid in } \langle W, v\rangle)\text{, iff for every } w \in W,\ w \Vdash_v \varphi.$$

As we explained above Definition 1.1.3, the set P of atomic formulas is a *parameter* in the definition of $\mathcal{L}_S$. So $\mathcal{L}_S$ is a function of P, and we write $\mathcal{L}_S^P$ to make this explicit. Thus general propositional logic is

$$\mathbf{L}_S = \left\langle \mathcal{L}_S^P : P \text{ is a set}\right\rangle.$$

$\mathbf{L}_S$ is an algebraizable (general) logic, cf. [And-Nem-Sai,01, p. 216] (where other nice properties of $\mathbf{L}_S$ are also stated and verified).

Example 1.2.2 (Modal logic **S5**). As in the previous example, a set P of atomic formulas is fixed. The set of connectives of *modal logic* **S5** is $\{\wedge, \neg, \Diamond\}$ with ranks 2, 1, 1. The set of formulas (denoted as $F_{\mathbf{S5}}$) of **S5** is defined to be $F\left(P, \{\wedge, \neg, \Diamond\}\right)$. We let $M_{\mathbf{S5}} \overset{\text{def}}{=} M_S$. The definition of $w \Vdash_v \varphi$ is the same as in the propositional case but we also have the case of $\Diamond$:

$$w \Vdash_v \Diamond\varphi \overset{\text{def}}{\Longleftrightarrow} (\exists w' \in W)\ w' \Vdash_v \varphi.$$

Then $mng_{\mathbf{S5}}\left(\varphi, \langle W, v\rangle\right) \overset{\text{def}}{=} \{w \in W : w \Vdash_v \varphi\}$, and the validity relation $\models_{\mathbf{S5}}$ are defined as in Example 1.2.1 above. Now modal logic **S5** with parameter P is $\mathbf{S5}^P \overset{\text{def}}{=} \langle F_{\mathbf{S5}}, M_{\mathbf{S5}}, mng_{\mathbf{S5}}, \models_{\mathbf{S5}}\rangle$, and modal logic **S5** in the general sense is $\langle \mathbf{S5}^P : P \text{ is a set}\rangle$. Again, this logic is an algebraizable logic (see [And-Nem-Sai,01, p. 219]).

Example 1.2.3 (First order logic with n variables $\mathbf{L}_n$).
Cf. [And-Nem-Sai,01, pp. 220–222].

(i) Let $n \in \omega$ be any natural number and let P be any set. The formulas of $\mathcal{L}_n^P$ are ordinary first order formulas with the restrictions that

1. the relation symbols come from P, all of them n-ary, and

2. we have only n variables all of them occurring in every atomic formula in a fixed order.

Thus, denoting the set of variables as $V \stackrel{\text{def}}{=} \{v_0, \ldots, v_{n-1}\}$, an atomic formula looks like $r(v_0, \ldots, v_{n-1})$ ($r \in P$). According to restrictions (1-2) above, $r(v_0, \ldots, v_{n-1})$ can be identified with r. So parameter P is regarded to be the set of atomic formulas of $\mathcal{L}_n^P$.

The set of connectives of $\mathcal{L}_n^P$ is $Cn_n \stackrel{\text{def}}{=} \{\wedge, \neg\} \cup \{\exists v_i : i < n\} \cup \{v_i = v_j : i, j < n\}$. The rank of $v_i = v_j$ is 0, $\wedge$ is binary, and the rest are unary. Now $F_n^P = F(P, Cn_n)$.

(ii) The class M_n of *models* $\mathfrak{M}$ of $\mathcal{L}_n$ is defined by

$$M_n \stackrel{\text{def}}{=} \{ \langle M, r^{\mathfrak{M}} \rangle_{r \in P} : M \text{ is a set, and for all } r \in P,\ r^{\mathfrak{M}} \subseteq {}^n \}.$$

(iii) While in **S5** the "basic semantical units" were the possible situations $w \in W$, in first order logic the basic semantical units are the evaluations $q \in {}^nM$ of the variables. To follow tradition, we write $\mathfrak{M} \models \varphi[q]$ instead of $\mathfrak{M}, q \Vdash \varphi$. We define $\mathfrak{M} \models \varphi[q]$ by recursion as follows.

- $\mathfrak{M} \models r[q] \quad \stackrel{\text{def}}{\iff} \quad q \in r^{\mathfrak{M}} \quad (r \in P)$.
- $\mathfrak{M} \models (v_i = v_j)[q] \quad \stackrel{\text{def}}{\iff} \quad q_i = q_j \quad (i, j < n)$.
- The conditions for $\neg$, $\wedge$, $\exists v_i$ are the usual ones, cf. e.g. [And-Nem-Sai,01, p. 221].

If $\mathfrak{M} \models \varphi[q]$ then we say that the *evaluation q satisfies φ in the model $\mathfrak{M}$*.

$$mng_n(\varphi, \mathfrak{M}) \quad \stackrel{\text{def}}{=} \quad \{q \in {}^nM : \mathfrak{M} \models \varphi[q]\},$$

$$\mathfrak{M} \models_n \varphi \quad \stackrel{\text{def}}{\iff} \quad \mathfrak{M} \models \varphi[q] \text{ for every } q \in {}^nM.$$

First order logic with n variables with parameter P is

$$\mathcal{L}_n^P \stackrel{\text{def}}{=} \langle F_n, M_n, mng_n, \models_n \rangle, \text{ and}$$

$$\mathbf{L}_n \stackrel{\text{def}}{=} \langle \mathcal{L}_n^P : P \text{ is a set} \rangle.$$

Our $\mathcal{L}_n$ might look somewhat unusual because we do not allow substitution of variables in atomic formulas $r_i(v_0 \dots)$. However, this does not restrict generality, because substitution is expressible by using quantifiers and equality – as Tarski discovered it in the 40's. Namely, if we want to substitute v_1 for v_0 in formula φ then the resulting formula is equivalent to $\exists v_0(v_0 = v_1 \land \varphi)$.

Next we define first order logic in a non-traditional form. Intuitive explanation for this form is given in detail in [Gab,94, item 2.15 and above it, pp. 409–410].

Example 1.2.4 (First order logic $\mathbf{L}_{\mathrm{FOL}}$, rank-free formulation).

(i) Let P be an arbitrary set. The set F^P_{FOL} of formulas of $\mathcal{L}^P_{\mathrm{FOL}}$ is defined similarly to F^P_n: the only difference is that, in case of F^P_{FOL}, we have an infinite set $V = \{v_i : i \in \omega\}$ of individual variables, and accordingly, we have ω-many quantifiers. Thus the set of connectives of $\mathcal{L}_{\mathrm{FOL}}$ is $Cn_\omega \overset{\mathrm{def}}{=} \{\land, \neg\} \cup \{\exists v_i : i \in \omega\} \cup \{v_i = v_j : i, j \in \omega\}$, and $F^P_{\mathrm{FOL}} \overset{\mathrm{def}}{=} F(P, Cn_\omega)$.

(ii) The class M^P_{FOL} of *models* of $\mathcal{L}^P_{\mathrm{FOL}}$ is

$$M^P_{\mathrm{FOL}} \overset{\mathrm{def}}{=} \big\{ \langle M, r^{\mathfrak{M}} \rangle_{r \in P} : M \text{ is a set, and for all } r \in P,\ r^{\mathfrak{M}} \subseteq {}^nM \text{ for some } n \in \omega \big\}.$$

(iii) We define $\mathfrak{M} \models \varphi[q]$ as follows:

- $\mathfrak{M} \models r[q] \overset{\mathrm{def}}{\iff} \langle q_0, \dots, q_{n-1} \rangle \in r^{\mathfrak{M}}$ for some $n \in \omega$ $(r \in P)$.
- The conditions for $(v_i = v_j)$, $\neg$, $\lor$, $\exists v_i$ $(i, j \in \omega)$ are just as in case of $\mathcal{L}^P_n$.

Now mng^P_{FOL} and $\models^P_{\mathrm{FOL}} \subseteq M^P_{\mathrm{FOL}} \times F^P_{\mathrm{FOL}}$ are defined as in case of $\mathcal{L}^P_n$. *First order logic* (in rank-free form) is

$$\mathcal{L}^P_{\mathrm{FOL}} \overset{\mathrm{def}}{=} \mathcal{L}_{\mathrm{FOL}} \overset{\mathrm{def}}{=} \langle F_{\mathrm{FOL}}, M_{\mathrm{FOL}}, mng_{\mathrm{FOL}}, \models_{\mathrm{FOL}} \rangle.$$

$\mathbf{L}_{\mathrm{FOL}}$ denotes the general logic version.

Both $\mathbf{L}_n$ and $\mathbf{L}_{\mathrm{FOL}}$ are algebraizable logics. However, in their traditional presentation, they would not be structural (the substitution property fails, because the atomic formulas like $r_2(v_0, v_2)$ and $r_2(v_6, v_7)$ are not independent). Detailed explanation and verification of these facts can be found in [And-Nem-Sai,01, sections 7.5–7.9 (pp. 220–237)]. Further references on the properties of $\mathcal{L}_n$ and $\mathcal{L}_{\mathrm{FOL}}$ are e.g. Andréka–Gergely–Németi [And-Ger-Nem,77], [And-Ger-Nem,73], Blok–Pigozzi [Blo-Pig,89, Appendix], Henkin–Monk–Tarski [Hen-Mon-Tar,85, §4.3], Henkin–Tarski [Hen-Tar,61], Németi–Andréka [Nem-And,94], Simon [Sim,91], Venema [Ven,91].

Further examples for logics in our sense are e.g. in [And-Nem-Sai,01], [And-Nem-Sai,09], [And-Nem-Sai,94a], [And-Kur-Nem-Sai,94], [Gab,94, Chapters 14, 15], and in the literature of Universal Logic.

3. Algebraization

Why to algebraize logics (or anything)? Some motivation is given in our Section 6.[7] A possible more substantial motivation starts out from the following two slogans: *"Abstract algebra (not to mention universal algebra or categories) is the mathematical theory of abstraction"*; and *"Abstraction is a necessary prerequisite for reasoning"*. Then we argue that, therefore, if our purpose with developing logic is to develop a science of reasoning as proposed by J-Y. Béziau[8], then an explicit study of the process of algebraizing (e.g. logics) is important.

Now we turn to introducing the algebraic machinery we intend to use for studying definability properties. This machinery is appropriate for studying, besides definability properties, a large variety of other logical properties as well, see e.g. [And-Nem-Sai,01], [Mik,95].

Definition 1.3.1 (Meaning algebra, Alg_m, Alg). Let $\mathcal{L} = \langle F, M, mng, \models\rangle$ be a compositional logic.

(i) First we turn every model into an algebra. Compositionality of $mng_{\mathfrak{M}}$ ensures that we can easily define an algebra of type Cn on the set $\{mng_{\mathfrak{M}}(\varphi) : \varphi \in F\}$ of meanings. This algebra is $mng_{\mathfrak{M}}(\mathfrak{F})$, it will be called the *meaning algebra of* $\mathfrak{M}$ and it will be denoted by $\mathfrak{Mng}\,(\mathfrak{M})$. In more detail, to any logical connective c of arity k we define a k-ary function $c^{\mathfrak{M}}$ on the meanings in $\mathfrak{M}$ by setting for all formulas $\varphi_1, \dots, \varphi_k$

$$c^{\mathfrak{M}}\big(mng_{\mathfrak{M}}(\varphi_1), \dots, mng_{\mathfrak{M}}(\varphi_k)\big) \stackrel{\text{def}}{=} mng_{\mathfrak{M}}\left(c(\varphi_1, \dots, \varphi_k)\right), \text{ and}$$

[7]Further motivation is in [And-Nem-Sai,01, Introduction of Chap.2 "Bridge..." p. 186, pp. 202–203], [And-Nem-Sai,94a], and in [Mik,95, Sec. 1.3 "Bridge...", pp. 18–29].

[8]E.g. in [Bez,10].

$$\mathfrak{Mng}\,(\mathfrak{M}) \stackrel{\text{def}}{=} \left\langle \{mng_{\mathfrak{M}}(\varphi) : \varphi \in F\}, c^{\mathfrak{M}} \right\rangle_{c \in Cn}.$$

(ii) $\mathsf{Alg}_m(\mathcal{L})$ denotes the class of all meaning algebras of $\mathcal{L}$, that is,

$$\mathsf{Alg}_m(\mathcal{L}) \stackrel{\text{def}}{=} \{mng_{\mathfrak{M}}(\mathfrak{F}) : \mathfrak{M} \in M\} = \{\mathfrak{Mng}\,(\mathfrak{M}) : \mathfrak{M} \in M\}.$$

(iii) Let $K \subseteq M$. Then for every $\varphi, \psi \in F$ we define

$$\varphi \sim_K \psi \stackrel{\text{def}}{\iff} (\forall \mathfrak{M} \in K)\, mng_{\mathfrak{M}}(\varphi) = mng_{\mathfrak{M}}(\psi).$$

Then $\sim_K$ is an equivalence relation, and is a congruence on $\mathfrak{F}$ by compositionality of $\mathcal{L}$. $\mathfrak{F}/{\sim_K}$ denotes the factor-algebra of $\mathfrak{F}$, factorized by $\sim_K$. It is called the *(semantical) Lindenbaum–Tarski algebra* of K. Now,

$$\mathsf{Alg}\,(\mathcal{L}) \stackrel{\text{def}}{=} \mathbf{I}\{\mathfrak{F}/{\sim_K} : K \subseteq M\}.$$

Thus $\mathsf{Alg}\,(\mathcal{L})$ is the class of isomorphic copies of the Lindenbaum–Tarski algebras of $\mathcal{L}$.

(iv) Let $\mathbf{L} = \langle \mathcal{L}^P : P \text{ is a set}\rangle$ be a general logic. Then

$$\mathsf{Alg}_m(\mathbf{L}) \stackrel{\text{def}}{=} \bigcup \{\mathsf{Alg}_m(\mathcal{L}^P) : P \text{ is a set}\}, \text{ and}$$

$$\mathsf{Alg}\,(\mathbf{L}) \stackrel{\text{def}}{=} \bigcup \{\mathsf{Alg}\,(\mathcal{L}^P) : P \text{ is a set}\}.$$

For logics of the form $\mathbf{L} = \langle \mathcal{L}^P : P \in \text{Vocabularies}\rangle$ (cf. the end of Section 1) we let $\mathsf{Alg}_m(\mathbf{L}) \stackrel{\text{def}}{=} \bigcup \{\mathsf{Alg}_m(\mathcal{L}^P) : P \in \text{Vocabularies}\}$ etc.

In the definition of $\mathsf{Alg}_m(\mathcal{L})$ above, it is important that $\mathsf{Alg}_m(\mathcal{L})$ is not an abstract class in the sense that it is not closed under isomorphisms. The reason for defining $\mathsf{Alg}_m(\mathcal{L})$ in such a way is that, since $\mathsf{Alg}_m(\mathcal{L})$ is the class of algebraic counterparts of the *models* of $\mathcal{L}$, we need these algebras as concrete algebras and replacing them with their isomorphic copies would lead to loss of information (in semantical – model theoretical matters). See e.g. the algebraic characterization of the *weak Beth definability property*, Theorem 1.5.1.

For a logic $\mathcal{L} = \langle F_{\mathcal{L}}, M_{\mathcal{L}}, mng_{\mathcal{L}}, \models_{\mathcal{L}}\rangle$, we let $Mng_{\mathcal{L}} \stackrel{\text{def}}{=} \{mng_{\mathfrak{M}} : \mathfrak{M} \in M_{\mathcal{L}}\}$. Similarly $Mng^P \stackrel{\text{def}}{=} Mng_{\mathcal{L}^P}$ if $\mathbf{L} = \langle \mathcal{L}^P : P \text{ is a set}\rangle$. If $\mathfrak{A}$ is an algebra and K is a class of algebras, then $Hom\,(\mathfrak{A}, \mathsf{K})$ denotes the class of all homomorphisms $h : \mathfrak{A} \to \mathfrak{B}$ with $\mathfrak{B} \in \mathsf{K}$. $\mathfrak{F}^P$ denotes the formula algebra of parameter P.

Proposition 1.3.2 (Characterization of structural logics). *Let $\mathcal{L}$ and* **L** *be a compositional logic and a compositional general logic, respectively. Then* (i)–(ii) *below hold.*

(i) *$\mathcal{L}$ has the semantical substitution property iff*

$$Mng_{\mathcal{L}} = Hom\left(\mathfrak{F}, \mathsf{Alg}_m(\mathcal{L})\right).$$

(ii) **L** *has the semantical substitution property iff*

$$Mng^P = Hom\left(\mathfrak{F}^P, \mathsf{Alg}_m(\mathbf{L})\right), \quad \textit{for all } P.$$

Theorem 1.3.3 (Connection between Alg_m and Alg).

(i) *Let $\mathcal{L}$ be a compositional logic. Then* $\mathbf{SP}\,\mathsf{Alg}\,(\mathcal{L}) = \mathbf{SP}\,\mathsf{Alg}_m(\mathcal{L})$.

(ii) *Let* **L** *be a structural general logic. Then* $\mathsf{Alg}\,(\mathbf{L}) = \mathbf{SP}\,\mathsf{Alg}_m(\mathbf{L})$.

We note that we also have that, for structural logics $\mathcal{L}$,

$$\mathsf{Alg}\,(\mathcal{L}) = \mathbf{SP}\,\mathsf{Alg}_m(\mathcal{L}) \cap \left\{\mathfrak{A} \,:\, |A| \leq |\mathfrak{F}|\right\}.$$

The classes of algebras corresponding to our example logics in Section 2 are summarized at the end of this section.[9] In the end of that summary we included the "abstract" class CA of cylindric algebras and its logical counterpart $\langle F_n, \vdash_n\rangle$ (briefly discussed below).

The claims of this section are proved in [And-Nem-Sai,01, items 43–44, p. 204].

To complete this section, let us say some words about the algebraization of the *syntax*. As we already pointed out, in this paper we concentrate on the semantical aspects of a logic. Accordingly, both $\mathsf{Alg}\,(\mathbf{L})$ and $\mathsf{Alg}_m(\mathbf{L})$ correspond to the semantical part $\langle \mathsf{Mod}, \models, mng\rangle$ of a logic **L**. Our present definition of $\mathsf{Alg}\,(\mathbf{L})$ admits a natural (and widely used) generalization that applies to the syntactical part $\langle F, \vdash\rangle$ as well. $\mathsf{Alg}\left(\langle F, \vdash\rangle\right)$ is the class of *"(syntactical) Lindenbaum–Tarski algebras"* of $\langle F, \vdash\rangle$. In defining $\mathsf{Alg}\left(\langle F, \vdash\rangle\right)$, instead of the semantical congruence relation $\sim_K$ (Def. 1.3.1 (iii)), we factor $\mathfrak{F}$ by the syntactical inter-derivability relation $\equiv_T \stackrel{\text{def}}{=} \left\{\langle\varphi,\psi\rangle \,:\, T,\, \varphi \vdash \psi \text{ and } T,\, \psi \vdash \varphi\right\}$ for all $T \subseteq F$. However, $\equiv_T$ may not be a congruence relation. This problem can be overcome in two different (equivalently good) ways:

[9] Cf. [And-Nem-Sai,01, Table 1 on p. 237].

(i) We insist on using as $\vdash$ only "local" derivation systems in the spirit of the Amsterdam logic school (Venema [Ven,91, Appendix B] contains a useful comparison of the local/global paradigms, cf. also e.g. Gabbay–Kurucz et al [Gab-Kur-Wol-Zak,03, p. 35 "local consequence"]);

(ii) We use the largest congruence relation $\sim_T \subseteq \equiv_T$ contained in the equivalence relation $\equiv_T$. Then $\mathsf{Alg}\left(\langle F, \vdash\rangle\right) \stackrel{\text{def}}{=} \{\mathfrak{F}/\sim_T\ :\ T \subseteq F\}$. This approach is taken by the Blok–Pigozzi school, cf. [And-Nem-Sai,94a, pp. 9–10], [Blo-Pig,89], [Fon-Jan,94]. (There $\sim_T$ is denoted as $\Omega(T)$.)

For more detail see e.g. [And-Nem-Sai,01, p. 167 ("connections with logic")], $\mathsf{Alg}^* S$ in [Hoo,01, Thm. 3.2.13 (p. 73)], Henkin et al. [Hen-Mon-Tar,85, Part II, Sec. 4.3 items 4.3.23–4.3.28 (pp. 159–161)], and Font–Jansana [Fon-Jan,94, pp. 58–60 and Sec. 2 (pp. 64–65)].

It is interesting to compare the algebraic forms RCA_n and Cs_n of the semantics of $\mathbf{L}_n$ (cf. the list below) with the algebraic form of its syntax (last line of the list). Let $\vdash_n$ be a "natural" Hilbert style inference system for $\mathbf{L}_n$. Then $\mathsf{Alg}\left(\langle F_n, \vdash_n\rangle\right)$ turns out to be the class CA_n of cylindric algebras (see [And-Nem-Sai,01, Def. 16, p. 161], cf. also [And-Nem-Sai,01, 167–168]). While $\mathsf{Alg}\,(\mathbf{L}_n)$ and $\mathsf{Alg}_m(\mathbf{L}_n)$ are classes of "concrete" algebras, CA_n is defined axiomatically (moreover, some elements of CA_n are not representable at all as "algebras of relations", cf. [Sim,thisVol]). We note that this $\vdash_n$ is necessarily incomplete for $\models_n$ by [And-Nem-Sai,01, items 48–52, pp. 208–210].

Summary of examples:

$\mathsf{Alg}\,(\mathbf{L}_S)$	$= \mathsf{BA}$	Boolean algebras
$\mathsf{Alg}_m(\mathbf{L}_S)$	$= \mathsf{SetBA}$	Boolean set algebras
$\mathsf{Alg}\,(\mathbf{S5})$	$= \mathsf{RCA}_1$	representable cylindric algebras of dimension 1
$\mathsf{Alg}_m(\mathbf{S5})$	$= \mathsf{Cs}_1$	cylindric set algebras of dimension 1
$\mathsf{Alg}\,(\mathbf{L}_n)$	$= \mathsf{RCA}_n$	representable cylindric algebras of dimension n
$\mathsf{Alg}_m(\mathbf{L}_n)$	$= \mathsf{Cs}_n$	cylindric set algebras of dimension n
$\mathsf{Alg}\,(\mathbf{L}_{\mathrm{FOL}})$	$= \mathbf{SP}\,\mathsf{Csf}_\omega$	$\mathsf{Csf}_\omega \stackrel{\text{def}}{=} \mathsf{Cs}_\omega^{\mathrm{reg}} \cap \mathsf{Lf}_\omega$
$\mathsf{Alg}_m(\mathbf{L}_{\mathrm{FOL}})$	$= \mathsf{Csf}_\omega$	
$\mathsf{Alg}\left(\langle F_n, \vdash_n\rangle\right)$	$= \mathsf{CA}_n$	cylindric algebras of dimension n

where $\mathsf{Cs}_\omega^{\mathrm{reg}}$: regular cylindric set algebras of dimension ω,
Lf_ω: locally finite cylindric algebras of dimension ω.

The definitions of the above classes of algebras (BA, RCA, Cs, $\mathsf{Cs}^{\mathrm{reg}}$, Lf, CA) can be found in any textbook on cylindric algebras, see the index of the present volume or [Hen-Mon-Tar,85], [Hen-Mon-Tar-And-Nem,81], [Mon,00], [And-Nem-Sai,01].

4. Definability

Now we turn to the algebraic characterization of some definability properties. Beth's definability properties of logics were defined e.g. in Barwise–Feferman [Bar-Fef,85]. Here we give the definitions in the framework outlined so far.

If $f : A \to B$ and $C \subseteq A$ then $f \restriction C$ denotes the *restriction* of f to C, that is, $f \restriction C \stackrel{\mathrm{def}}{=} \big\langle f(x) : x \in C \big\rangle = \big\{ \langle x, y \rangle \in f : x \in C \big\}$.

Definition 1.4.1 (Implicit definition, explicit definition). Let $\mathbf{L} = \langle \mathcal{L}^P : P \text{ is a set} \rangle$ be a general logic. Let $P \subseteq Q$ with $F^P \neq \emptyset$, and let $R \stackrel{\mathrm{def}}{=} Q \setminus P$.

- The set Σ of formulas *defines R implicitly in Q* iff any P-model can be extended to a Q-model of Σ at most one way, i.e., iff

$$\big(\forall \mathfrak{M}, \mathfrak{N} \in \mathsf{Mod}^Q(\Sigma)\big) \big(mng^Q_{\mathfrak{M}} \restriction F^P = mng^Q_{\mathfrak{N}} \restriction F^P \Longrightarrow \Longrightarrow mng^Q_{\mathfrak{M}} = mng^Q_{\mathfrak{N}} \big).$$

- Σ *defines R implicitly in Q in the strong sense* iff any P-model that in principle can, can indeed be extended to a Q-model of Σ, i.e., iff Σ defines R implicitly in Q and in addition we have

$$\big(\forall \mathfrak{M} \in \mathsf{Mod}^P \big(Th^Q \, \mathsf{Mod}^Q(\Sigma) \cap F^P \big)\big) \big(\exists \mathfrak{N} \in \mathsf{Mod}^Q(\Sigma)\big) \, mng^Q_{\mathfrak{N}} \restriction F^P = mng^P_{\mathfrak{M}} .$$

- Σ *defines R explicitly in Q* iff any element of R has an "explicit definition" that works in all models of Σ, i.e. iff

$$(\forall r \in R)(\exists \varphi_r \in F^P)\big(\forall \mathfrak{M} \in \mathsf{Mod}^Q(\Sigma)\big) \; mng^Q_{\mathfrak{M}}(r) = mng^Q_{\mathfrak{M}}(\varphi_r).$$

- Σ *defines R local-explicitly in Q* iff the above definition can vary from model to model, i.e., iff

$$(\forall r \in R)\big(\forall \mathfrak{M} \in \mathsf{Mod}^Q(\Sigma)\big)\,\big(\exists \varphi_r \in F^P\big)\; mng^Q_{\mathfrak{M}}(r) = mng^Q_{\mathfrak{M}}(\varphi_r).$$

Definition 1.4.2 (Beth definability properties)**.** Let **L** be a general logic.

- **L** *has the weak Beth definability property* (wBp for short) iff for all P, Q, R and Σ as in Definition 1.4.1,

$$\Sigma \text{ defines } R \text{ implicitly in } Q \text{ in the strong sense } \Longrightarrow \Sigma \text{ defines } R \text{ explicitly in } Q.$$

- **L** *has the (strong) Beth definability property* iff for all P, Q, R, and Σ as in Definition 1.4.1,

$$\Sigma \text{ defines } R \text{ implicitly in } Q \Longrightarrow \Sigma \text{ defines } R \text{ explicitly in } Q.$$

- **L** *has the local Beth definability property* iff for all P, Q, R and Σ as in Definition 1.4.1,

$$\Sigma \text{ defines } R \text{ implicitly in } Q \Longrightarrow \Sigma \text{ defines } R \text{ local-explicitly in } Q.$$

The wBp was introduced in Friedman [Fri,73], [Bar-Fef,85, Def. 7.3.1 (p. 73)], Hoogland [Hoo,01, 2.2.13 (p. 18)] (cf. also references of [Bar-Fef,85]) and has been investigated since then, cf. e.g. [Bar-Fef,85, pp. 73–76, 689–716], [And-Nem-Sai,01, pp. 211–215], [Hoo,01], [Nem-And,94], [Hod,93b].

Examples 1.4.3. $\mathbf{L}_S$, **S5**, and $\mathbf{L}_{\mathrm{FOL}}$ have the Beth property. We conjecture that $\mathbf{L}_2$ has wBp. Beth property fails in $\mathbf{L}_n$, $n > 1$. The wBp fails in $\mathbf{L}_n$, $n > 2$. Let $\mathbf{L}_{DL}$ be the fragment of $\mathbf{L}_S$ with connectives $\wedge$, $\vee$ but without $\neg$. Then wBp fails in $\mathbf{L}_{DL}$ (and so does Beth property). $\mathsf{Alg}\,(\mathbf{L}_{DL})$ = "distributive lattices". Higher order logics fail to have any of the three definability properties discussed here [Bar-Fef,85, pp. 32–34, 68–76] ([Bar-Fef,85, pp. 709–710] contain more material on definability properties).

Remark 1.4.4. In Definition 1.4.2 above we did not make cardinality restrictions on the set R. A natural version of Beth definability property is obtained from Definition 1.4.2 by requiring $|R| < \omega$: $\mathcal{L}$ has the *finite* Beth property iff for all P, Q, R, Σ with $|R| < \omega$ we have

$$\Sigma \text{ defines } R \text{ implicitly in } Q \text{ in the strong sense } \Longrightarrow \\ \Sigma \text{ defines } R \text{ explicitly in } Q.$$

Analogously one can obtain finite versions of the other definability properties (weak, local).

We note that a considerable part of the literature puts more emphasis on the finite Beth property than on the "infinite" one (see e.g. [Hen-Mon-Tar,85, Part II], [Mak,92], [Mak,97], [Nem,82], [Sai,90]) but, now, we feel that the infinite Beth property is at least as natural as the finite one. Actually, a third version of definability properties is often distinguished as well, namely, when we require that $|R|$ is exactly 1. Let us call this *"singleton Beth" property.* We think that it is for historical reasons that the greatest amount of attention was paid to this singleton version of Beth definability, cf. e.g. [Hen-Mon-Tar,85, Thm. 5.6.10, p. 259 in Part II]. The reason may have been the fact that in logics extending classical first order logic the finite versions and the singleton versions are equivalent.

In the present work we concentrate on the infinite versions. For future work, we think it will be fruitful to investigate the infinite versions and the finite versions in parallel fashion. (We feel that the singleton version is somewhat specialized.)

Definition 1.4.5 (Patchwork property)[10]**.** Let **L** be a general logic. **L** *has the patchwork property* iff any two "compatible" models have a common extension; formally, iff for all sets P, Q, and models $\mathfrak{M} \in M^P$, $\mathfrak{N} \in M^Q$ we have

$$\left(F^{P\cap Q} \neq \emptyset \text{ and } mng_{\mathfrak{M}}^{P\cap Q} = mng_{\mathfrak{N}}^{P\cap Q}\right) \quad \Longrightarrow \\ (\exists \mathfrak{P} \in M^{P\cup Q})\left(mng_{\mathfrak{P}}^{P} = mng_{\mathfrak{M}}^{P} \text{ and } mng_{\mathfrak{P}}^{Q} = mng_{\mathfrak{N}}^{Q}\right).$$

Recall that if K is a class of algebras, then by a morphism of K we understand a triple $\langle \mathfrak{A}, h, \mathfrak{B}\rangle$, where $\mathfrak{A}, \mathfrak{B} \in \mathsf{K}$ and $h : \mathfrak{A} \to \mathfrak{B}$ is a homomorphism. A morphism $\langle \mathfrak{A}, h, \mathfrak{B}\rangle$ is an *epimorphism of* K iff for every $\mathfrak{C} \in \mathsf{K}$ and every pair $f, k : \mathfrak{B} \to \mathfrak{C}$ of homomorphisms we have $f \circ h = k \circ h$ implies $f = k$. Typical examples of epimorphisms are the surjections. But for

[10]In simplifying the formulation of patchwork property as given in [And-Nem-Sai,01], here we used our condition (1.1.1) (close to the end of Section 1).

certain choices of K there are epimorphisms of K which are not surjective. Such is the case, e.g., when K is the class of distributive lattices.

Let $\mathsf{K}_0 \subseteq \mathsf{K}$ be two classes of algebras. Let $\langle \mathfrak{A}, h, \mathfrak{B} \rangle$ be a morphism of K. h is said to be K_0-*extensible* iff for every algebra $\mathfrak{C} \in \mathsf{K}_0$ and every surjective homomorphism $f : \mathfrak{A} \twoheadrightarrow \mathfrak{C}$ there exists some $\mathfrak{N} \in \mathsf{K}_0$ and $g : \mathfrak{B} \to \mathfrak{N}$ such that $\mathfrak{C} \subseteq \mathfrak{N}$ and $g \circ h = f$. It is important to emphasise here that $\mathfrak{C}$ is a concrete subalgebra of $\mathfrak{N}$ and *not* only is embeddable into $\mathfrak{N}$. A good picture for this definition is [Hoo,01, Fig. 3.6, p. 109] and [Nem-And,94, Sec. 33, pp. 429–431].

Theorem 1.4.6 (Characterization of Beth definability properties). *Let* $\mathbf{L}$ *be an algebraizable logic having the patchwork property.*

(i) $\mathbf{L}$ *has the Beth definability property iff all the epimorphisms of* $\mathsf{Alg}\,(\mathbf{L})$ *are surjective.*

(ii) $\mathbf{L}$ *has the local Beth definability property iff all the epimorphisms of* $\mathsf{Alg}_m(\mathbf{L})$ *are surjective.*

(iii) $\mathbf{L}$ *has the weak Beth definability property iff every* $\mathsf{Alg}_m(\mathbf{L})$*-extensible epimorphism of* $\mathsf{Alg}\,(\mathbf{L})$ *is surjective.*[11]

Proof. The proof of (i) is in Németi [Nem,82] and in Hoogland [Hoo,96]. A less general version of (i) is proved in [Hen-Mon-Tar,85, Part II, Thm. 5.6.10]. Part (ii) is due to Madarász. An early version of (iii) is in Sain [Sai,90], and the full version is proved in [Hoo,96]. ∎

In the formulation of Theorem 1.4.6 (ii),(iii) above, it was important that $\mathsf{Alg}_m(\mathbf{L})$ *is not an abstract class* in the sense that it is not closed under isomorphisms, since the definition of K-extensibility strongly differentiates isomorphic algebras.[12]

Theorem 1.4.6 has spectacular applications for (universal) logic. E.g. Madarász [Mad,12] gave a concrete characterization of those varieties of cylindric algebras (and their reducts) in which epimorphisms are surjective.

[11] Warning: There is a typo in [And-Nem-Sai,01, Thm. 58(iii), p. 213], namely, the last occurrence of $\mathsf{Alg}_m(\mathbf{L})$ should be $\mathsf{Alg}\,(\mathbf{L})$ (as in our Thm. 1.4.6(iii) here). We also note that an algebraic characterization of the projective Beth property (due to Craig 1957) is in Hoogland [Hoo,01, 3.4.3 (pp. 18, 87)].

[12] $\mathfrak{M} \cong \mathfrak{N} \quad \underset{\not\Leftarrow}{\Longrightarrow} \quad \mathfrak{Mng}\,(\mathfrak{M}) \cong \mathfrak{Mng}\,(\mathfrak{N})$.

This, according to Thm. 1.4.6(i), gives an intrinsic characterization of those logics **L** which have the Beth definability property (under some natural conditions on **L**). Besides Madarász and Németi, Steven Comer, András Simon, Maarten Marx, Eva Hoogland and others have important results in this area. (E.g., by a result of A. Simon, $\mathbf{L}_3$ fails to have wBp. We conjecture that this extends to $\mathbf{L}_n$ iff $n > 2$.)

Problem 1.4.7. It might be interesting to learn what is used in e.g. Institutions Theory in place of the patchwork property (we do not know). The patchwork property is needed in Thm. 1.4.6 above and Thm. 1.5.1 below, but we do not know to what extent this is an artifact of our present formalism.

Problem 1.4.8. What is the (essential) difference between infinite Beth and (infinite) projective Beth property? Is there a logic differentiating between them? (A distinguished one?) Projective Beth property is defined e.g. in Hoogland [Hoo,01, pp. 18, 86] and in [Bar-Fef,85, Def. 7.3.2 (p. 75)].

5. Category Theoretic Characterization of Weak Beth Definability

If K is a class of algebras, then $\mathbf{max}\,\mathsf{K}$ denotes the class of all $\subseteq$-maximal elements of K : $\mathbf{max}\,\mathsf{K} \stackrel{\text{def}}{=} \{\mathfrak{A} \in \mathsf{K} : (\forall \mathfrak{B} \in \mathsf{K})(\mathfrak{A} \subseteq \mathfrak{B} \Longrightarrow \mathfrak{A} = \mathfrak{B})\}$.

We will use the notions of "reflective subcategory" and "limits of diagrams of algebras" as defined in Mac Lane [McL,71]. Namely:

If $\mathbb{A}$ and $\mathbb{B}$ are categories, then $\mathbb{A}$ is called a *reflective subcategory* of $\mathbb{B}$ iff $\mathbb{A}$ is a full and isomorphism closed subcategory of $\mathbb{B}$, and for every $\mathfrak{B} \in \mathrm{Ob}\,\mathbb{B}$ there exists a morphism $\langle \mathfrak{B}, r, \mathfrak{A} \rangle$ of $\mathbb{B}$ with $\mathfrak{A} \in \mathrm{Ob}\,\mathbb{A}$ such that r is an $\mathbb{A}$-reflection. A morphism $\langle \mathfrak{B}, r, \mathfrak{A} \rangle$ is called an $\mathbb{A}$-*reflection* iff

$$\big(\forall(\mathfrak{B} \xrightarrow{\mathrm{a}} \mathfrak{C}),\ \mathfrak{C} \in \mathrm{Ob}\,\mathbb{A}\big)\,\big(\exists!\mathfrak{A} \xrightarrow{\mathrm{b}} \mathfrak{C})\big)\, b \circ r = a.$$

A *limit* of a diagram of algebras is a concept generalizing the direct product of algebras (in a limit, the projection functions must respect certain homomorphisms as well). These and related category theoretic concepts (like epimorphisms and reflected hull) are understood in the category $\mathsf{Alg}\,(\mathbf{L})$. (This will be relevant in our Corollary 1.5.2.)

Theorem 1.5.1 (Characterization of weak Beth definability property). *Let* $\mathbf{L}$ *be an algebraizable logic with patchwork property. Assume that every element of* $\mathsf{Alg}_m(\mathbf{L})$ *can be extended to a maximal element of* $\mathsf{Alg}_m(\mathbf{L})$*, that is, that* $\mathsf{Alg}_m(\mathbf{L}) \subseteq \mathbf{S}\,\mathbf{max}\,\mathsf{Alg}_m(\mathbf{L})$.

Then conditions (i)–(iii) *below are equivalent.*

(i) $\mathbf{L}$ *has the weak Beth definability property.*

(ii) *The smallest reflective subcategory of* $\mathsf{Alg}\,(\mathbf{L})$ *containing* $\mathbf{max}\,\mathsf{Alg}_m(\mathbf{L})$ *is the whole of* $\mathsf{Alg}\,(\mathbf{L})$.

(iii) $\mathbf{max}\,\mathsf{Alg}_m(\mathbf{L})$ *generates* $\mathsf{Alg}\,(\mathbf{L})$ *by taking limits of diagrams of algebras. That is, there is no limit-closed proper subclass separating these two classes of algebras.*

For the origins of this characterization of weak Beth definability property see [Sai,90, p. 223 and on]. A full proof is going to appear in [And-Nem-Sai,09].

Corollary 1.5.2. *If* $\mathbf{L}$ *satisfies the conditions of Theorem* 1.5.1 *then, denoting* $\mathbf{max}\,\mathsf{Alg}_m(\mathbf{L})$ *by* G*, we have the following implication.*

$$\mathbf{L} \text{ has wBp} \quad \Longrightarrow \quad \text{reflective hull}\,(\mathsf{G}) = \text{limit hull}\,(\mathsf{G})\ \big(= \mathsf{Alg}\,(\mathbf{L})\big).$$

Conjecture 1.5.3 (On the reverse direction of Cor. 1.5.2). Let $\mathbf{L}$ and G be as in Corollary 1.5.2. Then (i) and (ii) below hold.

(i) reflective hull (G) = limit hull (G) = $\mathsf{Alg}\,(\mathbf{L}) \Longleftrightarrow \mathbf{L}$ has wBp.

(ii) reflective hull (G) = limit hull $(\mathsf{G}) \not\Longrightarrow \mathbf{L}$ has wBp, i.e., there is an algebraizable logic $\mathbf{L}$ with reflective hull (G) = limit hull $(\mathsf{G}) \neq \mathsf{Alg}\,(\mathbf{L})$, and $\mathbf{L}$ fails to have wBp.

Proof idea. Let DL and BL denote the category of distributive lattices and that of Boolean lattices, respectively.

For (ii) choose $\mathsf{Alg}\,(\mathbf{L}) = \mathsf{DL}$ and $\mathsf{G} = \mathbf{max}\,\mathsf{Alg}_m(\mathbf{L}) = \mathsf{BL}$. Use Thm. 1.4.6 to show that some $\mathsf{Alg}_m(\mathbf{L})$ $(= \mathsf{BL})$-extendible epimorphisms are not surjective in DL. Conclude that wBp fails in $\mathbf{L}$ and that limit hull$(\mathsf{G}) \neq \mathsf{Alg}\,(\mathbf{L})$ $(= \mathsf{DL})$.

For (i): It seems that (i) is an immediate corollary of Thm. 1.5.1. ∎

For completeness, we note that there are further logic properties strongly related to definability. Such are e.g. *Craig's interpolation property* and its variants, see e.g. [And-Nem-Sai,01], [Hoo,01], and the so called *omitting types property* [And-Nem-Say,08]. Algebraic characterizations (similar in spirit to the present ones) are available e.g. in [And-Nem-Sai,01], [And-Nem-Say,08], [Hoo,01], [Mak,92], [Sai,90].[13]

In Institutions Theory, too, algebraic characterizations of Craig's interpolation and Beth's definability properties of logics are studied, cf. [Mos-Gog-Dia-Tar,05, pp. 123–124, items 4.11 and below 4.12], in a spirit similar to the present Andréka–Németi–Tarski approach. A similar remark applies to works of Blok–Pigozzi, Hodkinson, Hoogland, Maksimova, Marx, Diaconescu in the references.

6. Perspective

In this paper we arrived at *algebraic logic* from two different perspectives.

The first one is using *algebraic methods* in investigating logics. The point in this is that, in many cases, algebraic statements are easier to prove, with the advanced methods and techniques of this well-developed part of mathematics. (Works of Bjarni Jónsson, Willem J. Blok, Don Pigozzi might illustrate these methods.) The algebraic forms of logic statements are often even clearer, more transparent or elegant than the original logic statements. In the most fortunate cases, after having proved such an algebraic statement, one can conclude that the logic statement that we started with, holds. (Otherwise we may obtain some weaker results only, like implications in one direction only.) This way we build a *bridge* between the fields Logic and Algebra, for the benefit of both of them. Examples for such "bridge theorems" can be found in our sections 4 and 5. Cf. also [Mik,95, p. 18], [And-Nem-Sai,01, p. 186] for this terminology/metaphor.

Our second perspective for arriving at algebraic logic is investigating *definability*. In their pioneering paper [Blo-Pig,89], Blok and Pigozzi explain that definability is a corner stone of algebraic logic. To illustrate this, let us recall from any textbook on the subject (cf. e.g. [Mon,00], [Hen-Mon-Tar,85], [Hen-Mon-Tar-And-Nem,81], [And-Nem-Sai,09]) that the algebraic version

[13] Algebraic characterizations of further variants of Beth property are e.g. in [Hoo,01, Sec. 2.2.3 (pp. 17–19), subsections 3.3, 3.4].

$\mathfrak{Cs}(\mathfrak{M})$ of a model $\mathfrak{M}$ of first order logic is an algebra the universe of which is the collection $\{mng_{\mathfrak{M}}(\varphi) : \varphi \text{ a formula}\}$ of all *definable relations* of the model $\mathfrak{M}$; where $mng_{\mathfrak{M}}(\varphi) = \{k \in {}^{\omega}M : \mathfrak{M} \models \varphi[k]\}$.

In connection with definability, let us also remember that definability was one of Alfred Tarski's favourite subject already in the 1930's. In the paper [Tar,34] he formulated and started the project of bringing about a *definability theory.* The fact that, before exploring logic, Tarski did research in sciences and in the methodology of science, indicates that he might well have motivations coming from his scientific experience. In this line it is remarkable that Hans Reichenbach, in his book [Rei,20] (already in 1920), explains that definability is a basic factor in *relativity theory.* This idea appears already in Einstein's work, in 1905, but more implicitly than in [Rei,20]. Cf. also [And-Mad-Nem,01], [And-Mad-Nem,07], [Mad,02], [Sze,09].

Finally we would like to mention a new trend in definability theory, as follows.

Beth's famous theorem on definability was designed for defining new *relations* (new properties). Most of its later generalizations also concern defining new relations or new atomic formulas. That is, the tradition is defining *new properties* over existing entities. But there is a need for defining *new entities* over existing entities, as well. A number of authors, e.g. Wilfrid Hodges, Dale Myers, Anand Pillay and Saharon Shelah raised this problem. (See e.g. [Hod,88], where Hodges gives ideas on a possible solution as well.) Here is a formulation of the problem.

Problem 1.6.1. Find a concept of definability that concerns definability of *new (sets of) entities* as well (with a Beth type theorem). Extend this to (introducing a logic property in) the spirit of Universal Logic.

The paper Andréka–Madarász–Németi [And-Mad-Nem,01] gives a solution to the first part of this problem in that the authors prove a Beth type theorem for defining *new universes* in many-sorted first order logic. The trick is that they define new relations as well, together with the new universe(s); these new relations provide the connection between the old universes and the one(s) being defined.

Problem 1.6.2 below was motivated in [Nem-And,94] (cf. items 3.23–24 there). The problem was first posed in [Nem-And,94, p. 429, Problem 3.24] and is still open. It was re-stated and further motivated in [And-Nem-Sai,01, Problem 26 (p. 181), Problem 64 (p. 223), Thm. 52 (p. 211)], and more generally in Sec. 6.6 (pp. 222–226) there.

Problem 1.6.2. Is there a finitely axiomatizable quasi-variety $\mathsf{K} \subseteq \mathsf{RCA}_n$ such that $\mathbf{H}\mathsf{K} = \mathsf{RCA}_n$? ($\mathbf{H}\mathsf{K}$ denotes the class of all homomorphic images of algebras in K.)

In connection with our open problems we note the following recent news. The problem left open in the algebraic logic book Tarski–Givant [Tar-Giv,87, Sec. 3.10] was recently solved by Hajnal Andréka and István Németi [And-Nem,11a]. They proved that first order logic (and set theory) can be built up (interpreted or defined in a sense) in the equational theory of diagonal free cylindric algebras of dimension 3 (Df_3). Cf. the second problem posed in Sec. 3.10 of [Tar-Giv,87] (beginning with the last 3 lines of p. 89 and footnote 18 on p. 90).

Bibliography

[Aho-Sag-Ull,79] Aho, A. V., Sagiv, Y. and Ullman, J. D., Equivalences among relational expressions, *SIAM J. of Computing,* **8** (1979), 218–246.

[Ale,95] Alechina, N., *Modal quantifiers,* Doctoral Dissertation, University of Amsterdam, 1995. vi+121pp.

[Ale,00] Alechina, N., Functional dependencies between variables, *Studia Logica,* **66,2** (2000), 273–283.

[Ale-Lam,96] Alechina, N. and van Lambalgen, M., Generalized quantification as substructural logic, *J. Symbolic Logic,* **61,3** (1996), 1006–1044.

[Ame-Say,06] Amer, M. and Sayed Ahmed, T., Polyadic and cylindric algebras of sentences, *Math. Logic Quarterly,* **52** (2006), 444–449.

[And,87] Andréka, H., Boolean reducts of relation and cylindric algebras and the cube problem, *Proc. Amer. Math. Soc.,* **100** (1987), 148–153.

[And,91] Andréka, H., *Complexity of equations valid in algebras of relations,* Doctoral Dissertation with the Hungarian Academy of Sciences, Budapest, 1991.

[And,97a] Andréka, H., Complexity of equations valid in algebras of relations. I. Strong non-finitizability, *Ann. Pure and Appl. Logic,* **89** (1997), 149–209.

[And,97b] Andréka, H., Complexity of equations valid in algebras of relations. II. Finite axiomatizations, *Ann. Pure and Appl. Logic,* **89** (1997), 211–229.

[And,01] Andréka, H., A finite axiomatization of locally square cylindric-relativized set algebras, *Studia Sci. Math. Hungar.,* **38 (1–4)** (2001), 1–11.

[And-Ben-Nem,95a] Andréka, H., van Benthem, J. and Németi, I., Back and forth between modal logic and classical logic, *Logic J. IGPL,* **3** (1995), 685–720.

[And-Ben-Nem,95b] Andréka, H., van Benthem, J. and Németi, I., Submodel preservation theorems in finite variable fragments, in: *Modal Logic and Process Algebra. A Bisimulation Perspective* (Ponse, A., Rijke, M., Venema, Y., eds.), CSLI Lecture Notes No. 53, CSLI Publications (1995), 1–11.

[And-Ben-Nem,98] Andréka, H., van Benthem, J. and Németi, I., Modal Languages and Bounded Fragments of Predicate Logic, *J. of Philosophical Logic,* **27** (1998), 217–274.

[And-Blo-Nem-Pig-Sai,97] Andréka, H., Blok, W. J., Németi, I., Pigozzi, D. and Sain, I., Abstract Algebraic Logic, manuscript, incomplete.

[And-Com-Mad-Nem-Say,09] Andréka, H., Comer, S. D., Madarász, J. X., Németi, I. and Sayed Ahmed, T., Epimorphisms in cylindric algebras, *Algebra Universalis,* **61** (2009), 261–281.

[And-Com-Nem,83] Andréka, H., Comer, S. D. and Németi, I., Epimorphisms in cylindric algebras, manuscript (1983). Abstracted in: Abstr. Amer. Math. Soc., **4,7** (1983), 283.

[And-Ger-Nem,73] Andréka, H., Gergely, T. and Németi, I., Purely algebraic construction of first order logics, *Publications of Central Res. Inst. for Physics, Hungar. Acad. Sci., Budapest,* No. KFKI-73-71 (1973), 46 pp.

[And-Ger-Nem,77] Andréka, H., Gergely, T. and Németi, I., On Universal Algebraic Construction of Logics, *Studia Logica,* **36** (1977), 9–47.

[And-Giv-Mik-Nem-Sim,98] Andréka, H., Givant, S. R., Mikulás, Sz., Németi, I. and Simon, A., Notions of density that imply representability in algebraic logic, *Ann. Pure and Appl. Logic,* **91** (1998), 93–190.

[And-Giv-Nem,95] Andréka, H., Givant, S. R. and Németi, I., Perfect extensions and derived algebras, *J. Symbolic Logic,* **60** (1995), 775–796.

[And-Giv-Nem-Sim,97] Andréka, H., Givant, S. R., Németi, I. and Simon, A., Persistent properties and an application to algebras of logic, *Algebra Universalis,* **38** (1997), 141–149.

[And-Gol-Nem,98] Andréka, H., Goldblatt, R. and Németi, I., Relativised quantification: some canonical varieties of sequence-set algebras, *J. Symbolic Logic,* **63** (1998), 163–184.

[And-Hod-Nem,99] Andréka, H., Hodkinson, I. and Németi, I., Finite algebras of relations are representable on finite sets, *J. Symbolic Logic,* **64** (1999), 243–267.

[And-Jon-Nem,90] Andréka, H., Jónsson, B. and Németi, I., Relatively free relation algebras, in: *Algebraic Logic and Universal Algebra in Computer Science,* [Ber-Mad-Pig,90], 1–14.

[And-Jon-Nem,91] Andréka, H., Jónsson, B. and Németi, I., Free algebras in discriminator varieties, *Algebra Universalis,* **28** (1991), 401–447.

[And-Kur-Nem-Sai,94] Andréka, H., Kurucz, A., Németi, I. and Sain, I., Applying Algebraic Logic to Logic; A General Methodology, in: *Algebraic Logic and the Methodology of Applying it,* Tempus Summer School (Budapest, Hungary, 1994).

[And-Kur-Nem-Sai,95] Andréka, H., Kurucz, A., Németi, I. and Sain, I., General algebraic logic including algebraic model theory: an overview, in: *Logic Colloquium '92,* [Csir-Gab-Rij,95], 1–60.

[And-Kur-Nem-Sai-Sim,94] Andréka, H., Kurucz, A., Németi, I., Sain, I. and Simon, A., Exactly which logics touched by the dynamic trend are decidable? in: *Proceedings of 9th Amsterdam Colloquium,* 1993 (P. Dekker and M. Stokhof, eds.), ILLC, Department of Philosophy, University of Amsterdam (1994), 67–86.

[And-Kur-Nem-Sai-Sim,96] Andréka, H., Kurucz, A., Németi, I., Sain, I. and Simon, A., Causes and remedies for undecidability in arrow logics and in multi-modal logics, in: *Arrow Logic and Multi-modal Logic,* [Mar-Pol-Mas,96], 63–99.

[And-Mad-Nem,01] Andréka, H., Madarász, J. X. and Németi, I., *Definability of New Universes in Many-sorted Logic,* preprint (2001).

[And-Mad-Nem,02a] Andréka, H., Madarász, J. X. and Németi, I., Algebraic Logic, in: *Supplement III of Encyclopedia of Mathematics* (M. Hazewinkel, ed.), Kluwer Academic Publishers (2002), 31–34.

[And-Mad-Nem,02b] Andréka, H., Madarász, J. X. and Németi, I., Relational Algebras, in: *The Concise Handbook of Algebra* (A. V. Mikhalev, G. F. Pilz, eds.), Kluwer Akademic Publishers, Dordrecht, Boston, London, 2002. pp. 478–482.

[And-Mad-Nem,04] Andréka, H., Madarász, J. X. and Németi, I., Algebras of relations of various ranks, some current trends and applications, *J. of Relational Methods in Computer Science,* **1** (2004), 27–49.

[And-Mad-Nem,07] Andréka, H., Madarász, J. X. and Németi, I., Logic of space-time and relativity theory, in: *Handbook of Spatial Reasoning* (M. Aiello, J. van Benthem and I. Pratt, eds.), Springer, Dordrecht (2007), 607–711.

[And-Mik,93] Andréka, H. and Mikulás, Sz., The Completeness of the Lambek Calculus with Respect to Relational Semantics, *J. of Logic, Language and Information,* **3** (1993), 1–37.

[And-Mon-Nem,91a] Andréka, H., Monk, J. D. and Németi, I., Open problems, in: *Algebraic Logic,* [And-Mon-Nem,91b].

[And-Mon-Nem,91b] Andréka, H., Monk, J. D. and Németi, I. (eds.), *Algebraic logic* (Budapest, 1988), Colloq. Math. Soc. János Bolyai, 54, North-Holland, Amsterdam, 1991.

[And-Nem,75] Andréka, H. and Németi, I. A simple purely algebraic proof of the completeness of some first order logics, *Algebra Universalis,* **5** (1975), 8–15.

[And-Nem,80] Andréka, H. and Németi, I., On systems of varieties definable by schemes of equations, *Algebra Universalis,* **11** (1980), 105–116.

[And-Nem,84] Andréka, H. and Németi, I., On a problem of J. S. Johnson about representability of polyadic algebras, preprint (1984).

[And-Nem,85] Andréka, H. and Németi, I., On the number of generators in cylindric algebras, *J. Symbolic Logic,* **50**, (1985), 865–873.

[And-Nem,88] Andréka, H. and Németi, I., Relation algebraic conditions for representability of cylindric and polyadic algebras, preprint, Mathematical Institute Budapest, 1988. 46 pp.

[And-Nem,91] Andréka, H. and Németi, I., Representable positive cylindric algebras do not form a variety, manuscript, Mathematical Institute, Hungarian Academy of Sciences (1991).

[And-Nem,94] Andréka, H. and Németi, I., Simple proof of decidability of the universal theory of cylindric set algebras of dimension 2, in: *Algebraic Logic and the Methodology of Applying it,* Tempus Summer School (Budapest, Hungary, 1994).

[And-Nem,11a] Andréka, H. and Németi, I., Formalizing set theory in weak logics, searching for the weakest logic with Gödel's incompleteness property. Rényi Institute of Mathematics, Budapest, Preprint (2011), 26pp. arXiv:1111.0995v1

[And-Nem,11b] Andréka, H. and Németi, I., Reducing predicate logic to propositional logic, Part II. Notes for a talk in Department of Logic, ELTE Budapest, December 9, 2011. `http://www.math-inst.hu/pub/algebraic-logic/LapomDf2011b.pdf`

[And-Nem,thisVol] Andréka, H. and Németi, I., Reducing first-order logic to Df_3, free algebras, in this volume.

[And-Nem-Sai,79] Andréka, H., Németi, I. and Sain, I., Henkin-type semantics for program schemes to turn negative results to positive, in: *Fundamentals of Computation Theory'79* (Proc. Conf. Berlin, 1979) (L. Budach, ed.), Akademie Verlag (Berlin, 1979), 18–24.

[And-Nem-Sai,82] Andréka, H., Németi, I. and Sain, I., A complete logic for reasoning about programs via nonstandard model theory, Parts I–II, *Theoretical Computer Science,* **17(2–3)** (1982), 193–212 and 259–278.

[And-Nem-Sai,94a] Andréka, H., Németi, I. and Sain, I., Applying algebraic logic to logic, in: *Algebraic methodology and software technology* (AMAST'93) (M. Nivat, C. Rattray, T. Rus and G. Scollo, eds.), Proc. Twente, The Netherlands, June 1993), Springer-Verlag (London, 1994), 7–28. A more detailed version of this is in [And-Kur-Nem-Sai,94].

[And-Nem-Sai,94b] Andréka, H., Németi, I. and Sain, I., Craig property of a logic and decomposability of theories, in: *Proceedings of 9th Amsterdam Colloquium* (1993) (P. Dekker and M. Stokhof, eds.), ILLC, Department of Philosophy, University of Amsterdam (1994), 87–93.

[And-Nem-Sai,01] Andréka, H., Németi, I. and Sain, I., Algebraic Logic, in: *Handbook of Philosophical Logic,* Vol. 2, second edition, eds. D. M. Gabbay and F. Guenthner, Kluwer Academic Publishers (2001), 133–247.

[And-Nem-Sai,09] Andréka, H., Németi, I. and Sain, I., *Universal Algebraic Logic (Dedicated to the Unity of Science),* Studies in Universal Logic, Birkhäuser (Basel–Boston–Stuttgart), in preparation.

[And-Nem-Say,97] Andréka, H., Németi, I. and Sayed Ahmed, T., On neat reducts of algebras of logic, Presented in: *Logic Colloquium 1996.* Bull. Symbolic Logic, **3** (1997), 249.

[And-Nem-Say,98] Andréka, H., Németi, I. and Sayed Ahmed, T., Amalgamation in algebras of logic, Presented in the"Universal Algebra" conference held in Szeged, Hungary (1998).

[And-Nem-Say,08] Andréka, H., Németi, I. and Sayed Ahmed, T., Omitting types for finite variable fragments and complete representations of algebras, *J. Symbolic Logic,* **73** (2008), 65–89.

[And-Nem-Say,12] Andréka, H., Németi, I. and Sayed Ahmed, T., A non-representable infinite dimensional quasi-polyadic equality algebra with representable cylindric reduct, *Studia Sci. Math. Hungar.,* to appear.

[And-Nem-Tho,90] Andréka, H., Németi, I. and Thompson, R. J., Weak cylindric set algebras and weak subdirect indecomposability, *J. Symbolic Logic,* **55** (1990), 577–588.

[And-Sai,81] Andréka, H. and Sain, I., Connections between algebraic logic and initial algebra semantics of CF languages, in: *Mathematical Logic in Computer Science* (Proc. Coll. Salgótarján, 1978) (B. Dömölki and T. Gergely, eds.), Colloq. Math. Soc. J. Bolyai, 26, North-Holland (Amsterdam, 1981), 25–83.

[And-Say,99] Andréka, H. and Sayed Ahmed, T., Omitting types in logics with finitely many variables, Presented in: Logic Colloqium 1998, *Bull. Symbolic Logic,* **5** (1999), 88.

[And-Tho,88] Andréka, H. and Thompson, R. J., A Stone type representation theorem for algebras of relations of higher rank, *Trans. Amer. Math. Soc.,* **309** (1988), 671–682.

[And-Tuz,88] Andréka, H. and Tuza, Zs., Nonfinite axiomatizability of the polyadic operations in algebraic logic, *Abstr. Amer. Math. Soc.,* **9,6** (1988), 500.

[Ane-Hou,91] Anellis, I. H. and Houser, N., Nineteenth century roots of algebraic logic and universal algebra, in: *Algebraic Logic,* [And-Mon-Nem,91b], 1–36.

[Azr,79] Azriel, L., *Basic Set Theory,* Springer-Verlag, Berlin, Heidelberg, New York (1979).

[Bac,85] Bacon, J., The completeness of a predicate-functor logic, *J. Symbolic Logic,* **50** (1985), 903–926.

[Bal-Dwi,74] Balbes, R. and Dwinger, P., *Distributive Lattices,* University of Missouri Press, 1974.

[Banach,93] *Algebraic Methods in Logic and in Computer Science* (Warsaw, 1991), Banach Center Publ., 28, Polish Acad. Sci., Warsaw, 1993.

[Bar-Cat-Seg,11] Bárány, V., ten Cate, B. and Segoufin, L., *Guarded Negation,* Proceedings ICALP, Zurich (2011), 356–367.

[Bar-Got-Ott,10] Bárány, V., Gottlob, G. and Otto, M., *Querying the Guarded Fragment,* Proceedings of IEEE LICS (2010), 1–10.

[Bar-Jud,95] Bartoszynski, T. and Judah, H., *Set Theory. On the structure of the real line,* A. K. Peters, Ltd., Wellesley, MA, (1995).

[Bar,75] Barwise, J., *Admissible Sets and Structures,* Springer-Verlag, Berlin, 1975.

[Bar-Fef,85] Barwise, J. and Feferman, S. (eds.), *Model-Theoretic Logics,* Springer–Verlag (Berlin, 1985).

[Bar-Sel,95] Barwise, J. and Seligman, J., *Information Flow,* Cambridge University Press, Cambridge (1995).

[Bas,58] Bass, H., Finite monadic algebras, *Proc. Amer. Math. Soc.,* **9** (1958), 258–268.

[Ben,79] van Benthem, J., Canonical modal logics and ultrafilter extensions, *J. Symbolic Logic,* **44** (1979), 1–8.

[Ben,80] van Benthem, J., Some kinds of modal completeness, *Studia Logica,* **39** (1980), 125–141.

[Ben,83] van Benthem, J., *Modal Logic and Classical Logic,* Bibliopolis, Napoli (1983).

[Ben,88] van Benthem, J., A note on Jónsson's theorem, *Algebra Universalis,* **25** (1988), 391–393.

[Ben,91] van Benthem, J., *Language in Action, Categories, Lambdas and Dynamic Logic,* North-Holland, Amsterdam and MIT Press, Cambridge (Mass.) (1991).

[Ben,93] van Benthem, J., Modal Frame Classes Revisited, *Fundamenta Informaticae,* **18** (1993), 307–317.

[Ben,94] van Benthem, J., Dynamic Arrow Logic, in: *Logic and Information Flow* (J. van Eijck and A. Visser, eds.), The MIT Press, Cambridge (Mass.) (1994), 15–29.

[Ben,95] van Benthem, J., *What Makes Modal Logic Tick?,* manuscript, Institute for Logic, Language and Computation, University of Amsterdam (1995).

[Ben,96a] van Benthem, J., *Exploring logical dynamics,* in: *Studies in Logic, Language and Information,* CSLI Publications, Stanford, CA; FoLLI: European Association for Logic, Language and Information, Amsterdam, 1996. x+329 pp.

[Ben,96b] van Benthem, J., Complexity of contents versus complexity of wrappings, in: *Arrow Logic and Multi-modal Logic,* [Mar-Pol-Mas,96], 203–219.

[Ben,97a] van Benthem, J., *Dynamic Bits and Pieces,* Report LP-97-01, Institute for Logic, Language and Computation, University of Amsterdam (1997).

[Ben,97b] van Benthem, J., Modal Foundations for Predicate Logic, *Bull. of the IGPL,* **5** (1997), 259–286.

[Ben,01] van Benthem, J., Modal Logic in Two Gestalts, in: *Advances in Modal logic,* vol. II, [Rij-Seg-Wan-Zak,01], 73–100.

[Ben,05] van Benthem, J., Guards, Bounds, and Generalized Semantics, *J. of Logic, Language and Information,* **14** (2005), 263–279.

[Ben,11] van Benthem, J., *Logical Dynamics of Information and Interaction,* Cambridge University Press, Cambridge (2011).

[Ben,thisVol] van Benthem, J., Crs and guarded logics: a fruitful contact, in this volume.

[Ben-Bon,08] van Benthem, J. and Bonnay, D., Modal logic and invariance, *J. of Applied and Non-classical Logic,* **18** (2008), 153–173.

[Ben-Doe,83] van Benthem, J. and Doets, K., Higher-Order Logic, in: *Handbook of Philosophical Logic, vol. I.* (D. M. Gabbay and F. Guenthner, eds.), Reidel, Dordrecht (1983), 275–329.

[Ben-Mar,08] van Benthem, J. and Martinez, M.-C., The Stories of Logic and Information, in: P. Adriaans and J. van Benthem, eds., *Handbook of the Philosophy of Information,* Elsevier, Amsterdam (2008), 217–280.

[Ben-Pac,06] van Benthem, J. and Pacuit, E., The Tree of Knowledge in Action, *Proceedings Advances in Modal Logic,* ANU Melbourne (2006), 87–106.

[Ber,96] van den Berg, M., *The internal structure of discourse,* Dissertation, Institute for Logic, Language and Computation, University of Amsterdam (1996).

[Ber-Mad-Pig,90] Bergman, C., Maddux, R. D. and Pigozzi, D. (eds.), *Algebraic Logic and Universal Algebra in Computer Science,* Lecture Notes in Computer Science, 425, Springer, 1990.

[Bez,01] Bezhanishvili, G., Locally finite varieties, *Algebra Universalis,* **46** (2001), 531–548.

[Bez,02] Bezhanishvili, N., Varieties of two-dimensional cylindric algebras. Part I: Diagonal-free case, *Algebra Universalis,* **48** (2002), 11–42.

[Bez,04b] Bezhanishvili, N., Varieties of two-dimensional cylindric algebras II, *Algebra Universalis,* **51** (2004), 177–206.

[Bez,06] Bezhanishvili, N., *Lattices of intermediate and cylindric modal logics,* PhD thesis, ILLC, University of Amsterdam, 2006.

[Bez,thisVol] Bezhanishvili, N., Varieties of two-dimensional cylindric algebras, in this volume.

[Bez-Hod,04] Bezhanishvili, N. and Hodkinson, I., All normal extensions of $S5$-squared are finitely axiomatizable, *Studia Logica,* **78** (2004), 443–457.

[Bez-Mar,03] Bezhanishvili, N. and Marx, M., All proper normal extensions of $S5$-square have the polynomial size model property, *Studia Logica,* **73** (2003), 367–382.

[Bez,05] Béziau, J-Y. (ed.) *Logica Universalis. Towards a General theory of logic.* Birkhäuser Verlag, 2005.

[Bez,10] Béziau, J-Y., Preface: Is logic universal? *Logica Universalis,* **4** (2010), 161–162.

[Bir,84] Bíró, B., On isomorphic but not lower-base-isomorphic cylindric set algebras, *Bulletin Section of Logic, Łodź,* **13** (1984), 230–232.

[Bir,87a] Bíró, B., Isomorphic but not base-isomorphic base-minimal cylindric set algebras, *Algebra Universalis,* **24** (1987), 292–300.

[Bir,87b] Bíró, B., Isomorphism does not imply lower base-isomorphism for nonregular or nonlocally finite-dimensional cylindric algebras, *Comment. Math. Univ. Carolin.,* **28** (1987), 221–226.

[Bir,89] Bíró, B., Isomorphic but not lower base-isomorphic cylindric algebras of finite dimension, *Notre Dame J. of Formal Logic,* **30** (1989), 262–267.

[Bir,90] Bíró, B., On generalizations of a theorem of Vaught, *Notre Dame J. of Formal Logic,* **31** (1990), 330–336.

[Bir,92] Bíró, B., Non-finite-axiomatizability results in cylindric algebras, *J. Symbolic Logic,* **57** (1992), 832–843.

[Bir-She,88] Bíró, B. and Shelah, S., Isomorphic but not lower base-isomorphic cylindric set algebras, *J. Symbolic Logic,* **53** (1988), 846–853.

[Bla-Rij-Ven,01] Blackburn, P., de Rijke, M. and Venema, Y., *Modal Logic,* 53 in Cambridge Tracts in Theoretical Computer Science. Cambridge University Press, 2001.

[Blo,80] Blok, W. J., Pretabular varieties of modal algebras, *Studia Logica,* **39** (1980), 101–124.

[Blo-Hoo,06] Blok, W. J. and Hoogland, E., The Beth property in algebraic logic, *Studia Logica,* **83** (2006), 49–90.

[Blo-Pig,89] Blok, W. J. and Pigozzi, D., *Algebraizable Logics,* Memoirs of Amer. Math. Soc., **77** (1989), 396.

[Bor,88] Börner, F., *Operationen auf Relationen,* Dissertation, Karl-Marx-Universität Leipzig, (1988).

[Bor-Kra,99] Börner, F., *Krasneralgebren,* (German) [Krasner algebras] Habilitationsschrift, Universität Potsdam, Potsdam, 1999. Logos Verlag Berlin, Berlin, 2000. 194 pp.

[Bul-Hod] Bulian, J. and Hodkinson, I., Bare canonicity of representable cylindric and polyadic algebras, submitted.

[Buk,75] Bukhari, S. A., *A relation algebra model for data bases,* Research report CS-75-11, Department of Computer Science, University of Waterloo (1975).

[Bur,91] Burch, R. W., A Peircean reduction thesis, in: *The foundations of topological logic.* Philosophical Inquiries, 1. Texas Tech University Press, Lubbock, TX, 1991. xvi+152 pp.

[Bur,80] Burgess, J. P., Decidability for branching time, *Studia Logica,* **39** (1980), 203–218.

[Bur,03] Burgess, J. P., A remark on Henkin sentences and their contraries, *Notre Dame J. of Formal Logic,* **44(3)** (2003), 185–188.

[Bus,01] Van den Bussche, J., Applications of Alfred Tarski's ideas in database theory, in: *Computer Science Logic* (L. Fribourg, ed.), 15th International Workshop, CSL 2001. 10th Annual Conference of the EACSL, Paris, France, 2142 of *Lecture Notes in Computer Science,* 20–37. Springer Verlag (2001).

[Cam-Hod,01] Cameron, P. and Hodges, W., Some combinatorics of imperfect information, *J. Symbolic Logic,* **66** (2001), 673–684.

[Cat,05] ten Cate, B., *Model Theory for Extended Modal Languages,* Dissertation, Institute for Logic, Language and Computation, University of Amsterdam (2005).

[Cha-Kei,90] Chang, C. C. and Keisler, H. J., *Model theory.* North-Holland, 3rd edition (Amsterdam, 1990).

[Cha-Zak,97] Chagrov, A. and Zakharyaschev, M., *Modal Logic,* 35 of *Oxford Logic Guides,* Clarendon Press, Oxford, 1997.

[Chi-Tar,48] Chin, L. H. and Tarski, A., Remarks on projective algebras, *Bull. Amer. Math. Soc.,* **54** (1948), 80–81.

[Chi-Tar,51] Chin, L. H. and Tarski, A., Distributive and modular laws in the arithmetic of relation algebras, *University of California Publications in Mathematics, New Series,* **1** (1951), 341–384.

[Chu,77] Chuaqui, R., B., A semantical definition of probability, in: *Non-classical logics, Model Theory, and Computability,* Arruda, Costa and Chuaqui, eds., North Holland (1977).

[Cie,97] Ciesielski, K., *Set Theory for the Working Mathematician,* London Mathematical Society Student Texts, 39, 1997.

[Cig,91] Cignoli, R., Quantifiers on distributive lattices, *Discrete Mathematics,* **96(3)** (1991), 183–197.

[Cir,86a] Cirulis, J., An abstract description of data types and of varieties of data algebras (in Russian), in: *Algebra and discrete Mathematics, Latvian University, Riga,* (1986), 131–144.

[Cir,86b] Cirulis, J., Generalizing the notion of polyadic algebra, *Bulletin Section of Logic, Łodź,* **15** (1986), 2–7.

[Cir,88] Cirulis, J., Relational algebras in databases and generalized cylindric algebras. (In Russian), *Theoretical Questions of Programming, Latvian University, Riga,* (1988), 100–107.

[Cir,90] Cirulis, J., On heterogeneous quantifier algebras, *Mathematics, I* (Russian), 121–126, Latv. Univ. Zināt. Raksti, 552, Latv. Univ., Riga, 1990.

[Cir,91] Cirulis, J., An algebraization of first order logic with terms, in: *Algebraic Logic,* [And-Mon-Nem,91b], 125–146. *Acta Un. Latv.,* **595** (1994), 49–52.

[Cir,94a] Cirulis, J., Corrections to my paper: "An algebraization of first order logic with terms" in: *Algebraic Logic,* [And-Mon-Nem,91b], 125–146.

[Cir,94b] Cirulis, J., Abstract algebras of finitary relations: several non-traditional axiomatizations, *Mathematics,* 23–48, Latv. Univ. Zināt. Raksti, 595, Latv. Univ., Riga, 1994.

[Cir,94c] Cirulis, J., Beniaminov algebras revised: one more algebraic version of first-order logic, *Bulletin Section of Logic, Łodź,* **23** (1994), 140–145.

[Cir,03] Cirulis, J., Multi-algebras from the viewpoint of algebraic logic, *Algebra Discrete Math.,* 2003, 20–31.

[Cir,08] Cirulis, J., Quantifiers on multiplicative semilattices. I, *Contributions to general algebra,* **18** (Klagenfurt, 2008), 31–45.

[Cod,70] Codd, E. F., A relational model of data for large shared databanks, *Comm. of the ACM,* **13** (1970), 377–387.

[Cod,79] Codd, E. F., Extending the database relational model to capture more meaning, *ACM Trans. Database Syst.,* **4** (1979), 397–434.

[Com,69] Comer, S. D., Classes without the amalgamation property, *Pacific J. of Mathematics,* **28** (1969), 309–318.

[Com,72] Comer, S. D., A sheaf-theoretic duality theory for cylindric algebras, *Trans. Amer. Math. Soc.,* **169** (1972), 75–87.

[Com,84] Comer, S. D., Galois theory for cylindric algebras and its applications, *Trans. Amer. Math. Soc.,* **286** (1984), 771–785.

[Com,86] Comer, S. D., Combinatorial types, in: *Algebra, Combinatorics and Logic in Computer Science,* Vol. I, II (Győr, 1983), Colloq. Math. Soc. János Bolyai, 42, North-Holland (Amsterdam, 1986), 267–284.

[Com,91a] Comer, S. D., The representation of 3-dimensional cylindric algebras, in: *Algebraic Logic,* [And-Mon-Nem,91b], 147–172.

[Com,91b] Comer, S. D., A remark on representable positive cylindric algebras, *Algebra Universalis,* **28** (1991), 150–151.

[Com,93] Comer, S. D., On connections between information systems, rough sets and algebraic logic, in: *Algebraic Methods in Logic and in Computer Science,* [Banach,93], 117–124.

[Coo,03] Cooper, S. B., *Computability theory,* Chapman Hall/CRC Mathematics Series (2003).

[Cos,85] Cosmadakis, S., *Equational theories and database constraints,* Doctoral dissertation, Massachusetts Institute of Technology (1985).

[Cos,87] Cosmadakis, S., Database theory and cylindric lattices, in: *IEEE 28th Symposion on the Foundations of Computer Science,* (1987), 411–420.

[Cra,74] Craig, W., *Logic in Algebraic Form,* North-Holland, Amsterdam (1974).

[Cra,06] Craig, W., *Semigroups Underlying First-order Logic,* Mem. Amer. Math. Soc., **184** (2006), xxvi+263 pp.

[Cra-Vau,58] Craig, W. and Vaught, R. L., Finite axiomatizability using additional predicates, *J. Symbolic Logic,* **23** (1958), 289–308.

[Csir-Gab-Rij,95] Csirmaz, L., Gabbay, D. M. and de Rijke, M. (eds.), *Logic Colloquium '92* (Veszprém, 1992), Stud. Logic Lang. Inform., CSLI Publ., Stanford, CA, 1995.

[Dai,64] Daigneault, A., Freedom in polyadic algebras and two theorems of Beth and Craig, *Michigan Math. J.,* **1** (1964), 129–135.

[Dai-Mon,63] Daigneault, A. and Monk, J. D., Representation theory for polyadic algebras, *Fundamenta Math.,* **52** (1963), 151–176.

[Daw-Hel-Kol,95] Dawar, A., Hella, L. and Kolaitis, Ph. G., Implicit definability and infinitary logic in finite model theory, in: *Proc. 22nd Int. Coll. on Automata, Languages and Programming,* Lecture Notes in Computer Science 944, Springer Verlag (1995). 624–635.

[Dec,05] Dechesne, F., *Game, Set, Maths: Formal investigations into logic with imperfect information,* PhD thesis, Universiteit van Tilburg, 2005.

[Die,97] Diestel, R., *Graph theory,* 173 of Graduate Texts in Mathematics. Springer-Verlag (Berlin, 1997).

[Dis,90] Diskin, Z. B., Polyadic algebras for non-classical logics, Parts I–IV, *Latv. Mat. Ezhegodnik,* **35, 36** (1991).

[Dis,93] Diskin, Z. B., *Abstract Universal Algebraic Logic,* Univ. of Latvia, Riga. Research Report, 1993. 39 pp.

[Dis,94] Diskin, Z. B., When is a semantically defined logic algebraizable? *Mathematics,* 57–82, Latv. Univ. Zinãt. Raksti, 595, Latv. Univ., Riga, 1994.

[Dor,92] Đorđević, R. S., Analytic completeness theorem for absolutely continuous biprobability models, *Zeitschrift Math. Logik Grundlagen Math.,* **38** (1992), 241–246.

[Dor,93] Đorđević, R. S., Analytic completeness theorem for singular biprobability models, *Math. Logic Quarterly,* **39** (1993), 228–230.

[Dor-Ras,thisVol] Đorđević, R. S. and Rašković, M., Cylindric probability algebras, in this volume.

[Dor-Ras-Ogn,04] Đorđević, R. S., Rašković, M. and Ognjanović, Z., Completeness theorem for propositional probabilistic models whose measures have only finite ranges, *Arch. Math. Logic,* **43** (2004), 557–564.

[Dos,88] Došen, K., Second-order logic without variables, in: *Categorial Grammar* (W. Buszkowski, W. Marciszewski and J. van Benthem, eds.) Linguistic & Literary Studies in Eastern Europe, Amsterdam (1988), 245–264.

[Dun,89] Düntsch, I., Cylindric algebras and Codd's relational model of data (Part I), preprint, University of Brunei, Darussalam, Negara Brunei, Dept. Math., Dec. 1989.

[Dun,91] Düntsch, I., Cylindric lattices, manuscript, Univ. of Osnabrück, Germany (1991).

[Dun,93] Düntsch, I., A note on cylindric lattices, in: *Algebraic Methods in Logic and in Computer Science,* [Banach,93], 231–238.

[Dun,thisVol] Düntsch, I., Cylindric algebras and relational databases, in this volume.

[Dun-Mik,01] Düntsch, I. and Mikulás, Sz., Cylindric structures and dependencies in relational databases, *Theoret. Comput. Sci.,* **269** (2001), 451–468.

[Dun-Pra,09] Düntsch, I. and Pratt-Hartmann, I., Complex algebras of arithmetic, *Fund. Inform.,* **97** (2009), 347–367.

[Ebb-Flu,95] Ebbinghaus, H.-D. and Flum, J., *Finite Model Theory,* Springer, Berlin (1995).

[Ell,76] Ellentuck, E., Categoricity regained, *J. Symbolic Logic,* **41** (1976), 639–643.

[End,70] Enderton, H. B., Finite partially-ordered quantifiers, *Zeitschrift für Mathematische Logik und Grundlagen der Mathematik,* **16** (1970), 393–397.

[Erd,59] Erdős, P., Graph theory and probability, *Canad. J. Math.,* **11** (1959), 34–38.

[Erd-Fab-Lar,81] Erdős, P., Faber, V. and Larson, J., Sets of natural numbers of positive density and cylindric set algebras of dimension 2, *Algebra Universalis,* **12** (1981), 81–92.

[Esa-Mes,77] Esakia, L. and Meskhi, V., Five critical modal systems, *Theoria,* **43** (1977), 52–60.

[Fag,82] Fagin, R., Horn clauses and data dependencies, *J. of the ACM,* **29** (1982), 952–985.

[Fag-Var,86] Fagin, R. and Vardi, M. Y., The theory of data dependencies – a survey, in: *Mathematics of Information Processing* (M. Anshel and W. Gewirtz, eds.), AMS (1986), 19–71.

[Faj-Kei,96] Fajardo, S. and Keisler, H. J., Existence theorems in probability theory, *Advances in Math.,* **120** (1996).

[Fef-Kre,69] Feferman, S. and Kreisel, G., Persistent and Invariant Formulas for Outer Extensions, *Compositio Mathematica,* **20** (1969), 29–52.

[Fel,90] Feldman, N., Cylindric algebras with terms, *J. Symbolic Logic,* **55** (1990), 854–866.

[Fel,98] Feldman, N., The cylindric algebras of three-valued logic, *J. Symbolic Logic,* **63** (1998), 1201–1217.

[Fen,67] Fenstad, J. E., Representations of probabilities defined on first order languages, in: *Sets, Models and Recursion Theory,* North Holland (1967), 156–172.

[Fer,83] Ferenczi, M., Measures on cylindric algebras, *Acta Math. Hungar.,* **42** (1983), 3–17.

[Fer,86a] Ferenczi, M., On the connection of cylindrical homomorphisms and point functions for Crs_α's, in: *Lectures in Universal Algebras* (Proc. Coll. Szeged, 1983), Coll. Math. Soc. J. Bolyai, North Holland (1986), 123–141.

[Fer,86b] Ferenczi, M., Measures and measurable functions on cylindric algebras, Dissertation with The Hungarian Academy of Sciences, Budapest, 1986. In Hungarian.

[Fer,90] Ferenczi, M., On inducing homomorphisms between relation set algebras, *Algebra Universalis,* **27** (1990), 474–479.

[Fer,91] Ferenczi, M., Measures defined on free products of formula algebras and the analogies with cylindric homomorphisms, in: *Algebraic Logic,* [And-Mon-Nem,91b], 173–181.

[Fer,97] Ferenczi, M., On inducing cylindrical homomorphism by point function, *Bulletin Section of Logic, Łodź,* **26** (1997), 179–187.

[Fer,99] Ferenczi, M., On homomorphisms between relation algebras, *Algebra Universalis,* **27** (1999), 187–199.

[Fer,00] Ferenczi, M., On representability of neatly embeddable cylindric algebras, *J. of Applied and Non-classical Logic,* **10** (2000), 300–315.

[Fer,02] Ferenczi, M., *Mathematical Logic,* Műszaki Kiadó (2002), 350 pp (in Hungarian).

[Fer,05] Ferenczi, M., Probabilities on first order models, *Publ. l'Inst. Math.,* **78** (2005), 107–115.

[Fer,07a] Ferenczi, M., On cylindric algebras satisfying merry-go-round properties, *Logic J. IGPL,* **15** (2007), 183–197.

[Fer,07b] Ferenczi, M., Finitary polyadic algebras from cylindric algebras, *Studia Logica,* **87** (2007), 1–11.

[Fer,09] Ferenczi, M., On conservative extensions in logics with infinitary predicates, *Studia Logica,* **92** (2009), 121–135.

[Fer,10a] Ferenczi, M., Non-standard stochastics with a first order algebraization, *Studia Logica,* **95** (2010), 345–354.

[Fer,10b] Ferenczi, M., On the representability of neatly embeddable CA's by cylindric relativized algebras, *Algebra Universalis,* **63** (2010), 331–350.

[Fer,11] Ferenczi, M., Existence of partial transposition means representability in cylindric algebras, *Math. Logic Quarterly,* **57** (2011), 87–94.

[Fer,12a] Ferenczi, M., The polyadic generalization of the Boolean axiomatization of fields of sets, *Trans. Amer. Math. Soc.,* **364, 2** (2012), 867–886.

[Fer,12b] Ferenczi, M., The representations of polyadic-like equality algebras, *Arxiv*: 1104.1286, 2011.

[Fer,thisVol,a] Ferenczi, M., Probability measures and measurable functions on cylindric algebras, in this volume.

[Fer,thisVol,b] Ferenczi, M., A new representation theory: representing cylindric-like algebras by relativised set algebras, in this volume.

[Fer] Ferenczi, M., On the definitions and the representability of quasi-polyadic equality algebras, submitted.

[Fer-Pat-Ron,05] Ferenczi, M., Pataricza, A. and Rónyai, L. (eds.), *Formal Methods in Computing,* Kluwer Akadémiai Kiadó (2005), 425 pp.

[Fer-Sag,06] Ferenczi, M. and Sági, G., On some developments in the representation theory of cylindric-like algebras. Special issue dedicated to Walter Taylor, *Algebra Universalis,* **55** (2006), 345–353.

[Fin,74] Fine, K., An ascending chain of S4 logics, *Theoria,* **40** (1974), 110–116.

[Fin,75] Fine, K., Some connections between elementary and modal logic, in: *Proceedings of the Third Scandinavian Logic Symposium* (S. Kanger, ed.), North-Holland (Amsterdam, 1975), 15–31.

[Fle,93a] Fleischer, I., Completeness of the infinitary polyadic axiomatization, *Math. Logic Quarterly,* **39** (1993), 197–200.

[Fle,93b] Fleischer, I., The polyadic completion of a transformation algebra, *Proc. Amer. Math. Soc.,* **117** (1993), 511–514.

[Fon-Jan,94] Font, J. M. and Jansana, R., On the sentential logics associated with strongly nice and semi-nice general logics, *Bulletin of the IGPL,* **2** (1994), 55–76.

[Fon-Jan-Pig,97] Font, J. M., Jansana, R. and Pigozzi, D. (organizers), Workshop on Abstract Algebraic Logic (Abstracts of Talks). Centre de Recerca Matemática, Bellaterra (Spain), 1997.

[Fre,84] Fremlin, D. H., *Consequences of Martin's Axiom,* Cambridge University Press, 1984.

[Fri,73] Friedman, H. M., Beth's theorem in cardinality logics, *Israel Journal of Mathematics,* **14** (1973), 205–212.

[Gab,81] Gabbay, D. M., An irreflexivity lemma with applications to axiomatizations of conditions on linear frames, in: *Aspects of Philosophical Logic* (U. Mönnich, ed.), Reidel (Dordrecht, 1981), 67–89.

[Gab,94] Gabbay, D. M. (ed.), *What is a Logical System?* Oxford Science Publications, Clarendon Press, Oxford, 1994.

[Gab-Kur-Wol-Zak,03] Gabbay, D. M., Kurucz, A., Wolter, F. and Zakharyaschev, M., *Many-Dimensional Modal Logics: Theory and Applications,* 148 of Studies in Logic and the Foundations of Mathematics, Elsevier, 2003.

[Gab-She,98] Gabbay, D. M. and Shehtman, V., Products of modal logics, Part I. *Logic J. IGPL,* **6** (1998), 73–146.

[Gab-Kur-Wol-Zak,05] Gabelaia, D., Kurucz, A., Wolter, F. and Zakharyaschev, M., Products of 'transitive' modal logics, *J. Symbolic Logic,* **70** (2005), 993–1021.

[Gai,64] Gaifman, H., Concerning measures in first order calculi, *Israel J. of Math.,* **2** (1964), 1–18.

[Gal,75] Gallin, D., *Systems of Intensional and Higher-Order Modal Logic,* North-Holland, Amsterdam (1975).

[Geo,78] Georgescu, G., Extension of probabilities defined on polyadic algebras, *Bull. Math. Soc. Sci. Math. de Roumanie,* **22** (1978), 1–12.

[Geo,83a] Georgescu, G., Chang's modal operators in algebraic logic, *Studia Logica,* **42** (1983), 43–48.

[Geo,83b] Georgescu, G., On the Popper-Carnap probability functions in polyadic algebras, *Rev. Roumaine Math. Pures Appl.* **28** (1983), 583–585.

[Geo,84] Georgescu, G., Algebraic analysis of the logic with the quantifier "there exist uncountably many", *Algebra Universalis,* **19** (1984), 99–105.

[Giv,91] Givant, S. R., Tarski's development of logic and mathematics based on the calculus of relations, in: *Algebraic Logic,* [And-Mon-Nem,91b], 189–215.

[Giv,12] Givant, S. R., *Introduction to Relation Algebras.* Springer-Verlag, to appear.

[Giv-Ven,99] Givant, S. R. and Venema, Y., The preservation of Sahlqvist equations in completions of Boolean algebras with operators, *Algebra Universalis,* **41** (1999), 47–84.

[Gog-Bur,92] Goguen, J. and Burstall, R., Institutions: Abstract model theory for specification and programming, *J. Assoc. of Computing Machinery,* **39** (1992), 95–146.

[God,33] Gödel, K., Zum Entscheidungsproblem des logischen Funktionenkalküls, *Monatshefte für Mathematik und Physik,* **40** (1933), 433–443.

[Gol,84] Goldfarb, W., The unsolvability of the Gödel class with identity, *J. Symbolic Logic,* **49** (1984), 1237–1252.

[Gol,89] Goldblatt, R., Varieties of complex algebras, *Ann. Pure and Appl. Logic,* **44** (1989), 173–242.

[Gol,91] Goldblatt, R., On closure under canonical embedding algebras, in: *Algebraic Logic,* [And-Mon-Nem,91b], 217–229.

[Gol,95] Goldblatt, R., Elementary generation and canonicity for varieties of Boolean algebras with operators, *Algebra Universalis,* **34** (1995), 551–607.

[Gol,00] Goldblatt, R., Algebraic polymodal logic: A survey, *Logic J. IGPL,* **8**(4) (2000), 393–450.

[Gol-Hod-Ven,04] Goldblatt, R., Hodkinson, I. and Venema, Y., Erdős graphs resolve Fine's canonicity problem, *Bull. Symbolic Logic,* **10**(2) (2004), 186–208.

[Gol-Tho,74] Goldblatt, R. and Thomason, S., Axiomatic Classes in Propositional Modal Logic, in: J. Crossley, ed., *Algebra and Logic,* Springer Lecture Notes in Mathematics, **450** (1974), 163–173.

[Gra,99a] Grädel, E., Decision Procedures for Guarded Logics, in: *Automated Deduction – Proceedings CADE 16,* Lecture Notes in Computer Science 1632, Springer Verlag (Berlin, 1999), 31–51.

[Gra,99b] Grädel, E., On the Restraining Power of Guards, *J. Symbolic Logic,* **64** (1999), 1719–1742.

[Gra,99c] Grädel, E., The decidability of guarded fixed point logic, in: *Essays Dedicated to Johan van Benthem on the Occasion of his 50th Birthday* (J. Gerbrandy, M. Marx, M. de Rijke and Y. Venema, eds.), Amsterdam University Press, (1999), http://turing.wins.uva.nl/∼j50/cdrom/.

[Gra-Kol-Var,97] Grädel, E., Kolaitis, P. and Vardi, M., On the decision problem for two-variable first order logic, *Bull. Symbolic Logic,* **3** (1997), 53–69.

[Gra-Ros,99] Grädel, E. and Rosen, R., On preservation theorems for two variable logic, *Math. Logic Quarterly,* **45** (1999), 315–325.

[Gre,83] Grenander, U., *Probabilities on Algebraic Structures,* Wiley (1983).

[Gro-Sto,92] Groenendijk, J. and Stokhof, M.(1991), Dynamic predicate logic, *Linguistics and Philosophy,* **14**, 39–100.

[Gui,88] Guillaume, M., Constantes d'une algebre monadique libre sur une algebre de Boole et automorphismes préservant une partie génératrice de celle-ci, (French) [Constants of a free monadic algebra over a Boolean algebra and automorphisms preserving a set of generators of the Boolean algebra], *Methods and applications of mathematical logic* (Campinas, 1985), 45–53, Contemp. Math., 69, Amer. Math. Soc., Providence, RI, 1988.

[Gur,76] Gurevich, Y., The decision problem for standard classes, *J. Symbolic Logic,* **41** (1976), 460–464.

[Gur,84] Gurevich, Y., Towards logic tailored for computational complexity, in: *Computation and Proof Theory,* Lecture Notes in Mathematics 1104, Springer Verlag (1984), 175–216.

[Gye,11] Gyenis, Z., On atomicity of free algebras in certain cylindric-like varieties, *Logic J. IGPL,* **19** (2011), 44–52.

[Hac,01] Hacking, I., *An Introduction to Probability and Inductive Logic,* Cambridge Univ. Press (2001).

[Hal,56] Halmos, P., Algebraic logic II., Homogeneous, locally finite polyadic Boolean algebras, *Fundamenta Math.,* **43**, (1956), 255–325.

[Hal,57] Halmos, P., Algebraic logic IV. Equality in polyadic algebras, *Trans. Amer. Math. Soc.,* **86** (1957), 1–27.

[Hal,62] Halmos, P., *Algebraic Logic,* Chelsea Publishing Co., New York (1962).

[Hal-Var,89] Halpern, J. and Vardi, M., The Complexity of Reasoning about Knowledge and Time, I: lower bounds. *J. of Computer and System Sciences,* **38** (1989), 195–237.

[Han,88] Handschel, G., Eine Erweiterung der Booleschen Algebra., (German) [An extension of Boolean algebra], *Math. Semesterber.,* **35** (1988), 81–94.

[Han,92] Hansen, B., *Complications in making first-order logic nice.* Thesis (MA), Harvard Univ. (Dept. Math.), 1992.

[Han,95] Hansen, B., Finitizability questions for some reducts of cylindric algebras, in: *Logic Colloquium '92,* [Csir-Gab-Rij,95], 115–134.

[Hen,49] Henkin, L., Completeness in the Theory of Types, *J. Symbolic Logic,* **15** (1949), 81–91.

[Hen,61] Henkin, L., Some remarks on infinitely long formulas, in: *Infinitistic Methods: Proceedings of the Symposium on Foundations of Mathematics,* Warsaw, Pergamon Press (1961), 167–183.

[Hen,67] Henkin, L., Logical systems containing only a finite number of symbols. Les Presses de l'Université de Montréal, Montreal, Canada, 1967.

[Hen-Mon-Tar,85] Henkin, L., Monk, J. D. and Tarski, A., *Cylindric Algebras,* I–II., North Holland, Part I in 1971 and Parts I–II in 1985.

[Hen-Mon-Tar,86] Henkin, L., Monk, J. D. and Tarski, A., Representable cylindric algebras, *Ann. Pure and Appl. Logic,* **31** (1986), 23–60.

[Hen-Mon-Tar-And-Nem,81] Henkin, L., Monk, J. D., Tarski, A., Andréka, H. and Németi, I., *Cylindric Set Algebras,* Lecture Notes in Mathematics, 883, Springer-Verlag, Berlin (1981), vi + 323 pp.

[Hen-Tar,61] Henkin, L. and Tarski, A., Cylindric algebras, in: *Lattice theory, Proceedings of Symposia in Pure Mathematics* (R. P. Dilworth, ed.), 2. American Mathematical Society (Providence, R.I., 1961), 83–113.

[Hin,96] Hintikka, J., *The Principles of Mathematics Revisited,* Cambridge University Press, Cambridge, 1996.

[Hin-San,89] Hintikka, J. and Sandu, G., Informational independence as a semantical phenomenon, in: *Logic, Methodology and Philosophy of Science VIII* (J. E. Fenstad et al., eds.), 126 of Studies in Logic and the Foundations of Mathematics, North-Holland (1989), 571–589.

[Hin-San,97] Hintikka, J. and Sandu, G., Game-Theoretical Semantics, in: J. van Benthem and A. ter Meulen (eds.), *Handbook of Logic and Language,* Elsevier, Amsterdam (1997), 361–410.

[Hir,95] Hirsch, R., Completely representable relation algebras, *Bull. of the IGPL,* **3** (1995), 77–92.

[Hir,97] Hirsch, R., Expressive power and complexity in algebraic logic, *J. Logic Comput.,* **7** (1997), 309–351.

[Hir,07] Hirsch, R., Relation algebra reducts of cylindric algebras and complete representations, *J. Symbolic Logic,* **72** (2007), 673–703.

[Hir-Hod,97a] Hirsch, R. and Hodkinson, I., Step by step building representations in algebraic logic, *J. Symbolic Logic,* **62** (1997), 225–279.

[Hir-Hod,97b] Hirsch, R. and Hodkinson, I., Axiomatising various classes of relation and cylindric algebras, *Logic J. IGPL,* **5** (1997), 209–229.

[Hir-Hod,97c] Hirsch, R. and Hodkinson, I., Complete representations in algebraic logic, *J. Symbolic Logic,* **62** (1997), 816–847.

[Hir-Hod,01a] Hirsch, R. and Hodkinson, I., Connections between cylindric algebras and relation algebras, in: *Relational methods for computer science applications,* 239–246, Stud. Fuzziness Soft Comput., 65, Physica, Heidelberg, 2001.

[Hir-Hod,01b] Hirsch, R. and Hodkinson, I., Relation algebras from cylindric algebras. I., *Ann. Pure and Appl. Logic,* **112** (2001), 225–266.

[Hir-Hod,01c] Hirsch, R. and Hodkinson, I., Relation algebras from cylindric algebras. II., *Ann. Pure and Appl. Logic,* **112** (2001), 267–297.

[Hir-Hod,01d] Hirsch, R. and Hodkinson, I., Representability is not decidable for finite relation algebras, *Trans. Amer. Math. Soc.,* **353** (2001), 1403–1425.

[Hir-Hod,02a] Hirsch, R. and Hodkinson, I., *Relation Algebras by Games,* 147 of Studies in Logic and the Foundations of Mathematics, Elsevier, North-Holland, 2002.

[Hir-Hod,02b] Hirsch, R. and Hodkinson, I., Strongly representable atom structures of relation algebras, *Proc. Amer. Math. Soc.,* **130** (2002), 1819–1831.

[Hir-Hod,02c] Hirsch, R., Hodkinson, I., Relation algebra reducts of cylindric algebras and an application to proof theory, *J. Symbolic Logic,* **67** (2002), 197–213.

[Hir-Hod,09] Hirsch, R. and Hodkinson, I., Strongly representable atom structures of cylindric algebras, *J. Symbolic Logic,* **74** (2009), 811–828.

[Hir-Hod,thisVol] Hirsch, R. and Hodkinson, I., Completions and complete representations, in this volume.

[Hir-Hod-Kur,02a] Hirsch, R., Hodkinson, I. and Kurucz, A., On modal logics between $K \times K \times K$ and $S5 \times S5 \times S5$, *J. Symbolic Logic,* **67** (2002), 221–234.

[Hir-Hod-Mad,02b] Hirsch, R., Hodkinson, I. and Maddux, R. D., On provability with finitely many variables, *Bull. Symbolic Logic,* **8** (2002), 329–347.

[Hod,88] Hodges, W., Extending Beth's theorem, in: *Proceedings of the 6th Easter Conference in Model Theory,* Humboldt University (1988), 57–64.

[Hod,93a] Hodges, W., *Model Theory,* Cambridge University Press, 1993.

[Hod,97a] Hodges, W., Compositional semantics for a language of imperfect information, *Logic J. IGPL,* **5** (1997), 539–563.

[Hod,97b] Hodges, W., Some strange quantifiers, in: *Structures in Logic and Computer Science: A Selection of Essays in Honor of A. Ehrenfeucht* (J. Mycielski, G. Rozenberg and A. Salomaa, eds.), 1261 in Lecture Notes in Computer Science, 51–65. Springer, 1997.

[Hod,01] Hodges, W., Formal Features of Compositionality, *J. of Logic, Language and Information,* **10** (2001), 7–28.

[Hod,93b] Hodkinson, I., Finite variable logics, *Bull. European Assoc. Theoret. Comput. Sci.,* **51** (1993), 111–140. Updated version: http://www.doc.ic.ac.uk/~imh/papers/yuri.ps.gz

[Hod,97c] Hodkinson, I., Atom structures of cylindric algebras and relation algebras, *Ann. Pure and Appl. Logic,* (1997), 117–148.

[Hod,12] Hodkinson, I., A construction of cylindric and polyadic algebras from atomic relation algebras, *Algebra Universalis,* to appear.

[Hod-Mik,00] Hodkinson, I. and Mikulás, Sz., Axiomatizability of reducts of algebras of relations, *Algebra Universalis,* **43** (2000), 127–156.

[Hod-Ven,05] Hodkinson, I. and Venema, Y., Canonical varieties with no canonical axiomatization, *Trans. Amer. Math. Soc.,* **357** (2005), 4579–4605.

[Hoe-Rij,93] van der Hoek, W. and de Rijke, M., Generalized Quantifiers and Modal Logic, *J. of Logic, Language, and Information,* **2** (1993), 19–50.

[Hoo,96] Hoogland, E., Algebraic characterizations of two Beth definability properties. Master's thesis, University of Amsterdam, 1996.

[Hoo,00] Hoogland, E., Algebraic characterizations of various Beth definability properties, *Studia Logica,* **65** (2000), 91–112.

[Hoo,01] Hoogland, E., *Definability and interpolation (model theoretic investigations),* PhD thesis, Institute for Logic, Language and Computation, University of Amsterdam, 2001. ILLC Dissertation Series DS-2001-05, xii+209 pp.

[Hoo-Mar,02b] Hoogland, E. and Marx, M., Interpolation in the guarded fragment, *Studia Logica,* **70** (2002), 373–409.

[Hoo-Mar-Ott,99] Hoogland, E., Marx, M. and Otto, M., Beth Definability for the Guarded Fragment, in: *Logic for Programming and Automated Reasoning* (H. Ganzinger, D. McAllester and A. Voronkov, eds.), LPAR 6, Springer Lecture Notes in AI 1705, 273–285, Springer, Berlin (1999).

[Hoo,78] Hoover, D., Probability logic, *Ann. Math. Logic,* **14** (1978), 287–313.

[Hor-Tar,48] Horn, A. and Tarski, A., Measures in Boolean algebras, *Trans. Amer. Math. Soc.,* **64** (1948), 467–492.

[Iko-Ras-Mar-Ogn,07] Ikodinović, N., Rašković, M., Marković, Z. and Ognjanović, Z., *Measure Logic,* ECSQARU 2007, LNCS4724, Springer (2007), 128–138.

[Imi-Lip,84a] Imielinski, T. and Lipski, W., The relational model of data and cylindric algebras, *J. of Computer and System Sciences,* **28** (1984), 80–102.

[Imi-Lip,84b] Imielinski, T. and Lipski, W., On the undecidability of equivalence problems for relational expressions, in: *Advances in Database Theory Vol. 2* (H. Gallaire and J. M. Nicolas, eds.), 393–409. Plenum Press (1984).

[Imm-Koz,87] Immerman, N. and Kozen, D., Definability with bounded number of bound variables, in: *Proceedings 4th IEEE Symposium on Logic in Computer Science LICS'87,* Computer Society Press (1987).

[Jan-Khe,01] Janicki, R. and Khédri, R., On a formal semantics of tabular expressions, *Sci. Comput. Program.,* **39** (2001), 189–213.

[Jer,11] Jeřábek, E., The ubiquity of conservative translations. Preprint, arXiv: 1108.6263v1

[Joh,69] Johnson, J. S., Nonfinitizability of classes of representable polyadic algebras, *J. Symbolic Logic,* **34** (1969), 344–352.

[Joh,70] Johnson, J. S., Amalgamation of Polyadic Algebras, *Trans. Amer. Math. Soc.,* **17** (1970), 834–1970.

[Jon,62] Jónsson, B., Defining Relations for Full Semigroups of Finite Transformations, *Michigan Math. J.,* **9** (1962), 77–85.

[Jon-Tar,51] Jónsson, B. and Tarski, A., Boolean algebras with operators I. *American J. of Mathematics,* **73** (1951), 891–939.

[Jon-Tar,61] Jónsson, B. and Tarski, A., On two properties of free algebras, *Math. Scand.,* **9** (1961), 95–101.

[JSL,86] Issue devoted to papers on the work of A. Tarski, *J. Symbolic Logic,* **61** (1986).

[Kag-Qua,76] Kagan, J. and Quackenbush, R., Monadic algebras, *Reports on Mathematical Logic,* **7** (1976), 53–62.

[Kal,58] Kalman, J. A., Lattices with involution, *Trans. Amer. Math. Soc.,* **87** (1958), 485–491.

[Kan,89] Kanellakis, P. C., *Elements of relational database theory,* Technical Report CS-89-39, Department of Computer Science, Brown University (1989).

[Kei,63] Keisler, H. J., A complete first order logic with infinitary predicates, *Fundamenta Math.,* **52** (1963), 177–203.

[Kei,71] Keisler, H. J., *Model Theory for Infinitary Logic,* North-Holland, Amsterdam, 1971.

[Kei,77] Keisler, H. J., Hyperfinite model theory, in: R. O. Gandy and J. M. E. Hyland, eds., *Logic Colloquium 76,* North-Holland, Amsterdam,1977, 5–110.

[Kei,85] Keisler, H. J., Probability quantifiers, in: *Model Theoretic Logics* (J. Barwise and S. Feferman, eds.), Springer, 1985, 509–556.

[Kei,86] Keisler, H. J., A completeness proof for adapted probability logic, *Ann. Pure and Appl. Logic,* **31** (1986), 61–70.

[Ker,01] Kerdiles, G., *Saying it with pictures: a logical landscape of conceptual graphs,* Dissertation DS-2001-09, Institute for Logic, Language and Computation, University of Amsterdam (2001).

[Kha-Say,09a] Khaled, M. and Sayed Ahmed, T., Classes of algebras that are not closed under completions, *Bulletin Section of Logic, Łodź,* **38** (2009), 29–43.

[Kha-Say,09b] Khaled, M. and Sayed Ahmed, T., On complete representations of algebras of logic, *Logic J. IGPL,* **17** (2009), 267–272.

[Kha-Say,09c] Khaled, M. and Sayed Ahmed, T., Omitting types algebraically via cylindric algebras, *Int. J. Algebra,* **3** (2009), 377–390.

[Kha-Say,10] Khaled, M. and Sayed Ahmed, T., Vaught's theorem holds for L_2 but fails for L_n when $n > 2$, *Bull. Section of Logic,* **39** (2010), 15–22.

[Kik,10] Kikot, S., Axiomatization of modal logic squares with distinguished diagonal, *Mathematical Notes,* **88** (2010), 238–250.

[Kik-Kur,11] Kikot, S. and Kurucz, A., Undecidable two-dimensional products of modal logics with diagonal constant, submitted.

[Kis-Mar-Pro-Tho,83] Kiss, E. W., Márki, L., Pröhle, P. and Tholen, W., Categorial algebraic properties. Compendium on amalgamation, congruence extension, epimorphisms, residual smallness and injectivity, *Studia Sci. Math. Hungar.,* **18** (1983), 79–141.

[Kle,52] Kleene, S. C., *Introduction to Metamathematics,* Van Nostrand, 1952.

[Knu-Ron,83] Knuth, E. and Rónyai, L., Closed convex reference schemes. (A junction between computer science, cylindric- and partial algebras.) in: *Proc. IFIP Conf.,* Hungary (1983).

[Kur,95] Kurtonina, N., *Frames and labels, a modal analysis of categorial inference,* Dissertation, Institute for Logic, Language and Computation, University of Amsterdam (1995).

[Kur,00] Kurucz, A., On axiomatizing products of Kripke frames, *J. Symbolic Logic,* **65** (2000), 923–945.

[Kur,02a] Kurucz, A., Comparing decision problems for various paradigms of algebraic logic, *Algebra Universalis,* **47** (2002), 409–424.

[Kur,02b] Kurucz, A., $\mathbf{S5} \times \mathbf{S5} \times \mathbf{S5}$ lacks the finite model property, in: *Advances in Modal Logic* (F. Wolter, H. Wansing, M. de Rijke, and M. Zakharyaschev, eds.), 3 of *CSLI Lecture Notes,* (Stanford, 2002), 321–328.

[Kur,07] Kurucz, A., Combining modal logics, in: *Handbook of Modal Logic* (P. Blackburn, J. van Benthem and F. Wolter, eds.), 3 of Studies in Logic and Practical Reasoning, Elsevier (2007), 869–924.

[Kur,08] Kurucz, A., On axiomatizing products of Kripke frames, part II. in: *Advances in Modal Logic* (C. Areces and R. Goldblatt, eds.), 7 College Publications, (2008), 219–230.

[Kur,10] Kurucz, A., On the complexity of modal axiomatizations over many-dimensional structures, in: *Advances in Modal Logic* (L. Beklemishev, V. Goranko and V. Shehtman, eds.), 8 College Publications, (2010), 241–254.

[Kur,thisVol] Kurucz, A., Representable cylindric algebras and many-dimensional modal logics, in this volume.

[Kur-Nem,00] Kurucz, A. and Németi, I., Representability of pairing relation algebras depends on your ontology, *Fund. Inform.*, **44** (2000), 397–420.

[Kur-Nem-Sai-Sim,93] Kurucz, A., Németi, I., Sain, I. and Simon, A., Undecidable varieties of semilattice-ordered semigroups, of Boolean algebras with operators, and logics extending Lambek calculus, (English summary) *Bulletin of the IGPL,* **1** (1993), 91–98.

[Kut-Mos-Luc,10] Kutz, O., Mossakowski, T. and Lücke, D., Carnap, Goguen, and the hyperontologies: logical pluralism and heterogeneous structuring in ontology design, *Logica Universalis,* **4** (2010), 255–333.

[Lar,84] Larson, J. A., The number of finitely generated infinite cylindric set algebras of dimension two, *Algebra Universalis,* **19** (1984), 377–396.

[Lar,85] Larson, J. A., The number of one-generated cylindric set algebras of dimension greater than two, *J. Symbolic Logic,* **50** (1985), 59–71.

[Las-She,93] Laskowski, M. and Shelah, S., On the existence of atomic models, *J. Symbolic Logic,* **58** (1993), 1189–1194.

[Lit-Hid-Mik,09] Litak, T., Hidders, J. and Mikulás, Sz., Relational lattices: *An Introduction,* Preprint, 2009.

[Los,73] Łoš, J., Remarks on the foundations of probability, in: *Proc. Int. Congr. for Math. Stockholm,* 1962, Almquist and Wikselle (1973), 225–228.

[Lyn,50] Lyndon, R., The representation of relational algebras, *Annals of Mathematics,* **51** (1950), 707–729.

[Lyn,56] Lyndon, R., The representation of relation algebras, II. *Annals of Mathematics,* **63** (1956), 294–307.

[McL,71] Mac Lane, S., *Categories for the Working Mathematician,* Springer–Verlag (Berlin, 1971).

[Mad,12] Madarász, J. X., Epimorphisms in cylindric algebras. Preprint, Rényi Institute of Mathematics, Budapest. Submitted.

[Mad,98] Madarász, J. X., Interpolation and Amalgamation; pushing the limits. Part I, *Studia Logica,* **61** (1998), 316–345.

[Mad,02] Madarász, J. X., *Logic and Relativity (in the light of definability theory).* PhD thesis, ELTE, Budapest, 2002.

[Mad-Nem,01] Madarász, J. X. and Németi, I., Free Boolean algebras with closure operators and a conjecture of Henkin, Monk, and Tarski, *Studia Sci. Math. Hungar.*, **38** (2001), 273–278.

[Mad-Nem-Sag,97] Madarász, J. X., Németi, I. and Sági, G., On the finitization problem of relation algebras (completeness problem for the finite variable fragments), *Bulletin Section of Logic, Łodź,* **26** (1997), 139–143.

[Mad-Say,07] Madarász, J. X. and Sayed Ahmed, T., Amalgamation, interpolation and epimorphisms, *Algebra Universalis,* **56** (2007), 179–210.

[Mad-Say,09a] Madarász, J. X. and Sayed Ahmed, T., Neat reducts and amalgamation in retrospect, a survey of results and some methods. I. Results on neat reducts, *Logic J. IGPL,* **17** (2009), 429–483.

[Mad-Say,09b] Madarász, J. X. and Sayed Ahmed, T., Neat reducts and amalgamation in retrospect, a survey of results and some methods. II. Results on amalgamation, *Logic J. IGPL,* **17** (2009), 755–802.

[Mad-Say,thisVol] Madarász, J. X. and Sayed Ahmed, T., Amalgamation, interpolation and epimorphisms in algebraic logic, in this volume.

[MadR,92] Madarász, R. Sz., Some results on axiomatizability, *Zb. Rad. Prirod.-Mat. Fak. Ser. Mat.,* **22** (1992), 119–132.

[Mad,78] Maddux, R. D., Some sufficient conditions for the representability of relation algebras, *Algebra Universalis,* **8** (1978), 162–171.

[Mad,80] Maddux, R. D., The equational theory of CA_3 is undecidable, *J. Symbolic Logic,* **45** (1980), 311–316.

[Mad,82] Maddux, R. D., Some varieties containing relation algebras, *Trans. Amer. Math. Soc.,* **272** (1982), 501–526.

[Mad,83] Maddux, R. D., A sequent calculus for relation algebras, *Ann. Pure and Appl. Logic,* **25** (1983), 73–101.

[Mad,89a] Maddux, R. D., Nonfinite axiomatizability results for cylindric and relation algebras, *J. Symbolic Logic,* **54** (1989), 951–974.

[Mad,89b] Maddux, R. D., Canonical relativized cylindric set algebras, *Proc. Amer. Math. Soc.,* **107** (1989), 465–478.

[Mad,89c] Maddux, R. D., Finitary algebraic logic, *Z. Math. Logik Grundlag. Math.,* **35** (1989), 321–332.

[Mad,90] Maddux, R. D., A relation algebra which is not a cylindric reduct, *Algebra Universalis,* vol. **27** (1990), 279–288.

[Mad,91a] Maddux, R. D., Introductory course on relation algebras, finite-dimensional cylindric algebras, and their interconnections, in: *Algebraic Logic,* [And-Mon-Nem,91b], 361–392.

[Mad,91b] Maddux, R. D., The origin of relation algebras in the development and axiomatization of the calculus of relations, *Studia Logica,* Vol. **50**(3/4) (1991), 421–455.

[Mad,91c] Maddux, R. D., The neat embedding problem and the number of variables required in proofs, *Proc. Amer. Math. Soc.,* **112** (1991), 195–202.

[Mad,93] Maddux, R. D., Finitary algebraic logic. II., *Math. Logic Quarterly,* **39** (1993), 566–569.

[Mad,06] Maddux, R. D., *Relation Algebras.* Elsevier, 2006.

[Mak,87] Makkai, M., Stone duality for first order logic, *Adv. in Math.,* **65** (1987), 97–170.

[Mak-Rey,77] Makkai, M. and Reyes, G. E., First order categorical logic, in: *Model-theoretical methods in the theory of topoi and related categories,* Lecture Notes in Mathematics, Vol. 611. Springer-Verlag, Berlin–New York, 1977. viii+301 pp.

[Mak-Sai,89] Makowsky, J. A. and Sain, I., Weak second order characterizations of various program verification systems, *Theoretical Computer Science,* **66** (1989), 299–321.

[Mak,72] Maksimova, L. L., Pretabular superintuitionistic logics, *Algebra i Logika,* **11** (1972), 558–570.

[Mak,75] Maksimova, L. L., Pretabular extensions of Lewis's logic **S4**, *Algebra i Logika,* **14** (1975), 28–55.

[Mak,91] Maksimova, L. L., Amalgamation and interpolation in normal modal logics, *Studia Logica,* **50** (1991), 457–471.

[Mak,92] Maksimova, L. L., The Beth properties, interpolation, and amalgamability in varieties of modal algebras, *Soviet Math Dokl.,* **44** (1992), 327–331.

[Mak,97] Maksimova, L. L., On the Beth definability properties in varieties of modal algebras, in: *Workshop on Abstract Algebraic Logic (Abstracts of Talks),* [Fon-Jan-Pig,97], 71–75.

[Man,08] Mann, A. L., Perfect IFG-formulas, *Logica Universalis,* **2** (2008), 265–275.

[Man,09] Mann, A. L., Independence-friendly cylindric set algebras, *Logic J. IGPL,* **17(6)** (2009), 719–754.

[Man,thisVol] Mann, A. L., Cylindric set algebras and IF logic, in this volume.

[Mar,03] Mares, E., *Relevant Logic, a Philosophical Interpretation,* Cambridge University Press, Cambridge (2003).

[Mar-Ras-Dor,99] Marinković, S., Rašković, M. and Đorđević, R. S., Weak probability logic with infinitary predicates, *Publ. L'Inst. Math.,* **65 (75)** (1999), 8–16.

[Mar-Ras-Dor,01] Marinković, S., Rašković, M. and Đorđević, R., Weak probability polyadic algebras, *Facta Univ. Ser. Math. Inform.,* No. **16** (2001), 1–12.

[Mar-Ogn-Ras,04] Marković, Z., Ognjanović, Z. and Rašković, M., A probabilistic extension of intuitionistic logic, *Math. Logic Quarterly,* **49** (2004), 557–564.

[Mar-Mes,01] Martí-Oliet, N. and Meseguer, J., General Logics and Logical Frameworks, in: *Handbook of Philosophical Logic Vol. II,* Second edition (D. M. Gabbay and F. Guenthner, eds.), Oxford Science Publications, Kluwer Academic Publishers (2001), 355–391.

[Mar-Sol,70] Martin, D. A. and Solovay, R. M., Internal Cohen extensions, *Ann. Mathematical Logic,* **2** (1970), 143–178.

[Mar,95] Marx, M., *Arrow logic and relativized algebras of relations,* Dissertation, CCSOM and ILLC, University of Amsterdam (1995).

[Mar,97] Marx, M., Mosaics and cylindric modal logic of dimension two, in: *Advances in Modal Logic* (M. Kracht, M. de Rijke, H. Wansing and M. Zakharyaschev, eds.), 87 in CSLI Lecture Notes, 141–157. CSLI Publications, Stanford, 1997.

[Mar,00] Marx, M., Amalgamation in finite dimensional cylindric algebras, *Algebra Universalis,* **43** (2000), 41–49.

[Mar,01] Marx, M., Tolerance Logic, *J. of Logic, Language and Information,* **10** (2001), 353–373.

[Mar-Mik,99] Marx, M. and Mikulás, Sz., Decidability of cylindric set algebras of dimension two and first-order logic with two variables, *J. Symbolic Logic,* **64** (1999), 1563–1572.

[Mar-Mik-Nem,95] Marx, M., Mikulás, Sz. and Németi, I., Taming Logic, *Journal of Logic, Language, and Information* **4** (1995), 207–226.

[Mar-Mik-Nem-Sai,96] Marx, M., Mikulás, Sz., Németi, I. and Sain, I., Investigations in arrow logic, in: *Arrow Logic and Multi-modal Logic,* [Mar-Pol-Mas,96], 35–61.

[Mar-Pol-Mas,96] Marx, M., Pólos, L. and Masuch, M. (eds.), *Arrow Logic and Multimodal Logic* (1996). Studies in Logic Language and Information. CSLI Publications.

[Mar-Ven,97] Marx, M. and Venema, Y., *Multi-Dimensional Modal Logic,* Kluwer, Dordrecht (1997).

[Mar-Ven,07] Marx, M. and Venema, Y., Local variations on a loose theme: Modal logic and decidability, in: *Finite model theory and its applications,* Texts Theoret. Comput. Sci. EATCS Ser., pages 371–429. Springer, Berlin, 2007.

[Mas-Pol,92] Masuch, M. and Pólos, L. (eds)., *Logic at Work,* (Proc. Conf. Amsterdam, 1992).

[McC,97] McCune, W., Solution of the Robbins problem, *J. Automated Reasoning,* **19** (1997), 263–276.

[Met-Pao-Tsi,10] Metcalfe, G., Paoli, F. and Tsinakis, C., Ordered algebras and logic, *Probability, uncertainty and rationality,* 3–83, CRM Series, 10, Ed. Norm., Pisa, 2010.

[Mik,95] Mikulás, Sz., *Taming logics,* Dissertation, Institute for Logic, Language and Computation, University of Amsterdam (1995).

[Mil,80] Miller, A., Covering $^{\omega}2$ with ω_1 disjoint closed sets, in: *The Kleene Symposuim* (proceedings, Madison, Wisconsin, 1978). Studies in Logic and the Foundation of Mathematics, 101, North-Holland (Amsterdam, 1980), 415–421.

[Mil,82] Miller, A., A characterization of the least cardinal for which the Baire Category Theorem fails, *Proc. Amer. Math. Soc.,* **86** (1982), 498–502.

[Mon,61] Monk, J. D., *Studies in cylindric algebra,* Doctoral Dissertation, University of California, Berkeley, 1961.

[Mon,64] Monk, J. D., On representable relation algebras, *Michigan Mathematics Journal,* **11** (1964), 207–210.

[Mon,69] Monk, J. D., Nonfinitizability of classes of representable cylindric algebras, *J. Symbolic Logic,* **34** (1969), 331–343.

[Mon,70a] Monk, J. D., On equational classes of algebraic versions of logic I, *Mathematica Scandinavica,* **27** (1970), 53–71.

[Mon,70b] Monk, J. D., Completions of Boolean algebras with operators, *Mathematische Nachrichten,* **46** (1970), 47–55.

[Mon,76] Monk, J. D., *Mathematical Logic,* Springer-Verlag, Berlin, 1976.

[Mon,78] Monk, J. D., Omitting types algebraically, *Ann. Sci. Clermont Math.,* 16 (1978), 101–105.

[Mon,91a] Monk, J. D., Corrections for the books: Cylindric Algebras [Part I, 1971; Part II, 1985] and Cylindric Set Algebras (1981), in: *Algebraic Logic,* [And-Mon-Nem,91b], 719–722.

[Mon,91b] Monk, J. D., Remarks on the problems in the books: Cylindric Algebras [Part I, 1971; Part II, 1985] and Cylindric Set Algebras (1981), in: *Algebraic Logic,* [And-Mon-Nem,91b], 723–726.

[Mon,91c] Monk, J. D., Structure problems for cylindric algebras, in: *Algebraic Logic,* [And-Mon-Nem,91b], 413–429.

[Mon,93] Monk, J. D., Lectures on cylindric set algebras, in: *Algebraic Methods in Logic and in Computer Science,* [Banach,93], 253–290.

[Mon,00] Monk, J. D., An introduction to cylindric set algebras, *Logic J. IGPL,* **8** (2000), 449–494. Electronically available as: `http://www.jigpal.oupjournals.org`.

[Mon,70c] Montague, R., Pragmatics and Intensional Logic, *Synthese,* **22** (1970), 68–94.

[Mon-Aba-Sav-Sew,97] Monteiro, L. F., Abad, M., Savini, S. and Sewald, J., Free monadic Tarski algebras, *Algebra Universalis,* **37** (1997), 106–118.

[Mos-Gog-Dia-Tar,05] Mossakowski, T., Goguen, J., Diaconescu, R. and Tarlecki, A., What is a logic? in: *Logica Universalis. Towards a General Theory of Logic* (J-Y. Béziau, ed.), Birkhäuser Verlag (2005), 113–133.

[Mor,60] de Morgan, A., On the syllogism, no. iv, and on the logic of relations, *Transactions of the Cambridge Philosophical Society,* **10** (1860), 331–358.

[Mor,75] Mortimer, M., On languages with two variables, *Zeitchrift für mathematische Logik und Grundlagen der Mathematik,* **21** (1975), 135–140.

[Nat-Ros,93] Naturman, C. and Rose, H., Interior algebras: some universal algebraic aspects, *J. Korean Math. Soc.,* **30** (1993), 1–24.

[Nem,82] Németi, I., Surjectiveness of epimorphisms is equivalent to Beth definability property in general algebraic logic. Preprint, Mathematical Institute, Budapest, 1982.

[Nem,83] Németi, I., The class of neat-reducts of cylindric algebras is not a variety but is closed with respect to HP, *Notre Dame J. of Formal Logic,* **24** (1983), 399–409.

[Nem,84] Németi, I., Algebraic proofs of "$\mathfrak{Fr}_\beta \mathsf{CA}_\alpha$ ($\alpha \geq 4$) is not atomic". Manuscript, Mathematical Institute of the Hungarian Academy of Sciences, Budapest, 1984.

[Nem,85a] Németi, I., Cylindric-relativised set algebras have strong amalgamation, *J. Symbolic Logic,* **50** (1985), 689–700.

[Nem,85b] Németi, I., Logic with three variables has Gödel's incompleteness property – thus free cylindric algebras are not atomic, Mathematical Institute of the Hungarian Academy of Sciences, Preprint No. 49/1985, 1985. `http://www.math-inst.hu/~nemeti/NDis/NPrep85.pdf`

[Nem,86a] Németi, I., *Free algebras and decidability in algebraic logic,* Dissertation with the Hungarian Academy of Sciences, Budapest, 1986. In Hungarian. `http://www.math-inst.hu/~nemeti/NDis/NDis86.pdf`. English summary is `http://www.math-inst.hu/~nemeti/NDis/NSum.pdf`.

[Nem,86b] Németi, I., A nonrepresentable cylindric algebra with pairing functions, *Algebra Universalis,* **22** (1986), 117–119.

[Nem,87a] Németi, I., On varieties of cylindric algebras with applications to logic, *Ann. Pure and Appl. Logic,* **36** (1987), 235–277.

[Nem,87b] Németi, I., Decidability of relation algebras with weakened associativity, *Proc. Amer. Math. Soc.,* **100** (1987), 340–344.

[Nem,90] Németi, I., On cylindric algebraic model theory, in: *Algebraic Logic and Universal Algebra in Computer Science,* [Ber-Mad-Pig,90], 37–75.

[Nem,91] Németi, I., Algebraizations of quantifier logics, an introductory overview, *Studia Logica,* **50** (1991), 485–569. Updated version 12.1 (1997) is available from `ftp.math-inst.hu/pub/algebraic-logic/survey.ps`.

[Nem,95] Németi, I., Decidable versions of first order logic and cylindric-relativized set algebras, in: *Logic Colloquium '92,* [Csir-Gab-Rij,95], 177–241.

[Nem,96a] Németi, I., A fine-structure analysis of first-order logic, in: *Arrow Logic and Multi-modal Logic,* [Mar-Pol-Mas,96], 221–247.

[Nem,96b] Németi, I., Ontology can turn negative results to positive (an overview of recent results), *Bulletin Section of Logic, Łodź,* **25** (1996), 29–40.

[Nem,97] Németi, I., Strong representability of fork algebras, a set-theoretic foundation, *Logic J. IGPL,* **5** (1997), 3–23.

[Nem,11] Németi, I., Formalizing set theory in weak fragments of algebraic logic (updated in June 2011), 4pp. `http://www.math-inst.hu/~nemeti/NDis/formalizingsettheory.pdf`

[Nem-And,91] Németi, I. and Andréka, H., On Jónsson's clones of operations on binary relations, in: *Algebraic Logic,* [And-Mon-Nem,91b], 431–442.

[Nem-And,94] Németi, I. and Andréka, H., General Algebraic Logic: a Perspective on "What is logic", in: *What is a Logical System* (D. M. Gabbay, ed.) Oxford Science Publications, Clarendon Press (Oxford, 1994), 393–444.

[Nem-Sag,00] Németi, I. and Sági, G., On the equational theory of representable polyadic equality algebras, *J. Symbolic Logic,* **65** (2000), 1143–1167.

[Nem-Sai,00] Németi, I. and Sain, I., Editorial to the special issue of Logic J. IGPL on Algebraic Logic, *Logic J. IGPL,* **8** (2000), 379–381.

[Nem-Sai-Sim,95] Németi, I., Sain, I. and Simon, A., Undecidability of the equational theory of some classes of residuated Boolean algebras with operators, *Bulletin of the IGPL,* **3** (1995), 93–105.

[Nem-Say,01] Németi, I. and Sayed Ahmed, T., On neat reducts of algebras of logic, *Studia Logica,* **68** (2001), 229–262.

[Nem-Sim,97] Németi, I. and Simon, A., Relation algebras from cylindric and polyadic algebras, *Logic J. IGPL,* **5** (1997), 575–588.

[Nem-Sim,09] Németi, I. and Simon, A., Weakly higher order cylindric algebras ad finite axiomatizations of the representables, *Studia Logica,* **91** (2009), 53–62.

[Nem-Sim-Sai,92] Németi, I., Simon, A. and Sain, I., Beth properties on finite variable fragments of first order logics, Preprint, Mathematical Institute Budapest, 1992.

[New,87] Newelski, L., Omitting types and the real line, *J. Symbolic Logic,* **52** (1987), 1020–1026.

[Ogn-Per-Ras,08] Ognjanović, Z., Perović, A. and Rašković, M., Logics with the qualitative probability operator, *Logic J. IGPL,* **16** (2008), 105–120.

[Ogn-Ras,00] Ognjanović, Z. and Rašković, M., Some first order probability logic, *Theor. Comp. Sci.,* **247** (2000), 191–212.

[Ott,97] Otto, M., *Bounded variable logics and counting* – A study in finite models, 9 of Lecture Notes in Logic, Springer-Verlag, 1997.

[Pac-Sal,04] Pacuit, E. and Salame, S., Majority Logic, in: *Proceedings of the Ninth International Conference on Knowledge Representation (KR 2004),* (D. Dubois, Ch. Welty and M.-A. Williams, eds.), (2004), 598–605.

[Par-Sus,88] Parashchuk, S. D. and Sushchanski, V. I., Maximal subalgebras of complete Krasner algebras and $\mathscr{A}\mathscr{R}$-algebras over finite sets, (Russian) *Problems in group theory and homological algebra* (Russian), 28–46, Matematika, Yaroslav. Gos. Univ., Yaroslavl, 1988.

[Per-Ogn-Ras,08a] Perović, A., Ognjanović, Z. and Rašković, M., A Probabilistic Logic with Polynomial Weight Formulas, FoIKS 2008, LNCS 4932 (2008), 239–252.

[Per-Ogn-Ras,08b] Perović, A., Ognjanović, Z. and Rašković, M., How to restore compactness into probabilistic logics? JELIA 2008, LNCS 5293 (2008), 338–348.

[Pet,96] Petrovich, A., Distributive lattices with an operator, *Studia Logica,* **56** (1996), 205–224.

[Pet,99] Petrovich, A., Monadic De Morgan algebras, in: *Models, Algebras, and Proofs* (X. Caicedo and C. H. Montenegro, eds.), volume 203 of Lecture Notes in Pure and Applied Mathematics, 315–333. Marcel Decker, New York, 1999.

[Pig,71] Pigozzi, D., Amalgamation, congruence extension, and interpolation properties in algebras, *Algebra Universalis,* **1** (1971), 269–349.

[Pig,91] Pigozzi, D., Fregean algebraic logic, in: *Algebraic Logic and Universal Algebra in Computer Science,* [Ber-Mad-Pig,90], 473–502.

[Pig-Sal,93] Pigozzi, D. and Salibra, A., Polyadic algebras over nonclassical logics, in: *Algebraic Methods in Logic and in Computer Science,* [Banach,93], 51–66.

[Pin,73a] Pinter, C., Cylindric algebras and algebras of substitutions, *Trans. Amer. Math. Soc.,* **175** (1973), 167–179.

[Pin,73b] Pinter, C., A simple algebra of first order logic, *Notre Dame J. of Formal Logic,* **14** (1973), 361–366.

[Pla,90] Plattner, A., Zusammenhang zwischen der Theorie F der Faktorenimplikation und der Theorie der Zylinderalgebren, Reduktion der Vollständigkeit der Axiome von F. (German), Connection between the theory F of factorial implication and the theory of cylindric algebras; reduction of the completeness of the axioms of F, *Z. Math. Logik Grundlag. Math.,* **36** (1990), 561–572.

[Plo,91a] Plotkin, B. I., *Universal algebra, Algebraic logic, and Databases* Nauka, Fizmatlit (1991), (in Russian, English version is [Plo,94]).

[Plo,91b] Plotkin, B. I., Halmos (polyadic) algebras in the database theory, in: *Algebraic Logic,* [And-Mon-Nem,91b], 503–518.

[Plo,94] Plotkin, B. I., *Universal algebra, algebraic logic, and databases.* (English summary) Translated from the 1991 Russian original by J. Cirulis, A. Nenashev and V. Pototsky and revised by the author. Mathematics and its Applications, 272. Kluwer Academic Publishers Group, Dordrecht, 1994. xvii+438 pp.

[Plo,06] Plotkin, B. I., Algebraic geometry in first-order logic, (Russian) *Sovrem. Mat. Prilozh.,* **22**, Algebra i Geom. (2004), 16–62; translation in *J. Math. Sci. (N.Y.),* **137** (2006), 5049–5097.

[Pop,07] Popescu, A., Some algebraic theory for many-valued relation algebras, *Algebra Universalis,* **56** (2007), 211–235.

[Pra,95] Pratt, V., A Road Map of Some Two-Dimensional Logics, in: J. van Eijck, A. Visser, eds., *Logic and Information Flow,* The MIT Press, Cambridge (Mass.) (1995), 149–162.

[Pri,08] Priest, G., *An Introduction to Non-Classical Logic,* Cambridge University Press, Cambridge, 2nd edition, 2008.

[Pri,93] Priestley, H. A., Natural dualities for varieties of distributive lattices with a quantifier, in: *Algebraic Methods in Logic and in Computer Science,* [Banach,93], 291–310.

[Rad-Per-Ogn-Ras,08] Radojević, D., Perović, A., Ognjanović, Z. and Rašković, M., *Interpolative Boolean Logic,* AIMSA 2008, LNCS 5235 (2008), 209–219.

[Ras,86] Rašković, M., Completeness Theorem for Biprobability Models, *J. Symb. Log.,* **51** (1986), 586–590.

[Ras,87] Rašković, M., Weak completeness theorem for $L_{\mathcal{A}P\forall}$ logic, *Zb. Rad. (Kragujevac),* **8** (1987), 69–72.

[Ras-Dor,96] Rašković, M. and Đorđević, R. S., *Probability Quantifiers and Operators,* Series in pure and applied mathematics, Vesta, Belgrade, 1996.

[Ras-Dor,00] Rašković, M., and Đorđević, R. S., Cylindric probability algebras, *Publ. L'Inst. Math.,* **68** (2000), 20–36.

[Ras-Dor-Bra,97] Rašković, M., Đorđević, R. S. and Bradić, M., Weak cylindric probability algebras, *Publ. L'Inst. Math.,* **61** (1997), 6–16.

[Ras-Mar-Ogn,08] Rašković, M., Marković, Z. and Ognjanović, Z., A logic with approximate conditional probabilities that can model default reasoning, *Int. J. Approx. Reasoning,* **49** (2008), 52–66.

[Ras-Ogn-Mar,04] Rašković, M., Ognjanović, Z. and Marković, Z., A logic with conditional probabilities, JELIA 2004, LNCS 3229 (2004), 226–238.

[Rei,20] Reichenbach, H., *The Theory of Relativity and a Priori Knowledge,* University of California Press, 1960. Translated by M. Reichenbach from the original German edition published in 1920.

[Res,75] Resek, D., *Some results on relativized cylindric algebras,* PhD dissertation, Berkeley, 1975.

[Res-Tho,91] Resek, D. and Thompson, R. J., Characterizing relativized cylindric algebras, in: *Algebraic Logic,* [And-Mon-Nem,91b], 519–538.

[Rij,92] de Rijke, M., The modal logic of inequality, *J. Symbolic Logic,* **57** (1992), 566–584.

[Rij,93] de Rijke, M., *Extending modal logic,* Dissertation, Institute for Logic, Language and Computation, University of Amsterdam (1993).

[Rij-Seg-Wan-Zak,01] de Rijke, M., Segerberg, K., Wansing, H. and Zakharyashev, M. (eds.), *Advances in Modal logic, vol. II,* Uppsala 1998, CSLI Publications, Stanford (2001).

[Rij-Ven,95] de Rijke, M. and Venema, Y., Sahlqvist's theorem for Boolean algebras with operators with an application to cylindric algebras, *Studia Logica,* **54** (1995), 61–79.

[Ris-Lai,87] Riscos, A. and Laita, L. M., N-categories in logic, *Z. Math. Logik Grundlag. Math.,* **33** (1987), 507–516.

[Rog,09] Rognes, A., Turning decision procedures into disprovers, *Math. Logic Quarterly,* **55** (2009), 87–104.

[Ros,88] Rosenberg, S. M., On varieties of Halmos polyadic algebras (in Russian), *Latv. Mat. Ezhegodnik,* Vol. **32** (1988), 85–89.

[Ros,93] Rosenberg, S. M., Specialized relation algebras and Halmos algebras, (Russian) *Latv. Mat. Ezhegodnik,* No. **34** (1993), 219–229.

[Ryb,97] Rybakov, V. V., *Admissibility of Logical Inference Rules,* Studies in Logic and the Foundations of Mathematics, 136. North-Holland Publishing Co., Amsterdam, 1997. ii+617 pp.

[Sad-Pal-Ma,11] Sadrzadeh, M., Palmigiano, A. and Ma, M., Algebraic Semantics and Model Completeness for Intuitionistic Public Announcement Logic, in: *Proceedings LORI 3* (H. van Ditmarsch and J. Lang, eds.), Guangzhou, Springer Verlag (2011), 394–395.

[Sag,99a] Sági, G., *On the finitization problem of algebraic logic,* Ph.D. dissertation, Eötvös University, Budapest (1999).

[Sag,99b] Sági, G., Defining relations for finite, non-bijective selfmaps, *Semigroup Forum,* **58** (1999), 94–105.

[Sag,00] Sági, G., A completeness theorem for higher order logics, *J. Symbolic Logic,* **65** (2000), 857–884.

[Sag,01] Sági, G., Non-Computability of the Equational Theory of Polyadic Algebras, *Bulletin Section of Logic, Łodź* (2001), 155–165.

[Sag,02] Sági, G., A note on algebras of substitutions, *Studia Logica,* **72**, No. 2 (2002), 265–284.

[Sag,11] Sági, G., On nonrepresentable G-polyadic algebras with representable cylindric reducts, *Logic J. IGPL,* **19** (2011), 105–109.

[Sag,thisVol] Sági, G., Polyadic algebras, in this volume.

[Sag-She,06] Sági, G. and Shelah, S., Weak and strong interpolation for algebraic logics, *J. Symbolic Logic,* **71** (2006), 104–118.

[Sag-Szi,12] Sági, G. and Sziráki, D., Some variants of Vaught's conjecture from the perspective of algebraic logic, *Logic J. IGPL,* **20** (2012), 1–23.

[Sai,88] Sain, I., Concerning some cylindric algebra versions of the downward Löwenheim–Skolem theorem, *Notre Dame J. of Formal Logic,* **29** (1988), 332–344.

[Sai,89] Sain, I., Amalgamation, epimorphisms and definability properties in algebraic logic, *Bulletin Section of Logic, Łodź,* **18** (1989), 72–78.

[Sai,90] Sain, I., Beth's and Craig's properties via epimorphisms and amalgamation in algebraic logic, in: *Algebraic Logic and Universal Algebra in Computer Science,* [Ber-Mad-Pig,90], 209–226.

[Sai,94] Sain. I., On finitizing first order logic, *Bulletin Section of Logic, Łodź,* **23**, No. 2 (1994), 66–79.

[Sai,95] Sain, I., On the problem of finitizing first order logic and its algebraic counterpart (a survey of results and methods), in: *Logic Colloquium '92,* [Csir-Gab-Rij,95], 243–292.

[Sai,97] Sain, I., On characterizations of definability properties in abstract algebraic logic, in: *Workshop on abstract algebraic logic,* [Fon-Jan-Pig,97], 162–175.

[Sai,00] Sain, I., Searching for a Finitizable Algebraisation of First Order Logic, *Logic J. IGPL,* **8** (2000), 495–589.

[Sai,thisVol] Sain, I., Definability issues in universal logic, in this volume.

[Sai-Gyu,97] Sain, I. and Gyuris, V., Finite schematizable algebraic logic, *Logic J. IGPL,* **5** (1997), 699–751.

[Sai-Nem,99] Sain, I. and Németi, I., Fork algebras in usual and in non-well-founded set theories (an overview), in: *Logic at work,* Stud. Fuzziness Soft Comput., 24, Physica (Heidelberg, 1999), 669–694.

[Sai-Tho,91] Sain, I. and Thompson, R. J., *Strictly finite schema axiomatization of quasi-polyadic algebras,* in: *Algebraic Logic,* [And-Mon-Nem,91b], 539–571.

[Sal,91] Salibra, A., A general theory of algebras with quantifiers, in: *Algebraic Logic,* [And-Mon-Nem,91b], 573–620.

[Sam-Say,06] Samir, B. and Sayed Ahmed, T., Neat embeddings and amalgamation, *Bulletin Section of Logic, Łodź,* **35** (2006), 163–171.

[Sam-Say,07a] Samir, B. and Sayed Ahmed, T., A Neat embedding theorem for expansions of cylindric algebras, *Logic J. IGPL,* **15** (2007), 41–51.

[Sam-Say,07b] Samir, B. and Sayed Ahmed, T., Omitting types for first order logic with infinitary predicates, *Math. Logic Quarterly,* **53** (2007), 564–576.

[Sam-Say,08] Samir, B. and Sayed Ahmed, T., The class $\mathsf{Nr}_3\,\mathsf{CA}_k$ is not closed under completions, *Logic J. IGPL,* **16** (2008), 427–429.

[Say,01] Sayed Ahmed, T., The class of neat reducts is not elementary, *Logic J. IGPL,* **9** (2001), 593–628.

[Say,02a] Sayed Ahmed, T., *On neat reducts of algebras of logic,* Ph.D dissertation, Cairo University (2002).

[Say,02b] Sayed Ahmed, T., A model-theoretic solution to a problem of Tarski, *Math. Logic Quarterly,* **48** (2002), 343–355.

[Say,02c] Sayed Ahmed, T., Martin's axiom, omitting types, and complete representations in algebraic logic, *Studia Logica,* **72** (2002), 285–309.

[Say,02d] Sayed Ahmed, T., The class of 2-dimensional neat reducts of polyadic algebras is not elementary, *Fundementa Mathematicea,* **172** (2002), 61–81.

[Say,03] Sayed Ahmed, T., Neat embeddings, interpolation, and omitting types, an overview, *Notre Dame J. of Formal Logic,* **44** (2003), 157–173.

[Say,04a] Sayed Ahmed, T., On amalgamation of reducts of polyadic algebras, *Algebra Universalis,* **51** (2004), 301–359

[Say,04b] Sayed Ahmed, T., Tarskian Algebraic Logic, *J. on Relation Methods in Computer Science,* **1** (2004), 3–26.

[Say,05a] Sayed Ahmed, T., Algebraic logic, where does it stand today? *Bull. Symbolic Logic,* **11** (2005), 465–516.

[Say,05b] Sayed Ahmed, T., Amalgamation theorems in algebraic logic, an overview, *Logic J. IGPL,* **13** (2005), 277–286.

[Say,05c] Sayed Ahmed, T., Independence results in algebraic logic, *Logic J. IGPL,* **14** (2005) 87–96.

[Say,05d] Sayed Ahmed, T., On amalgamation of algebras of logic, *Studia Logica,* **81** (2005), 61–77.

[Say,06a] Sayed Ahmed, T., Omitting types for algebraizable extensions of first order logic, *J. of Applied and Non-classical Logic,* **15** (2005), 465–489.

[Say,06b] Sayed Ahmed, T., Some results on amalgamation in algebraic logic, *Logic J. IGPL,* **14** (2006), 623–627.

[Say,06c] Sayed Ahmed, T., The class of infinite dimensional neat reducts of quasi-polyadic algebras is not axiomatizable, *Math. Logic Quarterly,* **52** (2006), 106–112.

[Say,07a] Sayed Ahmed, T., A note on neat reducts, *Studia Logica,* **85** (2007), 139–151.

[Say,07b] Sayed Ahmed, T., An atomic theory with no prime models, *Australasian J. of Logic,* **5** (2007), 85–88.

[Say,07c] Sayed Ahmed, T., An interpolation theorem for first order logic with infinitary predicates, *Logic J. IGPL,* **15**(1) (2007), 21–32.

[Say,07d] Sayed Ahmed, T., Neat embedding is not sufficient for complete representations, *Bulletin Section of Logic, Łodź,* **36** (2007).

[Say,07e] Sayed Ahmed, T., On neat reducts and amalgamation, *Logic J. IGPL,* **15** (2007), 33–39.

[Say,08a] Sayed Ahmed, T., A note on atom structures of relation and cylindric algebras, *Bulletin Section of Logic, Łodź,* **37** (2008), 29–35.

[Say,08b] Sayed Ahmed, T., Amalgamation of reducts of polyadic algebras, a negative result, *Bulletin Section of Logic, Łodź,* **37** (2008), 37–50.

[Say,08c] Sayed Ahmed, T., On complete representability of reducts of polyadic algebras *Studia Logica,* **89** (2008), 325–332.

[Say,08d] Sayed Ahmed, T., The superamalgamation property via neat embeddings, and a problem of Henkin and Monk, *International J. of Algebra,* **2** (2008), 533–554.

[Say,08e] Sayed Ahmed, T., Variations on Martin's axiom and Omitting types from algebraic logic, lattice theory and topology, *Bulletin Section of Logic, Łodź,* **37** (2008), 103–114.

[Say,08f] Sayed Ahmed, T., Weakly representable atom structures that are not strongly representable, with an application to first order logic, *Math. Logic Quarterly,* **3** (2008), 294–306.

[Say,09a] Sayed Ahmed, T., A note on substitutions in representable cylindric algebras, *Math. Logic Quarterly,* **55** (2009), 280–287.

[Say,09b] Sayed Ahmed, T., Amalgamation for reducts of polyadic equality algebras, both a negative result and a positive result, *Int. J. Contemp. Math. Sci.,* **4** (2009), 139–160.

[Say,09c] Sayed Ahmed, T., On a Theorem of Vaught for first order logic with finitely many variables, *J. of Applied and Non-classical Logic,* **19** (2009), 97–112.

[Say,09d] Sayed Ahmed, T., On neat embedding of algebraisations of first order logic, *J. of Algebra, Number Theory, Advances and Applications,* **1** (2009), 113–125.

[Say,09e] Sayed Ahmed, T., On neat embeddings of cylindric algebras, *Math. Logic Quarterly,* **55** (2009), 666–668.

[Say,09f] Sayed Ahmed, T., On the amalgamation property for various algebraic logics, *Int. J. Algebra,* **3** (2009), 41–94.

[Say,09g] Sayed Ahmed, T., The amalgamation property, and a problem of Henkin Monk and Tarski, *J. of Algebra, Number Theory, Advances and Applications,* **1**(2) (2009), 127–141.

[Say,10a] Sayed Ahmed, T., The class of polyadic algebras has the superamalgamation property, *Math. Logic Quarterly,* **56** (2010), 103–112.

[Say,10b] Sayed Ahmed, T., On neat reducts of cylindric algebras, *Eur. J. Pure Appl. Math.,* **3** (2010), 853–880.

[Say,10c] Sayed Ahmed, T., Some results about neat reducts, *Algebra Universalis,* **63** (2010), 17–36.

[Say,11a] Sayed Ahmed, T., Classes of algebras without the amalgamation property, *Logic J. IGPL,* **19** (2011), 87–104.

[Say,11b] Sayed Ahmed, T., On the complexity of axiomatizations of the class of representable quasi-polyadic equality algebras, *Math. Logic Quarterly,* **57** (2011), 384–394.

[Say,thisVol,a] Sayed Ahmed, T., Neat reducts and neat embeddings in cylindric algebras, in this volume.

[Say,thisVol,b] Sayed Ahmed, T., Completions, complete representations and omitting types, in this volume.

[Say,a] Sayed Ahmed, T., On relation algebras with quasi-projections, preprint.

[Say,b] Sayed Ahmed, T., Varieties of algebras without the amalgamation property, *Logic J. of IGPl,* to appear.

[Say,c] Sayed Ahmed, T., Algebraic Logic, Book manuscript.

[Say,d] Sayed Ahmed, T., Algebras having the unique neat embedding property, submitted.

[Say,e] Sayed Ahmed, T., Non elementary classes in algebraic logic, submitted.

[Say,f] Sayed Ahmed, T., On representations of neat reducts, submitted.

[Say,g] Sayed Ahmed, T., On various axiomatizations of algebras of relations, preprint.

[Say-Nem,01] Sayed Ahmed, T. and Németi, I., On neat reducts of algebras of logic, *Studia Logica,* **68** (2001), 229–262.

[Say-Say,09] Sayed Ahmed, T. and Sayed Ahmed, A., On the amalgamation base of cylindric algebras, *Int. J. Contemp. Math. Sci.,* **4** (2009), 161–190.

[Sco,62] Scott, D., A decision method for validity of sentences in two variables, *J. Symbolic Logic,* **27** (1962), 477.

[Sco,08] Scott, D., The algebraic interpretation of quantifiers: intuitionistic and classical, *Andrzej Mostowski and foundational studies,* 289–312, IOS, Amsterdam, 2008.

[Sco-Kra,66] Scott, D., and Krauss, P., Assigning probabilities to logical formulas, in: *Aspects of inductive logic,* North Holland (1966), 219–264.

[Scr,51] Scroggs, S. G., Extensions of the Lewis system **S5**, *J. Symbolic Logic,* **16** (1951), 111–120.

[Seg,73] Segerberg, K., Two-dimensional modal logic, *J. of Philosophical Logic,* **2** (1973), 77–96.

[Ser,85] Serény, G., Compact cylindric set algebras, *Bulletin Section of Logic, Łodź,* **14** (1985), 57–64.

[Ser,86] Serény, G., Lower level connections between representations of relation algebras (abstract), *Bulletin Section of Logic, Łodź,* **15** (1986), 123–127.

[Ser,87] Serény, G., Finite models are one-generated, *Algebra Universalis,* **24** (1987), 193–195.

[Ser,91a] Serény, G., Neatly atomic cylindric algebras and isomorphisms, in: *Algebraic Logic,* [And-Mon-Nem,91b], 637–644.

[Ser,91b] Serény, G., *A non locally finite saturated cylindric set algebra,* preprint, 1991, 15 pp.

[Ser,93] Serény, G., Isomorphisms of finite cylindric set algebras of characteristic zero, *Notre Dame J. of Formal Logic,* **34** (1993), 284–294.

[Ser,97] Serény, G., Saturatedness in cylindric algebraic model theory, (English summary) *Logic J. IGPL,* **5** (1997), 25–48 (electronic).

[Ser,thisVol] Serény, G., Elements of cylindric algebraic model theory, in this volume.

[Ser-Sim,98] Serény, G. and Simon, A., Product topological spaces and cylindric set algebras, manuscript, 1998, 3 pp.

[She,78] Shehtman, V., Two-dimensional modal logics, *Mathematical Notices of the USSR Academy of Sciences,* **23** (1978), 417–424. (Translated from Russian).

[She,91] Shelah, S., On a problem in cylindric algebra, in: *Algebraic Logic,* [And-Mon-Nem,91b], 645–664.

[She,97] Sheepers, M., The least cardinal for which the Baire Category Theorem fails, *Proc. Amer. Math. Soc.,* **2** (1997), 579–585.

[Sik,50] Sikorski, R., On analogies between measures and homomorphisms, *Ann. Soc., Pol. Math.,* **23** (1950), 1–20.

[Sik,64] Sikorski, R., *Boolean Algebras,* Springer, 1964.

[Sil,94] Silionova, A., Substitutions in term algebras: equivalence of two axiomatizations, *Mathematics,* 53–56, Latv. Univ. Zināt. Raksti, 595, Latv. Univ., Riga, 1994.

[Sim,91] Simon, A., Finite schema completeness for typeless logic and representable cylindric algebras, in: *Algebraic Logic,* [And-Mon-Nem,91b], 665–670.

[Sim,96] Simon, A., What the finitization problem is not, in: *Algebraic Methods in Logic and in Computer Science,* [Banach,93], 95–116.

[Sim,99] Simon, A., *Non-representable algebras of relations,* PhD thesis, Hungarian Academy of Sciences, Budapest, 1999.

[Sim,07] Simon, A., Connections between quasi-projective relation algebras and cylindric algebras, *Algebra Universalis,* **56** (2007), 263–301.

[Sim,thisVol] Simon, A., Representing all cylindric algebras by twisting, in this volume.

[Slom,59] Slominski, J., *The Theory of Abstract Algebras with Infinitary Operations,* Rozprawy Matematyczne, **18** (1959), 67 pp.

[Ste,00] Stebletsova, V., *Algebras, Relations and Geometries (an equational perspective).* PhD Dissertation, University of Utrecht and University of Amsterdam, 2000.

[Ste-Ven,98] Stebletsova, V. and Venema, Y., Q-algebras, *Algebra Universalis,* **40** (1998), 19–49.

[Sze,09] Székely, G., *First-order logic investigation of relativity theory with an emphasis on accelerated observers,* PhD thesis, ELTE, Budapest, 2009.

[Tar,34] Tarski, A., Some methodological investigations on the definability of concepts (in Germain), in: *Collected Papers,* Volumes 1–4 (S. R. Givant and R. N. McKenzie, eds.), Birkhäuser (1986), 637–639.

[Tar,35] Tarski, A., Grundzüge der Systemenkalküls. Erster Teil, *Fundamenta Math.,* 25, (1935), 503–526. English translation in [A. Tarski, *Logic, Semantics, Metamathematics.* Papers from 1923 to 1938, edited by J. Corcoran, (1983)], 342–383.

[Tar,52] Tarski, A., A representation theorem for cylindric algebras. Preliminary report, *Bull. Amer. Math. Soc.,* **58** (1952), 65–66.

[Tar,53a] Tarski, A., A formalization of set theory without variables, *J. Symbolic Logic,* **18** (1953).

[Tar,53b] Tarski, A., An undecidable system of sentential calculus, *J. Symbolic Logic,* **18** (1953).

[Tar,54] Tarski, A., Contributions to the theory of models I, II. *Indagationes Mathematicae,* **16** (1954), 572–588.

[Tar,55] Tarski, A., Contributions to the theory of models III, *Indagationes Mathematicae,* **17** (1955), 56–64.

[Tar-Giv,87] Tarski, A. and Givant, S. R., *A formalization of set theory without variables,* in: *American Mathematical Society Colloquium Publications,* 41. American Mathematical Society, Providence, RI, 1987. xxii+318 pp.

[Tar-Tho,52] Tarski, A. and Thompson, F. B., Some general properties of cylindric algebras. Preliminary report, *Bull. Amer. Math. Soc.,* **58** (1952).

[Tha,96] Thalheim, B., An overview on semantical constraints for database models, in: *Proceedings of The 6th International Conference on Intellectual Systems and Computer Science* (1996), 1–10.

[Tho,79] Thompson, R. J., *Transformational structure of algebraic logics,* PhD thesis, University of California, Berkeley, 1979.

[Tho,90] Thompson, R. J., Noncommutative cylindric algebras and relativizations of cylindric algebras, in: *Algebraic Logic and Universal Algebra in Computer Science,* [Ber-Mad-Pig,90], 273–278.

[Tho,93] Thompson, R. J., Complete description of substitutions in cylindric algebras and other algebraic logics, in: *Algebraic Methods in Logic and in Computer Science,* [Banach,93], 327–342.

[Tsi,88] Tsirulis, Ya. P., Two generalizations of the notion of a polyadic algebra, (Russian) *Izv. Vyssh. Uchebn. Zaved. Mat.,* 1988, no. 12, 39–49; translation in *Soviet Math. (Iz. VUZ),* **32** (1988), 61–78.

[Vaa,07] Väänänen, J., *Dependence Logic: A New Approach to Independence Friendly Logic,* Cambridge University Press, Cambridge (2007).

[Vag,95] Vaggione, D. J., Free algebras in discriminator varieties, *Algebra Universalis,* **34** (1995), 391–403.

[Vak,96] Vakarelov, D., Many-dimensional arrow structures: arrow logics. II, in: *Arrow Logic and Multi-modal Logic,* [Mar-Pol-Mas,96], 141–187.

[Vel,96] Veltman, F., Defaults in Update Semantics, *J. of Philosophical Logic,* **25** (1996), 221–261.

[Ven,91] Venema, Y., *Many-dimensional modal logics,* Dissertation, Institute for Logic, Language and Computation, University of Amsterdam (1991).

[Ven,93] Venema, Y., Derivation rules as anti-axioms in modal logic, *J. Symbolic Logic,* **58** (1993), 1003–1034.

[Ven,94] Venema, Y., A Modal Logic for Quantification and Substitution, *Logic J. IGPL,* **2** (1994), 31–45.

[Ven,95a] Venema, Y., A modal logic of quantification and substitution, in: *Logic Colloquium '92,* [Csir-Gab-Rij,95], 293–309.

[Ven,95b] Venema, Y., Cylindric modal logic, *J. Symbolic Logic,* **60** (1995), 591–623.

[Ven,96] Venema, Y., A Crash Course in Arrow Logic, in: *Arrow Logic and Multi-modal Logic,* [Mar-Pol-Mas,96], 3–34.

[Ven,97a] Venema, Y., Atom structures and Sahlqvist equations, *Algebra Universalis,* **38** (1997), 185–199.

[Ven,97b] Venema, Y., Atom structures, in: *Advances in Modal Logic '96* (M. Kracht, M. de Rijke, H. Wansing and M. Zakharyaschev, eds.), 291–305. CSLI Publications (Stanford, 1997).

[Ven,98] Venema, Y., Rectangular games, *J. Symbolic Logic,* **63** (1998), 1549–1564.

[Ven,07] Venema, Y., Algebras and coalgebras, in: *Handbook of Modal Logic* (P. Blackburn, J. van Benthem and F. Wolter, eds.), Elsevier (2007), 331–426.

[Ven,thisVol] Venema, Y., Cylindric modal logic, in this volume.

[Ver,01] Verstik, A. M., Randomization of algebra and algebraization of probability, in: *Mathematics unlimited – 2001 an beyond,* ed. Engquist, Springer (2001).

[Vla,02] Vladimirov, D. A., *Boolean Algebras in Analysis,* Kluwer (2002).

[Vol,86a] Volkov, N. D., Halmos algebras and relation algebras (in Russian), *Latv. Mat. Ezhegodnik,* **30** (1986), 110–123.

[Vol,86b] Volkov, N. D., The transition from a relation algebra to a Halmos algebra, *Algebra and discrete mathematics: the theoretical foundations of software* (Russian), 43–53, Latv. Gos. Univ., Riga, 1986.

[Vol,88a] Volkov, N. D., Filters in relational algebras and in Halmos polyadic algebras (in Russian), *Latv. Mat. Ezhegodnik,* **32** (1988), 60–67.

[Vol,88b] Volkov, N. D., Operations of Halmos polyadic algebras in relational algebras (in Russian), *Latv. Mat. Ezhegodnik,* **31** (1988), 103–114.

[Vol,93] Volkov, N. D., On the equivalence of the categories of Halmos algebras and relation algebras, (Russian) *Latv. Mat. Ezhegodnik,* **34** (1993), 171–180.

[Wal,70] Walkoe, W. J., Jr., Finite partially-ordered quantification, *J. Symbolic Logic,* **35** (1970), 535–555.

[Wad,75] Wadge, W., A Complete Natural Deduction System for the Relation Calculus, Technical Report 5, University of Warwick (1975).

[Wil-Gab,03] Williamson, T. and Gabbay, D. M., (eds.) Combining Probability and Logic (4th de Morgan Workshop, London), *J. Appl. Logic,* **1** (2003).

[Wol,01] Woliński, A., Cylindric algebras for partial relational systems. Quasi cylindric algebras, in: *Relational methods for computer science applications,* 287–297, Stud. Fuzziness Soft Comput., 65, Physica, Heidelberg, 2001.

[Wu,10] Wu, J., Equationally distinct countable simple Q-relation algebras, *Algebra Universalis,* **63** (2010), 131–147.

[Yan,68] Yankov, V. A., The construction of a sequence of strongly independent superintuitionistic propositional calculi, *Soviet Mathematics Doklady,* **9** (1968), 806–807.

[Yan-Pap,82] Yannakakis, M. and Papadimitriou, C. H., Algebraic dependencies, *J. of Computer and System Sciences,* **25** (1982), 2–41.

[Zan,06] Zander, M. A., Cyclic Tarski algebras, *Bull. Austral. Math. Soc.,* **73** (2006), 147–158.

[Zla,84] Zlatos, P., Two notes on locally finite cylindric algebras, *Comment. Math. Univ. Carolin.,* **25** (1984), 181–199.

Index of Symbols

Index of Names and Subjects

MIX
Papier aus verantwortungsvollen Quellen
Paper from responsible sources
FSC® C105338

If you have any concerns about our products, you can contact us on
ProductSafety@springernature.com

In case Publisher is established outside the EU, the EU authorized representative is:
Springer Nature Customer Service Center GmbH
Europaplatz 3, 69115 Heidelberg, Germany

Printed by Libri Plureos GmbH
in Hamburg, Germany